Rarefied Gas Dynamics: Physical Phenomena

Edited by
E. P. Muntz
University of Southern California
Los Angeles, California

D. P. Weaver
Astronautics Laboratory (AFSC)
Edwards Air Force Base, California

D. H. Campbell
The University of Dayton Research Institute
Astronautics Laboratory (AFSC)
Edwards Air Force Base, California

Volume 117
PROGRESS IN ASTRONAUTICS AND AERONAUTICS

Martin Summerfield, Series Editor-in-Chief
Princeton Combustion Research Laboratories, Inc.
Monmouth Junction, New Jersey

Technical Papers selected from the Sixteenth International Symposium on Rarefied Gas Dynamics, Pasadena, California, July 10-16, 1988, subsequently revised for this volume.

Published by the American Institute of Aeronautics and Astronautics, Inc., 370 L'Enfant Promenade, SW, Washington, DC 20024-2518

American Institute of Aeronautics and Astronautics, Inc.
Washington, D.C.

Library of Congress Cataloging in Publication Data

International Symposium on Rarefied Gas Dynamics
(16th:1988:Pasadena, California)
Rarefied gas dynamics: physical phenomena/edited by E.P. Muntz, D.P. Weaver, D.H. Campbell.

p. cm. – (Progress in astronautics and aeronautics; v. 117)
"Technical papers selected from the Sixteenth International Symposium on Rarefied Gas Dynamics, Pasadena, California, July 10–16, 1988, Subsequently revised for this volume."
Includes index.
1. Rarefied gas dynamics – Congresses. 2. Gases, kinetic theory of – Congresses. 3. Aerodynamics – Congresses. I. Muntz, E. Phillip (Eric Phillip), 1934– . II. Weaver, D.P. III. Campbell, D.H. (David H.) IV. Title. V. Series.
TL507.P75 vol. 117 89–17616
[QC168.86]
629.1 s
[629.132'.3]
ISBN 0–930403–54–1

Symposium Advisory Committee

J. J. Beenakker (Netherlands)
G. A. Bird (Australia)
V. Boffi (Italy)
C. L. Burndin (UK)
R. H. Cabannes (France)
R. Campargue (France)
C. Cercignani (Italy)
J. B. Fenn (USA)
S. S. Fisher (USA)
W. Fiszdon (Poland)
O. F. Hagena (FRG)
F. C. Hurlbut (USA)
M. Kogan (USSR)
G. Koppenwallner (FRG)
I. Kuscer (Yugoslavia)
E. P. Muntz (USA)
R. Narasimha (India)
H. Oguchi (Japan)
D. C. Pack (UK)
J. L. Potter (USA)
A. K. Rebrov (USSR)
Yu. A. Rijov (USSR)
B. Shizgal (Canada)
Y. Sone (Japan)
J. P. Toennies (FRG)
Y. Yoshizawa (Japan)

Technical Reviewers

K. Aoki
G. Arnold
V. Boffi
J. Brook
R. Caflisch
D. H. Campbell
C. Cercignani
H. K. Cheng
N. Corngold
J. Cross
R. Edwards
D. Erwin
W. Fiszdon
A. Frohn
O. Hagena
L. J. F. Hermans
W. L. Hermina
E. L. Knuth
G. Koppenwallner
K. Koura
J. Kunc
H. Legge
E. P. Muntz
K. Nanbu
D. Nelson
J. Oguchi
J. L. Potter
A. Rebrov
Yu. A. Rijov
J. Scott
A. K. Sreekanth
B. Sturtevant
H. T. Yang

Local Organizing Committee

E. P. Muntz (Chairman)
D. P. Weaver
D. H. Campbell
R. Cattolica
H. K. Cheng
D. Erwin
J. Kunc
M. Orme
B. Sturtevant

Symposium Sponsors

Los Alamos National Laboratory
Strategic Defense Initiative Organization
University of Southern California
U. S. Army Research Office
U. S. Air Force Astronautics Laboratory

Table of Contents

Preface .... **xvii**

Chapter I. Inelastic Collisions **1**

Inelastic Collision Models for Monte Carlo Simulation Computation **3**
J. K. Harvey, *Imperial College, London, England, United Kingdom*

Null Collision Monte Carlo Method: Gas Mixtures with Internal Degrees of Freedom and Chemical Reactions **25**
K. Koura, *National Aerospace Laboratory, Chofu, Tokyo, Japan*

Nitrogen Rotation Relaxation Time Measured in Freejets **40**
A. E. Belikov, G. I. Sukhinin, and R. G. Sharafutdinov, *Siberian Branch of the USSR Academy of Sciences, Novosibirsk, USSR*

Rate Constants for R-T Relaxation of N_2 in Argon Supersonic Jets **52**
A. E. Belikov, G. I. Sukhinin, and R. G. Sharafutdinov, *Siberian Branch of the USSR Academy of Sciences, Novosibirsk, USSR*

Rotational Relaxation of CO and CO_2 in Freejets of Gas Mixtures **68**
T. Kodama, S. Shen, and J. B. Fenn, *Yale University, New Haven, Connecticut*

Diffusion and Energy Transfer in Gases Containing Carbon Dioxide **76**
J. R. Ferron, *University of Rochester, Rochester, New York*

Freejet Expansion of Heavy Hydrocarbon Vapor **92**
A. V. Bulgakov, V. G. Prikhodko, A. K. Rebrov, and P. A. Skovorodko, *Siberian Branch of the USSR Academy of Sciences, Novosibirsk, USSR*

Chapter II. Experimental Techniques **105**

Optical Diagnostics of Low-Density Flowfields **107**
J. W. L. Lewis, *University of Tennessee Space Institute, Tullahoma, Tennessee*

Electron Beam Flourescence Measurements of Nitric Oxide 133
R. J. Cattolica, *Sandia National Laboratories, Livermore, California*

Measurements of Freejet Densities by Laser Beam Deviation 140
J. C. Mombo-Caristan, L. C. Philippe, C. Chidiac, M. Y. Perrin, and J. P. Martin, *Laboratoire d'Energetique Moleculaire et Macroscopic Combustion du Centre Nationale de la Recherche Scientifique, Chatenay Malabry, France* and *Ecole Centrale Paris, Chatenay Malabry, France*

Turbulence Measurement of a Low-Density Supersonic Jet with a Laser-Induced Fluorescence Method 149
M. Masuda, H. Nakamuta, Y. Matsumoto, K. Matsuo, and M. Akazaki, *Kyushu University, Fukuoka, Japan*

Measurement of Aerodynamic Heat Rates by Infrared Thermographic Technique at Rarefied Flow Conditions 157
J. Allègre, X. Hériard Dubreuilh, and M. Raffin, *Société d'Etudes et de Services pour Souffleries et Installations Aérothermodynamiques (SESSIA), Meudon, France*

Experimental Investigation of CO_2 and N_2O Jets Using Intracavity Laser Scattering 168
R. G. Schabram, A. E. Beylich, and E. M. Kudriavtsev, *Stosswellenlabor, Technische Hochschule, Aachen, Federal Republic of Germany*

High-Speed-Ratio Helium Beams: Improving Time-of-Flight Calibration and Resolution 187
R. B. Doak and D. B. Nguyen, *AT&T Bell Laboratories, Murray Hill, New Jersey*

Velocity Distribution Function in Nozzle Beams 206
O. F. Hagena, *Kernforschungszentrum Karlsruhe, Karlsruhe, Federal Republic of Germany*

Cryogenic Pumping Speed for a Freejet in the Scattering Regime 218
J.-Th. Meyer, *DFVLR, Göttingen, Federal Republic of Germany*

Effectiveness of a Parallel Plate Arrangement as a Cryogenic Pumping Device 233
K. Nanbu, Y. Watanabe, and S. Igarashi, *Tohoku University, Sendai, Japan,* and G. Dettleff and G. Koppenwallner, *German Aerospace Research Establishment, Göttingen, Federal Republic of Germany*

Chapter III. Particle and Mixture Flows 245

Aerodynamic Focusing of Particles and Molecules in Seeded Supersonic Jets 247
J. Fernández de la Mora, J. Rosell-Llompart, and P. Riesco-Chueca, *Yale University, New Haven, Connecticut*

Experimental Investigations of Aerodynamic Separation of Isotopes and Gases in a Separation Nozzle Cascade 278
P. Bley and H. Hein, *Kernforschungszentrum Karlsruhe, Karlsruhe, Federal Republic of Germany,* and J. L. Campos, R. V. Consiglio, and J. S. Coelho, *Centro de Desenvolvimento da Tecnologia Nuclear, Belo Horizonte, Brazil*

General Principles of the Inertial Gas Mixture Separation 290
B. L. Paklin and A. K. Rebrov, *Siberian Branch of the USSR Academy of Sciences, Novosibirsk, USSR*

Motion of a Knudsen Particle Through a Shock Wave 298
M. M. R. Williams, *University of Michigan, Ann Arbor, Michigan*

Method of Characteristics Description of Brownian Motion Far from Equilibrium 311
P. Riesco-Chueca, R. Fernández-Feria, and J. Fernández de la Mora, *Yale University, New Haven, Connecticut*

Chapter IV. Clusters 327

Phase-Diagram Considerations of Cluster Formation When Using Nozzle-Beam Sources 329
E. L. Knuth and W. Li, *University of California, Los Angeles, California,* and J. P. Toennies, *Max-Planck-Institut für Strömungsforchung, Göttingen, Federal Republic of Germany*

Fragmentation of Charged Clusters During Collisions of Water Clusters with Electrons and Surfaces 335
A. A. Vostrikov, D. Yu. Dubov, and V. P. Gilyova, *Siberian Branch of the USSR Academy of Sciences, Novosibirsk, USSR*

Homogeneous Condensation in H_20 - Vapor Freejets 354
C. Dankert and H. Legge, *DFVLR Institute for Experimental Fluid Mechanics, Göttingen, Federal Republic of Germany*

Formation of Ion Clusters in High-Speed Supersaturated CO_2 Gas Flows 366
P. J. Wantuck, *Los Alamos National Laboratory, Los Alamos, New Mexico,* and R. H. Krauss and J. E. Scott Jr., *University of Virginia, Charlottesville, Virginia*

MD-Study of Dynamic-Statistic Properties of Small Clusters..........381
S. F. Chekmarev and F. S. Liu, *Siberian Branch of the USSR Academy of Sciences, Novosibirsk, USSR*

Chapter V. Evaporation and Condensation 401

Angular Distributions of Molecular Flux Effusing from a Cylindrical Crucible Partially Filled with Liquid..................................403
Y. Watanabe, K. Nanbu, and S. Igarashi, *Tohoku University, Sendai, Japan*

Numerical Studies on Evaporation and Deposition of a Rarefied Gas in a Closed Chamber...418
T. Inamuro, *Mitsubishi Heavy Industries, Ltd., Yokohama, Japan*

Transition Regime Droplet Growth and Evaporation: An Integrodifferential Variational Approach.......................434
J. W. Cipolla Jr., *Northeastern University, Boston, Massachusetts,* and S. K. Loyalka, *University of Missouri-Columbia, Columbia, Missouri*

Molecular Dynamics Studies on Condensation Process of Argon......439
T. Sano, *Tokai University, Kitakaname, Hiratsuka, Kanagawa, Japan,* and S. Kotake, *University of Tokyo, Hongo, Bunkyo-ku, Tokyo, Japan*

Condensation and Evaporation of a Spherical Droplet in the Near Free Molecule Regime................................447
J. C. Barrett and B. Shizgal, *University of British Columbia, Vancouver, British Columbia, Canada*

Theoretical and Experimental Investigation of the Strong Evaporation of Solids...460
R. Mager, G. Adomeit, and G. Wortberg, *Rheinisch-Westfälische Technische Hochschule Aachen, Aachen, Federal Republic of Germany*

Nonlinear Analysis for Evaporation and Condensation of a Vapor-Gas Mixture Between the Two Plane Condensed Phases. Part I: Concentration of Inert Gas ~ $O(1)$ 470
Y. Onishi, *Tottori University, Tottori, Japan*

Nonlinear Analysis for Evaporation and Condensation of a Vapor-Gas Mixture Between the Two Plane Condensed Phases. Part II: Concentration of Inert Gas ~ $O(Kn)$.......................492
Y. Onishi, *Tottori University, Tottori, Japan*

Author Index for Volume 117 514

List of Series Volumes...515

Other Volumes in the Rarefied Gas Dynamics Series..........522

Table of Contents for Companion Volume 116

Preface xix

Chapter I. Rarefied Atmospheres 1

Nonequilibrium Nature of Ion Distribution Functions in the High Latitude Auroral Ionosphere 3
B. Shizgal, *University of British Columbia, Vancouver, British Columbia, Canada,* and D. Hubert, *Observatoire de Meudon, Meudon,France*

VEGA Spacecraft Aerodynamics in the Gas-Dust Rarefied Atmosphere of Halley's Comet 23
Yu. A. Rijov, S. B. Svirschevsky, and K. N. Kuzovkin, *Moscow Aviation Institute, Moscow, USSR*

Oscillations of a Tethered Satellite of Small Mass due to Aerodynamic Drag 40
E. M. Shakhov, *USSR Academy of Sciences, Moscow, USSR*

Chapter II. Plasmas 53

Semiclassical Approach to Atomic and Molecular Interactions 55
J. A. Kunc, *University of Southern California, Los Angeles, California*

Monte Carlo Simulation of Electron Swarm in a Strong Magnetic Field 76
K. Koura, *National Aerospace Laboratory, Chofu, Tokyo, Japan*

Collisional Transport in Magnetoplasmas in the Presence of Differential Rotation 89
M. Tessarotto, *Università degli Studi di Trieste, Trieste, Italy,* and P. J. Catto, *Lodestar Research Corporation, Boulder, Colorado*

Electron Oscillations, Landau, and Collisional Damping in a Partially Ionized Plasma 102
V. G. Molinari and M. Sumini, *Università di Bologna, Bologna, Italy,* and B. D. Ganapol, *University of Arizona, Tucson, Arizona*

Bifurcating Families of Periodic Traveling Waves in Rarefied Plasmas 115
J. P. Holloway and J. J. Dorning, *University of Virginia, Charlottesville, Virginia*

Thruster Plume Impingement Forces Measured in a Vacuum Chamber and Conversion to Real Flight Conditions 226
A. W. Rogers, *Hughes Aircraft Company, El Segundo, California,* J. Allègre and M. Raffin, *Société d'Etudes et de Services pour Souffleries et Installations Aérothermodynamiques (SESSIA), Levallois-Perret, France,* and J.-C. Lengrand, *Laboratoire d'Aérothermique du Centre National de la Recherche Scientifique, Meudon, France*

Neutralization of a 50-MeV H⁻Beam Using the Ring Nozzle 241
N. S. Youssef and J. W. Brook, *Grumman Corporation, Bethpage, New York*

Chapter V. Tube Flow 255

Rarefied Gas Flow Through Rectangular Tubes: Experimental and Numerical Investigation 257
A. K. Sreekanth and A. Davis, *Indian Institute of Technology, Madras, India*

Experimental Investigation of Rarefied Flow Through Tubes of Various Surface Properties 273
J. Curtis, *University of Sydney, Sydney, New South Wales, Australia*

Monte Carlo Simulation on Mass Flow Reduction due to Roughness of a Slit Surface 283
M. Usami, T. Fujimoto, and S. Kato, *Mie University, Kamihama-cho, Tsu-shi, Japan*

Chapter VI. Expansion Flowfields 299

Translational Nonequilibrium Effects in Expansion Flows of Argon 301
D. H. Campbell, *University of Dayton Research Institute, Air Force Astronautics Laboratory, Edwards Air Force Base, California*

Three-Dimensional Freejet Flow from a Finite Length Slit 312
A. Rosengard, *Commissariat à l'Energie Atomique, Centre d'Etudes Nucléaires de Saclay, France*

Modification of the Simons Model for Calculation of Nonradial Expansion Plumes 327
I. D. Boyd and J. P. W. Stark, *University of Southampton, Southampton, England, United Kingdom*

Simulation of Multicomponent Nozzle Flows into a Vacuum 340
D. A. Nelson and Y. C. Doo, *The Aerospace Corporation, El Segundo, California*

Kinetic Theory Model for the Flow of a Simple Gas from a Two-Dimensional Nozzle 352
B. R. Riley, *University of Evansville, Evansville, Indiana,* and K. W. Scheller, *University of Notre Dame, Notre Dame, Indiana*

Transient and Steady Inertially Tethered Clouds of Gas in a Vacuum 363
T. L. Farnham and E. P. Muntz, *University of Southern California, Los Angeles, California*

Radially Directed Underexpanded Jet from a Ring-Shaped Nozzle 378
K. Teshima, *Kyoto University of Education, Kyoto, Japan*

Three-Dimensional Structures of Interacting Freejets 391
T. Fujimoto and T. Ni-Imi, *Nagoya University, Furo-cho, Chikusa-ku, Nagoya, Japan*

Flow of a Freejet into a Circular Orifice in a Perpendicular Wall 407
A. M. Bishaev, E. F. Limar, S. P. Popov, and E. M. Shakhov, *USSR Academy of Sciences, Moscow, USSR*

Chapter III. Atomic Oxygen Generation and Effects 127

Laboratory Simulations of Energetic Atom Interactions Occurring in Low Earth Orbit 129
G. E. Caledonia, *Physical Sciences, Inc., Andover, Massachusetts*

High-Energy/Intensity CW Atomic Oxygen Beam Source 143
J. B. Cross and N. C. Blais, *Los Alamos National Laboratory, Los Alamos, New Mexico*

Development of Low-Power, High Velocity Atomic Oxygen Source 156
J. P. W. Stark and M. A. Kinnersley, *University of Southampton, Southampton, England, United Kingdom*

Options for Generating Greater Than 5-eV Atmospheric Species 171
H. O. Moser, *Kernforschungszentrum Karlsruhe, Karlsruhe, Federal Republic of Germany*, and A. Schempp, *University of Frankfurt, Frankfurt, Federal Republic of Germany*

Laboratory Results for 5-eV Oxygen Atoms on Selected Spacecraft Materials 180
G. W. Sjolander and J. F. Froechtenigt, *Martin Marietta Corporation, Denver, Colorado*

Chapter IV. Plumes 187

Modeling Free Molecular Plume Flow and Impingement by an Ellipsoidal Distribution Function 189
H. Legge, *DFVLR, Göttingen, Federal Republic of Germany*

Plume Shape Optimization of Small Attitude Control Thrusters Concerning Impingement and Thrust 204
K. W. Naumann, *Franco-German Research Institute of Saint-Louis (ISL), Saint-Louis, France*

Backscatter Contamination Analysis 216
B. C. Moore, T. S. Mogstad, S. L. Huston, and J. L. Nardacci Jr., *McDonnell Douglas Space Systems Company, Huntington Beach, California*

Chapter VII. Surface Interactions 417

Particle Surface Interaction in the Orbital Context: A Survey 419
F. C. Hurlbut, *University of California at Berkeley, Berkeley, California*

Sensitivity of Energy Accommodation Modeling of Rarefied Flow Over Re-Entry Vehicle Geometries Using DSMC 451
T. J. Bartel, *Sandia National Laboratories, Albuquerque, New Mexico*

Determination of Momentum Accommodation from Satellite Orbits: An Alternative Set of Coefficients 463
R. Crowther and J. P. W. Stark, *University of Southampton, Southampton, England, United Kingdom*

Upper Atmosphere Aerodynamics: Gas-Surface Interaction and Comparison with Wind-Tunnel Experiments 476
M. Pandolfi and M. G. Zavattaro, *Politecnico di Torino, Torino, Italy*

Nonreciprocity in Noble-Gas Metal-Surface Scattering 487
K. Bärwinkel and S. Schippers, *University of Osnabrück, Osnabrück, Federal Republic of Germany*

Studies of Thermal Accommodation and Conduction in the Transition Regime 502
L. B. Thomas, C. L. Krueger, and S. K. Loyalka, *University of Missouri, Columbia, Missouri*

Large Rotational Polarization Observed in a Knudsen Flow of H_2-Isotopes Between LiF Surfaces 517
L. J. F. Hermans and R. Horne, *Leiden University, Leiden, The Netherlands*

Internal State-Dependent Molecule-Surface Interaction Investigated by Surface Light-Induced Drift 530
R. W. M. Hoogeveen, R. J. C. Spreeuw, G. J. van der Mee, and L. J. F. Hermans, *Leiden University, Leiden, The Netherlands*

Models for Temperature Jumps in Vibrationally Relaxing Gases 542
R. Brun, S. Elkeslassy, and I. Chemouni, *Université de Provence-Centre Saint Jérôme, Marseille, France*

Variational Calculation of the Slip Coefficient and the Temperature Jump for Arbitrary Gas-Surface Interactions 553
C. Cercignani, *Politecnico di Milano, Milano, Italy,* and M. Lampis, *Università di Udine, Udine, Italy*

Author Index for Volume 116 562

List of Series Volumes 563

Other Volumes in the Rarefied Gas Dynamics Series 570

Table of Contents for Companion Volume 118

Preface .... **xix**

Chapter I. Kinetic Theory **1**

Well-Posedness of Initial and Boundary Value Problems for the Boltzmann Equation **3**
R. E. Caflisch, *Courant Institute of Mathematical Sciences, New York University, New York, New York*

Stationary Flows from a Model Boltzmann Equation **15**
Y. Y. Azmy and V. Protopopescu, *Oak Ridge National Laboratory, Oak Ridge, Tennessee*

A Tensor Banach Algebra Approach to Abstract Kinetic Equations **29**
W. Greenberg, *Virginia Polytechnic Institute and State University, Blacksburg, Virginia,* and C. V. M. van der Mee, *University of Delaware, Newark, Delaware*

Singular Solutions of the Nonlinear Boltzmann Equation **39**
J. Polewczak, *Virginia Polytechnic Institute and State University, Blacksburg, Virginia*

Spatially Inhomogeneous Nonlinear Dynamics of a Gas Mixture **48**
V. C. Boffi, *University of Bologna, Bologna, Italy,* and G. Spiga, *University of Bari, Bari, Italy*

Diffusion of a Particle in a Very Rarefied Gas **61**
B. Gaveau, *Université Pierre et Marie Curie, Paris, France,* and M.-A. Gaveau, *Centre d'Etudes Nucléaires de Saclay, Gif-sur-Yvette, France*

Heat Transfer and Temperature Distribution in a Rarefied Gas Between Two Parallel Plates with Different Temperatures: Numerical Analysis of the Boltzmann Equation for a Hard Sphere Molecule **70**
T. Ohwada, K. Aoki, and Y. Sone, *Kyoto University, Kyoto, Japan*

Chapter II. Discrete Kinetic Theory **83**

Low-Discrepancy Method for the Boltzmann Equation **85**
H. Babovsky, F. Gropengiesser, H. Neunzert, J. Struckmeier, and B. Wiesen, *University of Kaiserslautern, Kaiserslautern, Federal Republic of Germany*

Investigations of the Motion of Discrete-Velocity Gases **100**
D. Goldstein, B. Sturtevant, and J. E. Broadwell, *California Institute of Technology, Pasadena, California*

Discrete Kinetic Theory with Multiple Collisions: Plane Six-Velocity Model and Unsteady Couette Flow **118**
E. Longo and R. Monaco, *Politecnico di Torino, Torino, Italy*

Exact Positive (2 + 1)-Dimensional Solutions to the Discrete Boltzmann Models 131
H. Cornille, *Physique Théorique CEN-CEA Saclay, Gif-sur Yvette, France*

Initial-Value Problem in Discrete Kinetic Theory 148
S. Kawashima, *Kyushu University, Fukuoka, Japan,* and H. Cabannes, *Université Pierre et Marie Curie, Paris, France*

Study of a Multispeed Cellular Automaton 155
B. T. Nadiga, J. E. Broadwell, and B. Sturtevant, *California Institute of Technology, Pasadena, California*

Direct Statistical Simulation Method and Master Kinetic Equation 171
M. S. Ivanov, S. V. Rogasinsky, and V. Ya. Rudyak, *USSR Academy of Sciences, Novosibirsk, USSR*

Fractal Dimension of Particle Trajectories in Ehrenfest's Wind-Tree Model 182
P. Mausbach, *Rheinisch Westfälische Technische Hochschule Aachen, Aachen, Federal Republic of Germany,* and *Eckard Design GmbH, Cologne, Federal Republic of Germany*

Scaling Rules and Time Averaging in Molecular Dynamics Computations of Transport Properties 194
I. Greber, *Case Western Reserve University, Cleveland, Ohio,* and H. Wachman, *Massachusetts Institute of Technology, Cambridge, Massachusetts*

Chapter III. Direct Simulations 209

Perception of Numerical Methods in Rarefied Gasdynamics 211
G. A. Bird, *University of Sydney, Sydney, New South Wales, Australia*

Comparison of Parallel Algorithms for the Direct Simulation Monte Carlo Method: Application to Exhaust Plume Flowfields 227
T. R. Furlani and J. A. Lordi, *Calspan Advanced Technology Center, Buffalo, New York*

Statistical Fluctuations in Monte Carlo Calculations 245
I. D. Boyd and J. P. W. Stark, *University of Southampton, Southampton, England, United Kingdom*

Applicability of the Direct Simulation Monte Carlo Method in a Body-fitted Coordinate System 258
T. Shimada, *Nissan Motor Company, Ltd., Tokyo, Japan,* and T. Abe, *Institute of Space and Astronautical Science, Kanagawa, Japan*

Validation of MCDS by Comparison of Predicted with Experimental Velocity Distribution Functions in Rarefied Normal Shocks 271
G. C. Pham-Van-Diep and D. A. Erwin, *University of Southern California, Los Angeles, California*

Direct Monte Carlo Calculations on Expansion Wave Structure Near a Wall 284
F. Seiler, *Deutsch-Französisches Forschungsinstitut Saint-Louis (ISL), Saint-Louis, France,* and B. Schmidt, *University of Karlsruhe, Karlsruhe, Federal Republic of Germany*

Chapter IV. Numerical Techniques 295

Numerical Analysis of Rarefied Gas Flows by Finite-Difference Method 297
K. Aoki, *Kyoto University, Kyoto, Japan*

Application of Monte Carlo Methods to Near-Equilibrium Problems 323
S. M. Yen, *University of Illinois at Urbana-Champaign, Urbana, Illinois*

Direct Numerical Solution of the Boltzmann Equation for Complex Gas Flow Problems 337
S. M. Yen and K. D. Lee, *University of Illinois at Urbana-Champaign, Urbana, Illinois*

Advancement of the Method of Direct Numerical Solving of the Boltzmann Equation 343
F. G. Tcheremissine, *USSR Academy of Sciences, Moscow, USSR*

New Numerical Strategy to Evaluate the Collision Integral of the Boltzmann Equation 359
Z. Tan, Y.-K. Chen, P. L. Varghese, and J. R. Howell, *The University of Texas at Austin, Austin, Texas*

Comparison of Burnett, Super-Burnett, and Monte Carlo Solutions for Hypersonic Shock Structure 374
K. A. Fiscko, *U. S. Army* and *Stanford University,* and D. R. Chapman, *Stanford University, Stanford, California*

Density Profiles and Entropy Production in Cylindrical Couette Flow: Comparison of Generalized Hydrodynamics and Monte Carlo Results 396
R. E. Khayat and B. C. Eu, *McGill University, Montreal, Quebec, Canada*

Chapter V. Flowfields 411

Direct Simulation of AFE Forebody and Wake Flow with Thermal Radiation 413
J. N. Moss and J. M. Price, *NASA Langley Research Center, Hampton, Virginia*

Direct Monte Carlo Simulations of Hypersonic Flows Past Blunt Bodies 432
W. Wetzel and H. Oertel, *DFVLR/AVA, Göttingen, Federal Republic of Germany*

Direct Simulation of Three-Dimensional Flow About the AFE Vehicle at High Altitudes 447
M. C. Celenligil, *Vigyan Research Associates, Inc., Hampton, Virginia,* J. N. Moss, *NASA Langley Research Center, Hampton, Virginia,* and G. A. Bird, *University of Sydney, Sydney, New South Wales, Australia*

Knudsen-Layer Properties for a Conical Afterbody in Rarefied Hypersonic Flow 462
G. T. Chrusciel and L. A. Pool, *Lockheed Missiles and Space Company, Inc., Sunnyvale, California*

Approximate Calculation of Rarefied Aerodynamic Characteristics of Convex Axisymmetric Configurations 476
Y. Xie and Z. Tang, *China Aerodynamics Research and Development Centre, Mianyang, Sichuan, People's Republic of China*

Procedure for Estimating Aerodynamics of Three-Dimensional Bodies in Transitional Flow 484
J. L. Potter, *Vanderbilt University, Nashville, Tennessee*

Drag and Lift Measurements on Inclined Cones Using a Magnetic Suspension and Balance 493
R. W. Smith and R. G. Lord, *Oxford University, Oxford, England, United Kingdom*

Three-Dimensional Hypersonic Flow Around a Disk with Angle of Attack 500
K. Nanbu, S. Igarashi, and Y. Watanabe, *Tohoku University, Sendai, Japan,* and H. Legge and G. Koppenwallner, *DFVLR, Göttingen, Federal Republic of Germany*

Direct Simulation Monte Carlo Method of Shock Reflection on a Wedge 518
F. Seiler, *Deutsch-Französisches Forschungsinstitut Saint-Louis (ISL), Saint-Louis, France,* H. Oertel, *DFVLR-AVA, Göttingen, Federal Republic of Germany,* and B. Schmidt, *University of Karlsruhe, Karlsruhe, Federal Republic of Germany*

Interference Effects on the Hypersonic, Rarefied Flow About a Flat Plate 532
R. G. Wilmoth, *NASA Langley Research Center, Hampton, Virginia*

Numerical Simulation of Supersonic Rarefied Gas Flows Past a Flat Plate: Effects of the Gas-Surface Interaction Model on the Flowfield 552
C. Cercignani and A. Frezzotti, *Politecnico di Milano, Milano, Italy*

Rarefied Flow Past a Flat Plate at Incidence 567
V. K. Dogra, *ViRA, Inc., Hampton, Virginia,* and J. N. Moss and J. M. Price, *NASA Langley Research Center, Hampton, Virginia*

Monte Carlo Simulation of Flow into Channel with Sharp Leading Edge 582
M. Yasuhara, Y. Nakamura, and J. Tanaka, *Nagoya University, Nagoya, Japan*

Structure of Incipient Triple Point at the Transition from Regular Reflection to Mach Reflection 597
B. Schmidt, *University of Karlsruhe, Karlsruhe, Federal Republic of Germany*

Author Index for Volume 118 608

List of Series Volumes 609

Other Volumes in the Rarefied Gas Dynamics Series 616

Preface

The 16th International Symposium on Rarefied Gas Dynamics (RGD 16) was held July 10–16, 1988 at the Pasadena Convention Center, Pasadena, CA. As anticipated, the resurging interest in hypersonic flight, along with escalating space operations, has resulted in a marked increase in attention being given to phenomena associated with rarefied gas dynamics. One hundred and seventy-three registrants from thirteen countries attended. Spirited technical exchanges were generated in several areas. The Direct Simulation Monte Carlo technique and topics in discrete kinetic theory techniques were popular subjects. The inclusion of inelastic collisions and chemistry in the DSMC technique drew attention. The Boltzmann Monte Carlo technique was extended to more complex flows; an international users group in this area was formed as a result of the meeting.

Space-related research was discussed at RGD 16 to a greater extent than at previous meetings. As a result of Shuttle glow phenomena and oxygen-atom erosion of spacecraft materials, the subject of energetic collisions of gases with surfaces in low Earth orbit made a strong comeback.

There were 11 excellent invited lectures on a variety of subjects that added significantly to the exchange of ideas at the symposium.

The Symposium Proceedings have, for the first time, been divided into three volumes. A very high percentage of the papers presented at the symposium were submitted for publication (practically all in a timely manner). Because of the large number of high-quality papers, it was deemed appropriate to publish three volumes rather than the usual two. An additional attraction is that the size of each volume is a little more convenient. Papers were initially reviewed for technical content by the session chairmen at the meeting and further reviewed by the proceedings' editors and additional experts as necessary.

In the proceedings of RGD 16, there are 107 contributed papers and 11 extended invited papers dealing with the kinetic theory of gas flows and transport phenomena, external and internal rarefied gas flows, chemical and internal degree-of-freedom relaxation in gas flows, partially ionized plasmas, Monte Carlo simulations of rarefied flows, development of high-speed atmospheric simulators, surface-interaction phenomena, aerosols and clusters, condensation and evaporation phenomena, rocket-plume flows, and experimental techniques for rarefied gas dynamics.

The Rarefied Gas Dynamics Symposia have a well-deserved reputation for being hospitable events. RGD 16 continued this admirable tradition. It is with much appreciation that we acknowledge Jan Muntz's contributions to RGD history along with Jetty and Miller Fong and Noel Corngold.

Many people helped make the symposium a success. Gail Dwinell supervised organizational details for the symposium as well as the pre-symposium correspondence. Eric Muntz was responsible for the computerized registration. Kim Palos, whose services were provided by the University of Dayton Research Institute, assisted with registration and retyped several manuscripts. Jerome Maes also retyped a number of manuscripts for these volumes. Nancy Renick and Marilyn Litvak of the Travel Arrangers of Pasadena were tireless in their efforts to assist the delegates.

E. P. Muntz
D. P. Weaver
D. H. Campbell
May 1989

Chapter 1. Inelastic Collisions

Inelastic Collision Models for Monte Carlo Simulation Computation

J. K. Harvey*
Imperial College, London, England, United Kingdom

Abstract

This paper is concerned with the use of the Monte Carlo direct simulation method to predict upper atmosphere flow fields and the choice of models to describe the molecular collisions is the central theme. The requirements for these to portray the scattering after collisions is examined and examples are cited where a sensitivity to the choice of post-collision particle trajectories has been observed. The range of models available to predict the internal energy exchange for non-reacting inelastic encounters is enumerated and the limitation imposed by the use of the currently favoured stochastic models for reacting collisions is discussed. The paper concludes with a brief description of a new approach to modelling gas-surface interaction in which non-reacting inelastic exchange of rotational and translational energy takes place.

Nomenclature

$\underline{c}$ = velocity
$\underline{c}_i$ = internal energy
g^* = normalised relative velocity
I_f = Indicator function
I_t = Transition probability
k_b = Boltzman constant
k_n = Knudsen number
N_m = Number of internal modes
R = $(\theta,\phi,\varepsilon)$

Presented as an Invited Paper

*Reader in Gas Dynamics, Department of Aeronautics.

T_s	=	Surface temperature
$\underline{x}$	=	position
Z	=	Impact parameter
γ	=	(number of degrees of freedom)/2
Γ	=	Gamma function
θ	=	angle measured with respect to surface normal
χ	=	deflection
Ω	=	scattering solid angle
$()_i$	=	internal
$()_s$	=	surface
$()_t$	=	translational

Introduction

The Monte Carlo Direct Simulation (MCDS) method pioneered by Bird[1] is now well established as the most practical numerical technique available for solving rarefied flow field problems where the Knudsen number, Kn, > O(0.01). The method assumes that the motion of an ensemble of several thousand "simulator" particles can adequately model that of the far larger number of molecules in the real flow. The particles move within a framework of cells in the physical space and carry with them the phase-space information of position, $\underline{x}$, velocity, $\underline{c}$, and internal energy, $\underline{e}_i$, The identity of individual species and/or quantum levels can also be retained if appropriate. The technique uses the concept of decoupling the convective and collision terms through suitable time discretization. The motion of the particles is advanced alternately with them moving freely through space for one time interval and then colliding with randomly chosen near neighbours for a second interval. During this later phase $\underline{c}$, $\underline{e}_i$ and, if reactions take place, the specie of each particle can alter. The changes in these parameters are dependent on the selection of a suitable collision model. An appreciation of some of the available choices for inelastic collisions is the subject of this paper.

Various criticisms of Bird's method have been voiced (Nanbu[2] and Yen[3]). It is argued that the technique does not solve the Boltzmann equation rigorously. The most serious objections are centred on the way the collision rate is computed during the molecular interaction phase. This is not the appropriate place to defend Bird's method except to say that when the calculations are performed carefully, the MCDS method appears to yield results that agree very precisely with experiments (Harvey[4]) for non-reacting hypersonic flows. Reliable evidence to validate the solutions for equivalent reacting flows is not available. Bird[5] has proposed two methods of assigning an appropriate collision frequency. Both appear to yield accurate answers and the choice between the two is dependent more on computation expediency and intuitive advantages than any obvious mathematical rigour. Bird's method of handling the collisions is not based on the same one particle concept of the Boltzmann collision integral but it is a plausible simulation that mimics the behaviour of pairs of the molecules in the real flow.

The MCDS scheme requires the molecular collisions, either between themselves or with any solid surface bounding the flow, to be modelled as accurately as possible. Most practical applications involve diatomic gases that can be excited vibrationally and rotationally and for which reactions, dissociation and ionisation are possible. The routines to compute the interactions will be called possibly millions of times during the course of a calculation and it is essential for them to be very brief. In choosing the algorithms a careful balance has to be struck between realism and extreme efficiency.

The event of powerful computers has heralded advances in the field of molecular dynamics. Accurate computations have been made (see for example Alder and Wainright[6] and the references of Page et al[7] in which collisions are dealt with in detail either quantum mechanically or semiclassically. In the latter a quantum mechanical calculation is made to determine the change in the populations of individual levels but the trajectories of approaching molecules are handled classically. In many of the examples in the literature the chemical reaction dynamics are of prime interest; hence fairly sweeping simplifications are often made. Frequently a non-rotating linear collision is assumed with the emphasis being placed on determining the behaviour of the vibrational quantum levels because these play a predominant role in the reaction processes. The full quantum mechanical calculations are regarded as precise benchmarks, and the semiclassical models need some "adjustments" to get them to agree and to force them to satisfy conservation of energy and detail balance.

It is completely out of the question to contemplate including these molecular dynamic methods into MCDS computations, since they require prodigious amounts of computing. However their relevance should not be discounted because they have the potential of yielding comprehensive collision cross-section data for a wide range of impact variables, in particular, for reacting species. Such information could be stored and interpolated within the MCDS program. For accurate prediction of rarefied reacting and radiating flows knowledge relating to individual excited quantum levels is needed. Such information is not generally available and one must look elsewhere for appropriate simpler inelastic collision models. In essence these deal with the interaction in two steps: (1) the scattering i.e., the determination of the post collision deflections, and (2) the computation of the energy and specie exchange during the interaction. We will examine each of these in turn.

Scattering Procedures

The main application of the MCDS method so far has been for spacecraft and generally it has been felt that not much emphasis needs to be placed on accurately determining the deflections caused by the collision. This is because computations that test the scattering models have detected very little sensitivity to the choice. Macrossan[8] reported that even when 90 deg was inadvertently added to each post collision angle, the results were tolerably close to the correct ones. The reason appears to be that for moderately dense situations, the orientation of most of the simulator collisions within the bulk of the disturbed flow is, to all intents and purposes, random. Thus the direction in which the particles move after a collision is of little consequence.

For polyatomic molecules it is expedient to calculate the scattering by representing each molecule with a single potential placed at the center of mass rather than accounting for each atom individually. Wu and Ohamua[9] and Pullin[10] have shown that the perturbation in direction from the monatomic trajectory due to internal energy exchange is of small order and thus can be justifiably neglected. The postcollision directions are thus computed as if the interaction were monatomic. The magnitudes of the particle velocities then have to be scaled by a common factor to compensate for the energy exchanged internally (generally a small fraction of the center-of-mass energy). This ensures energy conservation. The same procedure has been applied to chemically reacting collisions with dubious justification.

The most popular scattering model is Bird's variable hard sphere (VHS) model[11] for which the collision cross-section is reduced with increasing centre of mass energy. With an appropriate variation in the diameter a very close approximation to the observed change in viscosity with temperature can be obtained. The hard sphere isotropic scattering is accepted as being inappropriate but the model is very economical to compute. It is generally recommended for MCDS computations because it appears to give good flowfield results. We will examine if this is a universally valid conclusion.

The choices of pertinent scattering potentials include the inverse power (IP) family, of which the hard sphere and VHS models belong, and models which incorporate a long-range attractive force, i.e. the Lennard-Jones (6-12), Buckingham exp-6 and Morse models. Although examples from the former family have been most widely used in MCDS computations because of their simplicity, from the author's experience only a slight penalty in computing time is incurred if the latter more complex models are used via a look-up scheme. This involves storing an array of previously computed deflections and using a simple linear interpolation for each interaction.

As stated, in most situations the results are insensitive to the scattering model chosen. However two examples are cited here where noticeable differences are seen for different models and these serve to illustrate where caution should be exercised. The first is taken from Davies et al[12] in which the density profiles measured on a hollow cylinder in hypersonic flow are compared with MCDS calculations. Comparisons are made between the IP and Morse potential scattering.

It is clear from Figure 1 that the IP predictions give consistently wrong density gradients in the outer part of

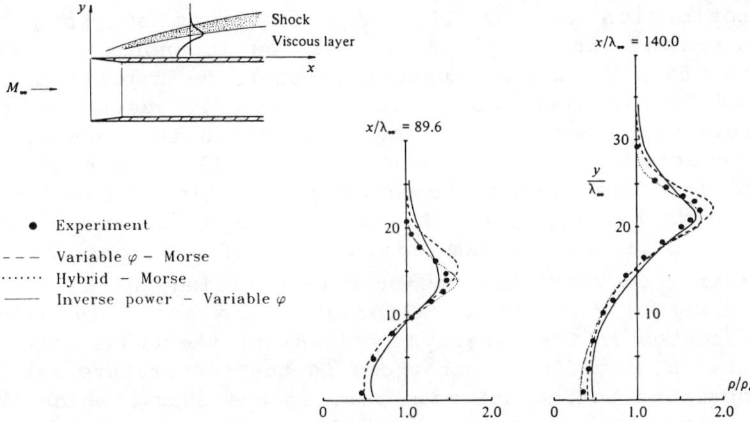

Fig. 1 Density profiles on a hollow cylinder at Mach 25 in Nitrogen.

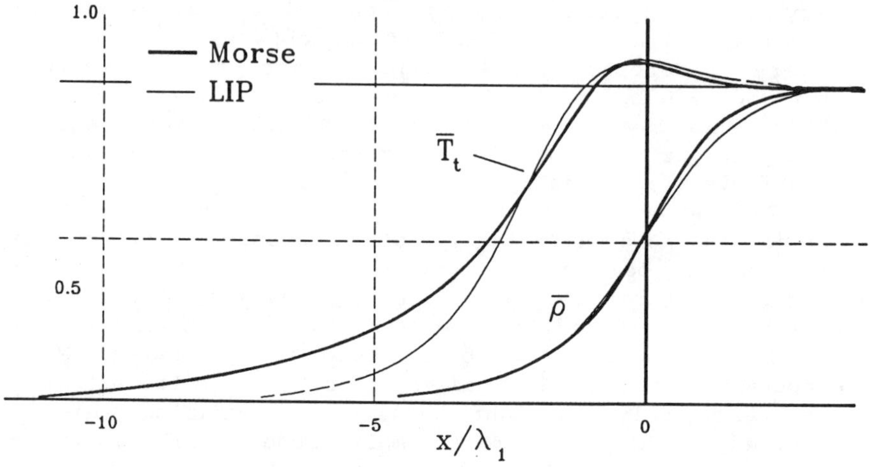

Fig. 2 Computed Mach 5 nitrogen normal shock wave profiles.

the flow, i.e. in the shock wave. The rest of the viscous layer is very well simulated. The shock is weak compared with that on a blunt body and it is a region where there are many primary freestream particles moving essentially in one direction into a cloud of slow "body reflected" particles. In this region the velocity distribution is markedly "two-stream" in character.

The second example, shown in Figure 2, presents the results of calculations made by Macrossan for a steady shock in nitrogen with a specularly reflecting downstream boundary that moves at the appropriate Rankine-Hugoniot velocity. The comparison is between the attractive Morse potential and LIP models. The latter is a very good approximation to the IP model that was suggested by Macrossan[8] in which a linear relationship between deflection, χ and the impact parameter, Z is presupposed. In both calculations a rapid rise is seen in the temperature almost before any increase in the density can be detected. This is due to a small number of hot particles from within the shock propagating forward into the lower density gas. Again we have a situation where there is a two-stream mixture of freestream almost uni-directed particles and ones originating in the denser hot body of the shock. Although there are only slight differences in the density profiles for the two scattering models, a significant departure in the temperature pattern is observed in the upstream part of the shock, which, had the flow been reacting, could have had significant repercussions. It should be noted that this normal shock is a stronger disturbance than the oblique shock of the previous example.

Figures 3 and 4 demonstrate how χ varies with Z for a variety of scattering models, all of which have the same viscosity cross sections. In the first figure the IP and Morse models are compared. For the former, χ, depends only on the single parameter, Z. The exponent for this model gives a good match for viscosity over an appropriate temperature range. The curve can be closely approximated by a straight line for most of its range which is the basis of Macrossan's LIP model[8]. For the Morse potential the deflections are also a function of the normalized particle relative velocity, g^*. For high missdistances inward deflections are possible due to the long-range attractive force but, as the relative velocity increases, this feature becomes less pronounced. Clearly such an effect will be missed in a calculation if the cutoff range of Z is set too low. From Figure 3 it is evident that any differences in the results between the IP model (or the very similar LIP model) and the Morse model are most likely to have arisen from collisions for which the miss distance is large.

Recalling molecular beam experiments, the scattering pattern of a stream of particles moving into a background gas is a function of the magnitude and not the sign of the collision deflection angles since the azimuthal orientation of encounters is random. The Morse model gives greater scattering than the IP models, in particular where the

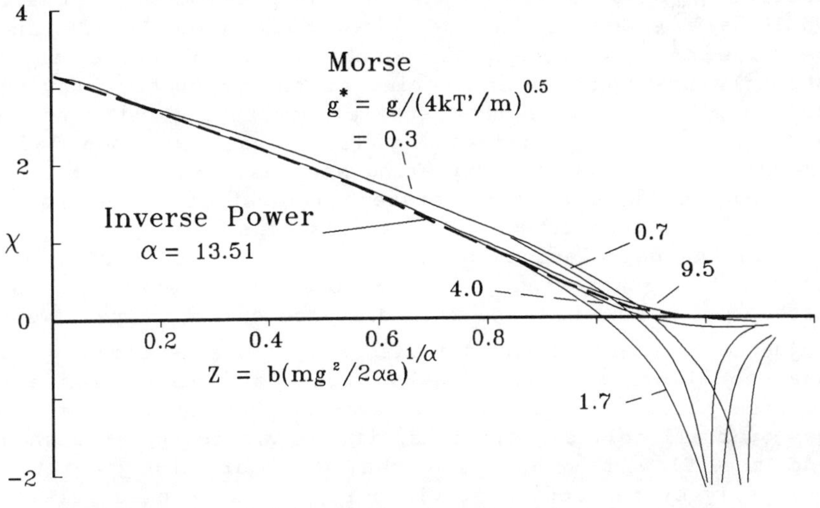

Fig. 3 Post collision deflections for inverse power and Morse potentials.

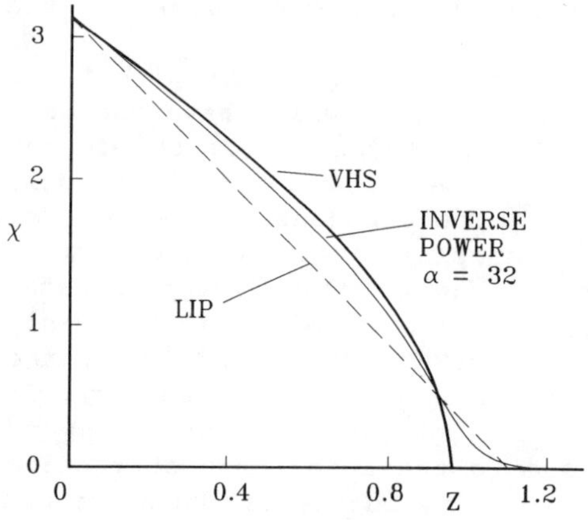

Fig. 4 Deflections for three further models.

inward deflections occur, and the pattern is different. The variations in the profiles seen in the shock waves in Figures 1 and 2 are thought to be attributable to this effect since everything else in the calculations was held constant. When there is a region where the orientation of one group of particles (for example, the freestream molecules in the edge of the disturbed flow) is not random there is evidence that the scattering model makes a difference to the computed results. Figures 3 and 4 show that deflection patterns of the much harder VHS model and the more realistic Morse model differ considerably and some caution should be exercised in the use of the former. There are examples in the literature (see Levine and Bernstein[13]) of chemical reactions where there is a strongly preferential scattering of the products. Having seen a sensitivity to the detailed scattering patterns in some regions of non-reacting flows caution should be exercise for flows with differing species. Where knowledge is available appropriate preferential scattering should be incorporated into future MCDS calculations.

In the calculations performed to test the sensitivity to the the scattering patterns, the parameters used for each model, including the miss distance cutoff values, were adjusted to ensure identical viscosity cross sections. The near-equilibrium Chapman-Enskog transport properties are derived through integrals that average the behavior of the gas over all collisions. Thus, if the viscosity is matched and the flow is one where the structure is generated primarily by the action of viscosity, it is not surprising that the effects of different scattering models are masked and the results appear very similar. This is not to say that in other highly non equilibrium situations where, the concept of a Chapman-Enskog viscosity is foreign, the results would be the same.

Energy Exchange

To complete the description of the collision, a model is required for the energy exchange process. Most of the flows of practical interest are of polyatomic gases and the requirement to include exchange with the vibrational and rotational internal modes presents a challenging problem due to the increase in the number of variables to be considered. The rotational coupling between colliding molecules is far stronger than the vibrational. Thus, for nonreacting hypersonic spacecraft, only rotational energy interchange is important except near the stagnation point where the gas can dwell long enough for the vibrational mode to be influential. However this is not so if chemical reactions take place because significant redistributions of the vibrational modes can occur during the collisions.

For nonreacting gases numerous models have be proposed to describe the internal energy exchange. These can be classed as classical, quantum mechanical, or stochastic, and examples of each type have been tried with varying success in MCDS calculations.

Examples in the first category are the models that were suggested by Parker[14] and Lordi and Mates[15] in their work on transport properties. Collisions are modeled deterministically with each atom being represented by a repulsive force potential that gyrates around the respective centres of the molecules as these approach and then move away from each other. Refinements to take some account of the long range attractive forces have been added. To reduce the number of variables to compute, "planar collisions" are assumed, i.e. only rotation in the collision plane is considered. This effectively disregards the orientation of the molecular rotation and a reduction in the spacing between atoms is made to compensate for this. Pullin[10] successfully used an extension of Parker's model in MCDS calculations. However he showed that it did not satisfy detailed balance and demonstrated that a MCDS computation for a relaxing gas did not tend toward equilibrium after a long time. This, in fact, is difficult to do rigorously, because the method exhibits random walk instabilities that manifest themselves if a relaxation simulation is allowed to run for too long. A good example of this is illustrated in

Figure 2 of Davis et al.[12] In most instances it is only possible to demonstrate analytically the failure of a model to satisfy detailed balance and not to quantify the extent. Failure in itself may be an excessively stringent criterion by which to judge a model which, in all events, will only be approximate in other respects.

Alternative classical formulations that could be used include rough and loaded spheres and axisymmetrical convex bodies. Pidduck[16] and Condiff et al.[17] have proposed rough sphere models but these are generally discredited because they give too high a rotational coupling if meaningful parameters are used for planar collisions. Melville's loaded sphere[18] and a variety of other earlier models in which the molecules are represented by ellipsoidal bodies of revolution or even disks have been used for the calculation of transport properties and for chemical kinetics. The chief objection to these is the patent lack of realism and the difficulty in handling efficiently the awkward geometry in a MCDS scheme. These models also fail to satisfy detailed balance.

Of the other classes of models, attempts to date to construct effective quantum mechanic models have not proved fruitful. Deiwart and Yoshikawa[19] and Itikawa and Yoshikawa[20] have used a formulation for inverse power molecules, but they were obliged to make such drastic simplifying assumptions that the resulting schemes are extremely limited and of dubious validity. There remains a vast chasm between the molecular dynamics models and practical algorithms suitable for MCDS which is certain to remain for a long time.

The third group of stochastic models has proved to be the most efficacious. These function by randomly selecting the post collision properties from a predetermined distribution which ensures that the fundamental principles for molecular collisions are satisfied on average. A paper by Larsen and Borgnakke[21] was a precursor of an important one published in 1975,[22] (B-L 2), which dealt with rotational exchange for a linear rigid molecule. In (B-L 2) the authors proposed a model in which for each interaction a fixed fraction of total internal energy is scattered elastically. The inelastic part of the exchange is modeled by redistributing the total center-of-mass energy randomly between the translational and rotational energies according to a parabolic probability. The summed resultant rotational energies are then randomly shared between the two molecules with a uniform distribution. The postcollision particle velocities, derived from monatomic scattering, then have to be uniformly scaled to ensure overall conservation of energy.

The (B-L 2) model is currently the most favored for MCDS calculations, although other statistical models have been proposed by Davis et al.[12], Pullin[23] and Kuscer[24] which are improvements. Pullin claims that (B-L 2) does not satisfy detailed balance exactly, and this is because the rate for forward collisions is not precisely that for the reverse collisions. Nor does it allow variation of viscosity with temperature whereas all of the three alternatives cited above do. The models in Davis' paper change the fraction of the energy scattered elastically as this energy increases, permitting a good fit to be obtained for the temperature variation of viscosity. The examples of predicted flowfields given in the paper show a very close agreement with experiments. The Pullin and Kuscer models have several disposable parameters that permit other transport properties of the gas to be fitted empirically. Although these more comprehensive stochastic models impose a computing penalty when compared with (B-L 2), they remove some of its mathematical restrictions when applied to multispecies gases (see Bird[11]). Their usefulness for MCDS computations should be explored.

Reactions

The reactions of primary interest in spacecraft flows are between nitrogen, oxygen and their products and, even though nitrogen is chemically relatively inactive, the temperatures are high enough for a number of different transitions to occur including some of the ionized states. Further reactions may take place between the air and rocket exhaust plume. The basis for the scheme commonly used in the MCDS computations for reacting flows is given in Bird[1,25] and Moss and Bird[26] and is based on the VHS-Borgnakke and Larsen stratagem which unfortunately imposes some unwelcome mathematical constraints. Reaction cross sections are chosen with an inverse dependence on the total center-of-mass energy. If this does not exceed a specified threshold, no reaction takes place. Parameters within the models are adjusted to give the desired temperature dependence for the reaction rates; this is analogous to determining the parameters for the nonreacting model by matching the viscosity. Again, by comparing only an integrated characteristic of the computational gas with its real counterpart, detailed differences can be overlooked. Ignoring ionic exchanges, there are two distinct categories of chemical interaction: (1) those involving molecules with large reaction cross sections, which are associated with a low threshold energy and low energy release during the reaction and (2) those with small cross sections, which implies a large threshold. The latter can be either exoergic or endoergic. Examples of both types of reactions occur in the spacecraft flow.

Bird was obliged to make sweeping simplifications in devising his scheme, and two aspects are worthy of further consideration. First, there is evidence that the products of many chemical reactions are preferentially scattered (see, for example, Levine and Bernstein[13]) either in the forward or backward directions depending on whether they are type 1 or type 2 reactions. Since, as noted above, there is evidence that the slightly different molecular scattering pattern for non-reacting gas models can influence the results of a simulation, the more pronounced preferential species scattering for reacting encounters could well be of greater significance. Intuitively one expects that where particular species migrate to will have an important influence on the development of a flow. This remains to be tested, but, at present, appropriate scattering data do not appear to be available for the relevant reactions.

Second, the dependence of the outcome of the reaction on the total center-of-mass energy and the way the post collision energy is allocated among the possible modes is significantly more complicated than the VHS-Borgnakke and Larsen scheme[26] admits. Reaction rates are influenced by the vibrational and even rotational levels of excitations. Stupochenko and Osipov[27] have demonstrated that it is the higher vibrational level and not the translational energy that determines dissociation rates. Thus the higher level populations can be depleted, distorting the Boltzmann distribution, and eventually slowing down the reaction rate even though the gas temperature is high. Coupled vibrational-dissociation has been reported by Carlson and Rieper[28] for N_2 and by Koshi[29] for O_2. Clearly, to handle such effects the MCDS programs will have to retain information on the population of many quantum levels for each specie and will require data on the reaction cross-section for many states, which will have to be derived from molecular dynamic calculations. In exoergic type (1) collisions in which an "early" reaction takes place a considerable fraction of the postcollision energy can be redistributed into the vibrational modes. This is illustrated in Figure 5, which shows a plot of potential energy contours and the "trajectory" of a typical reaction. The oscillating path signifies a product with high vibrational energy. In

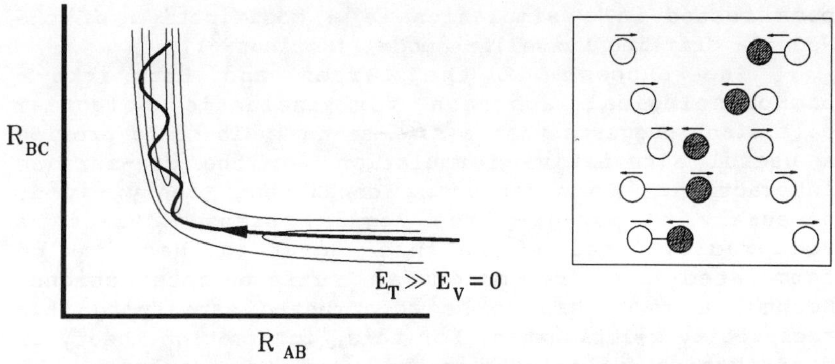

Fig. 5 A reaction leading to vibrational excitation.

reacting flows rapid changes in the levels of internal-mode excitation can be produced by a single collision which, in turn, can influence the reaction rate for next encounter. The importance of devising better schemes to realistically represent these features is self-evident. Phenomenological collision models derived from those of B-L, Pullin, and Kuscer are likely to be the most suitable for the purpose.

Surface Interactions

It is as important to model the interactions between gas molecules and a solid surface as accurately as those between the gas molecules themselves. For most engineering applications the distribution of particles reflected from "practical" solids appear to correspond closely to the diffuse/fully thermally accommodated pattern. There is evidence (Blanchard and Larman[30]) that is not quite consistent with observations from upper atmosphere flight trials. The effect of deviations from this assumption have been studies with MCDS calculations by applying the Maxwell boundary condition - a linear combination of fully accommodated diffuse and specular reflections - which is patently physically unrealistic. Numerous attempts have been made to find more satisfactory ways of predicting the gas-surface interaction which range from simple empirical to complex quantum lattice models. A brief review of them can be found in Agbormbai[31]. In general none of these models performs well and most are far too complicated for inclusion in MCDS calculations. They do not address the question of how to apportion the internal energy. Other than the Maxwell model, the only alternative that has been tested in a simulation is a modification of the Nochila drifting Maxwellian model (Hurlbut[32]).

The success of the Larsen and Borgnakke[21] phenomenological approach for inelastic molecular collisions suggests that a similar analysis could provide a useful alternative formulation for the gas-surface interaction. To construct a consistent theory, it is necessary to pursue three logical steps. First, a reciprocity/detailed balance equation has to be formulated for free-molecular surface interactions. Second, a model has to be constructed satisfying this reciprocity relationship. For this, information theory or distribution calculus can be used to determine the functional form of the distributions from which the post interaction properties are randomly sampled. As a final step it should be verified that the model accords with the H-theorem.

Agbormbai[31] has derived a generalized reciprocity equation for particles with N_m internal modes, each with $2\gamma_i$ degrees of freedom. These reflect from a surface which is composed of an array of atoms at a temperature, T_s, which oscillate with coupled simple harmonic motion. The equation, which is an extension of the Cercignani[33] reciprocity expression, is written in terms of the components of translational, internal, and surface-atom energies ε_t, ε_i, and ε_s respectively.

$$(\cos\theta')\epsilon_t'\epsilon_s'^{\gamma_s-1}\prod_{k=1}^{N_m}\epsilon_{ik}'^{\gamma_{ik}-1}\exp\left(-\frac{\epsilon_t'+\epsilon_s'+\sum_{k=1}^{N_m}\epsilon_{ik}'}{k_BT_s}\right)I_t(\mathbf{R}'\mid\mathbf{R}'')d\epsilon'd\Omega'd\epsilon''d\Omega''$$

$$= -(\cos\theta'')\epsilon_t''\epsilon_s''^{\gamma_s-1}\prod_{k=1}^{N_m}\epsilon_{ik}''^{\gamma_{ik}-1}\exp\left(-\frac{\epsilon_t''+\epsilon_s''+\sum_{k=1}^{N_m}\epsilon_{ik}''}{k_BT_s}\right)I_t(\mathbf{R}''\mid\mathbf{R}')d\epsilon''d\Omega''d\epsilon'd\Omega' \quad (1)$$

where θ is measured with respect to the surface normal and Ω is the scattering solid angle. The $I_t(\theta,\phi,\varepsilon)$ is a transition probability, analogous to the collision cross section in intermolecular collisions, which relates the pre- and postinteraction energies and trajectories.

After decoupling the scattering and the energy exchange, the evaluation of the transition probability can be avoided by using distribution calculus. This bypasses the detailed physics of the interaction but provides a family of possible distributions for the postinteraction properties that can be randomly sampled to determine the outcome of each collision. Agbormbai's formulation[31] is analogous to that of Pullin[23] for the molecular interaction, albeit different in mathematical detail. However the surface interaction problem is less well posed because of the absence of a convenient equation expressing momentum conservation. (The idealization of the surface and the existence of unspecified forces holding the solid lattice together precludes this; only the detailed balance and energy conservation equations are available.) The distribution functions include a number of disposable constants, which, when varied, give a smooth range of results ranging from fully diffuse to specular. These constants have to be set from experimental data. The results obtained for the angular scattering distributions are unrealistic and not useful, but the method provides plausible predictions within a rigorous framework for possible redistributions of the energy, including the internal modes.

We replace the transition probability with a density G(s), where s is a random correlating variable, i.e., as

$$I_t(\mathbf{R}' \mid \mathbf{R}'')d\epsilon''d\Omega'' = G(\mathbf{s}')d\mathbf{s}'$$

$$I_t(\mathbf{R}'' \mid \mathbf{R}')d\epsilon'd\Omega' = G(\mathbf{s}'')d\mathbf{s}''$$

where the single and double primes refer to pre- and post-collision states.

Writing $\gamma_t = 2$, $\xi = \varepsilon/k_bT_s$, and introducing the gamma distribution

$$Ga\langle x|\mu\rangle = \frac{1}{\Gamma(\mu)}x^{\mu-1}e^{-x}I_{f(|0,\infty|)}(x) \qquad (\mu > 0)$$

where $\Gamma(\mu)$ is the gamma and $I_f(x)$ the indicator functions, the reciprocity equation becomes

$$\cos\theta' Ga\langle\xi_t'|\gamma_t\rangle Ga\langle\xi_s'|\gamma_s\rangle \prod_{k=1}^{N_m} Ga\langle\xi_{ik}'|\gamma_{ik}\rangle G(\mathbf{s}')d\mathbf{s}'d\epsilon'd\Omega'$$

$$= -\cos\theta'' Ga\langle\xi_t''|\gamma_t\rangle Ga\langle\xi_s''|\gamma_s\rangle \prod_{k=1}^{N_m} Ga\langle\xi_{ik}''|\gamma_{ik}\rangle G(\mathbf{s}'')d\mathbf{s}''d\epsilon''d\Omega'' \tag{2}$$

The minus sign due to the reversal of normal velocity on reflection can be avoided by replacing the incident stream with a corresponding specularly reflected stream. Considering only one internal mode, the energy exchange component of both sides of the reciprocity equation take the form

$$dF = Ga\langle\xi_t|\gamma_t\rangle Ga\langle\xi_s|\gamma_s\rangle Ga\langle\xi_i|\gamma_i\rangle G_e(\mathbf{s}_e)d\mathbf{s}_e d\epsilon$$

Write

$$G_e(\mathbf{s}_e)d\mathbf{s}_e = \prod_{j=1}^{m} G_j(\mathbf{s}_j)d\mathbf{s}_j$$

and choose G_e to be the product of three Beta distributions

$$G_e(s_t, s_e, s_i) = \beta\langle s_t|\alpha_t\gamma_t, (1-\alpha_t)\gamma_t\rangle\beta\langle s_s|\alpha_s\gamma_s, (1-\alpha_s)\gamma_s\rangle\beta\langle s_i|\alpha_i\gamma_i, (1-\alpha_i)\gamma_i\rangle$$

Decomposing the energies into active and inactive fractions in the manner

$$\xi_t = \xi_{t1} + \xi_{t2}, \qquad s_t = \frac{\xi_{t1}}{\xi_{t1} + \xi_{t2}}$$

$$\xi_s = \xi_{s1} + \xi_{s2}, \qquad s_s = \frac{\xi_{s1}}{\xi_{s1} + \xi_{s2}}$$

$$\xi_i = \xi_{i1} + \xi_{i2}, \qquad s_i = \frac{\xi_{i1}}{\xi_{i1} + \xi_{i2}}$$

and recombining the active components yields

$$\xi_I = \xi_{i1} + \xi_{s1}, \qquad s_I = \frac{\xi_{i1}}{\xi_{i1} + \xi_{s1}}$$

$$\xi_a = \xi_{I1} + \xi_{t1}, \qquad s_a = \frac{\xi_{I1}}{\xi_{I1} + \xi_{t1}}$$

These transformations generate additional densities

$$G_i(\mathbf{s}_j)f_i(\xi_{j2}) = \beta\langle s_j|\alpha_j\gamma_j, (1-\alpha_j)\gamma_j\rangle Ga\langle \xi_{j2}|\alpha_j\gamma_j\rangle$$

Combining the transformations and identifying the right and left sides with the incidence and reflection respectively gives

$$\xi_t'' = (1-s_t')\xi_t' + (1-s_a')\xi_a$$
$$\xi_i'' = (1-s_i')\xi_i' + s_I's_a'\xi_a$$
$$\xi_s'' = (1-s_s')\xi_s' + (1-s_I')s_a'\xi_a$$

where

$$\xi_a = s_t'\xi_t' + s_i'\xi_i' + s_s'\xi_s' = s_t''\xi_t'' + s_i''\xi_i'' + s_s''\xi_s''$$

The associated correlation densities from which the random variables s are sampled are

$$G_e(\mathbf{s}_e) = \beta\langle s_t|\alpha_t\gamma_t, (1-\alpha_t)\gamma_t\rangle\beta\langle s_i|\alpha_i\gamma_i, (1-\alpha_i)\gamma_i\rangle\beta\langle s_s|\alpha_s\gamma_s, (1-\alpha_s)\gamma_s\rangle$$
$$\times\beta\langle s_I|\alpha_i\gamma_i, \alpha_s\gamma_s\rangle\beta\langle s_a|\alpha_i\gamma_i, +\alpha_s\gamma_s, \alpha_t\gamma_t\rangle$$

The adjustable α parameters represent the fraction of each energy mode that actually takes part in the exchange. When they are all zero there is no energy exchange and the process is elastic, corresponding to specular reflection. When they are unity, the process is fully inelastic with complete correlation between all the energy modes. This is close to, but not exactly, the fully accommodated limit. Further simplifications can be obtaned, which are analogous to Pullin's[23] restricted exchange and Borgnakke and Larsen's[26] molecular collision models. If the number of surface atoms is made indefinitely large, the vibrational energy of these becomes large and the correlation between their energy and those of each of the gas's modes disappears. Thus, considering each of the gas's modes separately one obtains

$$\xi_j{}'' = s'_j \xi'_j + \xi'_{j2}$$

$$s''_j = \frac{s'_j \xi'_j}{s'_j \xi'_j + \xi'_{j2}}$$

$$\xi''_{j2} = (1 - s'_j)\xi'_j$$

with

$$G_I(s_I)G_a(s_a) = \beta\langle s_I|\alpha_i\gamma_i, \alpha_s\gamma_s\rangle \beta\langle s_a|\alpha_i\gamma_i, +\alpha_s\gamma_s, \alpha_t\gamma_t\rangle$$

This model yields complete accommodation with the α's all set to unity.

Figures 6 and 7 show examples of the results of the energy distributions obtained with this model with all of the α's set equal to the values indicated. They are then equal to the accommodation coefficient. The plots illustrate that a smooth change from specular to diffuse reflections is obtained. The advantage of the model is that it provides a means of specifying, by changing the individual values of α, the degree of internal energy accommodation independently of the translational energy in situations where complete accommodation is not achieved. It is suggested that it is used with an independent scattering model such as Nochila's. Knowledge of how to set individual values of α remains to be discovered, as there is very little information in the literature on the accommodation of the separate modes. Measurements by Marsden[34] using an electron beam indicated that from uncleaned silver about the same 85% accommodation of both the translational and rotational modes was achieved in his experiment. 89% accommodation of the rotational mode was inferred from the experiments of Lewis[35]. Monte Carlo calculation for a spherical blunted slender 6 deg. cone using this model are presented in Figure 8. The bluntness ratio is 0.4. The differences between this stochastic and the Maxwell surface interaction model are, as expected, not marked

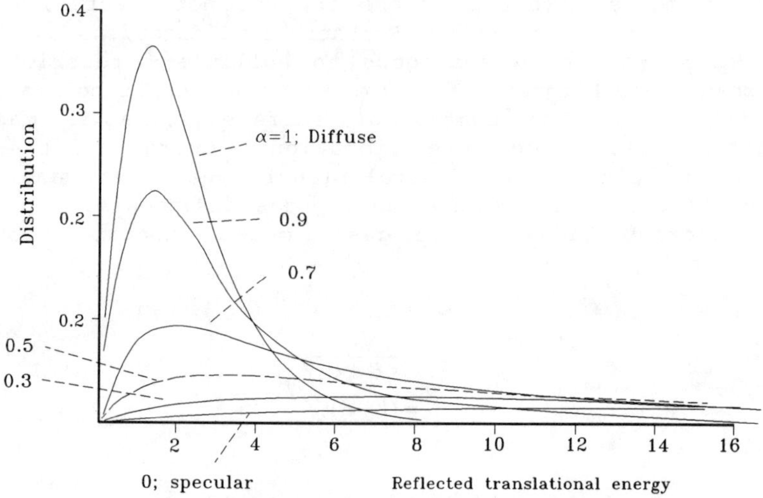

Fig. 6 The distribution of reflected translational energy for a thermal beam with a temperature ratio with respect to the surface of 20.

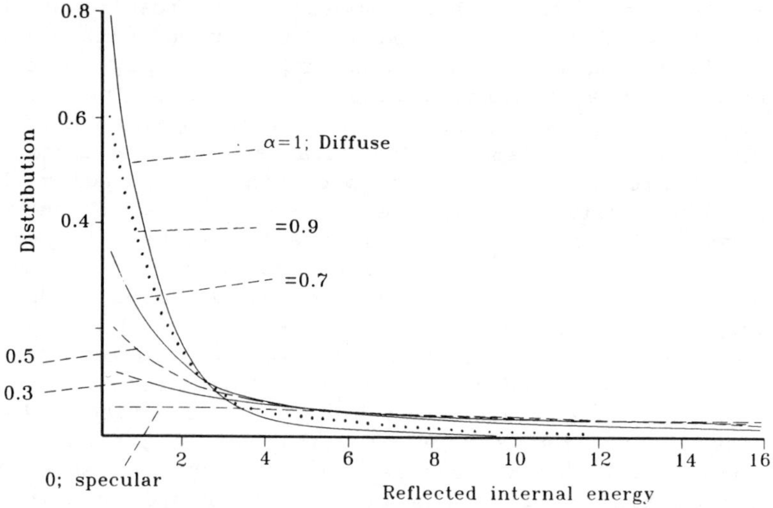

Fig. 7 The distribution of reflected internal energy for a thermal beam with a temperature ratio of 20.

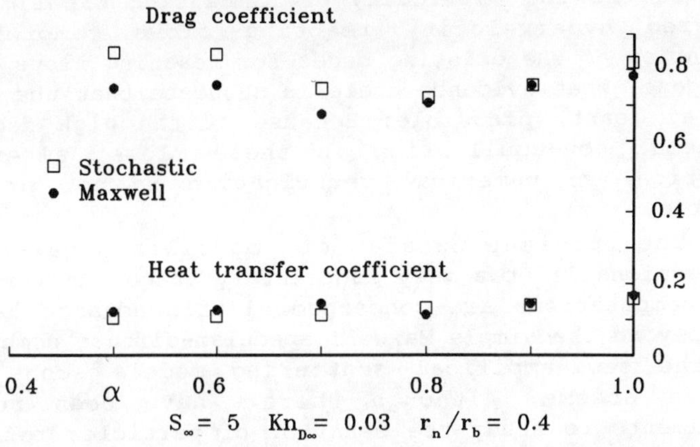

Fig. 8 Total heat transfer and drag coefficients for 5 deg. blunt cone with surface and free stream temperatures 60 K and 273 K respectively.

except for values of α below about 0.6. The differences in the two models at α=1 gives an indication of the computational uncertainty.

Concluding Remarks

The present state of modelling inelastic collision in relation to the MCDS method is that a range of viable models exist for encounters in which energy is exchanged with the internal modes or through chemical or ionic interactions. Of the available choices, the stochastic models are the most favoured. They have been shown to perform very adequately in predicting flows where there is only internal/translational energy exchange. Although the Borgnakke and Larsen 2nd model is the most widely used, the more sophisticated derivative such as Pullin's model, are likely to be more precise and may offer advantages when extending to reacting flows. Generally a very simplistic approach based on the (B-L 2) model has had to be adopted for collisions involving chemical exchange. Although conceptually better models could be devised using, for example, stored arrays of cross-sections for a wide variety of collision parameters, such information is not generally yet available and little progress has been made in this direction. The role of internal excitation in both the controlling of reaction rate and in the absorption of post collision energy has to be addressed in much more depth.

Due to the difficulty of simulating experimentally rarefied hypervelocity reacting flows comprehensive validation of the existing codes for reacting flows has not been done. What evidence there is suggests that the results are, at least, plausible. Because of the high degree of molecular non-equilibrium in these flows, alternative analytical or numerical verification is not presently possible.

The present state of modelling gas-surface interactions is at a very rudimentary stage. As far as the MCDS computations are concerned, little advance has been made beyond the simple Maxwell specular-diffuse combination and the semi-empirical scattering models such as the Nochila scheme. Although there have been numerous experiments to study the behavior of particles reflecting from a wide variety of different surfaces there is a dearth of information on scattering of neutral particle in the appropriate energy range for orbital flight. Setting up suitable experiments in a laboratory is physically impossible and there is an urgent need for space based tests to be done to provide basic information on internal mode and chemical behavior.

References

[1]Bird, G. A. Molecular Gas Dynamics, Oxford University Press.U.K. 1976.

[2]Nanbu, K., "Direct Simulation Scheme Derived from the Boltzmann Equation. (i) Mono-component Gases," Journal Physical Society of Japan, Vol. 45, 1980, pp. 2042-49.

[3]Yen, S. M., "Numerical Solution of the Nonlinear Boltzmann Equation for Non-equilibrium Gas flow Problems," Annual Review of Fluids Mechanics., Vol. 16, 1984, pp. 67-97.

[4]Harvey, J. K., "Direct Simulation Monte Carlo Method and Comparison with Experiment", Progress in Aeronautical Science, Thermophysical Aspects of Re-Entry Flows, Ed. Moss J. N. and Scott, C. D., Vol. 103, AIAA, New York, 1986, pp. 25-43.

[5]Bird, G. A. "The Perception of Numerical Methods in Rarefied Gas Dynamics," Paper presented at 16th Rarefied Gas Dynamics Symposium, Pasadena, Calif., July 1988.

[6]Alder, B. J. and Wainright, T. E., "Molecular Dynamics by Electronic Computers," Transport processes in Statistical Mechanics, Interscience, New York, 1958, pp. 97-131.

[7]Page, M., Oran, E. S., Boris, J. P., Miller, D. and Wyatt, R. E. "A Comparison of Quantum, Classical and Semi-classical Descriptions of a Model Collinear Inelastic

Collision of Two Diatomic Molecules," Naval Research Laboratories Report AD-A162357, 1985.

[8]Macrossan, M. N., "Diatomic Collision Models used in the Monte-Carlo Direct Simulation Method applied to Rarefied Hypersonic Flows," Ph.D. Thesis, Univ. of London. 1983.

[9]Wu, T-Y and Ohamua, T, The Quantum Theory of Scattering, Prentice Hall, Englewood Cuffs, NJ., 1962.

[10]Pullin, D. I.,"Rarefied Leading Edge Flow of a Diatomic Gas," Ph.D. Thesis, University of London. 1974

[11]Bird, G. A. "Monte-Carlo Simulation in an Engineering Context." Rarefied Gas Dynamics, Progress in Aeronautical Science, 74, AIAA, New York, 1980, pp. 113-139.

[12]Davis, J., Dominy, R. G., Harvey, J. K. and Macrossan, M. N. "Evaluation of some collision models used in the Monte Carlo calculation of diatomic rarefied hypersonic flows." Journal of Fluid Mechanics., Vol. 35, 1983, pp. 355-371.

[13]Levine, R. D. and Bernstein, R. B., "Molecular Reaction Dynamics and Chemical Reactivity," Oxford Univ. Press, 1987.

[14]Parker, J. G., "Rotational and Vibrational Relaxation in Diatomic Gases," Physics of Fluids, Vol. 2, 1959, pp. 449.

[15]Lordi, J. A. and Mates, R. E., "Rotational Relaxation in Nonpolar Diatomic Gases," Physics of Fluids, Vol. 13, No. 2, 1970, pp. 291.

[16]Pidduck, F. B., Proceedings of the Royal Society of London,, Series, A. Vol.A101, 1922, pp. 101-.

[17]Condiff, D. W., Lu, W.-K. and Dahler, J. S., Journal of Chemical Physics, 38, 1965, pp. 2963.

[18]Melville, W. K., "The Use of the Loaded Sphere Molecular Model for Computer Simulation of Diatomic Gases,". Journal of Fluid Mechanics., Vol. 51, 1972, pp. 571-.

[19]Deiwart, G. S. and Yoshikawa, K. K. "Analysis of a Semi-Classical Model for Rotational Transitional Probabilities." Physics of Fluids, Vol. 18, 1975, pp. 1085.

[20]Itikawa, Y. and Yoshikawa, K. K., "Monte-Carlo Calculations of Diatomic Gas flows," NASA, Ames Research. Center, D TN-8100, 1975.

[21]Larsen, P. I. and Borgnakke, C, "Statistical Collision Models for Simulating Polyatomic Gas with Restricted Energy Exchange," Rarefied Gas Dynamics, DFVLR-Press, paper A7., 1974.

[22]Borgnakke, C. and Larsen, P. S., "Statistical Collision Models for Monte-Carlo Simulation of Polyatomic Gas

Mixtures," Journal of Computational Physics, Vol. 18, 1975, pp. 405-420.

[23]Pullin, D. I. "Kinetic models for polyatomic molecules with phenomenological energy exchange," Physics of Fluids, Vol. 21, 1978, pp. 209-216.

[24]Kuscer. I. "A. Model for Monte Carlo Simulation of Rarefied Gas Flows", Journal of Fluid Mechanics., to be published 1988.

[25]Bird, G. A. "Simulation of Multi-Dimensional and Chemically Reacting Flows," Rarefied Gas Dynamics, Vol.1, Commissanat a l'Energie Atomiques, Paris, 1979, pp 365-388.

[26]Moss, J. N. and Bird, G. A., "Direct Simulation of Transitional Flows for Hypersonic Re-entry Conditions",Progress in Aeronautical Science, Vol. 96, AIAA, New York, 1985, pp. 113-139.

[27]Stupochenko, S. A., Losev, S. A. and Osipov, A. I. Relaxation Process in Shock Waves," Nanka, Moscow. 1965.

[28]Carlson, L. A. and Rieper, R. G., "Electron Temperature and Relaxation Phenomena behind Shock Waves," Journal of Chemical Physics, Vol. 57, 1972, pp. 760-766.

[29]Koshi, M et Al., "Dissociation of Nitric Oxide in Shock Waves," Symposium (International) on Combustion, Vol. 17, 1978, pp. 553-562.

[30]Blanchard, R. C. and Larman, K. T., "Rarefied Aerodynamics and Upper Atmosphere Flight Results from the Orbiter High Resolution Accelerometer Package Experiment," AIAA Paper 87-2366, 1987.

[31]Agbombia, A. A., "Gas Surface Interaction in Rarefied Hypersonic Flows," Ph.D. Thesis, Univ. of London. 1988.

[32]Hurlbut, F. C., "Sensitivity of Hypersonic Flow Over a Flat Plate to Wall/Gas Interaction Models using DSMC," AIAA 22nd. Thermophysics Conference, Paper No, AIAA-87-1545. 1987.

[33]Cercignani, C. Theory and Application of the Boltzmann Equation, Scottish Academic Press, London. 1975

[34]Marsden, D. J. "Measurement of Energy Transfer in Gas-Solid Interactions Using Electron Beam Excited Emission of Light," Rarefied Gas Dynamics, Ed. J. H. de Leeuw, Academic Press, New York, Vol. 2, 1971, pp. 329-583.

[35]Lewis, J. H. "Accommodation of Rotational Energy of a Diatomic Gas at the Surface of Slender Bodies in Rarefied Hypersonic Flow," Rarefied Gas Dynamics, Ed. Dino Dini, Editrice Technico Scientifica, Pisa, Italy, Vol. 1, 1971, pp. 329-337.

Null Collision Monte Carlo Method: Gas Mixtures with Internal Degrees of Freedom and Chemical Reactions

Katsuhisa Koura*
National Aerospace Laboratory, Chofu, Tokyo, Japan

Abstract

The null collision (NC) direct simulation and test particle Monte Carlo methods are extended to gas mixtures with internal degrees of freedom and chemical reactions. Quantitative comparisons between the NC and time counter (TC) direct simulation methods are made in the shock-wave structure and leading-edge flow for the simple gas of hard sphere molecules. It is indicated that the NC method is stronger to the small number of simulation molecules than the TC method.

Introduction

The null collision (NC) direct simulation Monte Carlo method developed for the simple gas[1] is extended to gas mixtures with internal degrees of freedom and chemical reactions and applied to the relaxation of a chemically reacting gas. The NC test particle Monte Carlo method for the simulation of trace molecules in the heat-bath gas is also extended to gas mixtures with internal degrees of freedom and chemical reactions. The NC test particle method is applied to the diffusion of trace hard sphere molecules in the heat-bath (hard sphere) gas and the diffusion coefficient obtained from the mean-square displacement is compared with the Chapman-Enskog solution.[2]

The NC and time counter (TC)[3] direct simulation methods are different in the method of evaluating the molecular collision number. Quantitative comparisons between the NC and TC methods are made in the shock-wave

* Chief of Rarefied Gas Dynamics Laboratory.

structure and leading-edge flow for the simple gas of hard sphere molecules. The convergence of the NC and TC results by the decrease in the time step and cell size and by the increase in the number of simulation molecules is studied. The convergent NC and TC results are compared with previous TC results.

Null Collision Direct Simulation Method

In the direct simulation Monte Carlo method, molecular motions and collisions are separately simulated during the time step Δt provided that Δt is small enough to satisfy the condition[4]

$$\Delta t << \min\,(\tau_m, \tau_c) \tag{1}$$

where τ_m and τ_c are the characteristic times for molecular motions and collisions, respectively, in a cell. The simulation of molecular collisions during Δt is carried out in each cell by disregarding molecular positions in the cell provided that the cell size is sufficiently small.

The NC method of simulating molecular collisions in a cell during Δt from time t to $t + \Delta t$ is straightforwardly extended to gas mixtures, where the molecular species I in the internal state i, I(i), undergoes the binary (reactive) collision

$$I(i) + J(j) \rightarrow K(k) + L(l) \tag{2}$$

and is described as follows:

1) A time interval Δt_c between successive collisions is assigned by the probability density function

$$p(\Delta t_c) = \nu_c \exp(-\nu_c \Delta t_c) \tag{3}$$

or $\Delta t_c = -\ln(R)/\nu_c$, where ν_c is the constant (between successive collisions) collision frequency in the cell given by

$$\nu_c = (nN/2)S_{max} \tag{4}$$

and R is a uniform random number over the range $0 < R < 1$, $n = \Sigma_{I(i)} n_{I(i)}$ and $N = \Sigma_{I(i)} N_{I(i)}$ are the total number density and number of simulation molecules, and $n_{I(i)}$ and $N_{I(i)}$ are the number density and molecular number of species I(i) in the cell, respectively. The S_{max} is a constant (between successive collisions) defined by

$$S_{max} = \max\,[S_{I(i)J(j)}(g)],\ g \leqq g_{max} \tag{5}$$

where

$$S_{I(i)J(j)}(g) = g\sigma_{I(i)J(j)}(g) \quad (6)$$

$$\sigma_{I(i)J(j)}(g) = \Sigma_{K(k)L(1)}\sigma_{I(i)J(j),K(k)L(1)}(g) \quad (7)$$

and g is the relative speed, g_{max} is the maximum value of g of the N molecules, and $\sigma_{I(i)J(j),K(k)L(1)}(g)$ is the total cross section for collision (2) and related to the differential cross section $\sigma_{I(i)J(j),K(k)L(1)}(g,\Omega)$ for the scattering into the solid angle Ω by

$$\sigma_{I(i)J(j),K(k)L(1)}(g) = \int\sigma_{I(i)J(j),K(k)L(1)}(g,\Omega)\, d\Omega \quad (8)$$

The null collision cross section $\sigma_{I(i)J(j),null}(g)$ [K(k)L(1) = null] for which no (real) collision occurs is defined by

$$S_{max} = g[\sigma_{I(i)J(j)}(g) + \sigma_{I(i)J(j),null}(g)] \quad (9)$$

2) A collision pair of species I(i) with velocity $\mathbf{v}$, [I(i),$\mathbf{v}$], and [J(j),$\mathbf{w}$] is selected from the N molecules by the probability density function

$$p\{[I(i),\mathbf{v}][J(j),\mathbf{w}]\} = y_{I(i)}y_{J(j)}P_{I(i)}(\mathbf{v})P_{J(j)}(\mathbf{w}) \quad (10)$$

where $y_{I(i)} = N_{I(i)}/N$ is the number fraction and $P_{I(i)}(\mathbf{v})$ is the velocity distribution function of species I(i) normalized as unity; i.e., a collision pair is randomly selected from the N molecules.

3) A molecular state after collision, K(k)L(1), is assigned by the probability

$$p[K(k)L(1)] = g\sigma_{I(i)J(j),K(k)L(1)}(g)/S_{max} \quad (11)$$

where K(k)L(1) includes the null collision. If K(k)L(1) = null, then the molecular states and velocities are not changed. If K(k)L(1) ≠ null, then the velocities after collision are calculated from the collision dynamics (momentum and energy conservation) by assigning a scattering solid angle Ω by the probability density function

$$p(\Omega) = \sigma_{I(i)J(j),K(k)L(1)}(g,\Omega)/\sigma_{I(i)J(j),K(k)L(1)}(g) \quad (12)$$

4) Procedures 1-3 are repeated until the summation of Δt_c exceeds Δt ($\Sigma\Delta t_c \leqq \Delta t$).

Equation (3) is rigorous provided that molecular collisions in a cell occur independently and uniformly (between successive collisions) consistent with the assumption of molecular chaos. Because the mean collision number in a cell during Δt is given by $\nu_c \Delta t$ (which is not an integer), the collision number during Δt may also be evaluated as $[\nu_c \Delta t] + 1$ with the condition that if $R \geqq \nu_c \Delta t / \{[\nu_c \Delta t] + 1\}$ then no collision occurs, where $[\nu_c \Delta t]$ is the maximum integer which does not exceed $\nu_c \Delta t$ and the constant (during Δt) value of ν_c is estimated at time t. The validity of this modified NC (MNC) method is verified by the NC method.

Null Collision Test Particle Method

In the NC test particle Monte Carlo method for the simulation of trace molecules in the heat-bath gas, all of the trace (simulation) molecules of the total number N are followed individually during the time step Δt because they do not mutually interact and the separation between molecular motions and collisions is unnecessary. (No condition is required for Δt.) The NC method of simulating the trace species I(i), which undergoes the binary (reactive) collision (2) with the heat-bath species J(j), is described as follows:

1) Each of trace molecules is followed in the subsequent procedures 2–5 until the summation of the collision time interval Δt_c exceeds Δt in procedure 3.

2) A time interval Δt_c between successive collisions of the trace molecule $[I(i), \mathbf{v}]$ with heat-bath molecules is assigned by the probability density function given by Eq. (3) with the constant (between successive collisions) collision frequency

$$\nu_c = nS_{max} \tag{13}$$

where n is the heat-bath number density and S_{max} is a constant (between successive collisions) defined by Eq. (5) with the effective maximum value of the relative speed, $g_{max} = |\mathbf{v}| + w_{max}$, between successive collisions, w_{max} being the effective maximum speed of heat-bath molecules.

3) When the summation of Δt_c is less than Δt ($\Sigma \Delta t_c \leqq \Delta t$), the velocity $\mathbf{v}$ and position $\mathbf{r}$ of the trace molecule are changed into the new velocity $\mathbf{v}'(= \mathbf{v})$ and position $\mathbf{r}'$ $(= \mathbf{r} + \mathbf{v}\Delta t_m)$, respectively, corresponding to the motion during the time interval $\Delta t_m = \Delta t_c$. When $\Sigma \Delta t_c > \Delta t$, $\mathbf{v}$ and $\mathbf{r}$ are changed into $\mathbf{v}'$ and $\mathbf{r}'$, respectively, corresponding to the motion during the remaining time interval $\Delta t_m = \Delta t_c$

$- (\Sigma\Delta t_c - \Delta t)$, and the simulation procedure for the trace molecule ends.

4) A collision partner $[J(j),\mathbf{w}]$ of heat-bath molecules is assigned by the probability density function

$$p[J(j),\mathbf{w}] = y_{J(j)}P_{J(j)}(\mathbf{w}) \tag{14}$$

i.e., $J(j)$ and $\mathbf{w}$ are chosen from $y_{J(j)}$ and $P_{J(j)}(\mathbf{w})$, respectively, which are assumed to be known.

5) A molecular state after collision, $K(k)L(l)$, is assigned by Eq. (11). If $K(k)L(l)$ = null, then $[I(i),\mathbf{v}]$ is not changed. If $K(k)L(l) \neq$ null, then the velocities after collision are calculated from the collision dynamics (momentum and energy conservation) by assigning a scattering solid angle Ω from Eq. (12) and $[I(i),\mathbf{v}]$ is replaced by that after collision.

Relaxation of a Chemically Reacting Gas

The NC and MNC direct simulation methods are applied to the relaxation of a chemically reacting gas where two molecular species A and B without internal degrees of freedom undergo the bimolecular reversible reaction

$$A + A \rightleftarrows B + B \tag{15}$$

without the heat of reaction. The reaction cross sections $\sigma_{r,AA}(\varepsilon)$ and $\sigma_{r,BB}(\varepsilon)$ for the forward ($A + A \rightarrow B + B$) and backward ($B + B \rightarrow A + A$) reactions, respectively, are chosen as the reactive hard sphere model

$$\sigma_{r,AA}(\varepsilon) = \sigma_A(1 - \varepsilon_A/\varepsilon), \quad \varepsilon \geqq \varepsilon_A \tag{16a}$$

$$\sigma_{r,AA}(\varepsilon) = 0, \quad \varepsilon < \varepsilon_A \tag{16b}$$

$$\sigma_{r,BB}(\varepsilon) = \sigma_B(1 - \varepsilon_B/\varepsilon), \quad \varepsilon \geqq \varepsilon_B \tag{17a}$$

$$\sigma_{r,BB}(\varepsilon) = 0, \quad \varepsilon < \varepsilon_B \tag{17b}$$

where $\varepsilon = \mu g^2/2$ is the relative energy, $\mu = m_A m_B/(m_A + m_B)$ is the reduced mass, m_A and m_B are the masses of species A and B, respectively, σ_A and σ_B are the constant cross sections, and ε_A and ε_B are the threshold energies. The mass conservation and microscopic reversibility require $m_A = m_B$ (= m), $\sigma_A = \sigma_B$ (= σ_r), and $\varepsilon_A = \varepsilon_B$ (= ε_r) for zero heat of reaction.[5] The elastic cross sections $\sigma_{e,AA}(\varepsilon)$, $\sigma_{e,AB}(\varepsilon)$, and $\sigma_{e,BB}(\varepsilon)$ for the A + A, A + B, and B + B collisions, respectively, are taken as the hard sphere

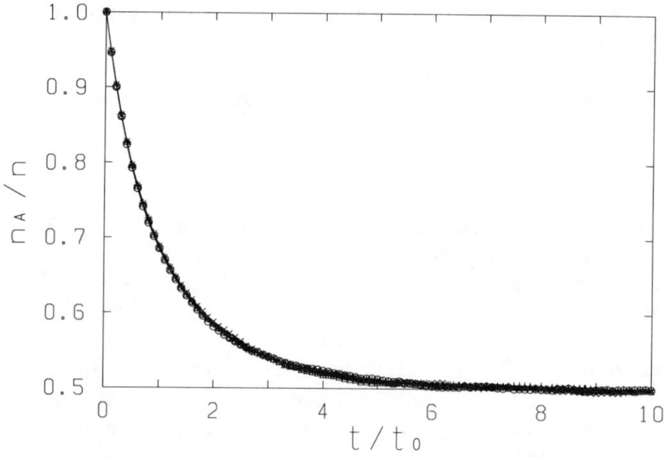

Fig. 1 Comparison of time evolution of reactant number fraction; o, NC ($\Delta t/t_0 = 0.1$); ×, NC ($\Delta t/t_0 = 10^{-5}$); Δ, MNC ($\Delta t/t_0 = 0.1$); +, MNC ($\Delta t/t_0 = 10^{-5}$).

model

$$\sigma_{e,AA}(\varepsilon) = \sigma_{e,AB}(\varepsilon) = \sigma_{e,BB}(\varepsilon) = \sigma_e \tag{18}$$

Both the elastic and reactive scatterings are taken as isotropic. The initial ($t = 0$) number fractions of species A (reactant) and B (product) are taken as $n_A/n = 1$ and $n_B/n = 0$, respectively. The initial velocity distribution function of species A is taken as the Maxwellian distribution at the (constant) temperature T. The total number of simulation molecules is taken as $N = 10^5$.

The time evolution of the reactant number fraction n_A/n ($n_B/n = 1 - n_A/n$) for $\sigma_r/\sigma_e = 1$ and $\varepsilon_r/\kappa T = 1$ is presented in Fig. 1 for the time steps $\Delta t/t_0 = 0.1$ and 10^{-5} ($= 1/N$), where κ is the Boltzmann constant, $t_0 = (nv_0\sigma_e)^{-1}$ is the elastic collision time, and $v_0 = (2\kappa T/m)^{1/2}$ is the most probable speed. The n_A/n monotonically approaches the equilibrium value of 1/2. The fact that the MNC results agree with the NC results within statistical fluctuations indicates that the MNC method is sufficiently accurate.

Diffusion of Hard Sphere Molecules

The NC test particle method is applied to the diffusion of trace hard sphere molecules in the heat-bath (hard sphere) gas, where heat-bath molecules are assumed to remain the equilibrium Maxwellian velocity distribution at the constant temperature T and have the same mass m as trace molecules. The Chapman-Enskog solution of the first

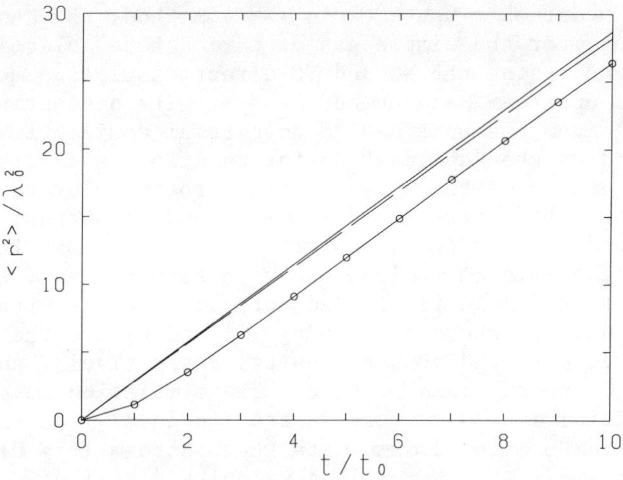

Fig. 2 **Time evolution of radial mean-square displacement (o) of trace hard sphere molecules in the heat-bath gas compared with the Chapman-Enskog solution of the first (- -) and second (——) approximations to the diffusion coefficient.**

(D_1) and second (D_2) approximations to the diffusion coefficient is known[2] for the hard sphere molecules. The initial (t = 0) positions and velocity distribution function of trace molecules are taken as the origin of coordinate, $\mathbf{r} = 0$, and the isotropic δ function distribution with the heat-bath speed $(3\kappa T/m)^{1/2}$, respectively. The number of trace molecules is taken as $N = 10^5$.

The time evolution of the radial mean-square displacement $<r^2>/\lambda_0^2$ of trace molecules is presented in Fig. 2 as compared with the two straight lines with the gradients of $6D_1/v_0\lambda_0 = (81\pi/32)^{1/2}$ and $6D_2/v_0\lambda_0 = (59/58) \times(81\pi/32)^{1/2}$, where $\lambda_0 = (n\sigma)^{-1}$ is the free path, n is the heat-bath number density, σ is the hard sphere cross section, $v_0 = (2\kappa T/m)^{1/2}$ is the most probable speed, and $t_0 = \lambda_0/v_0$ is the collision time. The steady-state gradient of $<r^2>/\lambda_0^2$ at $t/t_0 > 2$ is in excellent agreement with $6D_2/v_0\lambda_0$; the NC value of the diffusion coefficient obtained from $6D = d<r2>/dt$ is $D/v_0\lambda_0 = 0.477$, which agrees with $D_2/v_0\lambda_0 = 0.478$ within 0.2% ($D_1/v_0\lambda_0 = 0.470$).

Comparison between NC and TC Direct Simulation Methods

Shock-Wave Structure

To compare with the previous results obtained by Bird[6] using the TC direct simulation method and by Yen and Ng[7] using Nordsieck's Monte Carlo (NMC) method, the shock-wave structure for the simple gas of hard sphere molecules is calculated using the NC and TC direct simulation methods for the upstream Mach number $M_1 = 8$. The discontinuity surface from the upstream to downstream equilibrium states combined by the Rankine-Hugoniot relation is initially set at the origin of the x coordinate perpendicular to the surface. The upstream ($x/\lambda_1 \leqq -10$) and downstream ($x/\lambda_1 \geqq 10$) boundaries of the computation domain are set far enough from the origin to include the wave structure and the computation domain is divided into small cells with the same width Δx, where $\lambda_1 = (\sqrt{2}n_1\sigma)^{-1}$ and n_1 are the upstream mean free path and number density, respectively, and σ is the hard sphere cross section. The simulation molecules of the initial ($t = 0$) number N_0 are (uniformly) distributed in the computation domain with the upstream ($x < 0$) and downstream ($x \geqq 0$) Maxwellian velocity distributions. Influx molecules across the upstream and downstream boundaries are assigned by the boundary condition of the upstream and downstream Maxwellian distributions, respectively. When the stationary shock wave is established ($t/t_1 \geqq 2$), the wave properties are calculated at a convenient time interval and averaged until statistical fluctuations become sufficiently small, where $t_1 = \lambda_1/c_1$, $c_1 = (2\kappa T_1/m)^{1/2}$, and T_1 are the upstream collision time, most probable speed, and temperature, respectively, and m is the molecular mass. The shock wave appreciably moves from the origin for a small number of

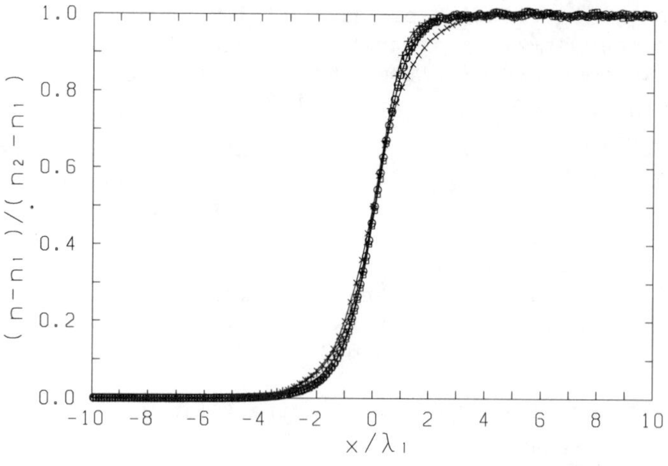

Fig. 3 Comparison of number density profile in the $M_1 = 8$ shock wave; o, NC ($\Delta t/t_1 = 10^{-4}$, $\Delta x/\lambda_1 = 0.1$); □, TC ($\Delta t/t_1 = 10^{-4}$, $\Delta x/\lambda_1 = 0.1$); ×, Bird (TC, Ref. 6); +, Yen and Ng (NMC, Ref. 7).

simulation molecules ($N_0 \leqq 10^4$) but for $N_0 = 10^5$ the movement is negligibly small; therefore, the results for $N_0 = 10^5$ are presented and the origin of coordinate is not adjusted. The characteristic times τ_m and τ_c are estimated as $\Delta x/u_1$ and λ_{12}/u_1, respectively, where u_1 is the upstream flow speed, $\lambda_{12} \sim (n_2\sigma)^{-1}$ ($\sim\lambda_2$) is the free path of upstream molecules for collisions with downstream molecules, and n_2 and λ_2 are the downstream number density and mean free path, respectively. Condition (1) leads to $\Delta t \ll \min(\Delta x, \lambda_2)/u_1$.

The convergence of the shock-wave structure by the decrease in the cell size $\Delta x/\lambda_1$ and time step $\Delta t/t_1$ is investigated over the ranges $0.1 \leqq \Delta x/\lambda_1 \leqq 1$ and $10^{-4} \leqq \Delta t/t_1 \leqq 1$. It is found that the sufficient convergence is obtained for $\Delta x/\lambda_1 \leqq 0.2$ ($\sim\lambda_{12}/\lambda_1$) and $\Delta t/t_1 \leqq 10^{-3}$ consistent with condition (1), $\Delta t/t_1 \ll 0.01$, for $M_1 = 8$.

The number density profile, $(n - n_1)/(n_2 - n_1)$, for $\Delta x/\lambda_1 = 0.1$ and $\Delta t/t_1 = 10^{-4}$ is presented in Fig. 3 as compared with the previous TC and NMC results. The present NC and TC profiles are in agreement but indicate a larger curvature in the upstream half of the shock wave than the previous TC and NMC curvatures and, in the downstream half, are closer to the NMC profile than to the previous TC profile.

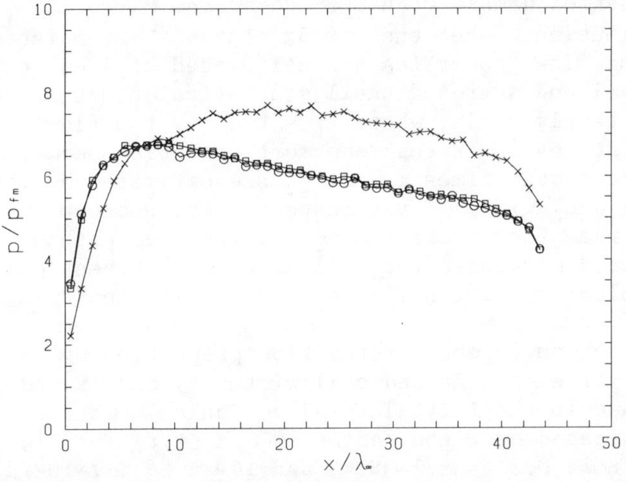

Fig. 4 Convergence of NC plate pressure distribution by the decrease in the time step for $N_0 = 10^5$, $\Delta y/\lambda_\infty = 0.05$, and $\Delta x/\lambda_\infty = 1$; o, $\Delta t/t_L = 10^{-4}$; □, $\Delta t/t_L = 10^{-3}$; ×, $\Delta t/t_L = 10^{-2}$.

Leading-Edge Flow

For comparison with the previous TC results obtained by Vogenitz et al.,[8] the leading-edge flow over the sharp flat plate with the finite length $L/\lambda_\infty = 44$ and diffuse reflection surface at the wall temperature $T_w = 0.08T_0$ for the freestream Mach number $M_\infty = 12.9$ is calculated using the NC and TC direct simulation methods for the simple gas of hard sphere molecules, where λ_∞ and T_0 are the freestream mean free path and total temperature, respectively. The origin of the x and y coordinates parallel and perpendicular to the plate, respectively, is taken as the leading edge of the plate. The computation domain is taken as a rectangle; the upstream $(x/\lambda_\infty \leqq -5)$ and top $(y/\lambda_\infty \geqq 20)$ boundaries are set far enough from the plate to include the plate disturbance and the boundary condition of the freestream Maxwellian velocity distribution is assigned to influx molecules across the boundaries. The rare boundary $(x/\lambda_\infty \geqq L/\lambda_\infty + 15)$ is set sufficiently far from the trailing edge of the plate and the streamwise gradient of the velocity distribution function of influx molecules across the rare boundary is assumed to be negligible. The bottom boundary ($y = 0$ except on the plate) is taken as the specular reflection wall. The computation domain is divided into rectangular cells with the same widths Δx and Δy in the x and y directions, respectively. The simulation molecules of the initial ($t = 0$) number N_0 are uniformly distributed in the computation domain with the freestream Maxwellian distribution. When the steady state is established $(t/t_L \geqq 1)$, the flow properties are calculated at a convenient time interval and averaged until statistical fluctuations become sufficiently small, where $t_L = L/c_\infty$ is the flow time over the plate by the freestream most probable speed c_∞. The characteristic times τ_m and τ_c are estimated as min $(\Delta x/u_\infty, \Delta y/c_\infty)$ and $\lambda_{bf}/c_\infty \sim \lambda_\infty/u_\infty$, respectively, where u_∞ is the freestream flow speed and λ_{bf} is the free path of body reflected molecules for collisions with freestream molecules; condition (1) leads to $\Delta t \ll \min(\Delta x/u_\infty, \Delta y/c_\infty, \lambda_\infty/u_\infty)$.

The convergence of the flow properties by the decrease in the time step Δt and cell widths Δy and Δx and by the increase in the initial total molecular number N_0 is investigated over the ranges $10^{-4} \leqq \Delta t/t_L \leqq 10^{-1}$, $0.05 \leqq \Delta y/\lambda_\infty \leqq 1$, $0.5 \leqq \Delta x/\lambda_\infty \leqq 4$, and $10^3 \leqq N_0 \leqq 7\times10^5$. The NC results of the plate pressure distribution, p/p_{fm}, normalized by the free-molecule value p_{fm} are presented in Figs. 4-7. It is indicated in Figs. 4-6 that the

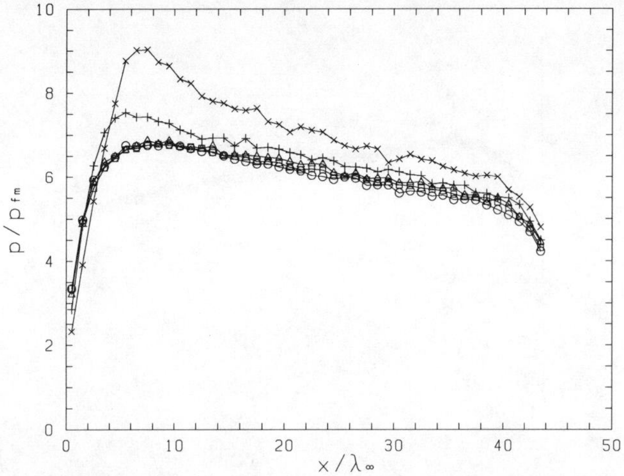

Fig. 5 Convergence of NC plate pressure distribution by the decrease in the cell width perpendicular to the plate for $N_0 = 10^5$, $\Delta t/t_L = 10^{-3}$, and $\Delta x/\lambda_\infty = 1$; o, $\Delta y/\lambda_\infty = 0.05$; □, $\Delta y/\lambda_\infty = 0.1$; Δ, $\Delta y/\lambda_\infty = 0.2$; +, $\Delta y/\lambda_\infty = 0.5$; ×, $\Delta y/\lambda_\infty = 1$.

sufficient convergence is obtained for $\Delta t/t_L \leqq 10^{-3}$ [consistent with condition (1), $\Delta t/t_L << 2\times10^{-3}$, for $L/\lambda_\infty = 44$], $\Delta y/\lambda_\infty \leqq 0.2$ ($\sim\lambda_{bf}/\lambda_\infty$), and $\Delta x/\lambda_\infty \leqq 2$. Figure 7 shows that the lower limit of N_0 is about 10^5, for which the lower limit of the average molecular number per cell, N_{min}, is evaluated to be about 5 because the total molecular number at the steady state is only about 10% larger than N_0. It is noted that the NC method yields persistent results even for the small molecular numbers $N_0 = 5\times10^4$ and 10^4.

Figure 8 presents the dependence of the TC results on N_0 and indicates that the lower limit of N_0 is about 3×10^5, for which N_{min} is evaluated to be about 15. It should be noted that N_{min} for the NC method is less than one third of that for the TC method and the computation time for the NC method is only 10-15% larger than that for the TC method for the same value of N_0. It has also been confirmed that the MNC method yields almost the same results as the NC method.

The convergent NC and TC results of p/p_{fm} for $N_0 = 7\times10^5$ are compared with the previous TC result [8] in Fig. 9. The present NC and TC results are in agreement but somewhat different from the previous TC result.

The NC results are also obtained for $M_\infty = 29$, $T_W = 0.5\times T_0$, and the diffuse reflection surface of the plate with

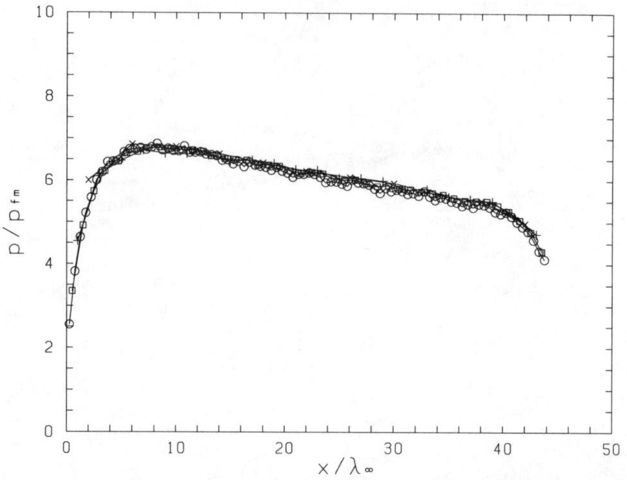

Fig. 6 Convergence of NC plate pressure distribution by the decrease in the cell width parallel to the plate for $N_0 = 10^5$, $\Delta t/t_L = 10^{-3}$, and $\Delta y/\lambda_\infty = 0.1$; o, $\Delta x/\lambda_\infty = 0.5$; □, $\Delta x/\lambda_\infty = 1$; +, $\Delta x/\lambda_\infty = 2$; ×, $\Delta x/\lambda_\infty = 4$.

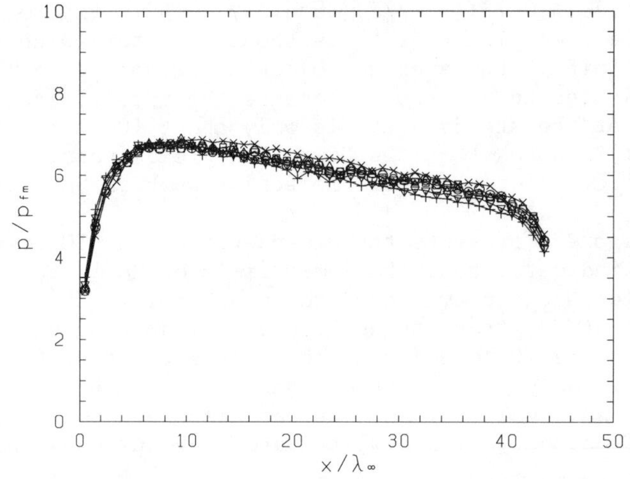

Fig. 7 Dependence of NC plate pressure distribution on the initial total molecular number N_0 for $\Delta t/t_L = 10^{-3}$, $\Delta y/\lambda_\infty = 0.05$, and $\Delta x/\lambda_\infty = 1$; o, $N_0 = 7\times10^5$; □, $N_0 = 5\times10^5$; Δ, $N_0 = 3\times10^5$; ∇, $N_0 = 10^5$; +, $N_0 = 5\times10^4$; ×, $N_0 = 10^4$.

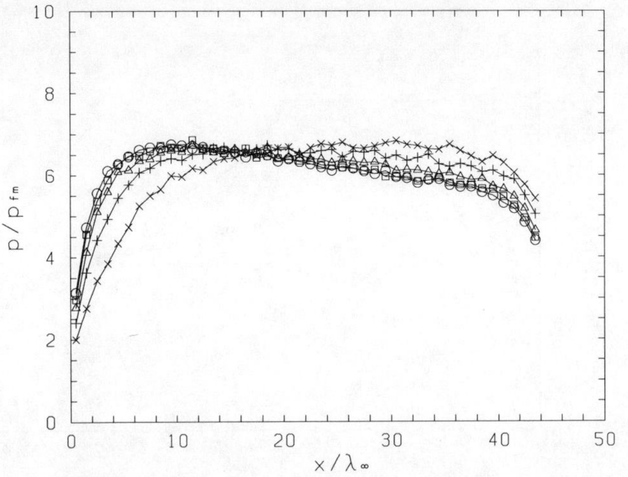

Fig. 8 Dependence of TC plate pressure distribution on the initial total molecular number N_0 for $\Delta t/t_L = 10^{-3}$, $\Delta y/\lambda_\infty = 0.05$, and $\Delta x/\lambda_\infty = 1$; o, $N_0 = 7\times10^5$; □, $N_0 = 5\times10^5$; Δ, $N_0 = 3\times10^5$; +, $N_0 = 2\times10^5$; ×, $N_0 = 10^5$.

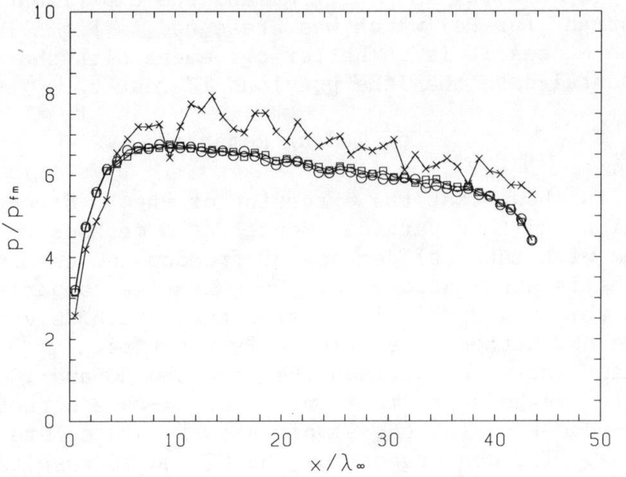

Fig. 9 Comparison of plate pressure distribution for $M_\infty = 12.9$ and $T_w = 0.08T_0$; o, NC ($N_0 = 7\times10^5$, $\Delta t/t_L = 10^{-3}$, $\Delta y/\lambda_\infty = 0.05$, $\Delta x/\lambda_\infty = 1$); □, TC ($N_0 = 7\times10^5$, $\Delta t/t_L = 10^{-3}$, $\Delta y/\lambda_\infty = 0.05$, $\Delta x/\lambda_\infty = 1$); ×, Vogenitz et al. (TC, Ref. 8).

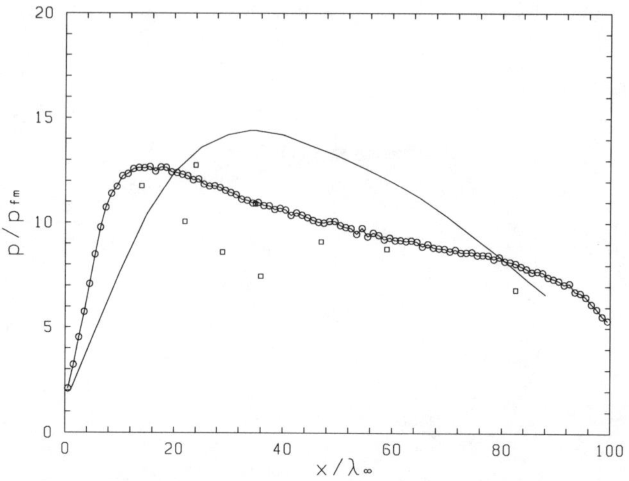

Fig. 10 Comparison of plate pressure distribution for $M_\infty = 29$ and $T_w = 0.5T_0$; o, NC ($N_0 = 7\times10^5$, $\Delta t/t_L = 10^{-3}$, $\Delta y/\lambda_\infty = 0.1$, $\Delta x/\lambda_\infty = 1$); ——, Vogenitz et al. [TC, Ref. 8 ($M_\infty = 29.2$, $T_w = 0.72T_0$)]; □, Horstman (experiment for He, Refs. 8 and 9).

the length $L/\lambda_\infty = 100$. The plate pressure p/p_{fm} for $N_0 = 7\times10^5$ is shown in Fig. 10 as compared with the previous TC result[8] ($M_\infty = 29.2$, $T_w = 0.72T_0$) and the experimental data of Horstman[9] for He (which was presented in Fig. 11 of Ref. 8). The NC result is in better agreement with the experimental data than the previous TC result.

Concluding Remarks

It is shown that the extension of the NC direct simulation and test particle Monte Carlo methods to gas mixtures with internal degrees of freedom and chemical reactions is straightforward. Quantitative comparisons between the NC and MNC direct simulation methods verify that the MNC method is sufficiently accurate.

Quantitative comparisons between the NC and TC direct simulation methods are made in the shock-wave structure and leading-edge flow for the simple gas of hard sphere molecules. The convergence of the NC and TC results by the decrease in the time step and cell size and by the increase in the number of simulation molecules is ascertained. The convergent NC and TC results are in agreement but somewhat different from the previous TC results. It is indicated in the leading-edge flow that the NC method is stronger to the small number of simulation molecules than the TC method.

References

[1]Koura, K., "Null-Collision Technique in the Direct-Simulation Monte Carlo Method," Physics of Fluids, Vol. 29, Nov. 1986, pp. 3509-3511.

[2]Chapman, S. and Cowling, T. G., The Mathematical Theory of Non-Uniform Gases, Cambridge Univ. Press, London, 1939.

[3]Bird, G. A., Molecular Gas Dynamics, Oxford Univ. Press, London, 1976.

[4]Koura, K., "Transient Couette Flow of Rarefied Binary Gas Mixtures," Physics of Fluids, Vol. 13, June 1970, pp. 1457-1466.

[5]Koura, K., "Nonequilibrium Velocity Distributions and Reaction Rates in Fast Highly Exothermic Reactions," Journal of Chemical Physics, Vol. 59, July 1973, pp. 691-697.

[6]Bird, G. A., "Aspects of the Structure of Strong Shock Waves," Physics of Fluids, Vol. 13, May 1970, pp. 1172-1177.

[7]Yen, S.-M. and Ng, W., "Shock-Wave Structure and Intermolecular Collision Laws," Journal of Fluid Mechanics, Vol. 65, Aug. 1974, pp. 127-144.

[8]Vogenitz, F. W., Broadwell, J. E., and Bird, G. A., "Leading Edge Flow by the Monte Carlo Direct Simulation Technique," AIAA Journal, Vol. 8, March 1970, pp. 504-510.

[9]Horstman, C. C., "Surface Pressures and Shock-Wave Shapes on Sharp Plates and Wedges in Low-Density Hypersonic Flow," Rarefied Gas Dynamics, Vol. I, edited by L. Trilling and H. Y. Wachman, Academic Press, New York, 1969, pp. 593-605.

Nitrogen Rotation Relaxation Time Measured in Freejets

A. E. Belikov,* G. I. Sukhinin,* and R. G. Sharafutdinov†
Siberian Branch of the USSR Academy of Sciences, Novosibirsk, USSR

Abstract

Rotational relaxation times for N_2 molecules in freejets of nitrogen over the temperature range of 6-360°K have been measured by the electron beam diagnostics method. This is the first time these relaxation times have been obtained for T = 6-80°K. Numerous comparisons are given between the results on relaxation times obtained in shock waves, ultrasonic measurements, transport coefficient measurements, etc, and the results of different theoretical calculations. In the temperature range of 80-360°K the obtained data are in agreement with experimental data of other authors. Comparisons with rotational relaxation times of nitrogen in argon are also presented.

Introduction

Rotational relaxation can be described theoretically using Wang Chang-Uhlenbeck's equations for a single-particle distribution function $f_j(\vec{r},t,\vec{v})$[1] where j is a rotational quantum number, $\vec{v}$ and $\vec{r}$ are velocity and coordinate, respectively, and t is the time. When the specific time of rotational relaxation is significantly greater than the specific time necessary for the equilibrium distribution of translational molecule energy, the distribution function f_j is represented in the form of the product $f_j = f(\vec{r},t,\vec{v})\, N_j(\vec{r},t)$. In this case relaxation of $N_j(\vec{r}, t)$ occurs at the equilibrium distribution of translational energy and is described by a system of master equations.[2] At present reliable information on rotational transition rate constants involved in these equations is absent, and this complicates the calculations of the level kinetics.

Very frequently the following relaxation equation is employed for a less detailed description of rotational mode relaxation:

$$dE_R/dt = (E_R - E_t)/\tau_R \qquad (1)$$

* Research Physicist, Institute of Thermophysics.
† Professor, Laboratory Chairman, Institute of Thermophysics.

which is valid for small deviations from equilibrium (E_R and E_t are the current and equilibrium value of rotational energy, respectively, and τ_R is the time of rotational relaxation).

The time of rotational relaxation has been determined in the past based on experiments of different types (ultrasonic, shock waves, thermotranspiration, etc.), including the measurements of parameters in a freejet.[3] For the interpretation of experiments it is usually assumed that Eq. (1) is valid for the entire flow, although in freejets considerable deviations from an equilibrium can be attained, which makes questionable the applicability of Eq. (1) over the entire region of parameters. The second shortcoming in the interpretation of available experimental data is that the characteristic collision number $Z_R = \tau_R/\tau_t$ (τ_t is the time of translational relaxation) is usually assumed to be constant for the entire flow and is selected out of the values measured in the jet, though it is well known that Z_R depends on the gas temperature, which, in its turn, changes significantly in the jet.

In the present paper the time of rotational relaxation in molecular N_2 is determined on the basis of measurements in supersonic freejets using the electron beam (EB) technique. The EB technique allows us to determine the populations of N_2 molecule rotational levels and rotational energy E_R based on intensities of rotational lines of the first negative system spectrum of nitrogen bands excited by an electron beam. With a small deviation from equilibrium, using the measured values of E_R the values of τ_R are determined with the help of Eq. (1) similar to the way it has been done in Ref. 4 for N_2 relaxation in Ar. The present method, in contrast to other experiments with freejets, allows us to determine the relaxation time as a function of temperature and to expand the region of τ_R determination down to temperatures of 10°K, which was inaccessible in previous investigations. With the increase of temperature range there appeared a possibility for more detailed analysis of theoretical calculations concerning rotational relaxation.

With a supersonic gas expansion out of a nozzle into a vacuum the gas temperature and density rapidly decrease. The decrease of the collision frequency results in constant violation of equilibrium between internal and translational degrees of molecular freedom. For low temperatures, vibrational degrees of freedom are not excited, and the energy exchange under collisions occurs only between rotational and translational degrees of freedom. If a gradual deviation from an equilibrium of rotational degrees of freedom is observed in the experiment, then one can obtain information concerning the rate of the relaxation process.

The gas flows into a vacuum chamber with a low but finite pressure. The flow parameters in a freejet supersonic core limited by a side shock wave and a Mach disk are similar to those for flow into vacuum. The core dimensions depends on the source parameters and pressure in a vacuum chamber. At low stagnation pressures the jet becomes diffusive and the molecules of a background gas and compressed layers behind the side waves penetrate into the core and destroy the flow. The application of such jets to investigate relaxation processes is difficult since, apart from gasdynamic expansion, one should take to account diffusion, scattering, etc. In the present study the penetration of a surrounding gas into the flow core was decreased by a high evacuation of the vacuum chamber by a booster pump and helium-cryogenic pumps. In those cases when the penetration effect was possible its influence was minimized by reducing the background gas pressure in the vacuum chamber.[5]

Another undesirable process resulting in a more complicated mechanism of energy exchange in the flow is that of condensation. At gas expansion a rapid decrease of temperature results in the possibility of gas condensation. The condensation effect on rotational relaxation can be limited if investigations are carried out at minimum gas densities, i.e., at minimum stagnation pressures. This condition contradicts the limiting condition of the background gas effect. Therefore, in the present study experiments were carried out with the nozzle size as large as possible with minimum stagnation pressures. The problem of data selection free from the effect of condensation and background gas is considered in more detail in the Refs. 5 and 6.

Experimental Technique

The experiments have been carried out in stationary freejets from sonic nozzles in the gasdynamic low-density wind tunnel[5] of the Institute of Thermophysics. Electron-vibrational rotational spectra of the first negative system of N_2 bands were recorded at different distances from the nozzle on the jet axis. These spectra were excited by a beam of electrons with the energies 10-15 KeV. The volume used for the emission selection was determined by the electron beam diameter (d_b = 2 mm) and size restriction with respect to the beam height (1-5 mm). The distance from the nozzle was changed by displacing the gas source with respect to a fixed electron beam. The range of stagnation temperatures T_o varied from 295 to 1000°K. The stagnation temperature was determined by a flow rate method and controlled by thermocouples. The problem of T_o determination is considered in more detail in Ref. 7, and the instrumentation is described in detail and the primary errors of the measurements are determined in Ref. 8.

To calculate the relaxation time by Eq. (1), it is necessary to obtain the dependence of rotational energy of the ground state $N_2(X\,^1\Sigma_g^+, v'' = 0)$ on the position in the jet. Rotational spectra of the first negative system of bands [process $N_2^+(B^2\Sigma_u^+, v' = 0) \rightarrow N_2^+(X^2\Sigma_g^+, v'' = 0) + h\upsilon_{1NS}$] excited by direct electron impact[8] [process $N_2(X\,^1\Sigma_g^+, v'' = 0) + e \rightarrow N_2^+(B^2\Sigma_u^+, v' = 0) + 2e$] were measured in the experiment. The emission intensity of rotational lines in a spectrum and the energy of rotational degrees of freedom in the state $N_2\,(X^1\,\Sigma_g^+, v = 0)$ are correlated as follows:[9]

$$E_R = \frac{\sum_{k'} I_{k'}\,(2k' + 1)\,(k' + 1)\vartheta}{\sum_{k'} I_{k'}\,(2k' + 1)/k'} - \Delta E \qquad (2)$$

where $\Delta E = 12.7°K$ is a rotational heating during the electron excitation of N_2 molecules into the state $N_2^+(B^2\Sigma_u^+, v' = 0)$. Because there exists a prohibition on transitions between N_2 ortho- and para-modifications, then the rotational energy was calculated separately for each modification.

Equation (1) for the central stream tube of a stationary freejet is converted into

$$\frac{u}{n\, d_*} \cdot \frac{d\, E_R}{d(x/d_*)} = - \frac{E_R - E_t}{n\tau_R} \tag{3}$$

where u and n are the gas velocity and density which are calculated on the basis of the isentropic relations for a diatomic gas ($\gamma = c_p/c_v = 1.4$), and d_* is the critical nozzle diameter. Equation (3) was employed to calculate the values of $n\tau_R$ based on the measurements of rotational energy E_R along the coordinate x/d_*. A smooth cubic spline was applied to determine the derivatives $dE_R/d(x/d_*)$. The experimental conditions are listed in Table 1.

Results

Figure 1 shows the values of $n\tau_R$ calculated with respect to the data of these experiments, with the dependance on translational temperature. Designations correspond to columns in Table 1. Points designated by numerals 1-6 represent the relaxation time in pure nitrogen. Data on the relaxation time of small N_2 admixture in Ar obtained in a similar manner in Ref. 4 (points on the diagram and lines 7-9 in Table 1) are shown here for comparison. Experiments 10 and 11 (Table 1) are inappropriate to determine $n\tau_R$ by the described method (see Ref. 4) due to large deviation from an equilibrium. The results of these experiments along with experiments 7-

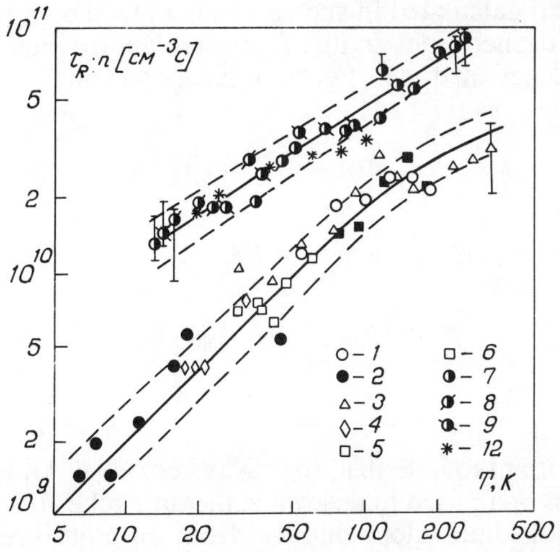

Fig. 1 Variation of relaxation time with temperature. See text for meaning of symbols.

Table 1 Experimental conditions

Run Number	T_o, °K	$n_o d_*$, 10^{-16} cm^{-2}	Range, x/d_*
1	880	11	0.8-9.7
2	293	74.9	15-35
3	665	2.9	0.8-9.8
4	293	9.9	0.7-10
5	293	8.1	0.9-5.8
6	860	5.8	1.4-8.2
7	990	70.1	0.8-13.7
8	900	26.1	0.8-9.7
9	730	12.1	0.8-10
10	730	6.6	0.8-6.8
11	670	2.9	0.8-5.8

9 were employed in Ref. 10 to determine the rotational relaxation rate constants $K_{ij}(T)$, which can be used to calculate $n\tau_R$ [11,12]

$$(n\tau_R)^{-1} = \sum_i \sum_{j>i} K_{ij}(T) N_i^*(T)\,(\varepsilon_i - \varepsilon_j)^2/(<\varepsilon^2> - <\varepsilon>^2)$$

where ε_i is the rotational energy of the ith level and

$$<\varepsilon^l> = \sum_i N_i^* \varepsilon_i^l .$$

The values of $n\tau_R$ calculated in such a manner are shown in Fig. 1 (point 12). Solid and dashed lines in this figure are approximate dependences with 95% confidence intervals. For N_2 relaxation in Ar,[4]

$$n\tau_R = 2.3\ 10^9\ T^{0.66}$$

and for relaxation in pure nitrogen (T < 200° K),

$$n\tau_R = 2.0\ 10^8\ T^{1.0} \tag{4}$$

where $n\tau_R$ is in cm^{-3} s.

It is important to note that, for the system N_2 + Ar, two essentially different methods were used to determine the time relaxation, based on the processing of data both close and far from an equilibrium, and both methods yielded the same result (Fig. 1).

Comparison of Results

Let us start the analysis of the obtained data with the comparison of rotational energy measured in the present study and those published earlier. In Ref. 3, 13-16 E_R was measured by a molecular beam technique. The molecular beam was taken from a supersonic jet with the help of a skimmer. The momenta of the molecule velocity distribution function, i.e., the mean velocity of the flow and translational temperature, were measured in this beam. Then the energy of rotational degrees of freedom was obtained from the conservation equation. The molecular beam was taken at a sufficiently large distance from the nozzle where the rotational relaxation is frozen. The method requires accurate measurements (in particular, the error of the rate measurement should not exceed 0.1%) as well as the applicability of the energy conservation equation for the axial stream tube (which may be questionable for highly rarefied jets). The validity of the conservation equation was verified in Ref. 14 by simultaneous measurements of E_R by the electron beam technique and the velocity distribution function by the time-of-flight method.

Figure 2 shows the values of the rotational energy measured in Refs. 3, 13-16 (designations 1-5, respectively), and our results (Ref. 6) obtained at the distance $x/d_* = 38$ from the nozzle. Under these conditions the rotational energy E_R is sufficiently close to its limiting frozen value and this justifies the comparison with the other studies in which E_R was determined at $x/d_* > 100$. The observable data agreement indicates the lack of discrepancy between the primary experimental results. Solid lines in Fig. 3 show the calculations of E_R using Eq. (3) with the relaxation time given by Eq. (4) curve 7, as well as with relaxation time variation within confidence range (curves 8 and 9). Calculated and experimental data agree only in the range of large values of the parameter $n_o d_*$. An agreement between calculations and experiment over the entire range can be attained by changing the

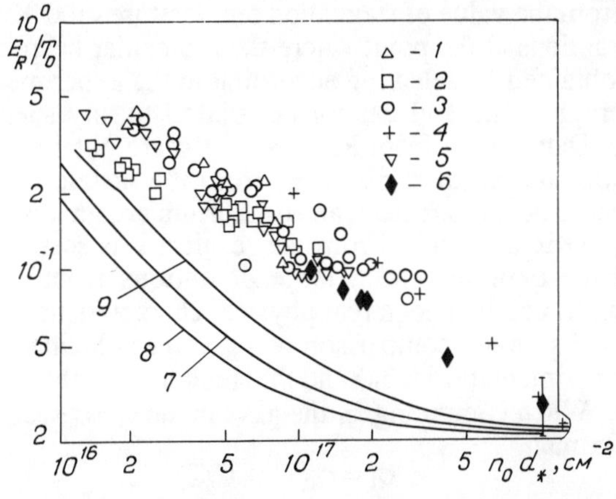

Fig. 2 Rotational energy variation with $n_o d_*$. See text for meaning of symbols.

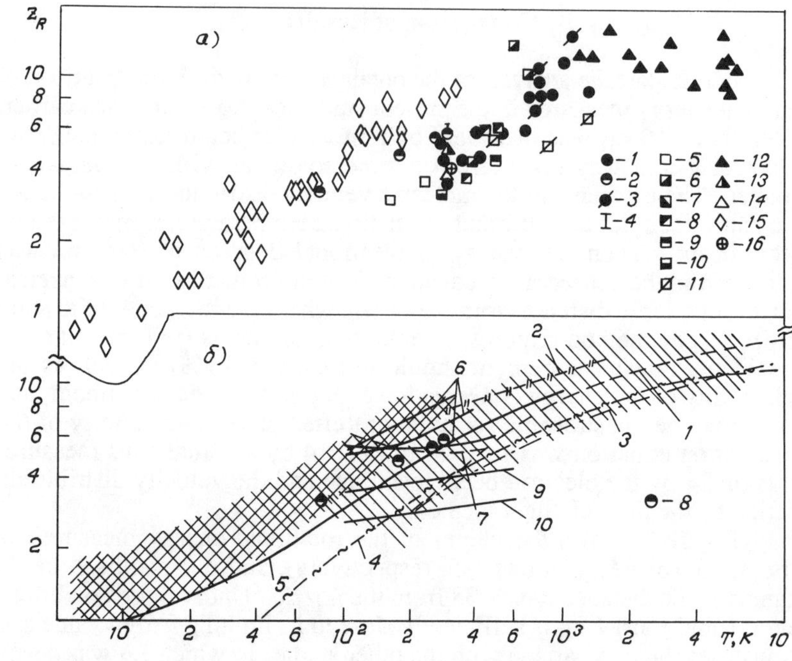

Fig. 3 Comparison of collision number (Z_R) variation with temperature for various experimental and theoretical results. See text for meaning of symbols.

value of Z_R, but, of course, only at the same distance from the nozzle according to the authors of Refs. 3, 13-16. However, this method to obtain Z_R yields inaccurate results for the following reasons. The value of Z_R depends on the gas temperature (see details below while considering Fig. 3). The gas temperature on the jet axis changes over very wide ranges, i.e., from the value of stagnation temperature (300°K and above) to the Kelvin fractions at the point where the molecular beam is taken. The value of Z_R obtained by selecting according to the experiment correspond to some average value and cannot be related to any specific value of temperature. Data on E_R from Refs. 3, 13-16 are obtained at rather low values of n_od_*, and at large distances from the nozzle, i.e., under such conditions where deviations from an equilibrium are great and the applicability of Eq. (1) is doubtful. From the results[4] it is seen that $n\tau_R$ determined from the experiments at large deviations from an equilibrium depends on n_od_*, i.e., it is not a real physical characteristic.

Figure 3 shows a comparison of Z_R values obtained in the present study with experimental (Fig. 3a) and theoretical (Fig. 3b) data available in the literature. When computing τ_R the gas kinetic cross-section was calculated by the formula

$$\sigma_t = \sigma_o (1 + C/T)$$

and $\sigma_o = 32.56 \text{ X } 10^{-16}$ cm^2 and C = 105 K.[17]

The results of ultrasonic measurements in Fig. 3a are designated by numerals 1, 2, and 3 (Refs. 18, 19, and 20, respectively). Numerous ultrasonic data obtained at room temperature and systematized in Ref. 19 are shown in Fig. 3a in the form of an interval restricting the limiting values of Z_R. They are designated by numeral 4. The values of Z_R calculated from the processing of experimental data on thermotranspiration are designated by numerals 5-11 (corresponding to Refs. 21-27). The values of Z_R obtained in shock waves are designated as 12-14 (Refs. 28-30) and our results as 15. The calculations of Z_R from the transfer coefficients, as more approximate, are not used for comparison, though they agree rather well, in terms of an order of magnitude, with those shown in Fig. 3a. In general, there is a good agreement between the results obtained by different methods, however, one can notice that data from acoustic and thermotranspiration measurements are somewhat lower than the rest. At low temperatures $Z_R \leq 1$. This may not correspond to reality due to the lack of reliable information on the value of the gas kinetic cross section at T<80° K.

Note that the results of ultrasonic measurements (points 2 and 16 in Fig. 3a) agree well with the results of our study, whereas there was not such an agreement for the mixture N_2 + Ar.[4] Since the number of rotational lines under investigation and the method of data processing were the same for pure N_2 and for the mixture N_2 + Ar, then one can suppose that the application of some N_2 admixture in Ar results in a slight sensitivity of the ultrasonic absorption coefficient under measurement in Ref. 31.

Figure 3b shows a comparison of experimental data with different theoretical dependences. The region of Z_R values obtained in the present study is shown by a cross-dashed line, and the results of the other works are shown by a conventional dashed line. The results of calculations performed within the framework of classical mechanics are shown by curves 1 (Ref. 32); 2 (Ref. 33); 3 (Ref. 28); 4 (Ref. 34); and 5 (Ref. 35). A more recent study[36] performed with the help of the classical trajectory method is shown by three curves - 6, differing in the values of the anisotropic part of interaction potential. Quite recently, the same authors have performed calculations[37] on the basis of more complicated expressions for the potential of molecule interaction.[38,39] Curves 9 and 10 in Fig. 3b correspond to those calculations. Calculations[40] performed by an analytical method for the conditions of the experiments in Ref. 19, which are in good agreement with the experiment, are presented by points 8. Results of the calculations in Ref. 41 are shown by curve 7.

Note that the difference between theoretical dependences Z_R (T) in the system $N_2 + N_2$ is approximately within the spread of the experimental data. The predictions of three-dimensional classical theories[32, 35] are the closest to the experimental data. For example, the authors of Ref. 36, having made the comparison over a narrow range of temperatures with data[13, 18, 19] (see Fig. 3a), chose the parameters a = 0.1 and b = 0.7 (designations of Ref. 36). This is the lower curve out of three designated by 6 in Fig. 3b. However, the curve for which a = 0.13, b = 0.5 shows better agreement with the total set of data.

In conclusion, let us note the main advantages of the method to determine rotational relaxation time developed in the present study in comparison with those developed earlier. A spectroscopic method to determine the energy of rotational degrees of freedom allows the possibility of determining the contribution of each rotational level to the relaxation time of the entire system. Such a possibility is absent in ultrasonic, thermotranspiration methods and those based on the measurements of the transport coefficient.

The relaxation time and the values measured directly in the experiments of different types are related via kinetic theory of the given phenomenon (propagation of ultrasound, flow of rarefied gas in a capillary, shock wave structure, etc.). It is evident that the interpretation of experimental data should include the analysis of correspondence between theoretical relations and conditions of the present experiments. In particular, one of the main assumptions of all the previously mentioned theories is that of small deviations from an equilibrium. While analyzing the data obtained in freejets there is a possibility of selecting conditions where the deviation from an equilibrium can be considered to be small and where the relaxation equation is applicable.[4] One can hardly expect that in thermomolecular flows large deviations from an equilibrium will occur, but the kinetic theory of this phenomenon is complicated and is still under development. The processing of experimental data on thermotranspiration with the help of new theories[42] yields higher values of Z_R in comparison with those shown in Fig. 3a.

With propagation of sound one cannot state that deviations from equilibrium are always small, therefore, one can agree with the authors of Ref. 32 that in ultra-acoustic measurements "deviations from an equilibrium especially at low temperatures (about 300°K) can be considerable". In strong shock waves there can, of course, be significant deviations from an equilibrium between rotational and translational degrees of freedom.

References

[1]Ferziger, J. H., and Kaper, H. G., Mathematical Theory of Transport Processes in Gases, North-Holland, Amsterdam, 1972.

[2]Kogan, M. N., Rarefied Gas Dynamics, Plenum Press, New York, 1969 (translated from Russian).

[3]Yamazaki, S., Taki, M., and Fujitani, Y., "Rotational Relaxation in Free Jet Expansion for N_2 from 300 to 1000 K.," Journal of Chemical Physics, Vol. 74, No. 8, 1981.

[4]Belikiv, A. E., Solovjev, I. Yu., Sukhinin, G. I., and Sharafutdinov, R. G., "Rotational Relaxation Time of N_2 in Ar," Journal of Applied Mechanics and Technical Physics, Vol. 28, No. 4, 1987.

[5]Borzenko, B. N., Karelov, N. V., et al., "Experimental Investigation of Nitrogen Rotational Level Populations in Free Jets," Journal of Applied Mechanics and Technical Physics, Vol. 17, No. 5, 1976.

[6]Zarvin A. E. and Sharafutdinov, R. G., "Rotational Relaxation in the Transient Regime of the Nitrogen Free Jets," Journal of Applied Mechanics and Technical Physics, Vol. 22, No. 6, 1981.

[7]Belikov, A. E., Dubrovskii, G. V. et al., "Rotational Relaxation of Nitrogen in Free Jet of Argon," Journal of Applied Mechanics and Technical Physics, Vol. 27, No. 6, 1986.

[8]Belikov, A. E., Karelov, N. V., Rebrov, A. K., and Sharafutdinov, R. G., "Measurements with Help of Electron Beam. The Role of Secondary Processes During Excitation of the N_2^+ $B^2S_u^+$ State," Diagnostics of Rarefied Gas Flows, Novosibirsk Institute of Thermophysics, Novosibirsk, USSR, 1975, p. 7-64 (in Russian).

[9]Belikov, A. E., Sedelnikov, A. I., Sukhinin, G. I., and Sharafutdinov, R. G., "Rotational Transitions in Ionization of Nitrogen to the N_2 ($B^2\Sigma_u^+$, v' = 0) State by Electron Impact," Preprint No. 149-86, Novosibirsk Institute of Thermophysics, Novosibirsk, USSR, 1986, (in Russian).

[10]Belikov, A. E., "Rotational Relaxation Rate Constants of Nitrogen in Argon," Kinetic and Gasdynamic Processes in Nonequilibrium Mediums, Moscow State University, Moscow, USSR, 1986 (in Russian).

[11]Sukhinin, G. I., "Generalized Relaxational Equations for Vibrational and Rotational Kinetics of Molecules in the Gas Flows," Journal of Applied Mechanics and Technical Physics, Vol. 29, No. 1, 1988.

[12]Sukhinin, G. I., "Relaxational Representation of Master Equations," Preprint No. 144-86, Novosibirsk Institute of Thermophysics, Novosibirsk, USSR, 1986, (in Russian).

[13]Gallagher, R. J. and Fenn, J. B., "A Free Jet Study of the Rotational Relaxation of Molecular Nitrogen from 300 - 1000 K," Rarefied Gas Dynamics, Proceedings of the 9th International Symposium, 1974, Vol. 2.

[14]Poulsen, P. and Miller, D. R., "The Energy Balance and Free Expansion of Polyatomics," Proceedings of the 10th International Symposium on Rarefied Gas Dynamics, AIAA, New York, 1976, Vol. 2, pp. 899-911.

[15]Bulk, U., Pauly, H., Pust, D., and Schleusener, J., "Molecular Beams from Free Jet Expansion of Molecules and Mixed Gases," Rarefied Gas Dynamics, Proceedings of the 9th International Symposium, 1974, Vol. 2, pp. B10-11.

[16]Brusdeylins, G. and Meyer, H. D., "Speed Ratio and Change of Internal Energy in Nozzle Beams of Polyatomic Gases," Rarefied Gas Dynamics, Proceedings of the 11th International Symposium, 1979, Vol. 2.

[17]Short Handbook of Physical and Chemical Quantities, Leningrad, Chemistry Publishing Co., 1974 (in Russian).

[18]Carnevale, E. H., Carey, C., and Larson, G., "Ultrasonic Determination of Rotational Collision Numbers and Vibrational Relaxation Times of Polyatomic Gases at High Temperatures," Journal of Chemical Physics, Vol. 47, No. 8, 1967.

[19]Prangsma, G. J., Alberga, A. H., and Beenakker, J. J. M., "Ultrasonic Determination of the Volume Viscosity of N_2, CO, CH_4 and CD_4 between 77 and 300 K," Physica, Vol. 64, No. 2.

[20]Winter, T. G. and Hill, G. U., "High Temperature Ultrasonic Measurement of Rotational Relaxation in Hydrogen, Deuterium, Nitrogen and Oxygen," Journal of the Acoustical Society of America, Vol. 42, No. 4, 1967.

[21]Ganzi, G. and Sandler, S. T., "Determination of Transport Properties from Thermal Transpiration Measurements," Journal of Chemical Physics, Vol. 55, No. 1, 1971.

[22]Annis, B. K. and Malinauskas, A. P., "Temperature Dependence of Rotational Collision Numbers from Thermal Transpiration," Journal of Chemical Physics, 1971, Vol. 54, N11.

[23]Malinauskas, A. P., Gooch, J. W., and Annis, B. K., "Rotational Collision Numbers of N_2, O_2, CO, and CO_2 from Thermal Transpiration Measurements," Journal of Chemical Physics, Vol. 53, No. 4, 1970.

[24]Healy, R. N. and Storvick, T. S., "Rotational Collision Number and Eucken Factors from Thermal Transpiration Measurements," Journal of Chemical Physics, Vol. 50, No. 3, 1969.

[25]Malinauskas, A. P., "Thermal Transpiration Rotational Relaxation Numbers from Nitrogen and Carbon Dioxide," Journal of Chemical Physics, Vol. 44, No. 3, 1966.

[26]Mason, E. A., "Molecular Relaxation Times from Thermal Transpiration Measurements," Journal of Chemical Physics, Vol. 39, No. 3, 1963.

[27]Butherus, T. F. and Storvick, T. S., "Rotational Collision Numbers and the Heat Conductivity of Nitrogen Gas from Thermal Transpiration Measurements to 1250 K," Journal of Chemical Physics, Vol. 60, No. 1, 1974.

[28]Bray, C. A. and Jonkman, R. M., "Classical Theory of Rotational Relaxation in Diatomic Gases," Journal of Chemical Physics, Vol. 52, No. 2, 1970.

[29]Linzer, M. and Hornig, D. F., "Structure of Shock Fronts in Argon and Nitrogen," Physics of Fluids, Vol. 6, No. 12, 1963.

[30]Robben, F. and Talbot, L., "Measurement of Shock Wave Thickness by the Electron Beam Fluorescence Method," Physics of Fluids, Vol. 9, No. 4, 1966.

[31]Kistemaker, P. G., and de Vries, A. E., "Rotational Relaxation Times in Nitrogen-Noble Gas Mixtures," Chemical Physics, Vol. 7, No. 2, 1970, pp. 371-382.

[32]Gerasimov, G. Ya. and Makarov, V. N., "On the Theory of Rotational Relaxation in Polyatomic Gases," Journal of Applied Mechanics and Technical Physics, Vol. 16, No. 1, 1975 (in Russian).

[33]Mason, E. A. and Monchick, L. "Heat Conductivity of Polyatomic and Polar Gases," Journal of Chemical Physics, Vol. 36, No. 6, 1962.

[34]Parker, J. G., "Rotational and Vibrational Relaxation in Diatomic Gases," Physics of Fluids, Vol. 2, No. 4, 1959, pp. 449-462.

[35]Lordi, J. A. and Mates, R. E., "Rotational Relaxation in Non-polar Diatomic Gases," Physics of Fluids, Vol. 13, No. 2, 1970.

[36]Nyeland, C. and Billing, G. D., "Rotational Relaxation of Homonuclear Diatomic Molecules by Classical Trajectory Computation," Chemical Physics, Vol. 30, No. 3, 1978.

[37]Nyeland, C., Poulsen, L. L., and Billing, G. D., "Rotational Relaxation and Transport Coefficients for Diatomic Gases: Computations on Nitrogen," Journal of Physical Chemistry, Vol. 88, 1984, pp. 1216-1221.

[38]Berns, R. M., and van der Avoird, A., "N_2-N_2 Interaction Potential from ab initio Calculations, with Application to the Structure of $(N_2)_2$," Journal of Chemical Physics, Vol. 72, 1980, pp. 6107-6116.

[39]van Hemert, M. C. and Berns, R. M., "Comparison of Electron Gas and ab initio Potentials for the N_2-N_2 Interactions. Application in the Second Virial Coefficient," Journal of Chemical Physics, Vol. 76, 1982, pp. 354-361.

[40]Turfa, A. P., Knaap, H. F. P., Thijsse, B. J., and Beenakker, J. J. M., "A Classical Dynamics Study of Rotational Relaxation in Nitrogen Gas," Physica, Vol. A 112, No. 1, 1982.

[41]Mukhametzyanov, R. E., "Calculations of Rotational Interaction Efficiency for Diatomic Molecules in the Low Temperature Plasma," Plasma Physics Proceedings of VI Allunion Conference, Leningrad, 1983, p. 135 (in Russian).

[42]Zhdanov, V. M., "Kinetic Phenomena in Rarefied Polyatomic Gas Flows in Channels," Rarefied Gas Dynamics Proceedings of the VIIIth Allunion Conference, Moscow, 1985, Vol. 2 (in Russian).

Rate Constants for R-T Relaxation of N_2 in Argon Supersonic Jets

A. E. Belikov,* G. I. Sukhinin,* R. G. Sharafutdinov†
Siberian Branch of the USSR Academy of Sciences, Novosibirsk, USSR

Abstract

This paper reports on the results of solving the inverse kinetic problem for relaxation of the rotational energy of small amounts of N_2 additive in Ar, using a one-dimensional master equation. The kinetics of rotational level relaxation was studied by the electron beam technique in a freejet. The populations were measured in five flow regimes ($d_* = 5$ mm, $T_0 = 670\text{-}990°K$, $p_0 = 4\text{-}140$ torr), where d_* is sonic nozzle diameter, T_0 is the stagnation temperature, and p_0 is the stagnation pressure. The translational temperature varied from 160°K to 10°K at different distances from the nozzle. The rate constants $K_{ij}(T)$ were determined using two well-known dependences on quantum numbers: the Exponential Gap Law and the Energy-Corrected Sudden Law. Results are compared with the available theoretical calculations.

Introduction

Rotational relaxation of molecules affects a number of phenomena occurring in gases, such as the diffusion, viscosity, and thermal conductivity of molecular gases, the broadening and shifting of the resolved lines of rotational spectra, the transformation of rotational energy in dense gases and shock waves, and the dispersion and absorption of ultrasonic waves. Information concerning cross sections and rate constants of inelastic rotational transitions is required to solve various problems of chemical kinetics, gasdynamics, and laser spectroscopy.

The theoretical calculation of rotational transition cross-sections, which is based on a numerical solution of Schrodinger's equation, has been performed for a few scattering channels and individual energies. Approximate methods of solving the problem, though less accurate, provide the possibility of its complete solution, with calculation of the complete set of rate constants and the times of rotational relaxation as well as

---

*Research Physicist, Institute of Thermophysics.

†Professor, Laboratory Chairman, Institute of Thermophysics.

clear physical interpretation of results and explanation of dependences on energy and quantum numbers.

In principle, rotational transition cross-sections can be measured using intersecting molecular beam techniques. However, difficulties arise in these experiments[1,2] in producing beams with the given distribution of rotational level populations and in detecting them after scattering. Additionally, these data need to be obtained over a wide range of energies to obtain the rate constants. Therefore, investigations of such type are few in number and, as far as the authors know, only two works[3,4] are devoted to the system Ar + N_2.

On the other hand, rotational relaxation times obtained from microscopic experiments (ultrasonic, thermotranspiration, transport coefficients, shock waves, freejets, rotational spectroscopy of dense gases) represent the values averaged not only in terms of translational energy, but also by rotational energy as well. In this case, as for any indirectly obtained information, the usefulness of the results depends on correctness and completeness of the theory relating the relaxation times with the dependences observed in the experiment. In this respect, experiments on the kinetics of rotational level population relaxation that have been conducted using laser[5] and electron beam technique[6] are certainly of great advantage because they yield detailed information on transition rate constants between the separate levels that are averaged only with respect to translational motion.

In this paper the kinetics of N_2 rotational level populations in an Ar jet is studied. Relaxation times and rate constants of R-T processes in this mixture are determined and comparison with other available experimental and theoretical results is made.

Rotational Relaxation of a Molecular Admixture in an Atomic Gas

Experimental results obtained in this work are interpreted using kinetic equations for the relaxation of a small molecular gas admixture in the flow of a monatomic gas ($n \ll n_a$)

$$\frac{\partial N_i}{\partial t} + \vec{u}\,\nabla N_i = n_a \sum_f (N_f K_{fi} - N_i K_{if}) \tag{1}$$

where $\vec{u}$ is the velocity of the atomic flow, N_a is the atom density, n_i is the number density of molecules occupying the ith rotational level, and $N_i = n_i/n$, where $n = \Sigma\, n_i$. In Eq. (1) K_{if} is the rate constant for the rotational transition from the level j_i to the level j_f initiated by collision

$$K_{if} = \frac{1}{n_i n_a} \iint f_i(\vec{c}_i)\, f_a(\vec{c}_a)\, v_i\, \sigma_{if}(v_i)\, d\vec{c}_i\, d\vec{c}_a \tag{2}$$

where $v_i = |\vec{c}_i - \vec{c}_a|$ is the relative velocity of atom and molecule before collision and $\sigma_{if}(v_i)$ the cross section for transition $j_i \rightarrow j_f$.

The solution of Eq. (1) is hindered by the fact that the rate constants [Eq. (2)] depend on the distribution functions f_i and f_a and one has to solve Wang Chang-Uhlenbeck's equations to obtain these distribution functions. However, in many cases the distribution of velocities is close to the equilibrium one (due to a rapid process of translational relaxation) and can be taken to be in the form of the local Maxwellian distribution.[7] In this case the rate constant [Eq. (2)] is reduced to a onefold integral:

$$K_{ji \to jf}(T) = \bar{v} \int_0^\infty \frac{E\, dE}{(k_B T)^2} \exp\left(-\frac{E}{K_B T}\right) \sigma_{ji \to jf}(E)$$

where $\bar{v}$ is the mean relative velocity and E the kinetic energy of relative motion before collision.

From the relation of microscopic reversibility for the cross sections of direct and reverse transitions,

$$E\,(2j_i + 1)\, \sigma_{ji \to jf}(E) = (E - \Delta E)\,(2j_f + 1)\, \sigma_{jf \to ji}(E - \Delta E)$$

the principle of detailed balance for the rate constants follows

$$N_i^* K_{ji \to jf}(T) = N_f^* K_{jf \to ji}(T) \tag{3}$$

where N_i^* is an equilibrium population of the j_i - level

$$N_i^* = \frac{g_i\,(2j_i + 1) \exp(-\varepsilon_i/k_B T)}{\sum_i g_i\,(2j_i + 1) \exp(-\varepsilon_i/k_B T)}$$

where $\varepsilon_i = B_v j_i(j_i + 1)$, B_v is the rotational constant for a given vibrational level, $\Delta E = \varepsilon_f - \varepsilon_i$, and g_i is the statistical weight related to the spin state of the molecule nuclei. For $^{14}N_2$ $g_i = 2/3$ at even j_i (orthonitrogen) and 1/3 for odd j_i (paranitrogen).

The kinetic equations [Eq. (1)] with due account of detailed balance (Eq. 3) are employed later in the paper to determine the rate constants $K_{ji \to jf}(T)$ on the basis of the known distributions of the gasdynamic parameters (n_a, $\vec{u}$, T) of the atomic flow and the experimentally measured populations $N_i(\vec{r},t)$.

When the population of rotational levels N_i are close to their equilibrium values N_i^*, for the mean rotational energy of the molecules one can write a simple relaxational equation

$$\frac{d E_R}{d t} = -\frac{1}{\tau_R}(E_R - \langle\varepsilon\rangle) \tag{4}$$

where $E = \Sigma\varepsilon_i N_i$ is the instantaneous value of rotational energy and $<\varepsilon> = \Sigma\varepsilon_i N_i^*$ the equilibrium value of rotational energy. In Eq. (4) τ_R is the specific time necessary to attain an equilibrium between rotational and translational degrees of freedom, i.e., the time of R-T relaxation

$$\tau_R^{-1} = n_a \sum_{j_i=0}^{\infty} \sum_{j_f>j_i} N_i^* K_{j_i\to j_f}(T) \frac{(\varepsilon_i-\varepsilon_f)^2}{<\varepsilon^2>-<\varepsilon>^2} \quad (5)$$

One-exponential relaxation [Eq. (4)] does not take place in the general case with arbitrary deviations of populations from equilibrium values. If the momenta of rotational energy are determined as $<\varepsilon_R^n> = \Sigma\varepsilon_i^n N_i$, n = 1, 2, ..., then in the general case they are all related to each other.[8]

Determination of Rate Constants and Rotational Relaxation Times from Measurements of Rotational Level Populations in Freejets

As mentioned earlier, information on the rate constants of rotational transitions can be obtained by studying the kinetics of the populations of molecular rotational levels in freejets. For this, it is necessary to measure the spatial-temporal dependences N_j and to obtain the distribution of gasdynamic parameters n_a, u_a, T in the gas jet and then solve the inverse problem for a system of kinetic equations [(Eq. (1)] in terms of the rate constants K_{ij}. This problem is solved most simply in the case of a stationary gas flow into a vacuum from a chamber in which the gas is characterized by the stagnation density n_o and the stagnation temperature T_o. In this case the system of kinetic equations [(Eq. (1)] is reduced to

$$\frac{u_a}{n_a d_*} \cdot \frac{d N_j}{d \tilde{x}} = \sum_{j'=0}^{j} K_{j'\to j} \left(N_{j'} - N_j \frac{N_{j'}^*}{N_j^*}\right) + \sum_{j'=j}^{j_r} K_{j\to j'} \left(N_{j'} \frac{N_j^*}{N_{j'}^*} - N_j\right) \quad (6)$$

where x is the distance from the nozzle along the jet axis, and $\tilde{x} = x/d_*$ where d_* is the nozzle diameter. The relaxational Eq. (4) for this case is of the form

$$\left(\frac{u_a}{d_*}\right)\left(\frac{dE_R}{d\tilde{x}}\right) = -\frac{E_R - E_t}{\tau_R} \quad (7)$$

where $E_t = k_B T$ is the equilibrium value of rotational energy at the position $\tilde{x}$.

In the present study the gasdynamic parameters n_a, u_a, T were determined from numerical calculations for a monatomic gas equilibrium isentropic expansion into a vacuum out of a sonic nozzle.[9] The applicability region of such an approximation is restricted to values of $n_o d_*$ above some minimum value, as well as the requirement for the molecule gas fraction not to exceed 7%.[10]

The experiments have been carried out in a low-density wind tunnel that consists of a vacuum chamber equipped with a system of gas supply which passes through a heated prechamber and exits through removable sonic nozzles, an evacuation system, and electron beam spectroscopic and other diagnostic instrumentation.[6] Vapor-oil and cryogenic pumps provided the evacuation of the vacuum chamber. The pressure level of the background gas in all measurements did not exceed 3×10^{-3} torr. An axisymmetric sonic nozzle of the diameter $d_* = 5.11$ mm served as the gasdynamic source. The design of the prechamber allowed operation at stagnation temperatures up to $T_o \sim 1300°K$. The temperature was checked by a Pt/Pt + Rh thermocouple.

The population of rotational levels in the N_2 ground state ($X\ ^1\Sigma_g^+$, $v'' = 0$) was determined by the electron beam diagnostic technique.[11] For this purpose nitrogen on the jet axis was excited by the electron beam into the positive ion state $N_2^+(B^2\Sigma_u^+, v' = 0)$ (electron energy ~ 10 KeV, current ~ 10 mA, electron beam diameter ~ 1 mm). The intensity of the resolved rotational lines of the R branch for the 0-0 band of the first negative system $N_2^+(B^2\Sigma_u^+, v' = 0) \rightarrow N_2^+(X^2\Sigma_g^+, v'' = 0)$ was then measured. The intensities $I_{j'}$ of the detected band lines are associated with the populations of rotational levels in the ground state $N_2(X\ ^1\Sigma_g^+, v'' = 0)$ as follows

$$I_{j'} = I \sum_j N_j P(j \rightarrow j') \tag{8}$$

where I is the normalization constant, and $P(j \rightarrow j')$ is the probability of the $j \rightarrow j'$ transition in nitrogen excited by electron impact. The experimental technique to determine the probability matrix $P(j \rightarrow j')$ is expounded in detail in Ref. 11. The errors of this technique and the methods used to minimize them are analyzed in that reference. The unknowns N_j were obtained from Eq. (8) with the help of an iterative procedure.

The main difficulty that arises when finding the rate constants lies in the fact that the system of equations [Eq. (1)] is of infinite order, whereas the number of measured N_j populations in Eq. (6) is finite ($0 \leq j \leq j_r$). It is clear that the break of the equation chain affects the accuracy of the determination of $K_{j \rightarrow j'}$ rate constants at $j' \sim j_r$. This problem is simplified when a priori information on the $K_{j \rightarrow j'}$ dependence on quantum numbers j and j' is available. For example, the rate constants can be given on the basis of physical considerations[11] in the form of a rather "flexible" functional dependence on j, j' with a small number of adjustable parameters. In this case a set of several experimental regimes at fixed $T(\tilde{x})$ is necessary only to decrease the statistical error in determining the desired parameters.

In this respect a question arises as to what criterion should be used to chose the model out of a set of models accounting for the experimental data. In our opinion, the coincidence of rotational relaxation time [Eq. (5)]calculated using the rate constant model and that measured experimentally [Eq. 7)] is convincing evidence in favour of the chosen model. It is not surprising that various models of the rate constants $K_{j \to j+\Delta}$ differing considerably in their Δ dependence do not describe satisfactorily the bulk of the experimental data with respect to N_j. Additionally, the relaxation times τ_R calculated within the framework of these models can vary significantly, since in Eq. (5) for the $K_{j \to j'}$ distribution the second moment $(\Delta E)^2$, which is sensitive to a great extent to the behavior of $K_{j \to j'}$ at $|j-j'|>>2$, is calculated.

Two models were applied for experimental data processing:[12] 1) the exponential gap law (EGL), which employs the adjustable relation

$$K_{j_i \to j_f}(T) = (2j_f + 1)\, A \cdot \exp(-\theta\, |\varepsilon_i - \varepsilon_f| / k_B T), \tag{9}$$

and 2) the energy-corrected sudden (ECS) law of recalculation with the basis vector of the rate constants (ECS-EP):

$$K_{j\text{-}0}(T) = A' \, [j(j+1)]^{-\gamma} \exp[-j(j+1)/j^*(j^*+1)] \tag{10}$$

In Eqs. (9) and (10) A, θ, A', γ, j^* are free parameters of the model that depend on temperature.

For both models the parameters are obtained by the functional minimization

$$Q = \sum_{a,j} \xi\, aj \left\{ \frac{u^a}{n^a d_*} \left(\frac{dN_j}{dx}\right)^a - \sum_{j'} K_{j' \to j} \left(N^a_{j'} - N^a_j \frac{N^*_{j'}}{N^*_j}\right) \right\}^2$$

where subscript a designates the number of the working regime ($n_o d_*$ and T_o), and ξ_{aj} are statistical weights characterizing the experimental accuracy with which the populations in each working regime are determined.

Table 1 Comparison of rotational rate constants

Transition	Our Results		Theories	
	EGL	ECS-EP	Ref. 17	Refs. 13, 14
2-0	0.9±0.2	1.1±0.3	3.6	1.6
4-2	3.3±0.9	2.9±0.7	7.8	3.1
6-4	4.3±1.1	3.1±1.0	7.5	3.7
8-6	4.5±1.3	2.4±1.4	7.1	4.2
10-8	4.2±1.6	1.5±1.4	6.2	4.5
4-0	0.5±0.1	0.5±0.1	1.6	0.49
6-2	1.3±0.5	1.4±0.4	3.0	1.0
8-4	1.3±0.7	1.5±0.7	2.7	1.3
10-6	0.9±0.6	1.2±0.9	2.2	1.5
6-0	0.2±0.1	0.3±0.1	0.44	0.25
8-2	0.4±0.3	0.8±0.3	0.73	0.55
10-4	0.3±0.2	0.8±0.5	0.62	0.72
8-0	0.06±0.05	0.2±0.1	0.08	0.16
10-2	0.08±0.06	0.5±0.2	0.12	0.36

In the temperature range 15-160°K we have obtained the following data

1) For EGL, $A = 2.0\ T^{-0.77}$ (10^{-10} cm^3/s) and $\theta = 0.2\ T^{0.3}$
2) For ECS-EP, $A' = 1.4\ T^{-1.0}$ (10^{-9} cm^3/s), $\gamma = 2.32\ T^{-0.35}$, and $1/[j^*\ (j^* + 1)] = 0.017\ T^{0.28}$

The rate constants obtained using these models at T = 40 K are presented in Table 1 for illustration. As seen in Table 1, the differences in $K_{j \to j'}$ values are beyond the confidence intervals only at large j and $|j-j'| \geq 8$. Similar agreement was found at other temperatures as well.

Comparison with Theoretical Calculations

As was mentioned above, numerous works on the calculation of inelastic rotational collision cross sections have been published for the collision pair Ar + N_2, whereas the number of papers devoted to the calculation of rate constants is limited.

The expressions for the rate constants within the framework of the generalized eikonal method have been obtained in Ref. 13. In Ref. 14 these state-to-state rates were used to describe the kinetics of populations in freejets. Analytic expressions for the rate constants of the form given in Ref. 14 are used in the present paper for comparison.

The results of the trajectory calculations of inelastic cross sections $\sigma_{j \to j'}$ (E) (see Ref. 15) were approximated (see Ref. 16) by different types of adjustable dependences. The rate constants were obtained by averaging the cross sections with locally Maxwellian distribution with respect to energy E. Particularly, an exponential model that corresponded better in the case of low energies was applied for comparison.[16] In Ref. 17 another method of inelastic cross-sectional calculation has been developed which is

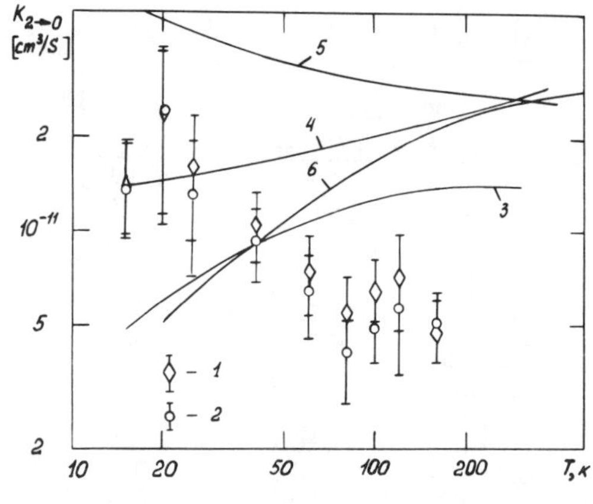

Fig. 1 Temperature dependence of the rate constant for low frequency rotational transitions. Symbols show present experimental results, curves are results of various theoretical analyses (see text).

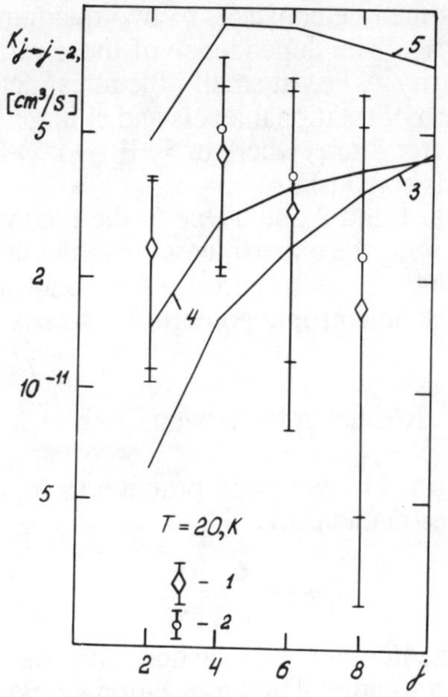

Fig. 2 Rate constants of two-quantum rotational transitions. Symbols show present experimental results, curves are results of various theoretical analyses (see text).

called the semiclassical approximation of centrifugal suddenness (SCS) and includes successive adiabatic correction of the sudden perturbation method. The rate constants calculated in Ref. 18 using this method[17] are listed in the fourth column of the table.

Figure 1 shows a typical temperature dependence of the rate constant for low-frequency rotational transitions that have been obtained in our experiments. As is seen, $K_{2\text{-}0}$ (T) decreases with increasing temperature in complete accordance with theoretical considerations[18] (curve 5). Curves 3 and 4 are plotted on the basis of the data from Refs. 16 and 14, respectively. If long-range attraction forces are neglected in the interaction potential, then short-range repulsive forces affect only the dynamics of strong collisions (close transits) for which the action increment increases with the energy increase. In this case the rate constant[18] increases with temperature (curve 6). For the collision pair Ar + N_2 the depth of the potential well of the isotropic part of the potential is ~ 120°K, therefore, at $T^* = k_B T/\varepsilon < 1$ this approximation is incorrect, and the neglect of attraction force effect, as done in Ref. 14, results in strong disagreement with experimental data. In Ref. 15 inelastic cross sections are calculated at sufficiently high energies of relative motion (768-12,280°K); therefore the calculation of the rate constants for the region of low temperatures performed on the basis of these data[16] is, in reality, a rough extrapolation, as shown by curve 3.

Figure 2 shows the rate constants of two-quantum rotational transitions j→j-2 at T = 20°K. The dependence of these rates on the quantum number j has a peculiarity[18], i.e., at small j the rates increase with increasing j due to degeneration of rotational levels and at large j the rates decrease due to the adiabatic factor 2 $\omega\tau_C$ where $\omega = 2B\,(j+1/2)$ (curve 5). Experimental data confirm this conclusion.

As seen in Figs. 1 and 2 and Table 1, the theoretical data[17] are in qualitative agreement with the experimental results but quantitatively are 1.5-2.5 times larger than the measured values. The anisotropy factor b_2 at the repulsive part of the anisotropic potential[19] is possibly somewhat too large.

Rotational Relaxation Time

In the literature it is a common practice to represent the data on rotational relaxation time in the form

$$Z_R = \frac{\tau_R}{\tau_t} = \frac{\sigma_t}{\sigma_R} \tag{11}$$

where Z_R is the average number of inelastic collisions necessary to attain on equilibrium between rotational and translational degrees of freedom; $\sigma_t = (n_o \bar{v}\,\tau_t)^{-1}$ is the effective cross section of translational relaxation. For Ar + N_2 we have

$$\sigma_t = \sigma_o\,(1 + \frac{C}{T}) \tag{12}$$

where σ_o = 30.2 Å^2 and C = 122°K. One should keep in mind that the reference values of the coefficients for Eq. (12) are obtained from analysis of experiments on heat and mass transfer that have been conducted at temperatures T > 80°K.

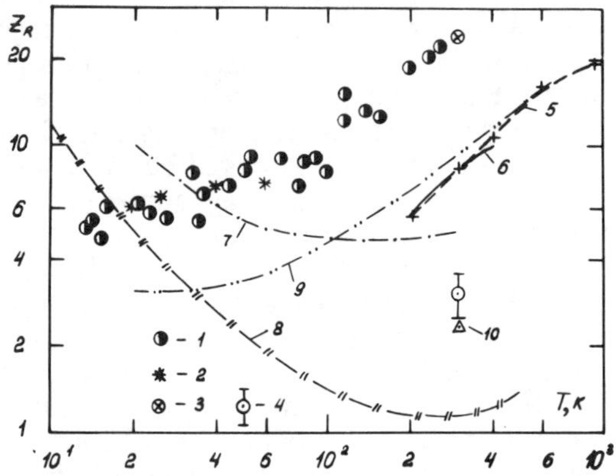

Fig. 3 Temperature dependence of collision number Z_R. See text for explanation of symbols.

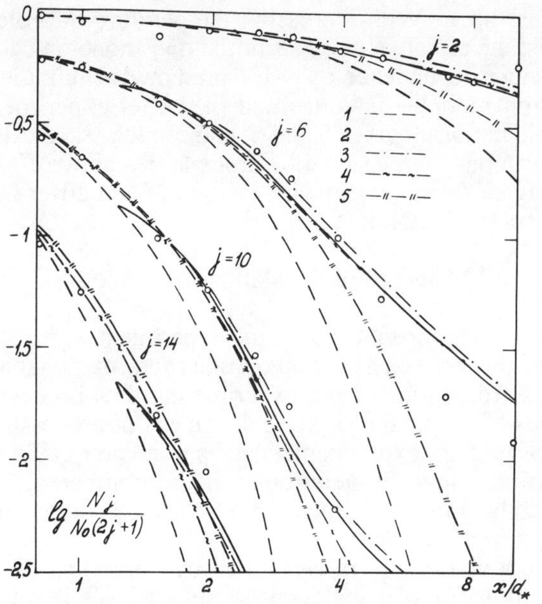

Fig. 4 Rotational populations at various positions in the jet for $T_0 = 730°K$, $p_0 d_* =$ 50 torr mm. See text for explanation of symbols.

Figure 3 shows the values of Z_R designated as follows: 1, calculated[20] using Eq. (4) after substitution:

$$E_R = \sum_i N_i \theta i (i + 1)$$

2,3, calculated by Eq. (5) using experimental constants of the present work and Ref. 21; 4, 10, data from ultrasonic measurements and trajectory calculations[22]; 5-9, calculations from Ref. 23, 24, 14, 16, and 18, respectively. It is seen that Z_R obtained by different methods (1 and 2 in Fig. 3) on the basis of rotational level population measurements coincide with each other. In Ref. 20 the rotational relaxation time was obtained from the experiments corresponding to the case of small deviations from an equilibrium, and Eq. (7), which is valid for this case, was employed for the data processing. In the present study this value was obtained as a convolution Eq. (5), and the rate constants were obtained more reliably from the experiments in which deviations from an equilibrium were not small. The coincidence of relaxation times τ_R obtained on the basis of Eqs. (5) and (7) testifies in favor of the chosen models for the rate constants.

As seen in Fig. 3, the average number of collisions Z_R increases with temperature for the case Ar + N_2. Such behavior is accounted for by the effect of the attractive forces that accelerate the motion of the molecules along classical trajectories, contributing to more effective R-T relaxation at lower temperatures. With increasing temperature the role of this effect decreases and the effective cross-sectional σ_{RT} monotonically decreases.[25] The temperature dependence σ_{RT}~1/T predicted within the framework of the j diffusion model[25] is confirmed in the jet experiments. Classical trajectory calculations (curves 5 and 6) and semiclassical calculation (curve 9) also confirm this conclusion. At temperatures above 300°K these theoretical data agree satisfactorily with each other and differ from the experimental results by a factor of 2-2.5.

Kinetics of Populations in a Free Jet

Earlier we compared theory and experiment with respect to the rate constants and the rate constant convolutions and rotational relaxation time. These values are obtained from experimental data processing. As additional sources of errors appear at all the stages of processing, comparison at the level of primary experimental data is more preferable. In this section the N_2 rotational level populations will be compared, i.e., the values obtained directly in the experiments in a freejet are compared to the populations obtained from the theory.

The populations were calculated by numerical integration of the kinetic equations [Eq. (6)] analogous to Ref. 26. While solving the kinetic equations with the rate constants [Eqs. (9) and (10)], one should note that they are obtained in the temperature range of 15-160°K. Therefore, the kinetics using these rates were calculated not from the critical nozzle but from a certain point in the flow where the gas temperature was close to 160°K.

Theoretical and experimental dependences are compared under the following conditions T_o = 730°K, p_od_* = 50 torr mm (Fig. 4), T_o = 900°K, p_od_* = 259 torr mm (Fig. 5). Experimental data are shown by dots in these figures. Curve 1 is the calculation of equilibrium flow of a small nitrogen admixture in argon, and curves 2 and 3 are the results of a calculation with the rate constants from Refs. 32 and 33, respectively. Theoretical data are shown by curve 4 calculated with the rate constants from Ref. 14 and in curve 5 from Ref. 18.

From the analysis of the data shown in Figs. 4 and 5 it follows that curves 2, 3 and 4 reproduce the experimental data rather well . The coincidence of curves 2 and 3 with the experiment and between each other is not surprising, since the solution of the direct problem with these rate constants only indicates that the inverse problem has been solved correctly. The calculation with the theoretical rate constants obtained in Ref. 18 predicts a population distribution that is closer to the equilibrium one than that obtained in the experiment. The coincidence of theoretical curve 4 with the experiment is somewhat unexpected. Moreover, the rate constants predicted by the theory[14] have an opposite temperature dependence compared with the experimental data (Fig. 1). The explanation of these results is as follows. The solution of kinetic equation system [(Eq. (6)] is sensitive not only to the assignment of initial conditions but also to the local values of the rate constants. Therefore, there is an agreement with respect to the

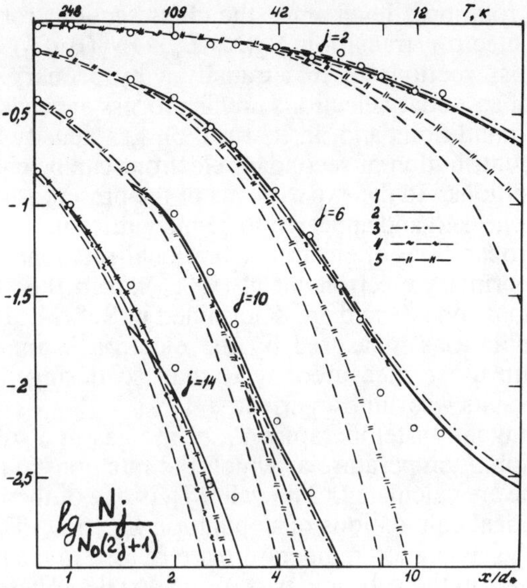

Fig. 5 Rotational populations at various positions in the jet for $T_0 = 900°K$, $p_0 d_* =$ 259 torr mm. See text for explanation of symbols.

population distribution in the temperature interval where the rate constants[14] intersect with the experimental ones.

Discussion

As is clear from the preceding paragraph, none of the theories used for comparison describes the entire bulk of experimental data on rotational relaxation time, rate constants and population kinetics. Let us briefly discuss the possible reasons for the discrepancy between experiment and theory.

In the experiment one can question the correctness of measuring rotational level populations and stagnation temperature, and, in its interpretation, the assumption of the applicability of gasdynamic and kinetic models. Uncertainty in the theoretical calculation of rate constants occurs both due to the model character of the potential surface and the possible discrepancy between the region of the theory applicability and the temperature range in which comparison with the experiment is made.

The problem concerning reliable measurements of rotational temperature and level population by the electron beam technique has been discussed for a rather long period of time (see Ref. 27). The main aspects of this problem are related to the role of secondary electrons and the difference in the probabilities of electron-vibrational-rotational transitions from those assumed initially in the Muntz dipole model.[28] In Ref. 27 it has been shown that secondary electrons change the distribution of the populations of $N_2^+(B^2\Sigma_u^+)$ rotational level since the cross sections for rotational level excitation at electron transition $N_2(X\ ^1\Sigma_g^+) \rightarrow N_2^+(B^2\Sigma_u^+)$ differ from the excitation cross sections of this transition by primary electrons. The contribution of secondary electrons in this process depends substantially on experimental conditions and, in its turn, on gas density and observation region. The contribution of secondary electrons can be checked by changing these conditions. In the experiments of the present study this contribution has been checked and appeared to be insignificant.

The probabilities of electron-vibrational-rotational transitions upon excitation by primary electrons of $N_2^+(B^2\Sigma_u^+)$ from the state $N_2(X\ ^1\Sigma_g^+)$ differ from dipole ones[28] and are determined in Ref. 11. In the same work the level populations measured by the electron beam technique were compared with those measured by Raman scattering. The agreement between the results was almost perfect.

The stagnation temperature T_o of the gas in a source defines the local translational temperature at which the rate constants and rotational relaxation time are calculated. The value T_o is one of the initial conditions for the theoretical calculations of population kinetics. The difficulties in determining the stagnation temperature are related to the effect of friction and heat transfer in the subsonic part of the nozzle. These difficulties are intensified by the low level of the gas density in the source due to the necessity to obtain a nonequilibrium in the jet. An analysis of different methods of determining the correct T_o has shown[14] that the flow rate method is the best one. A good agreement between measured and calculated populations of rotational levels in the equilibrium part of flows near the nozzle (small x/d_*, large $n_o d_*$) confirms the accuracy of the T_o determination (Fig. 5). Nevertheless, the possibility that the applied method of stagnation temperature measurement introduces a pronounced error in all the values obtained from the experiment should not be excluded.

The complexity of the formation process of the freejet out of a sonic nozzle at low Reynolds numbers results in one more uncertainty, i.e., a discrepancy between the gas dynamic model assumed in the calculations and the real flow.[26] According to Ref. 29, for nozzles made in the form of an opening in a thin wall at Re > 100, the flow on the axial stream tube is close to the calculated one. The nozzles of the present study are close in geometry to those given in Ref. 28 and the condition Re > 100 was constantly satisfied. This problem is considered in more detail in Ref. 26.

In the present study Maxwellian distribution in velocities of the relative motion of atoms and molecules is assumed to describe the kinetics of populations. This simplification results in the system [Eq. (6)] of kinetic equations with the transition rates not depending on time. But, as experiments show[30,31] that not only a Boltzmann distribution of populations but also a Maxwellian distribution in velocities are violated. The violation of equilibrium distributions occurs in close conditions[31] that are evidently related to the closeness of rotational and translational relaxation times. For the system $N_2 + N_2$ at low temperatures the value Z_R can be even smaller than 1.[32] Moreover, recent experiments[33] show that the velocity distribution function depends on rotational quantum number. All of this testifies to the necessity of describing the kinetics of rotational level populations within the framework of Wang Chang-Uhlenbeck's equations for the function $f_i(\vec{c}_i, \vec{r}, t)$ but the solution of these equations has not been obtained so far for arbitrary deviations from an equilibrium. For small N_2 admixture in Ar the minimum value is $Z_R \sim 5$ (Fig. 3); therefore, the effect of translational nonequilibrium for this system is expected to be smaller. According to the estimates[14] in all of the experiments of the present study, the region of Maxwellian distribution violation is beyond the measurement region. Additionally, model calculations[34] show that the contribution of translational nonequilibrium on the kinetics of rotational level populations is small.

As far as the theory[18] is concerned, the region of its applicability is defined by inequality $k_BT \geq \langle\Delta E\rangle$, where $\langle\Delta E\rangle$ is the average value of rotational energy increment per collision. This condition is satisfied beginning at 20°K. The assumption of a classical trajectory evidently becomes invalid at lower temperatures. The condition of quasiclassical relative motion is also violated.

The potential surface[19] is a model one, and the rate constant calculations performed with it are expected to have only qualitative agreement with the experiment. Theoretical results are the most sensitive to the value of the anisotropy factor at the repulsive part of the potential b_2. For example, the effective section of rotational relaxation σ_{RT} at high temperatures depends on b_2 squared.[25] This means that theoretical and experimental agreement between the data on Z_R can be attained by decreasing the coefficient b_2 (by ~ 1.5 times). The authors [18] do not attach great importance to this fact as the real potential surface for the system Ar + N_2 seems to differ in an essential way from the model one. Theory [17] allows one to effectively optimize the parameters of the potential surface, but to do this correctly a reliable bulk of experimental data should be available for different observations sensitive to both the isotropic and the anisotropic part of the potential surface.

References

[1]Faubel, M., "Vibrational and Rotational Excitation in Molecular Collisions," Advances in Atomic and Molecular Physics, Vol. 19, 1983, p. 345.

[2]Toennies, J. P., "Progress in Understanding Intermolecular Forces and the Prediction of Gas Flow Phenomena," Annual Review of Physical Chemistry, Vol. 27, 1976, p. 225.

[3]Scott, P. B., Mincer, T. R., and Muntz, E. P., "Nitrogen Rotational Excitation by Collisions with Argon-Observations and Comparison with Theory," Chemical Physics Letters, Vol. 22, 1973, p. 71.

[4]Van Den Biesen, J. J. H., Treffers, M. A., and Van Den Meijdenberg, C. J. , Physica, Vol. 116 A, 1982, p. 101.

[5]Lang, N. C., Polanyi, J. C., and Wanner, I., "Laser Fluorescence Studies of HF Rotational Relaxation," Chemical Physics, Vol. 24, 1977, p. 219.

[6]Borzenko, B. N., Karelov, N. V., Rebrov, A. K., and Sharafutdinov, R. G., "The Experimental Studying of Populations of Rotational Levels of Molecules in N_2 Free Jets," Zhurnal Prikladnoe Mekhaniki i Tekhnicheskoi Fiziki, No. 5, 1976, p. 20 (in Russian), English translation: Journal of Applied Mechanics and Technical Physics, Vol. 17, 1976.

[7]Ferziger,J. H. and Kaper, H. G., Mathematical Theory of Transport Processes in Gas, North-Holland, Amsterdam, 1972.

[8]Sukhinin, G. I., "Generalized Relaxation Equations for Vibrational and Rotational Kinetics of Molecules in Gas Flows," Journal of Applied Mechanics and Technical Physics, Vol. 29, No. 1, 1988.

[9]Skovorodko, P. A., in Rarefied Gas Dynamics, Nauka, Novosibirsk, 1976 (in Russian).

[10]Belikov,A. E., Zarvin, A. E., Karelov, N. V., Sukhinin, G. I., and Sharafutdinov, R. G., "Distrubance of Boltzmann's Rotational Distribution in Nitrogen Free Jets," Journal of Applied Mechanics and Technical Physics, Vol. 25, 1984, p. 12.

[11]Belikov, A. E., Sedlnikov, A. I., Sukhinin, G. I., Sharafutdinov, R. G., "Rotational Transitions in Processes of Ionization of N_2^+ ($B^2\Sigma u$, v' = 0) by Electron Impact," Journal of Applied Mechanics and Technical Physics, Vol. 29, No. 4, 1988.

[12]Brunner, T. A. and Pritchard, D. E., "Fitting Laws for Inelastic Collisions," Dynamics of the Excited State, Wiley, New York, 1982, pp. 589-641.

[13]Dubrovskii, G. V., Pavlov, V. A., and Muhametzianov, R. A., Inzhenernofizicheskii zhurnal, No. 47, 1984, p. 300 (in Russian).

[14]Belikov, A. E., Dubrovskii, G. V., and Zarvin, A. E., "Rotational Relaxation of Nitrogen in Ar Free Jet," Journal of Applied Mechanics and Technical Physics, Vol. 25, 1984, p. 19.

[15]Pattengill, M. D. and Bernstein, R. B, "Surprisal Analysis of Classical Trajectory Calculations of Rotationally Inelastic Cross Sections for the Ar + N_2 System; Influences of Potential Energy Surface," Journal of Chemical Physics, Vol. 65, 1976, pp. 4007-4015.

[16]Koura, K., "Rotational Distribution of N_2 in Ar Free Jet", Physics of Fluids, Vol. 24, 1981, pp. 401-405.

[17]Storozhev, A. V. and Strekalov, M. L., Chim. Fizika, 1988, in Russian (to be published).

[18]Belikov, A. E., Burshtein, A. I., Dolgushev, S. V., et al., "Rate Constants and Times of Rotational Relaxation of N_2 in Ar," Preprint No. 188-88, Institute of Thermophysics, Novosibirsk, 1988.

[19]Pattengill, M. D., La Budde, P. A., Bernstein, R. B., and Curtiss, C. F., "Molecular Collisions. XVI. Comparison of GPS with Classical Trajectory Calculations of Rotational Inelasticity for the Ar-N_2 System," Journal of Chemical Physics, Vol. 55, 1971, pp. 5517-5522.

[20]Belikov, A. E., Soloviev, I. Yu., Sukhinin, G. I., and Sharafutdinov, R. G., "The Rotational Relaxation Time of N_2 in Ar," Journal of Applied Mechanics and Technical Physics, Vol. 28, No. 4, 1987.

[21]De-Pristo, A. E. and Rabitz, H., "Scaling Theoretic Decombination of Bulk Relaxation Data: State-to-State Rates from Pressure-Broadening Linewidth," Journal of Chemical Physics, Vol. 68, 1978, pp. 1981-1987.

[22]Kistmaker, P. G. and de Vries, A. E. "Rotational Relaxation Times in Nitrogen-Noble Gas Mixtures," Chemical Physics, Vol. 7, 1975, pp. 371-382.

[23]Gelb, A. and Kapral, R., "Rotational Relaxation in the Ar-N_2 System," Chemical Physics Letters, Vol. 17, 1972, pp. 397-400.

[24]Russel, J. D., Berstein, R. B., and Curtiss, C. F., "Transport Properties of Gas of Diatomic Molecules. VI Classical Trajectory Calculations of the Rotational Relaxation Times of Ar-N_2 System," Journal of Chemical Physics, Vol. 57, 1972, pp. 3304-3307.

[25]Strekalov, M. L. and Burshtein, A. I., "Theory of Vibrational Line Width in Dense Gases," Chemical Physics, Vol. 82, 1983, p. 11.

[26]Skovorodko, P. A., and Sharafutdinov, R. G., "Kinetics of Populations at Rotational Levels in N_2 Free Jet," Journal of Applied Mechanics and Technical Physics, Vol. 22, No 5, 1981.

[27]Belikov, A. E., Karelov, N. V., Rebrov, A. K., and. Sharafutdinov, R. G, in The Diagnostics of Rarefied Gas Flows, Institute of Thermophysics, Novosibirstk, 1979, pp. 7-63 (in Russian).

[28]Muntz, E. P., "Static Temperature Measurements in a Flowing Gas," Physics of Fluids, Vol. 5, 1962, pp. 80-90.

[29]Ashkenas, H. and Sherman, F. S., "The Structure and Utilization of Supersonic Free Jets in Low Density Wind Tunnels," Proceedings of the 4th Symposium (International) on Rarefied Gas Dymanics, Vol. 2, 1966, pp. 84-105.

[30]Anderson, J. B. and Fenn, J. B., "Velocity Distribution in Molecular Beam from Nozzle Sources," Physics of Fluids, Vol. 8, 1965, pp. 780-787.

[31]Zarvin, A. E. and Sharafutdinov, R. G., "The Rotational Relaxation at Transition Regime Free Jets of Nitrogen," Journal of Applied Mechanics and Technical Physics, Vol. 22, No. 5, 1981.

[32]Belikov, A. E., Soloviev, I. Yu., Sukhinin, G. I., and Sharafutdinov, R. G., "The Rotational Relaxation Time for Nitrogen," Journal of Applied Mechanics and Technical Physics, Vol. 29, No. 4, 1988.

[33]Belikov, A. E., Khmel, S. Ya., and Sharafutdinov, R. G. "The Study of Velocity Distribution by t-o-f Photoelectronic Technique," Proceedings of the 8th Symposium (National) on Rarefied Gas Dynamics, Moscow, 1985, pp. 116-120 (in Russian).

[34]Koura, K., "Transition and Rotational Relaxation," Journal of Chemical Physics, Vol. 65, 1976, pp. 2156-2160.

Rotational Relaxation of CO and CO_2 in Freejets of Gas Mixtures

Takeshi Kodama,* Shida Shen,† and John B. Fenn‡
Yale University, New Haven, Connecticut

Abstract

By means of Fourier transform infrared spectrometry in the emission mode, the terminal rotational energy content has been determined for CO and CO_2 in free jet expansions of various gas mixtures. The vibrational excitation required for radiation was achieved by heating the source gas or by exciting it with a corona discharge. Characteristic rotational relaxation rates were determined for each species. The rate for CO_2 is always higher than that for CO, and there seems to be very little coupling between the rotational modes of the two species.

Introduction

In recent studies of the exhaust plume from a novel microjet burner exhausting into vacuum, emission spectra obtained by Fourier transform emission spectrometry (FTIS) revealed a terminal rotational energy content that was substantially lower for CO_2 than for CO.[1] Possible explanations include 1) a substantial departure from equilibrium in the reaction products before expansion so that the CO started at a higher rotational temperature than the CO_2 and 2) the rotational relaxation during expansion in the jet of gas mixture was not as fast for CO as for CO_2. Previous studies have shown that in free jets containing either species alone, or in admixture with He or Ar, the relaxation rate is always slower for CO.[2] There have been no results reported for jets in which both species are present. Consequently, it seemed appropriate to carry out some experiments with mixtures of the

---

*Visiting Scientist; currently at Komatsu Technical Research Center, Japan.

† Graduate Student; currently Research Scientist, Department of Chemical Engineering.

‡ Professor Emeritus, Department of Chemical Engineering.

two in order to determine to what extent coupling between their rotational modes might effect the relaxation process.

Apparatus and Procedures

The methods and equipment have been previously described in detail.[3] A free jet from a heatable source with a nozzle diameter of 0.15 mm is located near the focal point of a CaF_2 lens with a diameter and focal length of 100 mm. An opposing aluminized front surface mirror approximately doubles overall collection efficiency. The center of the effective field of view, an ellipsoidal volume with major and minor axes of roughly 50 and 5 mm, is 120 or more nozzle diameters downstream. Reported data were obtained under conditions for which variations in axial and radial distance as well as in background pressure (normally less than 5x10-5 torr) made no significant differences in the spectra. Thus, we believe that collected radiation was mostly from jet gas in its terminal state.

The source temperature could be raised to 1100 K by a resistance heater in order to produce enough radiation from the jet gas for analysis by the spectrometer (Nicolet FTIR-1500). It has been well established that vibrational relaxation in CO and CO_2 is sufficiently incomplete in jets from these small nozzles so that the terminal gas will radiate if the source gas is sufficiently excited. In the present experiments we also used a corona discharge in the source gas to obtain vibrational excitation without substantial increases in the rotational and translational temperatures. The discharge current was kept constant at 300 μ A. The voltage drop across the corona varied between about 1100 and 1800 V, depending on source gas density and composition. On the basis of very crude approximations of the amount of terminal vibrational energy in the jet molecules, we arrive at a crude estimate of about 150 K for the increase in effective stagnation temperature due to the corona discharge. Source gas for mixtures was prepared by filling an evacuated container to the appropiate partial pressure of each component from cylinders of reagent grade commercial gases (Matheson). Mixing was assured by natural convection due to gentle heating of the bottom of the container after it was filled. Most spectra were the result of 200 scans at a resolution of 0.1 cm-1 and required about an hour.

Results and Discussion

Figure 1 shows a spectrum for an equimolar mixture of CO and CO_2 that has been heated to 900 K before expansion. Figure 2 shows a spectrum for the same mixture excited in the source by a corona

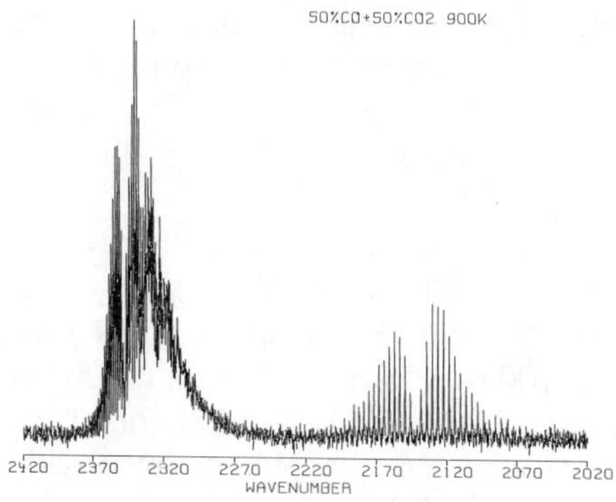

Fig. 1. Spectrum of radiation from the terminal state of a free jet of equimolar CO-CO_2 from a source at 900 K and 300 torr. The CO_2 band is on the left.

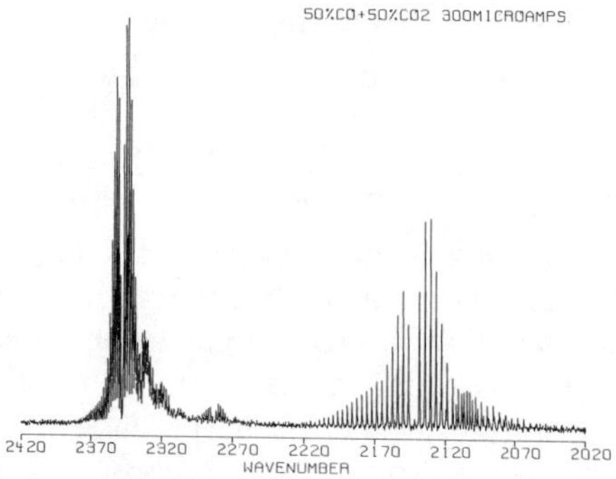

Fig. 2. Spectrum for conditions of Fig. 1 except that the source gas has been excited by a corona discharge.

discharge. Both show clearly separated and resolved bands for CO_2 on the left and CO on the right. The thermal spectrum for CO_2 is much more congested, especially in the P branch which includes, within the rotational envelope of the fundamental transition in the asymmetric stretch mode, the combination bands for which the bending and symmetric stretch modes are excited. The discharge spectrum for CO_2 in Fig. 2 is much "cleaner." Because the rotational temperature is much lower, combination bands are largely outside of the fundamental rotational envelope and the overall resolution is much better. The appearance of a new band about 70 wave numbers to the red of the CO_2 band center that we have observed previously is noteworthy. Careful analysis, which will be reported elsewhere, leaves no doubt that it is due to $^{13}CO_2$ which cannot be discerned in thermal spectra because it is embraced within the much broader rotational envelope.

When CO_2 is dispersed in helium or argon, the intensity of this new band reaches 15% of the intensity of the band due to the lighter isotope, even though its relative abundance is only 1%. We believe that this enhanced intensity is due to vibrational pumping of a ground state molecule of the heavier isotope by collision with an excited lighter one. The excess 70 wave numbers of energy go into rotation and/or translation of the colliding partners. Because the temperature in the jet is so low this excess energy in translation and rotation is rapidly dissipated. Consequently, there are no collision partners of the lighter isotope available that have the energy necessary for the reverse process to take place. Thus, the energy transfer from the lighter to the heavier isotope is along a one-way street.

In the CO band of the thermal spectrum in Fig. 1, the rotational lines are more clearly resolved because the rotational energy levels are more widely spaced in CO than in CO_2. Thus, it is possible, though not very clearly in this particular spectrum, to distinguish a band due to excitation of CO to the $v = 2$ level. A vibrational "temperature" can be obtained from the relative intensity of the two bands and turns out to be equal to the source temperature. This equality clearly supports the assumption that vibrational energy does not relax during the free jet expansion. The rotational lines in the discharge spectrum of Fig. 2 show a much more jagged profile than their counterparts in Fig. 1. This irregularity is much more prominent for spectra of CO alone or in admixture with helium or argon. Careful analysis of these spectra clearly reveals a Boltzmann distribution of vibrational energy over levels up to $v = 6$, corresponding to vibrational temperatures above 3000 K in some cases. It turns out that some of the rotational lines from these various vibrational levels are effectively coincident within the resolution of our spectrometer and thus superpose to produce the irregularity in the rotational envelope for CO in Fig. 2. If we use argon as carrier gas to achieve very low terminal rotational temperatures (because of its high specific heat ratio), the rotational band is greatly narrowed and the radiation from the higher vibrational levels is clearly revealed. Figure 3 shows such a discharge spectrum obtained with CO in argon. In this case the terminal rotational temperature is so low that rotational lines corresponding only to the lowest two or three levels can be seen.

It is important for our present purpose to consider some further aspects of the behavior of CO_2 as revealed by its discharge spectrum in Fig. 2. Unlike the spectrum for CO it shows no appreciable contribution from any vibrational levels of the asymmetric stretch mode above $v = 1$. All of the radiation on the red side of the rotational envelope of the fundamental seems to be accounted for by the combination bands at the $v = 1$ level. It is well known that when CO_2 is excited by a discharge under static conditions in a cell, the spectrum shows appreciable populations at vibrational levels up to as high as v = 16 or higher.[4] We have never seen radiation from these higher levels in all the CO_2 emission spectra we have ever obtained from free jets, whether the source gas has been excited by heating or by discharge. We conclude that there must be extremely rapid relaxation of these higher vibrational levels by some sort of a resonant exchange during the free jet expansion, the energy being dissipated among other modes. A consequence of this conclusion is that the effective specific heat ratio and source temperature are not well defined for free jet expansion of CO_2 that has been excited by a corona discharge. Therefore, we are unable to arrive at meaningful rotational relaxation rates under these conditions.

As set forth in the Introduction, the original primary objective of this study was to determine whether the rotational relaxation of CO and CO_2 were at all coupled when they were present together in a freely expanding jet. Our approach was to take advantage of the dependence of the rotational collision number Zr on terminal rotational temperature that emerged from numerical integrations for free jet expansion of a relaxing gas. Zr is a convenient measure of relaxation rate and represents the number of collisions that the average molecule in a population undergoes while the difference between rotational and vibrational temperatures in that population decreases to l/e of its initial value. Such calculations were first carried out by Gallagher and Fenn[5] and then systematically extended over a range of specific heat ratios by Quah et al.[6] More recently, they were refined by Labowsky et al [7] who carried out calculations by the method of characteristics for nonequilibrium free jet expansions. We used the latter results which were only slightly different from those obtained by Quah et al.

The first step in arriving at a value for Zr is to obtain the terminal rotational "temperature" from the spectrum. Because the terminal distribution of rotational energy frequently departs markedly from a Boltzmann distribution, one cannot obtain a true rotational temperature. Instead, we arrive at a pseudo terminal rotational temperature by the following strategem. We find that the observed distributions in the thermal spectra can be well represented by the two temperature model originally proposed by Venkateshan et al.[3] It pictures the terminal gas as comprising two separate populations of molecules, each with its own Boltzmann distribution of rotational energy and thus possessing a meaningful rotational temperature. We represent the total rotational energy as being equal to the sum of the rotational energy in each population, i.e., $R(C_1Tr_1 + C_2Tr_2)$ where the R is the molar rotational heat capacity, C the mole fraction, and Tr the rotational temperature. The subscripts 1 and 2 refer to the two sub populations. We assume that both CO and CO_2 behave as linear

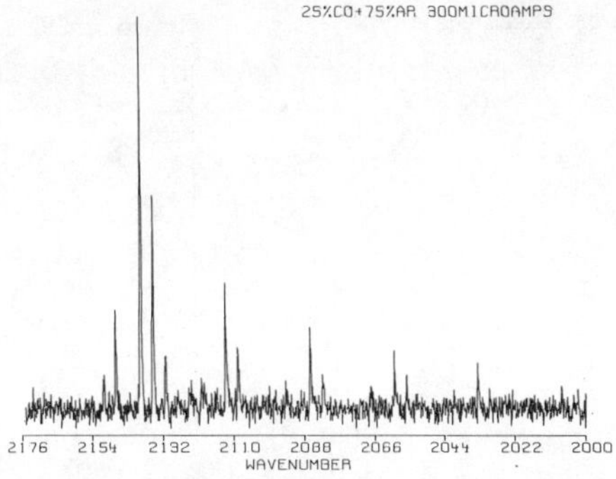

Fig. 3. Spectrum of radiation from the terminal state of a free jet of 25% CO in argon from source gas at 1000 torr and 300 K that has been excited by a corona discharge.

molecules with two rotational degrees of freedom and that the vibrational energy does not relax enough to participate in the expansion. In other words we assume that Cp/Cv is 7/5. For several reasons these assumptions do not hold when the vibrational excitation is by corona discharge in the source. In the first place the rotational distributions are so non-Boltzmann that the two-temperature model cannot provide a good description of the actual distribution. Figure 4 shows an example of the marked departure from a Boltzmann distribution that we have observed. In the second place, as we pointed out earlier, it appears that a substantial amount of energy in the higher vibrational levels of CO_2 in the source does relax during the expansion so that the effective value of Cp/Cv is not well defined and must be somewhat less than 7/5. As the results of Quah et al. show, the value of Zr depends strongly on this specific heat ratio.[5] In the third place, when discharge excitation is used in source gas that is typically at room temperature, especially when the CO_2 is in admixture with He or Ar, the jet temperatures become low enough to promote condensation of both CO_2 and CO so that effective values of T_0 and Cp/Cv become pretty much indeterminate. Even for mixtures with CO as the major component Cp/Cv was apparently low enough to bring about some condensation of CO_2. For these reasons we gave up trying to extract meaningful values of Zr when the source gas was excited by a discharge.

Figure 5 shows Zr values for CO and CO_2 mixtures expanded from source gas at 1000 Torr and 900 K. The values for CO show a larger, less systematic dependence on composition than do those for CO_2. There is no obvious evidence of any strong positive or negative correlation or coupling between the two species. There does seem to

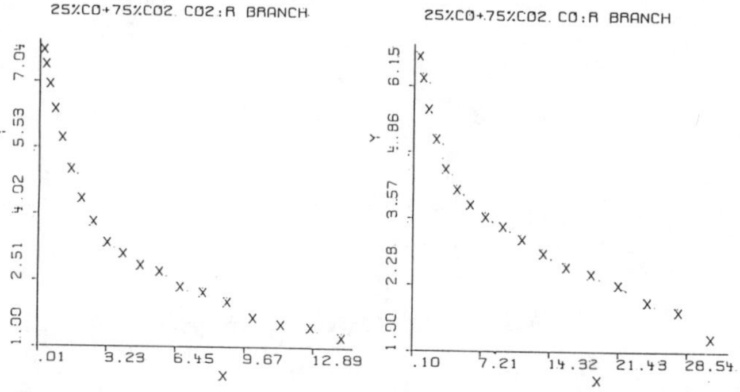

Fig. 4. Boltzmann plot of the rotational line intensities in the spectrum of radiation from the terminal state of a free jet of 25% CO_2 in CO from source gas at 300 torr and 300 K that has been excited by a corona discharge. The departure from a Boltzmann distribution of rotational energy is extreme.

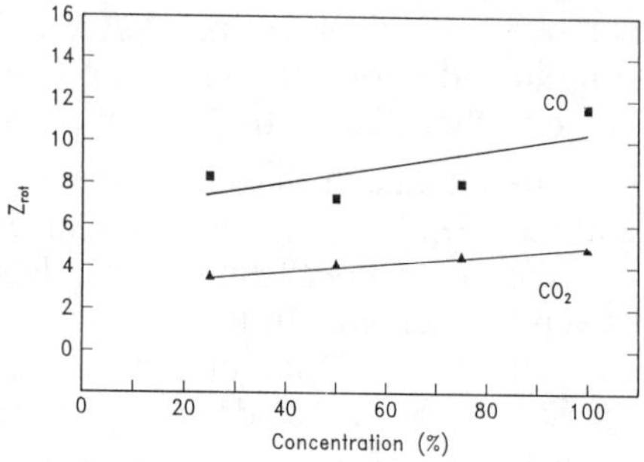

Fig. 5. Values of Zr for CO and CO_2 in free jet expansions of their mixtures of varying proportions.

be a slight indication, stronger in CO than in CO_2, that for each species the effective rotational transfer cross sections are larger in collisions between like rather than unlike partners. In any event, it is clear that CO_2 relaxes more rapidly than CO. Even when we cannot extract a very meaningful value of Zr, as in the case of discharge excitation, we always find that CO_2 relaxes rotationally more rapidly than CO, no matter what the initial temperature, pressure or source gas composition. Experiments have covered temperatures from 500 to 1100 K, source pressures from 300 to 1000 Torr, and source gases comprising CO or CO_2 at various concentrations in N2 and Ar. In no case have we observed any effects that could be attributed to an appreciable coupling of rotational modes in different species.

Acknowledgments

This research was supported in part by the U.S. Air Force Office of Scientific Research. Appreciation is also due to Komatsu, Ltd. for funding and participation of Takeshi Kodama from its Technical Research Center at Hiratsuka-shi near Tokyo.

References

[1]Groeger, W. and Fenn, J. B., "Microjet Burners for Molecular Beam Sources and Combustion Studies" Scientific Instruments, Sept. 1988, pp.1971-79.

[2]Ryali, S. B., Venkateshan, S. P., and Fenn, J. B., "Terminal Distributions of Rotational Energy in Free Jets of CO and CO_2," Rarefied Gas Dynamics, (edited by H. Oguchi), Univ. Tokyo Press, Tokyo, 1984, pp. 567-576.

[3]Venkateshan, S. P., Ryali, S. B., and Fenn, J. B., "Terminal Distributions of Rotational Energy in Free Jets of CO_2 by Infrared Emission Spectrometry," Journal of Chemical Physics, Vol. 77, Sept. 1982, pp. 2599-2606.

[4]Bailly, D., Farrenq, R., Guelachvili, G., and Rossetti, C., "CO_2 Analysis of Emission Fourier Spectra in the 4.5 μm Region: Rovibrational Transitions," Journal of Optical Spectroscopy, Vol. 90, Jan. 1981, pp. 74-105.

[5]Gallagher, R. J. and Fenn, J. B., "Relaxation Rates from TOF Analysis of Molecular Beams," Journal of Chemical Physics, Vol. 60, May 1974, pp. 3487-3499.

[6]Quah, C. G. M., Fenn, J. B., and Miller, D. R., "Internal Energy Relaxation Rates from Observations on Free Jets," Rarefied Gas Dynamics (edited by R. Compargue), Commissariat a l'Energie Atomique, Paris, 1979, Vol. II. pp. 885-898.

[7]Labowsky, M., Ryali, S. B., Fenn, J. B., and Miller, D. R., "Flowfield Calculations in Nonequilibrium Free Jets by the Method of Characteristics," Progress in Astronautics and Aeronautics: Rarefied Gas Dynamics , Vol. 74, edited by S. S. Fisher, AIAA, New York, 1981, pp. 695-709.

Diffusion and Energy Transfer in Gases Containing Carbon Dioxide

John R. Ferron*

University of Rochester, Rochester, New York

Abstract

Diffusion of vibrational energy is expected to be an important part of transport phenomena in polyatomic gases. It is difficult to estimate the contribution with the aid of current theories, suitably separating it from the more evident influences of rotational energy. Phenomenological models of vibrational energy transfer are explored here as a means of representing vibrational energy transfer explicitly. These lead to predictions of internal diffusivities that are in approximate accord with pressure, temperature, and concentration dependence that can be deduced from laser excitation experiments and from measured transport properties. Examples are given for pure carbon dioxide and for mixtures with argon.

Introduction

The transport properties of polyatomic gases were first given explicit expressions by Wang Chang and Uhlenbeck[1] by means of their approximate, semiclassical treatment. The contributions of translational motion and of internal energy are separately represented in these expressions, and multiple internal states may be included. Coupling among states occurs because the various energies are utilized in the evaluation of the collision integrals.

Such evaluation is complicated, and simplifying assumptions and substitutions are required for development of practicable formulas. Mason, Monchick, and co-workers[2-4] provided these in their derivation of expressions for the thermal conductivity of pure gases[2] and in their

---

*Professor, Department of Mechanical Engineering.

construction of a polyatomic theory for mixtures.[3,4] Wang Chang and Uhlenbeck had included relaxation terms for exchange of energy among modes. Mason and Monchick represented these by the approximate form

$$\frac{C_{int}}{z_{int}} = \frac{C_{rot}}{z_{rot}} + \frac{C_{vib}}{z_{vib}} \tag{1}$$

where

$$C_{int} = C_{rot} + C_{vib} \tag{2}$$

is the sum of rotational and vibrational contributions to heat capacity and z_{rot} and z_{vib} are, respectively, average numbers of collisions required for exchange of a quantum of rotational and vibrational energy. Equation 1 defines the internal relaxation number, z_{int}.

Mason and Monchick also found it expedient to represent certain integrals by diffusion coefficients D_{rot}, D_{vib}, and D_{int}, related by

$$\frac{C_{int}}{D_{int}} = \frac{C_{rot}}{D_{rot}} + \frac{C_{vib}}{D_{vib}} \tag{3}$$

representing the ability of a gas or gas mixture to transmit internal energy by diffusional processes.

To use the theory to predict thermal conductivity, one must supply several physical properties, of which the relaxation numbers and internal diffusivities are the most difficult to evaluate. Rotational relaxation numbers for certain pure materials may be obtained from ultrasonic and other experiments.[5] Ordinarily, however, mixture values must be estimated from pure-component data. Vibrational relaxation numbers are commonly two or three orders of magnitude larger than those for rotational relaxation.[5] As a result the final term of Eq. 1 is customarily omitted, and except for the influence of C_{int}, computed values depend wholly on rotational relaxation.

The internal diffusivities have not been independently measured. In[1] Mason and Monchick's first illustrations of their theory, they supposed that a single coefficient, D_{int}, might be used; that is, D_{rot} and D_{vib} were taken to be equal. They then used measured values of the self-diffusion coefficient to represent D_{int}. In the theory the thermal conductivity is taken as the sum of a translational term and an internal term. The latter part has rotational and vibrational portions, but the average diffusivity D_{int} is applied to both of these.

Ahtye[6] objected to this because both rotational and vibrational relaxation were governed by the same characteristic collision number, and diffusion of both forms of internal energy had the same coefficient. He proposed an approximate, qualitative extension of the Wang Chang-Uhlenbeck theory that results in separate contributions to thermal conductivity from translational, rotational, and vibrational modes. He used the self-diffusion coefficient for D_{rot} and inferred values of D_{vib} from experimental thermal conductivities with the aid of his theory. For carbon dioxide at 300-1500°K and 1 atm pressure, D_{vib} was found to be 60-70% of the self-diffusion coefficient. This fraction is consistent with measurements of the laser-excited diffusivity (at about 300°K) of $CO_2(001)$, the asymmetric-stretch mode, to which we refer in more detail later. [Here we use the notation $CO_2(v_1,v_2,v_3)$ to suggest the quantum number of the three normal modes of vibration in carbon dioxide.]

Up to the present time a procedure like that of Ahtye, utilizing experimental data for thermal conductivity, has been the best that can be applied to determination of D_{vib}. The objective of this work is an independent estimate. We use a phenomenological method that makes use of laser excitation results for diffusion and energy transfer.

Vibrational Relaxation and Internal Diffusivities

Irradiation of a sample of carbon dioxide by a Q-switched CO_2-laser pulse produces a sudden increase in the concentration of the 001 level, that representing asymmetric stretching, in the CO_2 sample. Subsequent decay to the ground state can be monitored by infrared fluorescence. The technique has been used extensively for study of vibrational-vibrational transitions,[8] as a sensitive method for quantitative analysis of carbon dioxide in gases[9] and, most pertinent to the present discussion, for measurement of the diffusion coefficient of the vibrationally excited carbon dioxide.[7,10,11]

Decay of the excited state may be considered to occur primarily by three paths: quenching collisions with eventual transfer of energy to the translational mode of ground-state CO_2; diffusion of excited molecules to the wall of the cell, where de-exitation presumably occurs by collision with CO_2 adsorbed on the wall; and spontaneous radiation, with self-absorption followed by reradiation or de-exitation by the two diffusional mechanisms. Radiation is a small part of the total rate, although it is sufficient to monitor the experiment, and at low-to-moderate pressures self-absorption requires only a small correction.[10,11]

Experiments[7,11] were carried out at constant temperature (about 300°K) and at pressures ranging from 1 mTorr to about 10 Torr. Various cell sizes and wall materials were used in order to establish an appropriate diffusion-reaction model and to ascertain that accommodation at the cell wall is independent of cell material, hence determined by adsorbed CO_2. At pressures above about 1 Torr, collisions are sufficiently numerous that local gradients in concentration of the excited CO_2 are eliminated. Hence, a diffusion coefficient for $CO_2(001)$ is obtained in the limit of zero pressure.

Analysis of the experimental data was originally carried out[7,11] in terms of transient radial diffusion and homogeneous decay of $CO_2(001)$ to the translational mode, in a cylindrical system. Decay was sufficiently rapid that a single eigenfunction, in the form $J_0(ar/r_o)\exp(-bt)$, could represent concentration of $CO_2(001)$ for large time t. Here r is the radial coordinate, $0 \le r \le r_o$, with r_o the radius of the sample cell. Total decay rate is b, corrected for radiation and related to the eigenvalue a by

$$b = b' + a^2 D_o / r_o^2 \tag{4}$$

where b′ is the collisional decay rate.

The boundary condition at the cell wall is given by

$$aJ_1(a) - (vr_o/2D_o)(1-f)J_0(a)/(1+f) = 0 \tag{5}$$

where $v = (8k_B T/m\pi)^{1/2}$ is the arithmetic mean speed at temperature T for m = 44.01, the molecular mass of CO_2; k_B is the Boltzmann constant; and f is the probability of molecular reflection at the cell wall. The quantities J_0 and J_1 are Bessel functions of the first kind.

Decay rate b′ is proportional to pressure and is obtained from the slope of total decay rate as a function of pressure at relatively high pressure, above 1 Torr. The experimental value[7,11] for CO_2 is about 350 $sec^{-1}\cdot Torr^{-1}$, characteristic of intramolecular energy transfer between vibrational modes.[12]

Diffusivity D_0 may be supposed to be inversely proportional to pressure,

$$D_0 = D_1/P \tag{6}$$

where P is in atmospheres and D_1 stands for the value of $CO_2(001)$ diffusivity expected at 1 atm.

One may use the observed total decay rate as a

function of pressure, cell size, and material in the following way. Assume a value of D_1 and determine the eigenvalue a from Eq. 4. Use this in Eq. 5 to obtain $vr_o(1-f)/(1+f)$, which should be independent of pressure and proportional to r_o. Adjust D_1 to satisfy these conditions. Reflection probability f would in this way be obtained for various cell materials. In practice f was 0.78 ± 0.08 for cell surfaces of Pyrex, brass, Mylar, and Teflon[7] in one test and 0.6 ± 0.05 for Pyrex in the other.[11] One value of D_1 was 0.07, later[10] modified to 0.066; the other[11] was 0.075 ± 0.005 cm^2/s. The self-diffusion coefficient D_{11} of CO_2 at 1 atm and 313°K is 0.119; thus D_1 is about 60% of D_{11}.

Phenomenological Model for D_{vib}

The analysis we have outlined leads to a diffusivity for the transient excited state, $CO_2(001)$, based on its progress to a static surface where it may decay to $CO_2(000)$. The parameter D_{vib} has a more complex meaning in its expression of the effectiveness of CO_2 in transferring vibrational energy. When there are intramolecular vibrational-vibrational (V-V) transitions during collision and when energy is exchanged in vibrational-translational (V-T) pathways, the gas has transferred energy as surely as in the case in which an excited molecule makes its way along a trajectory of finite length.

We need to simulate these complex mechanisms by a more detailed model than that used for $CO_2(001)$ diffusion alone without burdening the analysis unduly. Fortunately, a few steps of V-V transition may control the pertinent rates,[8,13] and we are encouraged to construct a model with a small number of active vibrational levels.

For pure CO_2 this might in simplest form go as follows. We represent decay of $CO_2(001)$ by a single V-T path:

$$CO_2(001) + CO_2(000) \rightarrow CO_2(000) + CO_2^* \tag{7}$$

where the components of a mixture are $CO_2(000)$, the ground state, component 1, $CO_2(001)$, the excited state, component 2, and CO_2^*, a trace of ground-state material considered to carry the translational energy (of the order of 2-3 kcal/mole), component 3.

We imagine an experiment like that previously described, occurring in the radial direction of a cylindrical system. The conservation equations for the three components might be written as

$$\partial x_1/\partial t = 0 \tag{8a}$$

$$c(\partial x_2/\partial t) = -r^{-1}[\partial(rN_2)/\partial r] - k_v Pcx_2 \tag{8b}$$

$$c(\partial x_3/\partial t) = -r^{-1}[\partial(rN_3)/\partial r] + k_v Pcx_2 \tag{8c}$$

where x_1, x_2, and x_3 are the respective mole fractions, c is the (constant) molar density, $c=P/RT$, with R the gas constant, and k_v ($atm^{-1}\cdot s^{-1}$) is the rate constant for V-T decay.[14] The molar fluxes have the Fickian forms

$$N_i = -cD_{im}(\partial x_i/\partial r) + x_i(N_1+N_2+N_3) \tag{9}$$

with $i=1,2,3$, where D_{im} is an effective diffusivity of component i in the mixture. We suppose that x_2 and x_3 are small compared to x_1 and that $N_1 = 0$. We identify D_{2m} with D_0, the value for $CO_2(001)$, available from the laser excitation data in the limit of zero pressure. And we set $D_{3m} = D_{11}$, the self-diffusion coefficient of CO_2. Finally we define D_{vib} by

$$N_2 + N_3 = -cD_{vib}[\partial(x_2+x_3)/\partial r] + (x_2+x_3)(N_2+N_3) \tag{10}$$

With the assumptions listed, Eqs. 9 and 10 resolve to

$$D_{vib}[\partial(x_2+x_3)/\partial r] = D_0(\partial x_2/\partial r) + D_{11}(\partial x_3/\partial r) \tag{11}$$

Now we may go on to study the one-dimensional boundary-value problem in more detail. For the second component, $CO_2(001)$, we find for constant density,

$$\frac{\partial x_2}{\partial t} = \frac{D_1}{Pr}\frac{\partial}{\partial r}\left(r\frac{\partial x_2}{\partial r}\right) - k_v Px_2 \tag{12}$$

with a given uniform initial state, symmetry about $r=0$, and the wall condition, at $r=r_o$,

$$\frac{D_1}{P}\frac{\partial x_2}{\partial r} + \frac{vx_2}{2} = f_2\left(-\frac{D_1}{P}\frac{\partial x_2}{\partial r} + \frac{vx_2}{2}\right) \tag{13}$$

where f_2 is the surface reflection probability of the excited species.

A solution may be developed in terms of a Fourier-Bessel series, but we propose to utilize the solution as an asymptote in time. Hence, a single eigenfunction is used, with

$$x_2(r,t) = J_0(h_2^{1/2} r)e^{-bt}$$

where

$$h_2 = (P/D_1)(b-k_vP) \tag{14}$$

and b is, as before, the total decay rate less that from radiation. The wall condition resolves to Eq. 5 with the eigenvalue $a=h_2^{1/2} r_0$ and $f=f_2$.

We may now go on to substitute the solution in the third part of Eq. 8, solving for $x_3(r,t)$ and using the wall condition of Eq. 5 with D_{11} and f_3 in place of D_0 and f. The initial eigenvalue is the same provided that

$$\frac{1}{D_{11}}\left(\frac{1-f_3}{1+f_3}\right) = \frac{1}{D_0}\left(\frac{1-f_2}{1+f_2}\right) \tag{15}$$

Because D_{11} is greater that D_0, we find f_3 less that f_2; accommodation of the ground-state CO_2 is better than that of the vibrationally excited form. The solution may be written as

$$x_3(r,t) = k_vPx_2(r,t)(1-e^{-h_3t})/h_3$$

where

$$h_3 = (D_{11}/D_0)(b-k_vP)-b \tag{16}$$

With the two solutions available we next return to Eq. 11, substitute for the gradients, and average the result with respect to time, evaluating the average in the limit of infinite time. In this way we form the identity

$$b/D_{vib} = (b-k_vP)/D_0 + k_vP/D_{11} \tag{17}$$

In the limit of zero pressure Eq. 17 yields $D_{vib} = D_0$, consistent with the hypothesis of the model, that $CO_2(001)$ decays by interaction with the cell wall for these conditions. If, however, the wall reflection coefficient is unity, total decay is determined by collisional quenching, $b-k_vP=0$ and we obtain $D_{vib} = D_{11}$.

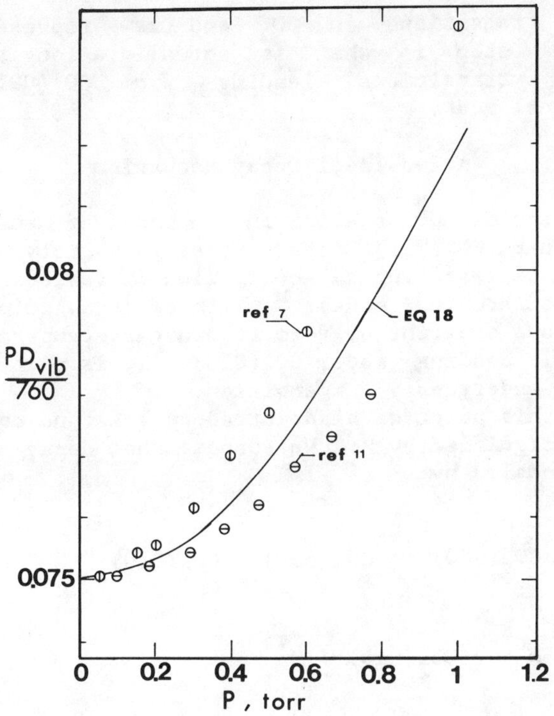

Fig. 1 Pressure dependence of the vibrational diffusivity at about 300°K. Experimental data are based on laser excitation in carbon dioxide.

At pressures greater than 1 Torr, $b = k_v P$ is inversely proportional to pressure.[7] Equation 17 resolves in this case to

$$D_{vib} = \frac{1}{P}\left(\frac{1 + 0.298P^2}{13.333 + 2.502P^2}\right) \tag{18}$$

where P is in Torr. Figure 1 illustrates this form along with data from Refs. 7 and 11. A value of b at 1 Torr has been chosen to fit the line shown. The low-pressure limit, $D_1 = 0.075$, is taken from Ref. 11. At 1 atm D_{vib} is essentially equal to D_{11}.

For the calculations of Fig. 1 we have used $k_v = 193.5$ $s^{-1} \cdot Torr^{-1}$, based on the V-T rate.[12,14] This makes sense for our model of Eq. 7, involving no explicit intramolecular vibrational exchange. The data[7,11] suggest that $k_v = 350$ $s^{-1} \cdot Torr^{-1}$ is better, however. This value is associated[12] with V-V transitions in CO_2 and may represent a rate-controlling step in what is actually a long and complex chain of transitions leading from $CO_2(001)$ to the translational mode.

A Two-level Decay Mechanism

To provide more realism in the model of decay, we next suppose that $CO_2(001)$ first takes a variety of paths to $CO_2(nmo)$, representing a collection of various bending or symmetric-stretching modes.[13] These decay quickly to the most populous equilibrium form at ordinary temperatures, the fundamental bending mode, $CO_2(010)$. It is this state that eventually undergoes V-T transition.

At this point we also introduce a second component, a rare gas, signified by M. We suppose that decay of $CO_2(001)$ might be modeled by

$$CO_2(001) + CO_2(000) \xrightarrow{k_1} CO_2(nmo) + CO_2(010) \tag{19a}$$

$$CO_2(001) + M \xrightarrow{k_2} CO_2(nmo) + M^* \tag{19b}$$

$$CO_2(nmo) + CO_2(000) \xrightarrow{k_3} 2CO_2(010) \tag{19c}$$

$$CO_2(010) + CO_2(000) \xrightarrow{k_4} 2CO_2^* \tag{19d}$$

As before, the asterisk identifies a trace of component considered to carry the excess translational energy arising from the overall collisional decay.

At the temperature at which we wish to use the model, the population of $CO_2(010)$ and other excited states is substantial.[6] This equilibrium content may also participate in step four of the mechanism of Eq. 19. In defining the trace diffusivity of $CO_2(010)$, we must account for this. We define D_1^* by

$$\frac{1}{D_1^*} = \frac{x_1}{D_{11}} + \frac{x_2}{D_{12}} \tag{20}$$

Components 1 and 2, with mole fractions x_1 and x_2, are $CO_2(000)$ and M, respectively.

The trace diffusivity of M is given by

$$\frac{1}{D_2^*} = \frac{x_1}{D_{12}} \quad \frac{x_2}{D_{22}} \qquad (21)$$

where D_{12} and D_{22} are binary diffusivity of the mixture and self-diffusion coefficient of M. The diffusivity of CO_2(001) is defined as D_{00} in

$$\frac{1}{D_{00}} = \frac{x_1}{D_{01}} + \frac{x_2}{D_{02}} \qquad (22)$$

where $D_{01} = D_0$ is the zero pressure limit for CO_2 with a trace of M and D_{02} is the corresponding value for a trace of CO_2 in M.

The vibrational diffusivity also has two contributions, weighted by the respective mole fractions,

$$\frac{1}{D_{vib}} = \frac{x_1}{D_{1,vib,1}} + \frac{x_2}{D_{1,vib,2}} \qquad (23)$$

where $D_{1,vib,1} = D_{vib}$ when $x_1 = 1$, that is, for pure CO_2. This is to be compared with the same quantity shown, for example, in Fig. 1.

The processes of Eq. 19 have various rates, identified by k_1, k_2, k_3, and k_4, each in units of $s^{-1} \cdot Torr^{-1}$, as with k_v of the previous model. The third step is rapid,[13] and we suppose that k_3 is much larger than the other rates. This makes a simplification possible; we will assume that the mole fraction of CO_2(nmo) has a constant small value given by $x_3(k_1x_1+k_2x_2)/k_3x_1$, where x_3 is the mole fraction of CO_2(001). This is a steady-state approximation obtained by setting the time derivative to zero in the rate equation for mole fraction of CO_2(nmo).

As with the previous model we solve first for $x_3(r,t)$, and we obtain the same single-eigenfunction form as that obtained before, $J_0(ar/r_o)\exp(-bt)$, valid for large times. This is used in the diffusion-reaction equations for the remaining transient components, CO_2(010), CO_2^*, and M^*. As before we take $x_1+x_2=1$, other concentrations being regarded as small.

A weighted combination of gradients, analogous to Eq. 11, leads to D_{vib} The gradients are averaged over infinite time, as before. We finally obtain

$$\frac{A}{D_{vib}} = \frac{B}{D_{00}} + \frac{C}{D_1^*} + \frac{E}{D_2^*}$$

$$A = [b-(1-\phi)k_1x_1-k_2x_2](b-k_1x_1) + (yb+2k_1x_1+k_2x_2)[b-(1-\phi)k_1x_1-k_2x_2] \tag{24a}$$

$$B = (b-k_1x_1-k_2x_2)[(1+y)b+(1+\phi)k_1x_1] \tag{24b}$$

$$C = k_1x_1\phi(yb+2k_1x_1+k_2x_2) \tag{24c}$$

$$E = k_2x_2[b-(1-\phi)k_1x_1-k_2x_2] \tag{24d}$$

where $\phi = k_{VT}/k_1$ and $k_{VT}\ s^{-1}\cdot Torr^{-1}$ is the measured rate for the V-T step.[13] The equilibrium mole fraction of all vibrational states is $y = c_{vib,1}/c_{v,1}$, where $c_{v,1}$ is heat capacity at constant volume of carbon dioxide. Values of the rates are $k_1 = 350P\ s.^{-1}$ and, for example, $k_2 = 53.5P\ s.^{-1}$ for CO_2-argon,[12] with P in Torr.

At P=0 we find $D_{vib} = D_{00}$, in agreement with the previous result for pure CO_2. For mixtures D_{vib} is the mole-fraction-weighted harmonic mean of D_{01} and D_{02}.

When wall reflection is complete, total decay rate is $b = k_1x_1 + k_2x_2$. This leads to

$$\frac{b(y+2)}{D_{vib}} = \frac{b(y+2) - k_2x_2}{D_1^*} + \frac{k_2x_2}{D_2^*} \tag{25}$$

If, in addition, P=0, then $k_2x_2 = 0$, and $D_{vib} = D_1^*$, this time the harmonic mean of D_{11} and D_{12}.

For high pressure we again find $b = k_1x_1 + k_2x_2$, and Eq. 25 holds. We may write this in the form

$$\frac{(y+2)(x_1+Kx_2)}{D_{vib}} = \frac{x_1(y+2)+Kx_2(y+1)}{D_1^*} + \frac{Kx_2}{D_2^*} \tag{26}$$

where $K = k_2/k_1$ and Eqs. 20, 21 and 23 define the diffusivities in terms of mole fraction of CO_2 (x_1) and rare gas ($x_2 = 1-x_1$).

Comparison with Experiment

As we remarked earlier, experimental thermal conductivities provide the only readily available means for evaluating internal diffusivities. Here we use Mason and Monchick's theory in its nonlinear form (Ref. 3, eqs. 93 and

Table 1 Comparison of Linear and Nonlinear Estimates of Thermal Conductivity

Rotational relaxation numbers assumed		Calculated thermal conductivity 10^5 cal/cm•s•°K	
z_{11}	z_{12}	Linear Model	Nonlinear Model
1	1.7	3.60	4.14
2	2	3.99	4.19
4	4	4.18	4.25
8	8	4.28	4.30
10	10	4.30	4.31
20	20	4.34	4.34

94). This is preferred for CO_2, which has rotational relaxation numbers of the order of 1-2.

The linear form, first order in reciprocal rotational relaxation numbers, may disagree with the nonlinear form by 10% or more, especially for high CO_2 content. Moreover, convergence of linear to nonlinear results may require unrealistic rotational relaxation numbers. Table 1 illustrates for a mixture of CO_2 (x_1=0.965) and argon (x_2=0.035) at 296°K. The experimental thermal conductivity is 4.14×10^{-5} cal/cm•s•°K.[15] This value is attained by the nonlinear model for the initial relaxation number but not by the linearized form (nor is convergence found) until relaxation numbers are larger than are to be expected.

In general we have used for z_{11} the estimate[16]

$$z_{11} = 22.53[1+\pi^{3/2}/2T^{*1/2} + (2+\pi^2/4)/T^* + (\pi/T^*)^{3/2}]^{-1} \qquad (27)$$

where T^*=T/251.2. For z_{12} we take a value about twice z_{11}. The diffusion coefficients are smoothed values from Refs. 17 and 18. The ratio of vibrational relaxation rates for CO_2 and argon may be given temperature dependence by[12]

$$K = \exp(-12.533/T^{1/3}) \qquad (28)$$

Pure Carbon Dioxide

Exact matching of a variety of literature values of thermal conductivity of CO_2 leads to an estimate of D_{int} for

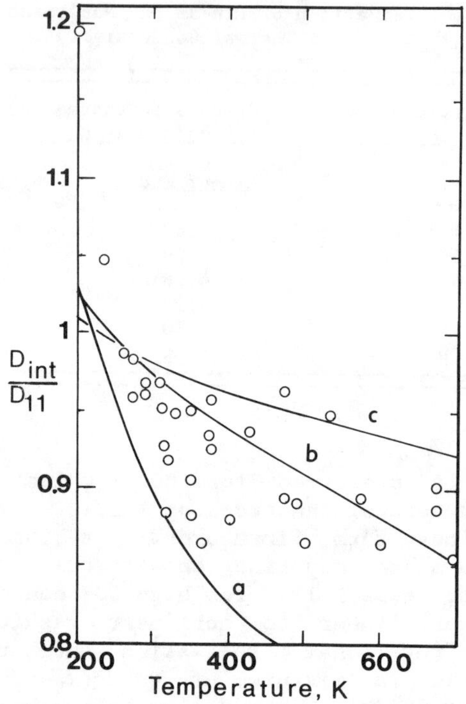

Fig. 2 Temperature dependence of the internal diffusivity of carbon dioxide at 1 atm. The points are values calculated from experimental thermal conductivities. Curve a is calculated according to the method of Ahtye.[6] Cruves b and c are determined by Eq. 26 with $x_1 = 1$, $x_2 = 0$. D_{rot} is estimated as equal to the experimental self-diffusion coefficient D_{11} in curve b. For curve c, D_{rot} is set equal to D_{11} determined by Lennard-Jones parameters obtained from experimental viscosities and second virial coefficients.

each experimental point. We may compare by calculating D_{vib} from Eq. 26, setting $D_{rot} = D_{11}$ and using Eq. 3 for D_{int}. Figure 2 illustrates the results. Also shown is the slightly higher line produced when D_{11} is estimated by the parameters of the Lennard-Jones potential obtained by fitting experimental viscosities and second virial coefficients. Finally, we show an estimate based on the recommendations of Ahtye,[6] which seems to provide a lower bound for the data. We have used experimental data of Ref. 16 in Fig. 2; but we have not included their smoothed curve for D_{int}/D_{11}, which is based on temperatures above 300°K and appears to yield an unrealistic low-temperature limit.

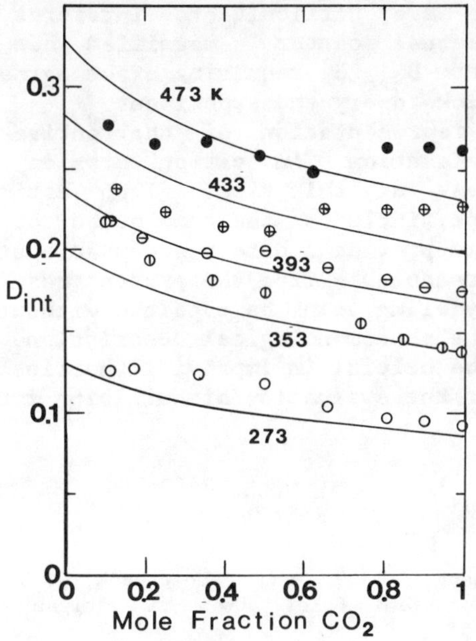

Fig. 3 Concentration dependence of the internal diffusivity for mixtures of carbon dioxide and argon. The points are values obtained from experimental thermal conductivities. The lines are calculated by Eq. 26. D_{11} is taken as equal to the experimental binary diffusion coefficient D_{12}.

Carbon Dioxide and Argon

Figure 3 is a similar calculation for CO_2-argon mixtures based on the thermal conductivities of Ref. 15. The agreement between values based on experiment and those calculated from Eq. 26 is reasonable in trend and magnitude. The experimental data are quite good, and the Mason-Monchick theory can be fit within 0.7% precision on the average.

Discussion

The phenomenological models of Eqs. 7 and 19 represent two steps in a sequence of possible approximations for the complex V-V and V-T interactions characterizing the diffusion of internal energy in carbon dioxide.

The two models are in agreement with respect to low-pressure diffusional phenomena. The two-level version is more successful in matching predictions from experiment and the Mason-Monchick theory, however. Agreement with both temperature dependence and the effects of concentration in CO_2-argon mixtures is good in terms of both magnitude and

trend. Details are difficult to interpret, of course, because experimental scatter is magnified when raw data are used to predict D_{int} by requiring exact agreement between the Mason-Monchick theory and experiment.

Accurate representation of the entire pathway of vibrational relaxation in carbon dioxide is far too ambitious a task at this time. Flynn[8] estimates that as many as 45 steps, including the same number of experimental parameters, would provide a barely acceptable start.

It seems reasonable from the present results to expect that better modeling can be obtained without excessively complicating the phenomenological description. Use of such models should be helpful in improving theories of transport phenomena and in the evaluation of collision integrals.

References

[1]Wang Chang, C. S. and Uhlenbeck, G. E., "Transport Phenomena in Polyatomic Gases," Univ. of Michigan, Ann Arbor, Rep. CM-681, July 1951.

[2]Mason, E. A. and Monchick, L., "Heat Conductivity of Polyatomic and Polar Gases," Journal of Chemical Physics, Vol. 36, March 1961, pp. 1622-1633.

[3]Monchick, L., Yun, K. S., and Mason, E. A., "Formal Kinetic Theory of Transport Phenomena in Polyatomic Gas Mixtures," Journal of Chemical Physics, Vol. 39, Aug. 1963, pp. 654-669.

[4]Monchick, L., Pereira, A. N. G., and Mason, E. A., "Heat Conductivity of Polyatomic and Polar Gases and Gas Mixtures," Journal of Chemical Physics, Vol. 42, May 1965, pp. 3241-3256.

[5]Amme, R. C., "Vibrational and Rotational Excitation in Gaseous Collisions," Advances in Chemical Physics, edited by J. W. McGowan, Wiley, New York, 1975, Vol. 28, pp. 171-265.

[6]Ahtye, W. F., "Thermal Conductivity in Vibrationally-Excited Gases," Journal of Chemical Physics, Vol. 57, Dec. 1972, pp. 5542-5555.

[7]Kovacs, M., Rao, D. R., and Javan, A., "Study of Diffusion and Wall De-excitation Probability in 00°1 State in CO_2," Journal of Chemical Physics, Vol. 48, June 1968, pp. 3339-3340.

[8]Flynn, G. W., "Collision-Induced Energy Flow between Vibrational Modes of Small Polyatomic Molecules," Accounts of Chemical Research, Vol. 14, Nov. 1981, pp. 334-341.

[9]Herman, I. P., Marietta, R. P., and Javan, A., "The Laser-stimulated Reaction: $NO_2 + CO \rightarrow NO + CO_2$," Journal of Chemical Physics, Vol. 68, Feb. 1968, pp. 1070-1087.

[10]Kovacs, M. A. and Javan, A., "Pressure Dependence of the Decay Rate for Imprisoned 4.35-micron Spontaneous Emission in CO_2," Journal of Chemical Physics, Vol. 50, May 1969, pp. 4111-4112.

[11]Margottin-Maclou, M., Doyenette, L., and Henry, L., "Relaxation of Vibrational Energy in CO, HCl, CO_2 and N_2O," Applied Optics, Vol. 10, Aug. 1971, pp. 1768-1779.

[12]Yardley, J. T. and Moore, C. B., "Intramolecular Vibration-to-Vibration Energy Transfer in Carbon Dioxide," Journal of Chemical Physics, Vol. 46, June 1967, pp. 4491-4495.

[13]Moore, C. B., "Vibration→Vibration Energy Transfer," Advances in Chemical Physics, edited by I. Prigogine and S. A. Rice, Wiley, New York, 1973, Vol. 23, pp. 41-83.

[14]Cottrell, T. L. and Day, M. A., "The Effect of Noble Gases on Vibrational Relaxation in Carbon Dioxide," Molecular Relaxation Processes, Academic, London, 1966, pp. 253-256.

[15]Barua, A. K., Manna, M., and Mukhopadhyay, P., "Thermal Conductivity of Argon-Carbondioxide and Nitrogen-Carbondioxide Gas Mixtures," Journal of the Physical Society of Japan, Vol. 25, Sept. 1968, pp. 862-867.

[16]Millat, J., Mustafa, M., Ross, M., Wakeham, W. A., and Zalaf, M., "The Thermal Conductivity of Argon, Carbon Dioxide and Nitrous Oxide," Physica, Vol. 145A, Oct. 1987, pp. 461-497.

[17]Pakurar, T. A. and Ferron, J. R., "Diffusivities in the System: Carbon Dioxide-Nitrogen-Argon," Industrial and Engineering Chemistry Fundamentals, Vol. 5, Nov. 1966, pp. 553-557.

[18]Ferron, J. R., Kerr, D. L., and Chatwani, A. U., "Diffusion of Carbon Dioxide at High Temperatures," Proceedings of the 10th Materials Research Symposium, edited by J. W. Hastie, National Bureau of Standards Spec. Pub. 561, Washington, D. C., 1979, pp. 1373-1403.

Freejet Expansion of Heavy Hydrocarbon Vapor

A. V. Bulgakov,* V. G. Prikhodko,† A. K. Rebrov, ‡
and P. A. Skovorodko †
Siberian Branch of the USSR Academy of Sciences, Novosibirsk, USSR

Abstract

This paper is devoted to experimental and theoretical investigations of heavy hydrocarbon freejet flow. The experiments were performed in a low-density wind tunnel. A hot-wire anemometer measuring technique using a probe with perpendicular and parallel orientation to the streamline direction has been developed. For a molecular gas the difference in accommodation coefficients for translational and internal energies on the surface of the hot-wire probe is taken into account. In methodical experiments the accommodation coefficients for translational ($\alpha' = 0.93$) and rotational ($\alpha'' = 0.79$) energies have been determined for a nitrogen freejet. The possibility of measuring the density of hydrocarbon vapor based on electron-beam-excited radiation in the x-ray and optical spectrum ranges is shown. The distributions of density, speed ratio, and recovery temperature in a mineral oil vapor freejet from a sonic nozzle were measured by electron beam and hot-wire anemometer diagnostic techniques.

The computation of the flow parameters was performed in the framework of a simple model for internal energy relaxation of vapor molecules. Based on the comparison between the theoretical and experimental results, a slow relaxation of vapor internal energy ($Z \sim 10^3$) is revealed that significantly affects the flow parameters and results in condensation of vapor during its expansion in the freejet.

Introduction

Flows of liquid vapors with large molecular weights are widely used in various technological processes. For example, the basic element of vacuum diffusion pumps is the vapor jet of working fluid (mineral oils, ethers, etc.).[1] The possible use of such flows is also connected with the development of the jet membrane method for gasdynamic separation of heavy gases and isotope mixtures.[2]

---

*Research Physicist, Institute of Thermophysics.
†Senior Research Physicist, Institute of Thermophysics.
‡Professor, Laboratory Chairman, Institute of Thermophysics.

For the designing of modern devices employing liquid vapors, as well as for the improvement of existing technical devices using heavy molecular weight vapors, it is desirable to be able to compute the flow parameters in an expansion of these vapors. For this purpose information on the relaxation properties of large molecules is necessary. For complex polyatomic molecules, particularly for heavy hydrocarbon vapors, such data are practically absent at the present time. This lack of information may be caused by the considerable difficulties involved in employing flow diagnostics to experimental investigations of these large molecules.

This paper deals with experimental and theoretical investigations of a mineral oil vapor freejet. The purpose is to reveal the features of internal energy relaxation under the expansion of molecular gas with large number of internal degrees of freedom.

A vapor freejet expanding into vacuum from a sonic nozzle was used to allow a gasdynamic investigation of the expansion process under nonequilibrium conditions employing the standard methods of rarefied gas diagnostics. The application of electron beam diagnostics is limited to density measurements due to complete spectral nonresolution of the fluorescence excited by electrons. The basic experimental results have been obtained with the help of specially developed hot-wire anemometer diagnostics. To analyze the experimental results, computations of freejet parameters for relaxing gas flow were made.

Experimental Equipment and Procedure

The experiments were performed in a low-density wind tunnel at the Institute of Thermophysics of the Siberian Branch of the USSR Academy of Sciences.[3] The wind tunnel was equipped with an oil boiler together with a built-in heater located directly in the oil. The vapor formed in the boiler was supplied through a throttle valve and steam pipeline with thermal isolation to a nozzle stagnation chamber. All of the elements of vapor supply (from the throttle valve to the nozzle) were equipped with external heaters to compensate for heat losses and to preheat the vapor. The working medium was continuously circulated during the experiment to maintain a constant fractional composition. The pressure was measured with the help of a deformation vacuum gage (operating range 10-30000 Pa). The vapor pressure in the boiler as well as in the stagnation chamber (p_o) was measured using the technique described in Ref. 4. The temperature in the critical points of the vapor supply elements was measured and controlled by thermocouples. Superheated vapor was used in the experiments; the vapor temperature in the stagnation chamber was kept 20-30° C higher than that in the boiler.

The jet was formed by a sonic nozzle (orifice) with the diameter $d_* = 10$ mm. The vapor of mineral oil BM-5 was used as a working gas [the mean chemical formula is $C_{33}H_{62}$, the molecular weight is 458, the equilibrium ratio of specific heats $\gamma \sim 1.01$ (for T = 500 K), and the gas kinetic diameter of the molecule is $\delta \sim 9$ Å (at T = 500 K)].[5] The background condition for the vapor freejet was either vacuum ($p_\infty \sim 10^{-2}$ Pa) or flooded space formed by nitrogen or helium. For comparative purposes a nitrogen freejet flow was also investigated.

A hot-wire anemometer with perpendicular and parallel orientation of the probe to the streamline direction was placed in the jet field. The hot-wire probe used was constructed of 8 μm diam., 2-3 mm length gilt tungsten wire welded to steel support needles with a 0.2 mm diam. at the junction with the wire. The temperature of the supports was measured by thermocouples welded close to the tips of the needles. The probe resistance was measured using a constant-current bridge circuit with a digital voltmeter. To determine the thermal losses in the hot-wire probe, the method described in Ref. 6 was used. The temperature coefficient of probe resistance was measured using calibration in a thermostat.

The hot-wire anemometer technique is widely used for the diagnostics of rarefied gas flows.[7-9] To obtain information on the flow two values are used, the coefficient of heat transfer h between the hot-wire probe and the flow, and the wire adiabatic temperature (recovery temperature) T_r, for which the dependences on thermodynamic parameters, geometry of flow-probe interaction, and accommodation coefficients are known in the literature.

The measuring technique utilizing hot-wire anemometer in gases of high molecular weight is described by one of the authors in Ref. 10. For a probe with perpendicular orientation to the streamline direction, the values of h and T_r are connected with the flow parameters by the following relations, which are correct for free molecular flow around the probe:

$$h_\perp = Anu/\varphi(S) \tag{1}$$

$$T_{r\perp} = \frac{T}{4\alpha' + \upsilon}\left(\alpha'(5 + 2S^2) + \upsilon\frac{T_{in}}{T} - \frac{\alpha'}{1 + S^2(1 + I_o/I_1)}\right) \tag{2}$$

and for the probe with parallel orientation to the streamline direction

$$h_{\parallel} = An\sqrt{\pi kT/2m} \tag{3}$$

$$T_{r\parallel} = T\frac{2\alpha'(S^2 + 2) + \upsilon\frac{T_{in}}{T}}{4\alpha' + \upsilon} \tag{4}$$

where $A = L \cdot d \cdot R\,(4\alpha' + \upsilon)/2$, L and d are the wire length and diameter, R = k/m is the gas constant, α' is the accommodation coefficient for translational energy, $\upsilon = \alpha'' \cdot j$ is the product of the mean accommodation coefficient for internal energy and the number of internal degrees of freedom per molecule, n is the number density of molecules, T is the translational temperature, T_{in} is the temperature of the internal degrees of freedom, u is the velocity, $S = u/(2kT/m)^{1/2}$ is the speed ratio, φ (S) is the known function of S, m is the mass of the molecule, and I_o and I_1 are the modified Bessel functions of argument $S^2/2$.

Based on the measured values of h and T_r the flow parameters can be determined as follows. The value of S can be directly found from the ratio $h_{\perp}/h_{11} = 2S/\sqrt{(\pi)}\ \varphi(S)$. Then the value of n·u can be determined from Eq. (1), if the accommodation coefficients α' and α" are known. Density measurements, obtained by other diagnostics technique (e.g., by electron beam diagnostics) allows the velocity and the translational temperature to be determined. The data for T_r in accordance with Eqs. (2) and (4) contain the information about the value of T_{in}. The accommodation coefficients α' and α" can be determined from Eqs. (1), (2), and (4) based on the values of h and T_r measured in a standard flow with known flow parameters.

It should be noted that utilizing the hot-wire anemometer in gases with high molecular weight have some advantages because the measured value of h in accordance with Eqs. (1) and (3) is proportional to the number of internal degrees of freedom.

The vapor density at the point under study was measured using measurement of fluorescence excited in the flow by an electron beam (beam current ~ 1 mA, beam diameter ~ 2 mm, electron energy ~ 10 keV). The x-ray bremsstrahlung was used for density measurements since the intensity of such radiation does not depend on chemical bonds in a molecule and is proportional to the concentration of atom targets and the charge squared of the atom nucleus. That is why the intensity of x-ray bremsstrahlung is high for gases with high molecular weight (e.g., in BM-5 vapor this intensity is 13 times higher than in nitrogen for the same concentration of molecules).

Induced radiation of BM-5 vapor in the optical spectral range is characterized by very weak intensity (the detected fluorescence was of continuous spectrum with the maximum intensity being in the wavelength range 405-420 nm). That is why the optical radiation was used only for the flow visualization and for measuring the background molecule concentration in the jet under the vapor expansion into flooded space. The data acquisition system for electron beam measurements is described in Ref. 11.

Theoretical Model

Together with the experimental investigations, computations of the flow in a jet expanding into vacuum from an axisymmetric sonic nozzle were carried out. To compute a steady two-dimensional flow in a jet issuing from a sonic nozzle, a numerical method based on a coordinate system formed by streamlines and lines normal to them[12] was used. The basic system of gasdynamic equations in natural coordinates is as follows:

$$\rho \cdot u \cdot A = \text{const} \tag{5}$$

$$\rho u \frac{\delta u}{\delta s} + \frac{\delta p}{\delta s} = 0 \tag{6}$$

$$\rho u^2 \frac{\delta\varphi}{\delta s} + \frac{\delta p}{\delta n} = 0 \tag{7}$$

$$i + u^2/2 = \mathrm{const} \tag{8}$$

where s is the distance along the streamline, n is the distance normal to the streamline, p is the pressure, ρ is the density, A is the streamtube area, i is the specific enthalpy, and φ is the flow angle with respect to the symmetry axis.

To describe the relaxation of the vapor molecule internal energy a simple model was considered. It was assumed that besides three translational degrees of freedom the molecule had a constant number j of internal degrees of freedom which is not dependent on temperature . In this case the specific enthalpy of the gas is expressed as follows:

$$i = 5RT/2 + jRT_{in}/2 \tag{9}$$

The relaxation process is described by the relaxation equation:

$$dT_{in}/dt = u \cdot \delta T_{in}/\delta s = (T - T_{in})/\tau \tag{10}$$

where τ is the relaxation time.

To close the system of Eqs. (5-10) the equation of state was used:

$$p = \rho RT \tag{11}$$

The value of j can be determined from the expression for the equilibrium ratio of specific heats

$$\gamma = (j + 5)/(j + 3) \tag{12}$$

where j is equal to 197 for $\gamma = 1.01$.

The relaxation time τ was assumed to be proportional to the average time τ_t between the collisions of molecules:

$$\tau = Z\tau_t \tag{13}$$

$$\tau_t = \pi \mu/4p \tag{14}$$

where μ is the viscosity of the vapor.

Based on the results of Ref. 5, where the viscosity of mineral oil BM-5 vapor was determined, we find the following expression for τ_t

$$\tau_t \cdot n = 1.8 \times 10^9 \ \mathrm{cm^{-3}\,s} \tag{15}$$

The system of Eqs. (5-11) together with the expressions for j and τ were solved numerically as applied to experimental conditions. The corresponding computational method is described in Ref. 13, where the analysis of the relaxing gas flow in a jet from a sonic nozzle for j = 2 is also performed.

Results and Discussion

Figure 1 shows the results of electron beam measurements of axial density distributions in the vapor freejet with the stagnation parameters p_o = 350 Pa, T_o = 540°K. Points 1 were obtained for a jet expanding into vacuum, and points 2 correspond to a jet expanding into flooded space formed by helium with the pressure p_∞ = 3.1 Pa. The computational results obtained for equilibrium flow are shown by the solid line. The density distribution is typical for freejets of such type. The experimental results are in good agreement with the computational ones both for a jet expanding into vacuum for all values of x/d_* and for the jet core in a jet expanding into flooded space. In the latter case the experimental results distinctly reveal the appearance of the Mach disk.

The density of helium penetrating into the jet is shown by points 3. The helium density was measured based on the radiation intensity of the

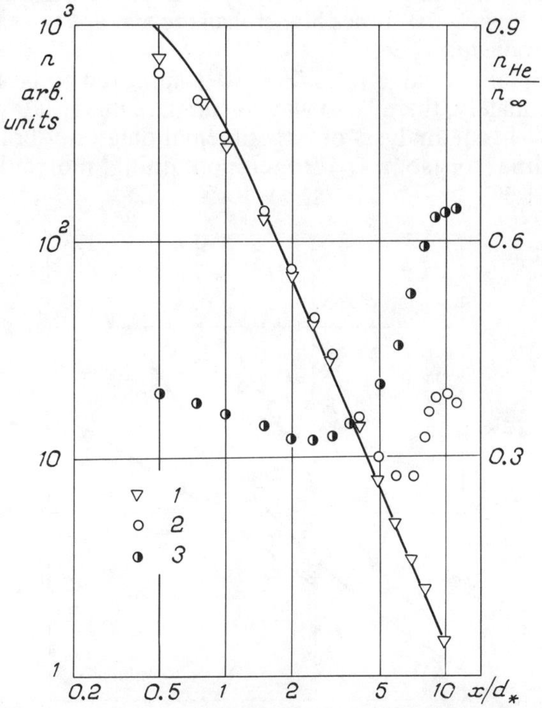

Fig. 1 Axial density distributions in mineral oil vapor freejet. Vapor density: 1, p_∞ = 10^{-2} Pa; 2, p_∞ = 3.1 Pa (helium backgound); solid line, theory; 3, helium density.

line with wavelength 501.6 nm (3^1p - 2^1s transition). The data were obtained using the fact that the contribution of BM-5 vapor excited by the electron beam to the radiation at $\lambda = 501.6$ nm is practically negligible. As can be seen from Fig. 1, a high penetration of helium into the core of sufficiently dense vapor jet is observed, even at $x/d_* = 0.5$ $n_{He} \approx 0.38 n_\infty$. The presence of a minimum in the helium density distribution may be explained by helium penetration into the jet through the barrel shock waves.

The experimental distributions of speed ratio along the flow axis for nitrogen and vapor jets which have been obtained using hot-wire anemometer are shown by points in Fig. 2. The data for nitrogen are in a good agreement with the computational results for jet flow of a perfect gas with $\gamma = 1.4$ (curve 1), which proves the correctness of the diagnostic method. For the vapor jet ($p_o = 350$ Pa, $T_o = 540°K$) at $x/d_* < 2$ the values of S are noticeably higher than those for equilibrium flow with $\gamma = 1.01$ (curve 2). This result is evidence that slow relaxation of internal energy of vapor molecules is occurring in the jet.

The computational distribution of $S(x/d_*)$ for nonequilibrium flow with j = 197 and relaxation time $\tau \cdot n = 1.9 \times 10^{12}$ cm^{-3} s is also shown in Fig. 2 (curve 3). As one can see, for $x/d_* < 2$ this distribution is in satisfactory agreement with the experimental one. The determined relaxation time is approximately 10^3 times higher than the average time τ_t between the collisions of molecules [Eq. 15].

The obtained value of Z ($Z = 10^3$) appeared to be unexpectedly high. Unfortunately, there is no way to compare this result with available data. In Ref. 14 an analysis of experimental data on vibrational energy relaxation time for some hydrogen-containing molecules has been

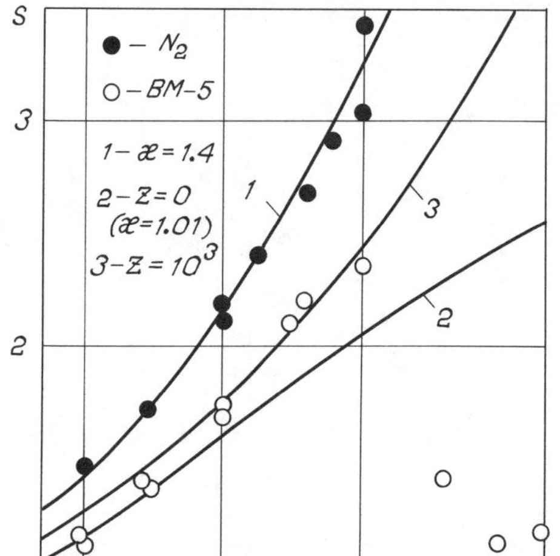

Fig. 2 Experimental (points) and theoretical (lines) results for speed ratio distributions along the jet axis.

performed. As a result, an empirical Lambert-Salter correlation was found

$$\log_{10} Z_{10} = X\upsilon_{min} \tag{16}$$

where υ_{min} is the frequency of the lowest vibrational mode of a polyatomic molecule, Z_{10} is the collision number for a 1– 0 transition, and X is a constant. The value of Z is connected with the value of Z_{10} by the relation[14]

$$Z = Z_{10} \cdot C_{\Sigma}/C_{min} \tag{17}$$

where C_{Σ} is the total vibrational specific heat and C_{min} the contribution due to the lowest mode. The lack of information on the values of C_{min} and υ_{min} for BM-5 vapor molecule does not allow a calculation of the value of Z_{10} for discovering whether the value $Z = 10^3$ corresponds to the Lambert-Salter correlation. Some experimental results testifying to a slow relaxation of vapor internal energy will be given below.

Figure 3 represents the computational axial distributions of T and T_{in} for the experimental conditions. The distribution of saturation temperature T_s is also shown. As can be seen, the slow relaxation of internal energy results in its freezing and produces a fast decrease of translational temperature. It also causes supersaturation in the flow that can result in condensation. The abrupt decrease of experimental values of S at $x/d_* > 2$ (Fig. 2) can be explained by the condensation process in the vapor flow. This effect can be caused by the flow heating due to vapor condensation and by the change of heat-transfer character due to interaction of condensed particles with the probe. The experimental data on the recovery temperature behavior in the flow are in agreement with this conclusion.

Let us analyze expressions (2) and (4) for recovery temperature from the viewpoint of nonequilibrium flow diagnostics. With the same accommodation of translational and internal energies on the probe surface ($\alpha' = \alpha''$)

1) The recovery temperature does not depend on the value of the accommodation coefficient;

2) The $T_{r\perp}$ does not depend on internal energy relaxation and $T_{r\shortparallel}$ depends slightly only through the value of translational temperature in the flow. This follows from the energy balance equation:

$$(5 + j)\, kT_o = mu^2 + 5kT + jkT_{in} \tag{18}$$

or

$$(5 + j)\, T_o/T = 2S^2 + 5 + jT_{in}/T \tag{19}$$

Specifically, for the hypersonic flow ($S^2 \gg 1$)

$$T_{r\perp} = \frac{5 + j}{4 + j} T_o \tag{20}$$

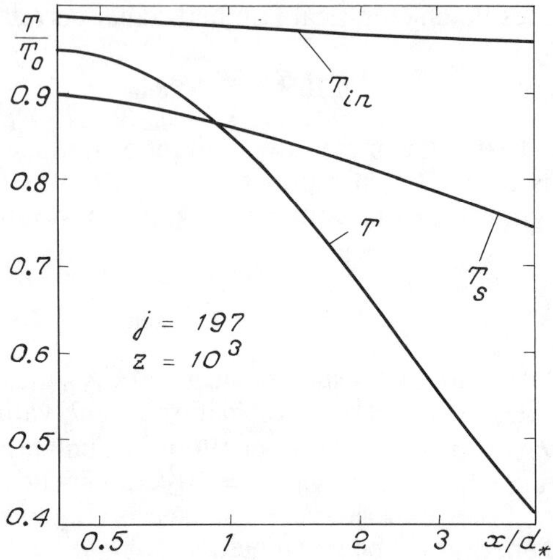

Fig. 3 Computational axial distributions of traslational (T) and internal (T_{in}) temperatures in mineral oil vapor freejet (T_s, saturation temperature).

$$T_{r\shortparallel} = T_o + \frac{T_o - T}{4 + j} \tag{21}$$

In the case of vapor with a large number of internal degrees of freedom, $T_r \approx T_o$.

The analysis of data on recovery temperature is substantially complicated when $\alpha' \neq \alpha''$. If the difference between α' and α'' is not taken into account, errors in the interpretation of the measured values of T_r in the molecular gas may arise. Smaller accommodation of internal energy compared to translational energy ($\alpha'' < \alpha'$) would result in the increase of T_r, and freezing of internal energy decreases T_r at $\alpha'' < \alpha'$ and increases it at $\alpha'' > \alpha'$.[10, 15, 16]

Our data for T_r in a nitrogen freejet under equilibrium conditions ($T_{in} = T$) have shown that the accommodation coefficient for rotational energy is less than that for the translational mode (Fig. 4). Experimental conditions are as follows: $d_* = 10$ mm, $T_o = 289°K$, $p_o = 500\text{-}4000$ Pa (in Fig. 4, points 1, 2, 3 correspond to $p_o = 3560$, 1320, 734 Pa, respectively). The apparatus and hot-wire anemometer measuring technique are analogous to the ones previously mentioned.

Dashed lines in Fig. 4 correspond to calculation of $T_{r\perp}$ and $T_{r\shortparallel}$ by formulas (2) and (4) for N_2 equilibrium flow at $\alpha' = \alpha''$. The fact that the measured values of T_r lie higher than the calculations for $\alpha' = \alpha''$ seems to be connected with smaller accommodation of rotational energy. The calculation by formulas (2) and (4) for $\alpha''/\alpha' = 0.85$ (solid lines) is in a good agreement with the experiment. Using the measured results for h in

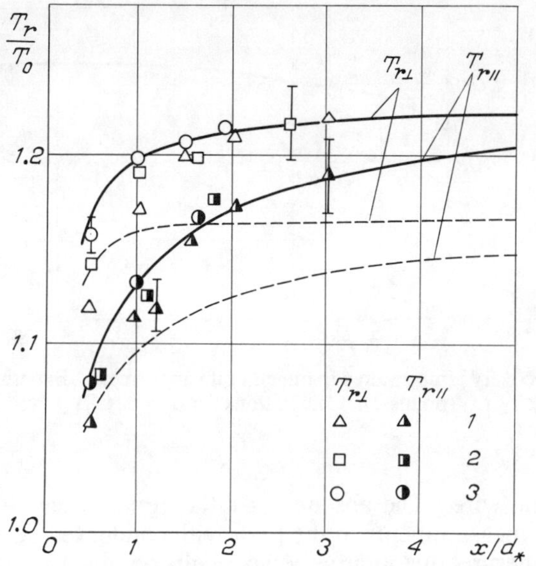

Fig. 4 Recovery temperatures for equilibrium flow in nitrogen freejet. Points, experimental results at different p_0; dashed lines, calculations for $\alpha = \alpha''$; solid lines, calculaitons for $\alpha' = 0.93$, $\alpha'' = 0.79$.

nitrogen jet according to $2\alpha' + \alpha'' = 2.65$, we have determined the absolute values of accommodation coefficients on the surface of the probe: $\alpha' = 0.93$, $\alpha'' = 0.79$. Note that the data of Ref. 15 concerning the measurements of T_r in H_2 jets also indicate that $\alpha'' < \alpha'$.

For a gas of high molecular weight, and in particular, for the oil vapor, the accommodation coefficient for internal energy can be evaluated based on the measured value of h in the flow with known parameters n, u, S. The determined value $4\alpha' + j \cdot \alpha''$ is approximately equal to $j \cdot \alpha''$ at arbitrary α' when $j >> 1$. Using the experimental data for h, n, S and computed value of u, the accommodation coefficient for internal energy of BM-5 vapor molecules on the probe surface was found to be equal to 0.5.

The accommodation coefficient for translational energy cannot be deduced from the measured data for recovery temperature due to the high uncertainty in such parameters as j, T, and T_{in}. Therefore, the following analysis of data for T_r is performed on the assumption that $\alpha' = \alpha''$.

Figure 5 shows the measured axial distributions of the recovery temperatures $T_{r\perp}$ and $T_{r\parallel}$ (points 1 and 2, respectively) in the vapor jet. The solid line is the calculation of $T_{r\perp}$ for the case when $\alpha' = \alpha''$ based on the computational results for vapor jet flow, and the dashed line shows the same data for $T_{r\parallel}$. At small values of x/d_* the experimental and theoretical results are in qualitative agreement, i.e., recovery temperature increases due to the increase of speed ratio. A noticeable decrease of measured T_r values at $x/d_* \geq 1.5$ may be attributed to the evaporation of condensed particles at their interaction with the probe surface that should result in

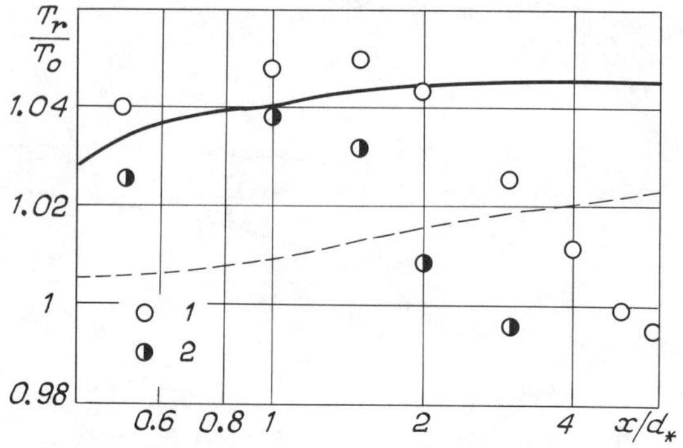

Fig. 5 Recovery temperatures in mineral oil vapor freejet. Experiments: $T_{r\perp}$ (points 1), $T_{r\parallel}$ (points 2). Calculations for $\alpha' = \alpha''$: $T_{r\perp}$ (solid line), $T_{r\parallel}$ (dashed line).

cooling of the wire. The absence of a theoretical treatment of hot-wire anemometer measurements in the flow with condensed particles makes it difficult to interpret quantitatively the results obtained at sufficiently high values of x/d_*.

The experimental results for both speed ratio (Fig. 2) and recovery temperature (Fig. 5) show the appearance of condensate in the freejet of mineral oil vapor. These results give evidence of slow relaxation of internal energy of vapor molecules, since for the equilibrium flow, which is close to the isothermal one, condensation in the flow should not be observed.

Conclusion

It has been established that the molecules of heavy hydrocarbon vapor are characterized by the slow relaxation of the internal energy, which can considerably influence the parameters of freejet flow, causing, in particular, condensation phenomenon. These peculiarities should be taken into account when analyzing the operation of the technical devices using a freejet expansion of vapor.

References

[1]Rebrov, A. K., "Studies on Physical Gas Dynamic of Jets as Applied to Vacuum Pumps," Proceedings of the Fifteenth Symposium (International) on Rarefied Gas Dynamics, Vol. 2, Teubner, Stuttgart, FRG,1986, pp. 455-473.

[2]Brook, J. W., Calia, V. S., Muntz, E. P., Hamel, B. B., Scott, P. B., and Deglow, T. L., "Jet Membrane Process for Aerodynamic Separation of Mixtures and Isotopes," Journal of Energy, Vol. 4, Sept.-Oct. 1980, pp. 199-208.

[3]Bochkarev, A. A., Kosinov, V. A., Prikhodko, V. G., and Rebrov, A. K., "Structure of a Supersonic Jet of an Argon-Helium Mixture in a Vacuum," Journal of Applied Mechanics and Technical Physics, Vol. 11, Sept.-Oct. 1970, pp. 857-861.

[4]Manov, M. G., "Pressure Measurement in High-Vacuum Pumps," Pribory i Tekhnika Experimenta, Sept.-Oct. 1959, pp. 99-102 (in Russian).

[5]Belikov, A. E., Bulgakov, A. V., Rebrov, A. K., Skovorodko, P. A., and Stankus, N. V., "Computational and Experimental Investigation of Vacuum Oil Vapour Jet Flows," Gasdynamics of Jet Vacuum Pumping Processes, edited by S. S. Kutateladze, Institute of Thermophysics, Novosibirsk, USSR, 1985, pp. 53-101 (in Russian).

[6]Lord, R. G., "Hot-Wire End-Loss Correction in Low Density Flows," Journal of Physics E: Scientific Instruments, Vol. 7, 1974, pp. 56-60.

[7]Gusev, V. N., Nikolsky, Yu. V., and Chernikova, L. G., "Experimental Investigation of Hypersonic Jets with the Help of Hot-Wire Anemometer," Uchenye Zapiski TsAGI, Vol. 3, No. 5, 1972, pp. 33-39 (in Russian).

[8]Mutoo, S. K. and Brundin, C. L., "Near Wake Flow Field Measurements Behind Spheres in Low Reynolds Hypersonic Flow," Proceedings of the Ninth Symposium (International) on Rarefied Gas Dynamics, Vol. 1, DFVLR, Porz-Wahn, FRG, 1974, pp. B.10-1-B.10-10.

[9]Gottesdiner, L., "Hot Wire Anemometry in Rarefied Gas Flow," Journal of Physics E: Scientific Instruments, Vol. 13, 1980, pp. 908-913.

[10]Bulgakov, A. V., "Application of Hot-Wire Anemometry for High-Molecular Low-Density Flows," Molecular Physics of Nonequilibrium Systems, edited by Ye. V. Baklanov, Institute of Thermophysics, Novosibirsk, USSR, 1984, pp. 20-27 (in Russian).

[11]Experimental Methods in Rarefied Gas Dynamics, edited by S. S. Kutateladze, Institute of Thermophysics, Novosibirsk, USSR, 1974 (in Russian).

[12]Boynton, F. P. and Thomson, A., "Numerical Computation of Steady, Supersonic, Two-Dimensional Gas Flow in Natural Coordinates," Journal of Computational Physics, Vol. 3, March 1969, pp. 379-398.

[13]Skovorodko, P. A., "Rotational Relaxation upon Expansion of a Gas into a Vacuum," Fluid Mechanics - Soviet Research, Vol. 6, Sept.-Oct. 1977, pp. 116-133.

[14]Lambert, J. D., Vibrational and Rotational Relaxation in Gases, Clarendon, Oxford, UK, 1977.

[15]Legge, H., "Recovery Temperature Determination in Free Molecular Flow of a Polyatomic Gas," Proceedings of the Fourteenth Symposium (International) on Rarefied Gas Dynamics, Vol. 1, Univ. of Tokyo Press, Tokyo, 1984, pp. 271-278.

[16]Bulgakov, A. V. and Prikhodko, V. G., "Diagnostics of Rarefied Gases Using Hot-Wire Anemometer," Journal of Applied Mechanics and Technical Physics, Vol. 27, Sept.-Oct. 1986, pp. 749-752.

Chapter 2. Experimental Techniques

Optical Diagnostics of Low-Density Flowfields

J. W. L. Lewis*
University of Tennessee Space Institute, Tullahoma, Tennessee

Abstract

A review is presented of the use of specific optical diagnostic techniques for the study of low-density, gas flowfields. Of specific interest to this review are the optically active techniques of electron beam fluorescence, laser-induced fluorescence, laser-Rayleigh and laser-Raman scattering. Typical results and regimes of applicability of these methods of measurement are presented.

Introduction

The rich heritage of the study of low-density vapors and gases using optical techniques can be traced at least to the late nineteenth century and the fabrication of incandescent bulbs. From the observation of the sharpness of the deposition on the interior of the bulb's wall, the straight-line trajectory of the evaporated, filament material was inferred.[1] In less than 20 years the natural radiation optical source of Ref. 1 had been replaced by sodium resonance radiation to enable Dunoyer[2] to visualize a sodium beam. From the sharpness of the atomic beam's spatial profile, Dunoyer inferred the shortness of the excitation lifetime of the sodium atom. During the next twenty-odd years, the optical study of low-density flowfields and molecular beams became more quantitative, and, as an example, Koenig and Ellet[3] made use of the resonance radiation of cadmium and the 4-mm-long afterglow of an irradiated, cadmium beam to determine that the radiative lifetime of the excited level of cadmium was on the or-

Presented as an Invited Paper.

*Professor of Physics, Center for Laser Applications.

der of 1μs. It was during this same period that the molecular beam laboratory of Stern was established at Hamburg, and the subsequent research of Stern and his co-workers contributed significantly to the development of quantum mechanics. The last of the 30 reports from this laboratory that were published during the 1926-1933 period was Frisch's, "The Experimental Verification of the Einstein Radiation Recoil,"[4] which investigated Stern's suggestion that the deflection of an atomic beam could provide evidence of the momentum of the photon. A quotation at the end of this 1933 paper is noteworthy and seems to have near-timeless relevance: "It would certainly be possible to do more careful experiments .... Unfortunately, for reasons beyond our control, the experiments had to be stopped prematurely."

This early work, beginning nearly 100 years ago and extending to approximately 50 years in the past, shows the close relationship that has existed between the research of physics and low-density gas flows and gives some perspective to the current optical measurements of low-density flowfields.

In general, there are two classes of optical measurements of low-density flowfields. The first class requires the use of optical measurements to define the flowfield's properties and to interpret or validate aspects of the gasdynamics study. The second major class makes use of the low-density flowfield to accomplish specific optical measures or studies. Clearly, these two classes are not mutually exclusive. The subsequent material in this brief review of the subject will concentrate on the former category.

A review of the research reported at this and previous symposia show that the nature and complexity of the gasdynamic problems have changed dramatically from those studied a half-century ago, and a primary difference is the detailed description of the flowfield that can be provided by optical, flow-diagnostic techniques. The low-density, expansion flowfields that are currently of interest are quite varied in their demands on the flow-diagnostic techniques. As examples, depending on the gas source conditions, the flowfield can vary in characterization from continuum to free molecular flow, it can be comprised of reacting gas phase species, and it can exhibit internal energy-state relaxation phenomena. Further possible complications for description of the flowfield may include flowfield-surface interactions, homogeneous or heterogeneous condensation, or the more simple but equally demanding two-phase, nonreacting expansions. Additionally, both continuous- and pulsed-source flows are encountered in practice.

The parameters that are generally required to characterize the low-density flowfield can be more easily specified. These requirements can include spatially and temporally (if necessary) resolved measurements of species densities, velocities, internal energy-state enthalpies, and translational-mode enthalpy or temperature(s). For two-phase flows, this list is increased to include the mass fraction, particulate size distribution function, and, possibly, the velocities of the particulate constituents of the flowfield. Finally, and quite often, the most immediate requirement is to provide visualization of the flowfield. Considering the types of flowfield that are encountered in practice and the parameters that are desired, it is not surprising that there is no one optical measurement technique that satisfies all, or even most, of the needs. However, despite the often encountered need for the use of more than one optical measurement technique to define particular low-density flowfields, there is one inescapable conclusion: the more hostile the flowfield environment, the more attractive the use of optical measurement methods.

The subsequent sections of this review will focus on several general types of optical measurements that have been used successfully for the study of low-density flowfields. As shown in Fig. 1, these techniques rely on the injection of a well-defined beam of "particles" to induce the excitation of specific optical spectra that provide information concerning the desired flowfield property. The initial differentiation of these techniques is the nature of the injected beam, i.e.,

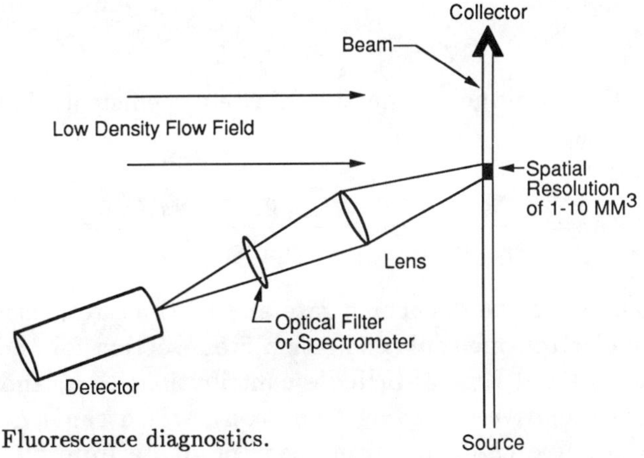

Fig. 1 Fluorescence diagnostics.

charged particles, such as electrons or photons. The specific optical measurements that will be addressed will include electron beam fluorescence, laser-induced fluorescence, and laser-scattering methods, and examples of the successful use and difficulties or uncertainties encountered in the use of each technique will be presented.

Electron Beam Fluorescence

For nearly 30 years the electron beam fluorescence (EBF) technique has been the most general and nearly universal nonintrusive diagnostics method for low-density flowfield studies. The basis for the quantitative use of EBF for the measurement of local values of number density and temperatures was established by the pioneering work of Schumacher and Gadamer[5] and Muntz[6,7]. These and other early EBF studies of low-density gases are described by Muntz[8].

As a summary description, and as sketched in Fig. 1, a multikilovolt electron beam is injected into the low-density environment of number density $\underset{\sim}{n}$, traverses a path length L and a total columnar number density $\tilde{L} = n \cdot L$, and the fluorescence that results from the electron-molecule interactions is detected optically; the observed fluorescence provides a measure of the local number density, the number distribution of molecules among the internal-mode energy levels, and the translational temperature for some species. The cylindrical beam of cross-sectional area A_B, current I, and energy E excites a fraction of the N_g ground state species in the observed volume $A_B \cdot L$ to produce a number density n_i of species in the ith excited level. The subsequent, fluorescent photon emission rate S_{ij} for the radiative transition between levels i and j is, for an observed beam length L,

$$S_{ij} = A_{ij} \cdot n_i \cdot A_B \cdot L \tag{1}$$

where A_{ij} is the spontaneous emission Einstein coefficient. The number density n_i is

$$n_i = \frac{N_i}{A_B \cdot L} = \frac{\left[\sigma_{gi} + \sum_j \sigma_{gj}\beta_{ji}\right] n_g(I/e)}{\left[(1/\tau_i) + k_i n_g\right] A_B} \tag{2}$$

where k_i and τ_i are the quenching rate and lifetime, respectively, of level i. The electron-molecule excitation cross section for the g to i transition is σ_{gi}, and Eq. (2) includes contributions to n_i and therefore S_{ij} from radiative cascading transitions, which usually are not well defined. These cascading transitions originate from all higher-

Table 1 Electron Beam Divergence Parameters for N_2

E keV	$A \times 10^{28}$ cm^3	$B \times 10^{28}$ cm
20	4.00	2.94
40	1.00	1.43
100	0.262	0.87

Beam Divergence: $\theta_{1/2} = A\tilde{L}^{3/2} / (1 + B\tilde{L}^{1/2})$; $\tilde{L} = n \times L$.

lying levels j that have been excited with excitation cross section σ_{gj} and that decay to level i with the radiative branching factor B_{ji}.

A spatially and energetically well-defined electron beam requires comparatively low values of beam divergence $\Theta_{1/2}$ and stopping power $dE/d\tilde{L}$. Table 1 gives the results obtained by Center[9] for the half-current divergence angle $\Theta_{1/2}$ and the stopping power for electrons in N_2. From these results, it is seen that high beam energies are desired. To satisfy these requirements and yet obtain the largest fluorescence signals, the excitation transitions of interest are dipole-allowed and obey the Bethe-Born relation $\sigma_{\text{exc}} = (C'/E)\ln(CE)$; the maximum values of σ_{exc} typically occur in the energy range of 100-300 eV and have maximum values on the order of 10^{-17} cm^2. The 10 keV, electron-excited, optical-emission cross section $\sigma(0,0)$ for the $(0,0)$ band of the N_2^+ First Negative system $[N_2^+(1-)]$ is 9.83×10^{-19}cm^2; the optical emission cross section is obtained from the level excitation cross section by replacing the branching factor (A_{ij}/A_i) by unity. For reference,[10,11]

$$\sigma(0,0) = \left[(1.69 \pm 0.02)^{\times 10^{-15}} / E\right] \ln\left[(0.035 \pm 0.009)E\right] \tag{3}$$

where $\sigma(0,0)$ is in cm^2 and E is in eV.

Muntz[6] proposed that the rotation-vibration band intensities of the N_2^+ First Negative system could be related to the number densities $n_g(v,k)$ of the individual vibrational and rotational levels v and k, respectively, of the N_2 electronic ground state $X^1\sum_g^+$. This relation was based on the use of vibrational band strengths and dipole transition, rotational line-strength factors for the $N_2 X^1\sum_g^+ \rightarrow N_2^+ B^2\sum_u^+$ transition. Use of this model enabled successful EBF measurements of the rotational (T_R) and vibrational (T_V) temperatures and number density of N_2 in low-density flows.[8] However, deviations from the dipole excitation model, as manifested by non-Boltzmann rotational distributions, were observed at low density for low values of T_R, and these deviations persisted as the density and T_R increased.[12] Coe, et al[13] proposed the dipole-quadrupole excitation model to explain the discrepant non-Boltzmann distribution, and better agreement was obtained. More recently, the subject of the excitation model has been revisited by Baronavski, et al[14] who combined the use of electron beam excitation and laser-induced fluorescence (LIF) detection and observation of the ground state N_2 and $N_2^+ X^2 \sum_u^+$ species. These results indicated disagreement with both the Muntz and Coe models. A resolution of this discrepancy is desirable.

Measurement of T_v of N_2 using EBF is performed routinely[8,15] using the ratios of fluorescence signals of the $v = 0$ and 1 vibrational levels of the $N_2^+ B^2 \sum_u^+$ ion. However, for excitation of N_2 at 300 K, anomalously intense, fluorescent of the transitions First Negative system $[N_2^+(1-)]$ have been observed from vibrational levels $v \geq 2$.[16] Lewis, et al[11] measured the relative excitation cross sections of this system for thermal equilibrium N_2 for 300 K $\leq T \leq$ 900 K, and anomalously large, temperature-dependent excitation cross sections were found for $N_2^+ B^2 \sum_u^+$, $v \geq 2$. Further study of this problem is required before routine use can be made of branching transitions from $v \geq 2$ of $N_2^+ B^2 \sum_u^+$ for measurements of T_v in nonequilibrium, low-density flowfields.

The increased importance of collisional quenching effects for the higher-density regions of the flow is indicated by Eq.(2). An order-of-magnitude estimate of the magnitude of the quenching term of Eq.(2) can be obtained using the elastic collision cross sections. In particular, for the binary collisions of excited neutral and ground state neutral molecular species at temperature T, the hard-sphere rate constant $k_i(HS)$ can be used, and, assuming a reduced mass $\overline{M}$ and hard-sphere cross section $\sigma_i^2(HS)$,

$$k_i(HS) = \sigma_i^2(HS) \cdot (8RT/\pi\overline{M})^{1/2} \quad (4)$$

where R is the universal gas constant.

For binary collisions of a molecule of charge e and a neutral molecule of polarizability α, the charge-induced dipole model is appropriate, and for this case[17]

$$k_i(\text{ion}) = 2\pi e(\alpha/\bar{M})^{1/2} \quad (5)$$

Assuming typical values of these parameters, one finds that $k_i(HS) \leq 10^{-10}$ cc/s and $k_i(\text{ion}) \geq 10^{-10}$ cc/s. For a radiative lifetime $\tau_i \simeq 10^{-7}$ s, it is seen that a 10% quenching correction can be expected for hard-sphere collisions for $n \geq 10^{16}$ cc^{-1} and for $n \leq 10^{16}$ cc^{-1} for excited ionic species. An increasing degree of sophistication and accuracy of data interpretation acknowledge the possible dependence of k_i on the electronic state, the vibrational level, and the rotational-vibrational level[18] of the excited state. Unfortunately, very little rotationally resolved quenching data exist for either pure or mixed gases. One such measurement was reported by Lewis and Price[19] for the $N_2^+(1-)$ system at 300 and 94 K. Figure 2 shows the result for k_i at $T = 94$ K for the resolved R branch of the $(0,0)$ band of this system. The significant variation with rotational quantum number K is evident, as is the difference in k_i for the two spin states of $N_2^+ B^2 \sum_u^+$. Of equal interest is the fact that k_i increases by an approximate factor of two as T decreases from 300 to 94 K. Since the ion-neutral, elastic-collision rate constant is temperature independent, the results of Ref. 19 allow one to infer that the probability of transfer increases with decreasing T, which is characteristic of near-resonant, energy transfer processes.[20] It is clear that further data of this type, for

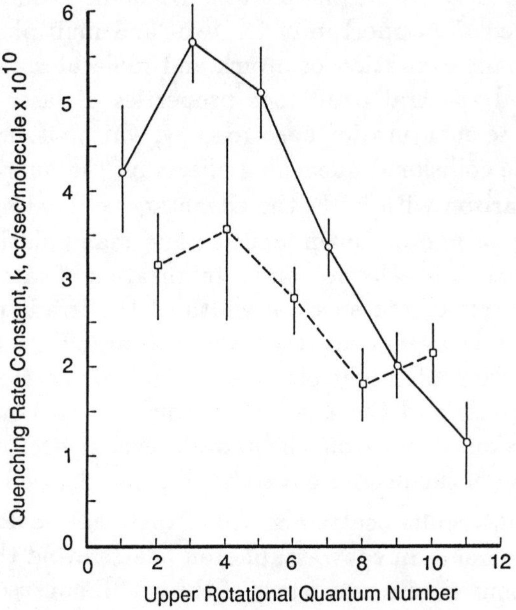

Fig. 2 Quenching rate constants for $N_2^+(1-)(0,0)$ R-branch rotational lines at 94 K and N_2 number densities less than 4×10^{-16} cc^{-1}.

both pure and mixed gases, are needed for the confident application of EBF to higher-density flowfield regions.

The concentration of the previous material on N_2 and its EBF spectra is, of course, due to the common use of N_2 for low-density studies and the resulting extensive investigations of electron-N_2 interactions. However, EBF has been used successfully for the study of low-density gaseous species and mixtures.[8] As examples, in a series of studies Petrie and co-workers applied EBF to NO,[21] O_2, and O[22] flowfields; a CO_2 jet-expansion was studied by Beylich[23] and CO fluorescence in a supersonic expansion was reported in Ref. 24.

Laser-Induced Fluorescence

The basic principle of LIF is the use of laser-excited, radiative emission to provide local measures of the number density distribution functions. Before the introduction of laser sources, photon-induced absorption and fluorescence studies of low-density gases were limited, in general, to the use of one-photon, resonance radiation excited by lamp sources of low spectral irradiance. The introduction of tunable lasers of high spectral irradiance has resulted not only in greater sensitivity for the use of one-photon, resonant radiation, but also it has provided the opportunity for two- and multiphoton resonant and nonresonant excitation of atoms and molecules. Moreover, the tunability and spectral irradiance properties of laser sources have enabled the use of saturation spectroscopy, which either minimizes or eliminates the collisional quenching effects on the fluorescent signal.

In comparison with EBF, the advantages offered by LIF include the simplicity of photon beam focusing and manipulation, the specificity of excited state selection and minimization of cascade-radiation effects by control of the spectral width of the irradiance. Furthermore, completely absent are the complicating effects of EBF, such as the production of Auger electrons, autoionizing transitions, secondary electrons, and the beam-spreading due to both elastic and inelastic collisions result from the kilovolt level of electron energy and the interaction of the beam's electrons with the charged electrons and nuclei of the molecular scatterers. All of these effects depend linearly upon beam current. In contrast, photon beams avoid the majority of these deleterious effects as a result of the small, nonresonant photon-molecule cross sections. Moreover, two- and multiphoton excitation phenomena are nonlinear processes, and, as a result, the presence

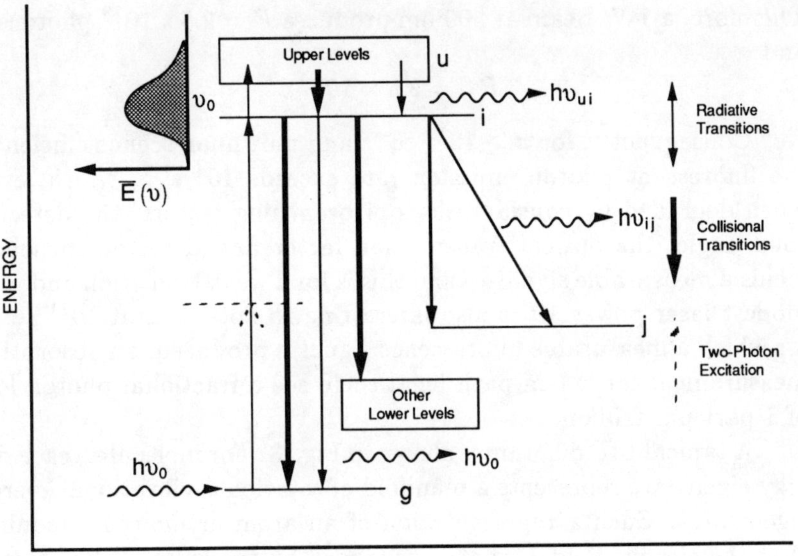

Fig. 3 Laser-induced fluorescence transition diagram.

and strength of these phenomena can be continuously adjusted by varying the incident beam flux.

The sensitivity of LIF for the detection of low-density gaseous species is easily seen by a simple order-of-magnitude computation. An incident laser beam of power P_o, wavelength λ_o is focused to an area A_B to produce a photon flux F/A_B and interacts with the atomic species of number density n with a photon-atom cross section σ to produce an excitation rate R_E. The ratio of the numbers of excited atoms and incident photons per unit length of the observed beam is

$$R_E/F = n \cdot \sigma \tag{6}$$

Assuming a comparatively weak transition, let $\sigma \doteq 10^{-25}$ cm^2,

$$R_E/F \doteq 10^{-25} \cdot n \tag{7}$$

The photon rate F is given by

$$F \doteq 5 \times 10^{15} \cdot \lambda_o \cdot P_o \tag{8}$$

where F is in photons/s, λ_0 is in nm, and P_0 is in watts.

Therefore, a 1-W beam at 500 nm produces $F \doteq 2.5 \times 10^{18}$ photons/s and

$$R_E \doteq 2.5 \times 10^{-7} n \tag{9}$$

Consequently, for $n \geq 10^{11}$ cc^{-1} and unit fluorescence efficiency, the fluorescent photon emission rate exceeds 10^4 s^{-1}, which, even when degraded by considerations of branching factors, the detector solid angle, the optical transmission factor and detector efficiency, yields a measurable signal — and this is for a weak transition and very modest laser power. It is also interesting to note that at 10^{11} cc^{-1}, for which a measurable fluorescence signal is produced, an absorption measurement for a 1-m path length will see a fractional photon loss of 1 part per trillion.

A typical LIF diagram is shown in Fig. 3. For molecules, each energy eigenstate represents a manifold of rotational-vibrational energy eigenstates. Such a representation of an atom or molecule requires that the electric field $\overline{E}$ that is produced by the laser at the atom is sufficiently small to allow the use of the field-free energy eigenfunctions. This criterion, for low-lying states of excitation, is simply that $\overline{E}$ is much, much less than the intra-atomic field $\overline{E}_{\text{at}}$; i.e.,

$$\overline{E} \ll \overline{E}_{\text{at}} \doteq 5 \times 10^9 \text{ V/cm}$$

where $\overline{E}$ can be computed in terms of the power density $\overline{P}$ according to

$$\overline{E} \doteq 27.3(\overline{P})^{1/2} \tag{10}$$

where $\overline{E}$ is in V/cm^2 and $\overline{P}$ is in W/cm^2.

Consequently, if $\overline{E}$ must be less than $10^{-2}\ \overline{E}_{\text{at}}$, [25]

$$\overline{P} \leq 3.3 \times 10^{12} \text{ W/cm}^2$$

a value that is exceeded by existing amplified picosecond laser systems. This critical field strength is reduced by many orders of magnitude for resonant interactions,[26] and these effects, such as the Bloch-Siegert effect, are excluded from this discussion.

The LIF diagram of Fig. 3 shows the representative, radiative, and collisional transition for the "$g - i - j$" three-state model. Determination of the number density distribution function of the atomic and molecular species is accomplished by relating the measured radiation emission from excited-state i to the ground state number density by means of an excitation-de-excitation model that includes the properties of the excitation source and the de-excitation mechanisms. The principles of LIF can be adequately demonstrated without algebraic complexity with a two-state model; i.e., radiative excitation and emission are limited to the states, or levels, g and i. Since many applications of LIF use pulsed sources, a time-dependent model is appropriate. The laser source is assumed to be a rectangular pulse of duration τ_L with a spectral irradiance $E^L(\nu)$, an integrated irradiance $\hat{E}^L$, and an effective spectral irradiance $E^L_{eff}(\nu)$ that is defined by

$$E^L_{eff}(\nu) \equiv \int E^L(\nu) L(\nu) d\nu, \tag{11}$$

where $L(\nu)$ is the absorption, line-profiie function.

Degeneracy factors g_g and g_i exist for the two levels and, for later use, $\bar{g} = g_i/g_g$ and $\hat{g} = \bar{g}/(1+\bar{g})$. The field-free characteristic times τ_i^R and τ_i^o for level i are defined in terms of the spontaneous emission coefficient A_{ig} and the collisional decay rate $k_i\tau_i n_T = Q_i$, which equals the reciprocal, collisional, characteristic time τ_i^c, according to the relations

$$1/\tau_i^R = A_{ig} = A_i \tag{12}$$

and

$$\begin{aligned} 1/\tau_i^o &= A_i + Q_i \\ &= (1/\tau_i^R) + (1/\tau_i^c). \end{aligned} \tag{13}$$

Using the rate equation approach and assuming $n_i(t=0) = 0$, one finds[27]

$$n_i(t) = H \cdot n_T \cdot \hat{g}\big[1 - \exp(-t/\tau_i)\big] \qquad t < \tau_L \tag{14}$$

$$= H \cdot n_T \cdot \hat{g}\big[1 - \exp(-\tau_L/\tau_i)\big] \qquad t = \tau_L \tag{15}$$

$$= H \cdot n_T \cdot \hat{g}\big[1 - \exp(-\tau_L/\tau_i)\big]$$

$$\exp\big[-(t-\tau_L)/\tau_i^o\big], \; t > \tau_L \tag{16}$$

where the factor H is

$$H = S_I/(1+S_I) \tag{17}$$

and the saturation factor S_I is

$$S_I = (1+\overline{g})(B_{ig}/c)E^L_{eff}(\nu)\tau_i^o \tag{18}$$

where B_{ig} is the Einstein stimulated emission coefficient.
The characteristic time τ_i is given by

$$\tau_i = \tau_i^o/(1+S_I) \tag{19}$$

The steady-state values of number density n_i^{SS} for the general, weak-field, and saturated cases are, respectively,

$$n_i^{SS} = \left[S_I/(1+S_I)\right]\hat{g}\cdot n_g \tag{20}$$

$$= \hat{g}(B_{ig}/c(\cdot E^L_{eff}(\nu)\tau_i^o \cdot n_g \tag{21}$$

$$= \hat{g}\cdot n_g \tag{22}$$

The spontaneous emission corresponding to the weak-field excitation case of Eq. (21) is a function of τ_i^o and is therefore implicitly dependent on the collisional decay channels. The number density n_g represents, for molecules, the number density of the ground state g rotation-vibration level (v_1, J_1), and n_i, the value for level $o(v_1', J_1')$ of level i. If, as proposed in Ref. 28 and shown in Fig. 4, excitation occurs sequentially with the wavelengths λ_1 and λ_2 to the same (v_1', J_1') level from different (v_1, J_1) levels of g, the ratio of the two fluorescence signals varies as the ratio of densities $n_g(v_1, J_1)/n_g(v_1', J_1')$, which is a function of T_R and T_V. Therefore, for $v_1 = v_1'$, T_R is obtained, and for $J_1 = J_1'$, T_V is obtained. The use of the ratio of the fluorescence signals improves the measurement accuracy by formally eliminating the lifetime parameters from the equation. However, since collisional quenching exists in the gas, the fluorescence efficiency of each signal is reduced accordingly and the precisions of the measured signals and their ratio are degraded. Of course, the measurement of the density will still be affected by collisional quenching effects, and a summary of quenching effects on LIF for combustion environments is given by Crosley.[29]

Saturated LIF is achieved when $S_I \gg 1$ and, as Eq.(22) shows, for $S_I \gg 1$, the spontaneous emission rate is directly proportional to the ground state density n_g. Therefore, if saturation is achieved,

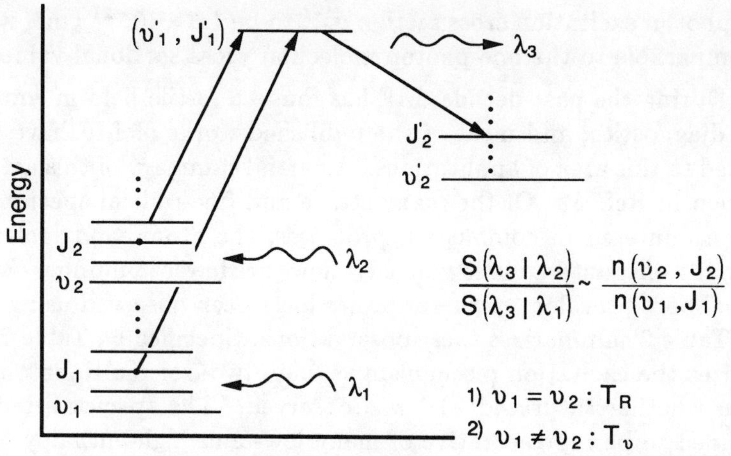

Fig. 4 LIF probe.

both density and temperatures can be obtained without corrections for collisional quenching.

The large spectral irradiance of laser sources makes possible the use of one-, two- or multiphoton excitation of atoms and molecules. Spectrally integrated, one-photon, absorption cross sections ($\sigma_a^{(1)}$) are given by Measures[30] for various atomic and molecular transitions. As listed in the work of Measures, values of $\sigma_a^{(1)}$ for resonant absorption of atomic species are on the order of 10^{-11} to 10^{-15} cm^2, and for typical molecular transitions, $\sigma_a^{(1)} \sim 10^{-19}$ to 10^{-25} cm^2. Two-photon cross sections ($\sigma_a^{(2)}$) can be defined, using the Beer-Lambert equation, in terms of the two-photon absorption coefficient $\alpha^{(2)}$ and the photon irradiance $\hat{F} = F/A_B$:

$$\begin{aligned} -\mathrm{d}F &= n \cdot \alpha^{(2)} \cdot \hat{F}^2 \cdot \mathrm{d}x \\ &= \sigma_a^{(2)} \cdot \hat{F} \cdot \mathrm{d}x \end{aligned} \tag{23}$$

so that the two-photon cross section is given by

$$\sigma_a^{(2)} = \alpha^{(2)} \hat{F} \tag{24}$$

Values of $\alpha^{(2)}$ are in the range of 10^{-49} to 10^{-51}cm^2/(photon/s$\cdot$ cm^2). As an example, a 1-mJ, 1-ps pulsed laser at 300 nm with a power density of 10^{11} W/cm^2 has $\hat{F} = 1.5 \times 10^{29}$ photons/(s $\cdot$ cm^2), and, assuming $\alpha^{(2)} = 10^{-50}$ cm^2/[photon/(s $\cdot$ cm^2)],[31] one finds the two-photon excitation cross section $\sigma_a^{(2)}$ to be 1.5×10^{-21} cm^2, which is comparable to the one-photon molecular cross sectional values.

During the past decade, LIF has found a fertile field in combustion diagnostics, and most of the published works of LIF have been related to this area of applications. A partial summary of this activity is given in Ref. 32. Of the many stable and free-radical species that are encountered in combustion processes, the atoms and molecules normally encountered in low-density flows are fewer in number. Many of the more prevalent of these species have been observed using LIF, and Table 2 summarizes these observations. Specifically, Table 2 categorizes the excitation mechanism as one-, two-, or multiphoton and notes whether saturated LIF was observed. The species listed are intended to be representative of many low- and high-enthalpy flows, and the atomic species N_a represents merely one of the entire family of alkali atoms that are easily observed using LIF. Included in Table 2 is an appropriate reference for each entry.[33–40]

Rayleigh and Raman Scattering

Diganostics accessibility to the higher-density, near-field regions of expansion flowfields is provided by laser-Rayleigh (RyS) and laser-

Table 2 LIF Partial Summary

Species	Saturation	Single-photon	Two- or Multi-photon	Ref.
H	•		•	33
N			•	34
O			•	34
Na	•	•		35
OH	•	•	•	36
H_2		•	•	37
N_2			•	38
O_2		•		39
CO			•	40
NO		•	•	28
H_2O				

• Observed

IL-0528a

spontaneous Raman scattering (RS). The interaction of the incident laser-beam photons with the atomic and molecular species results in a RyS signal with wavelength equal to that of the incident laser beam, and a signal strength proportional to the total gas density. In contrast, RS components are shifted to longer and shorter wavelengths, which are the Stokes and anti-Stokes components, respectively. As described in detail in Refs. 41-43, the RS wavelength, or frequency, displacements are characteristic of the internal, energy-level structure of each molecular species, and, as a result, the technique exhibits molecular specificity. Furthermore, the strength of the RS signal is proportional to the number of molecular scatterers in the internal energy level that corresponds to the specific wavelength-shifted, spectral feature. Consequently, RS provides a method of determining local values of species densities and internal-mode distribution functions. In contrast to fluorescence excitation, both RyS and RS are scattering processes that occur on sub-picosecond time scales and, as a result, collisional quenching does not affect the measurements.

Figure 5 shows a general experimental configuration for the use of laser RyS and RS for the study of flowfield expansions. The laser beam is injected into the flow, and the scattered signal is collected, dispersed in wavelength, and detected by either one or more photomultiplier tubes or an intensified, array-tube detector. Shown in Fig. 5 is a polarization rotator that uses to maximum advantage

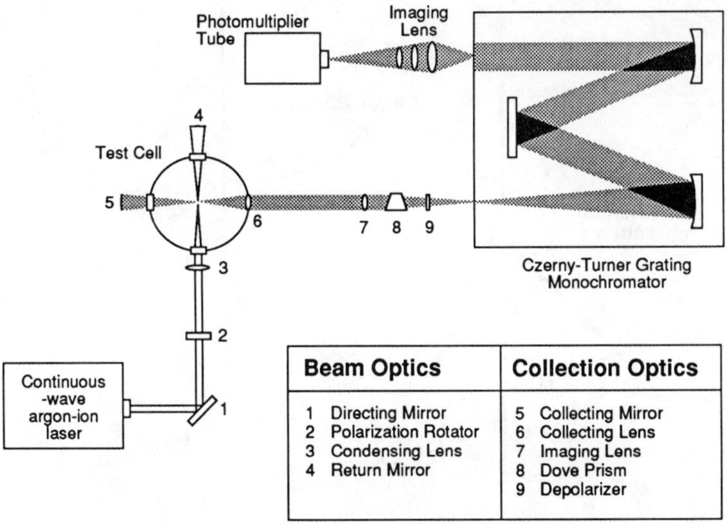

Fig. 5 Experimental configuration.

the polarization dependence of the RyS and RS cross sections. Since the RyS and RS total cross sections are small ($\leq 10^{-27}$ cm^2),[27] also shown is the possible enhancement of sensitivity by multipassing the incident laser beam and by increasing the detection solid angle by collection of photons scattered at the supplementary angle of the detection system.

The properties of RyS and RS for the measurement of density and temperatures of the flowfield are summarized in the following sections.

Rayleigh Scattering

Atomic or molecular species of number density n are assumed to be located at the origin of the coordinate system of Fig. 6. The incident laser beam of polarization direction $\hat{y}$ and incident direction of propagation $\hat{z}$ is assumed to have power P_o, wavelength λ_o, focal volume area A_B, and length L at the scattering location. Photons scattered in the $\hat{x}$ direction are detected, and polarization-resolved RyS signals are observed. The molecular scattering species are characterized by the diagonalized, optical, polarizability tensor[27]

$$S(\|) = C\sigma(\|) \cdot n \cdot L \cdot (P_o/hc)\lambda_o \tag{25}$$

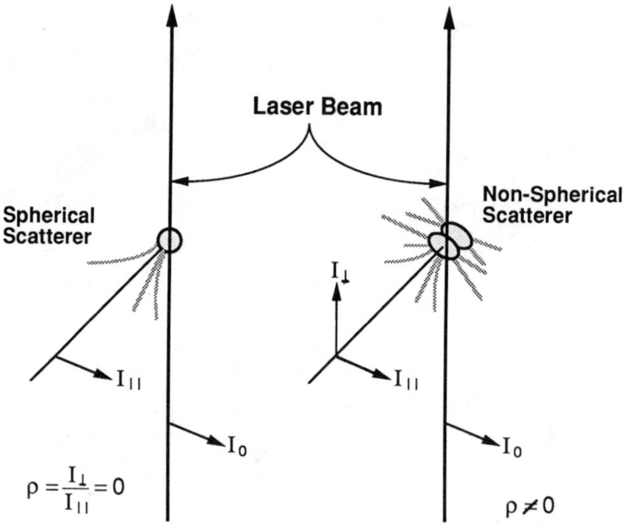

Fig. 6 Polarization diagram.

and

$$S(\perp) = C\sigma(\perp) \cdot n \cdot L \cdot (P_o/hc)\lambda_o, \tag{26}$$

where C represents the polarization-independent solid angle and sensitivity factors and $\sigma(\|)$ and $\sigma(\perp)$ are the polarization-specific cross sections; therefore, for low-density samples, the RyS components are a measure of the number density of the gas. Furthermore, it can be shown that[27]

$$\sigma(\|) = C'\left[\alpha^2 + (4\beta^2/45)\right] \tag{27}$$

and

$$\sigma(\perp) = C'(\beta^2/15) \tag{28}$$

where

$$\alpha = (1/3)(\alpha_1 + \alpha_2 + \alpha_3) \tag{29}$$

and

$$\beta^2 = (1/2)\left[(\alpha_1 - \alpha_2)^2 + (\alpha_1 - \alpha_3)^2 + (\alpha_2 + \alpha_3)^2\right] \tag{30}$$

Using the α_i values of Baas and van den Hout,[44] it can be shown that, in general,

$$4\beta^2/(45\alpha^2) \leq 10^{-2}$$

Consequently, reasonable accuracy is maintained if the second term of Eq.(27) is neglected.

For a gas mixture expansion flowfield, it can be shown that a nondimensional RyS signal $\hat{S}$ can be defined for the parallel polarization component

$$\hat{S}(\|) = S(\|)/[K \cdot n_{To} \cdot \sigma_{N_2}(\|)] = \hat{n}_T \sum_i X_i[\sigma_i(\|)/\sigma_{N_2}(\|)], \tag{31}$$

where X_i is the mole fraction of species i, n_{To} is the total reservoir number density, K is the collection of all unimportant parameters, and $\sigma_{N_2}(\|)$ has been used as a reference RyS cross section. Finally, $\hat{n}_T = n_T/n_{To}$, the spatially varying number density ratio for the expansion flow. Therefore, for chemically frozen flows for which the composition is unchanging, $\hat{S}(\|)$ provides a direct measure of the density ratio of the expansion flow. For a single-gas expansion, a non-dimensional RyS signal is defined to be equal to $\hat{n} = n/n_o$, and

RyS provides a direct measure of the local number density ratio for an expansion flowfield. Figure 7 shows the use of RyS for the measurement of $\hat{n}$ of N_2 and CO_2 conical nozzle expansions;[45] comparison of the results with the method-of-characteristics (MOC) prediction is seen to be excellent, and the minimum absolute number density obtained with RyS was on the order of 10^{14} cc^{-1}.

The detection of condensation onset and cluster growth has been one fruitful application of RyS.[45–52] As shown in Ref. 52, the growth of molecular clusters, which possess larger polarizabilities than the monomer species, will be manifested by a deviation of $\hat{S}(||)$ from the isentropic expansion's RyS signal, as given by Eq.(31). This behavior is shown in Fig. 8 for an expansion of N_2 for which this deviation, and the condensation onset and growth are obvious.

Raman Scattering

Although both atomic and molecular species exhibit Raman-type scattering processes, the spontaneous rotational and vibrational Raman scattering (RRS and VRS, respectively) of molecules is more

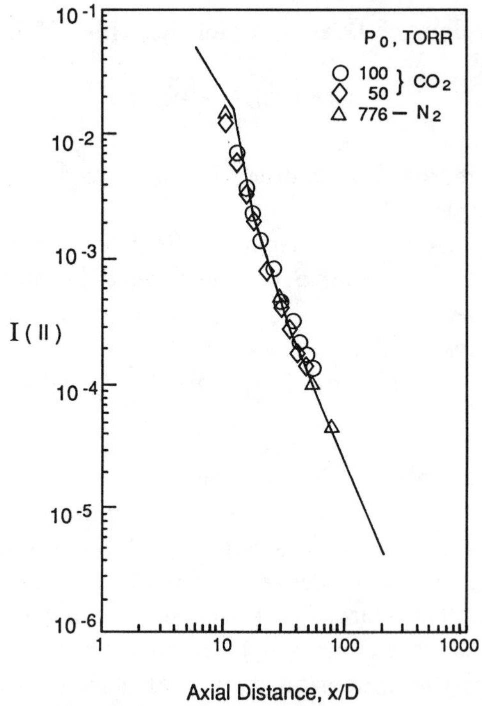

Fig. 7 Axial variation of scattering function I_s for pure gas expansions.

pertinent for the study of low-density flowfields. Characteristic frequency shifts, relative to the laser excitation frequency, are determined by the characteristic rotational and vibrational energies of the molecules. For typical molecules (Ref. 42), in cm^{-1} units, the characteristic rotational displacements of the Raman-shifted lines are on the order of $10 - 100\ cm^{-1}$. The characteristic vibrational displacements vary between 500 and 3000 cm^{-1}. This difference in scale between RRS and VRS displacements also occurs for the characteristic cross sections: for RRS, $\sigma^{RRS} \cong 10^{-28}\ cm^2$ and $\sigma^{VRS} \cong 10^{-30}\ cm^2$. Both RRS and VRS are applicable for measurement of the species number density $n \geq 10^{17}\ cc^{-1}$. For gas expansions comprised of one to a few molecular species, the much larger value of σ^{RRS} relative to σ^{VRS} is strong indication for the use of RRS for low-density expansions. However, even with the selection of RRS, the very small value of σ^{RRS} has prompted some investigators to locate the flowfield within the laser cavity. Such intracavity measurements can enhance the scattering signal by over two orders of magnitude.[53] The difficulty of implementation of this technique for general use should be noted. The following discussion will focus exclusively on the more general applications of RRS.

Figure 8 shows the axial variation of the measured n and T_R for a conical nozzle expansion of N_2; also shown is the theoretical prediction, and the agreement is excellent. The physical basis for the results of Fig. 8 is the interrogation by the incident photons of the population of the individual molecular rotational levels. The identification of the interrogated level is provided by the wavelengths of the Stokes and anti-Stokes photons.

The results of Fig. 8 are typical for RRS studies using a visible laser with average power on the order of 1 W, a double spectrometer to provide the required rejection of the nearby RyS, or Mie scattered, signal, and an efficient detector capable of single-photon detection. The low collection and transmission efficiency of the double spectrometer and the comparatively high resolution that is required for spectrally resolved measurements combine to limit the use of RRS to flow regions for which $n_{min} \geq 10^{16}\ cc^{-1}$. Replacing the visible laser with a commercially available, high-power, excimer laser with ultraviolet wavelength will decrease this limit by an order of magnitude; i.e., $n_{min} \geq 10^{15}\ cc^{-1}$. Furthermore, for the research environment the location of the flow within the cavity of a CW, visible laser of output power on the order of 1 W yields $n_{min} \geq 10^{14}\ cc^{-1}$.

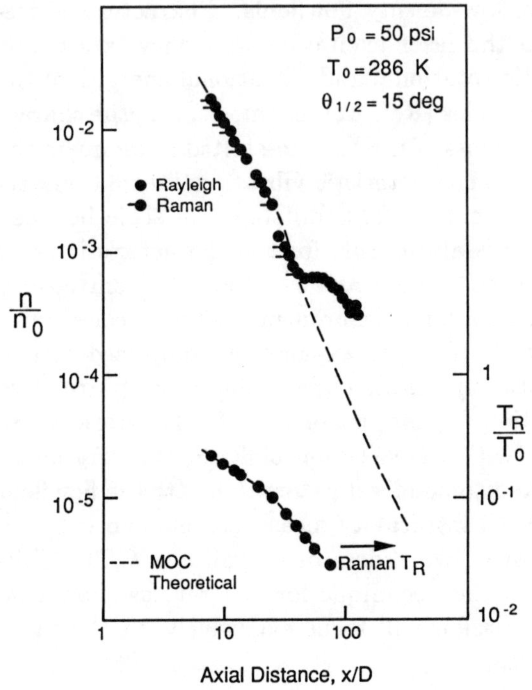

Fig. 8 N_2 conical nozzle expansions.

Table 3 Order-of-Magnitude Relations for N_2

	Photons (Molecule/cc) • Joule • SR • MM
Rayleigh	3×10^{-11}
Rotational Raman	10^{-12}
Vibrational Raman	10^{-14}
$\lambda = 300$ nm	

For low-density applications involving most diatomic and linear triatomic molecules, the use of Fabry-Perot dispersion, as suggested by Barrett[54–56] is an attractive alternative for expansion flows of single gases and mixtures. As discussed by Barrett, replacement of the spectrometer with a Fabry-Perot interferometer increases the étendue of the system and exploits the coincidence in frequency space of the comb filter, equally spaced, transmission peaks of the Fabry-Perot and the RRS lines with the near-equal spacing of four times the

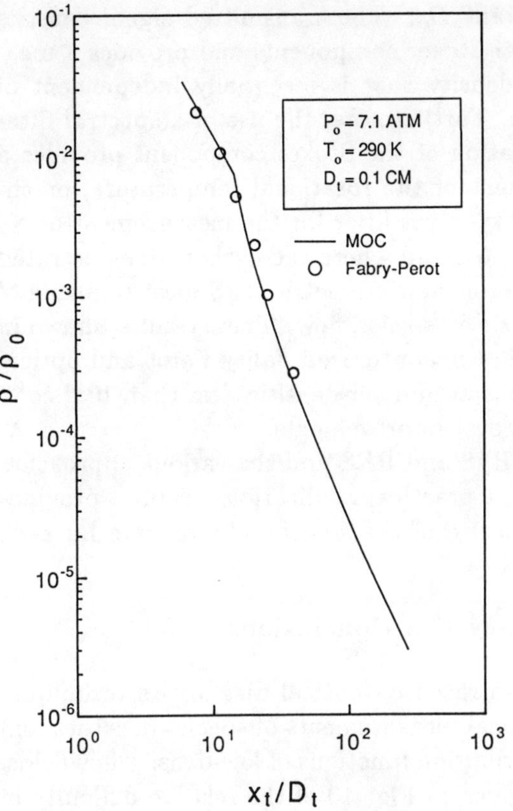

Fig. 9 Axial variation of N_2 density.

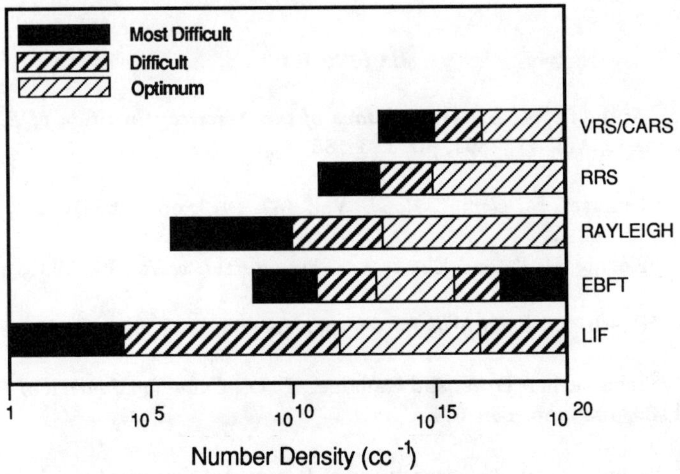

Fig.10 Diagnostics application regimes.

molecular rotational constant B_v. The transmitted signal consists of both the Stokes and anti-Stokes components and provides a measure of the total species density that is essentially independent of the rotational distribution. Furthermore, the use of a spectral filter for the selective discrimination of the Stokes component provides a method for the measurement of the rotational temperature, or enthalpy. The use of the Fabry-Perot filter for the measurement of N_2 density is described in Ref. 57, and where the authors demonstrated the application of Fabry-Perot interferometric RRS measurements of N_2 density in a conical nozzle expansion flow. These results, shown in Fig. 9, were obtained with a nonoptimized Fabry-Perot and optical system, and it is estimated that number densities less than 10^{14} cc^{-1} could be obtained with modest improvements.

Finally, summarizing RyS and RRS and the various approaches that are available for routine practical applications, Table 3 provides an order-of-magnitude measure of the specific photon rate for each technique for N_2.

Summary and Conclusions

This review has concentrated on optical diagnostics techniques that are available for practical measurements of species densities and internal mode number distribution functions of low-density flowfields. A subjective summary is given in Fig. 10 of the relative difficulty of application of each technique for the various density regimes that are encountered in low-density flows.

References

[1]Anthony, W. A., *Transactions of the American Institute of Electrical Engineers,* Vol. 11, 1894, pp. 133–185.

[2]Dunoyer, L., *Compt. Rend.,* Vol. 157, 1913, pp. 1068–1070.

[3]Koenig, H. D. and Ellett, A., *Physics Review,* Vol. 37, 1931, p. 1685.

[4]Frisch, R., *Zeitschrift fur Physik,* Vol. 86, 1933, pp. 42–48.

[5]Schumacher, B. W. and Gadamer, E. O., *Canadian Journal of Physics,* Vol. 36, 1958, pp. 659–671.

[6]Muntz, E. P., "Measurement of Rotational Temperature, Vibrational Temperature, and Molecule Concentration in Non-Radiating Flows of Low

Density Nitrogen," *University of Toronto Institute of Aerophysics Report 71*, Apr. 1961.

[7]Muntz, E. P., "Static Temperature Measurements in a Flowing Gas," *The Physics of Fluids*, Vol. 5, Jan. 1962, pp. 80–90.

[8]Muntz, E. P., "The Electron Beam Fluorescence Technique," *AGARDograph 132*, Dec. 1968.

[9]Center, R. E., *Physics of Fluids*, Vol. 13, 1970, pp. 79–88.

[10]Aarts, J.F.M., DeHeer, F.J., and Vroom, D.A., *Physica*, Vol. 40, 1968, pp. 197–206.

[11]Lewis, J.W.L., Price, L.L., and Powell, H.M., *Physics Review A*, Vol. 11, Apr. 1975, pp. 1214–1222.

[12]Coe, D. C., "An Experimental Study of Non-Equilibrium Free Jet Expansions of Nitrogen," *University of California Report*, FM-77-2, Nov. 1977.

[13]Coe, D., Robben, F., Talbot, L., and Cottolica, R., "Measurement of Nitrogen Rotational Temperatures Using the Electron Beam Fluorescence Technique," *Rarefied Gas Dynamics, Vol II.*, edited by R. Campargue, CEA, Paris, France, 1979, pp. 907–918.

[14]Baronavski, A. P., Helvajian, H., and DeKoven, B. M., "Direct Rotational Population Distributions of N_2 and N_2^+ in Pulsed Supersonic Beams," *SPIE Proceedings*, Vol. 482, 1984, pp. 31-35.

[15]Lewis, J.W.L. and Williams, D. W., "Vibrational Temperature Measurements Using the Electron Beam," *AIAA Journal*, Vol. 7, June 1969, pp. 1202–1204.

[16]Pendleton, W. R. and O'Neil, R., *Journal of Chemical Physics*, Vol. 56, No. 12, June 1972, pp. 6260–6262.

[17]McDaniel, E. W., Cermák, V., Dalgarno, A., Ferguson, E. E., and Friedman, L., *Ion-Molecule Reactions.* Wiley-Interscience, New York, 1970.

[18]Mitchell, K. B., "Fluorescence Efficiencies and Collisional Deactivation Rates for N_2 and N_2^+ Bands, Excited from Deposition of Soft X-Rays," *Los Alamos Scientific Laboratory*, LA-4248, Nov. 1969.

[19]Lewis, J.W.L. and Price, L. L., "Collisional Quenching of Atomic and Molecular Nitrogen: I. Experimental Results," *AEDC-TR-75-151*, Dec. 1975.

[20]Yardley, J. T., *Introduction to Molecular Energy Transfer*, Academic Press, New York, 1980, pp. 205–234.

[21]Petrie, S. L. and Komr, J. J., "Application of the Electron Beam Fluorescence Technique to the Measurement of the Properties NO in Non-Equilibrium Flows," *AFFDL-TR-72-144*, Dec. 1972.

[22]Petrie, S. L., Boiarski, A. A., and Lazdinis, S. S., "Electron Beam Studies of the Properties of Molecular and Atomic Oxygen in Non-Equilibrium Flows," *6th Aerodynamic Testing Conference*, AIAA Paper 71-271, March 1971.

[23]Beylich, A. E., "Experimental Investigation of Carbon Dioxide Jet Plumes," *The Physics of Fluids*, Vol. 14, No. 5, May 1971, pp. 898–905.

[24]Dagdigian, P. J. and Doering, J. P., "Electron Impact Ionization-Excitation of CO in a Supersonic Beam," *Journal of Chemical Physics*, Vol. 78, No. 4, Feb. 1983, pp. 1846–1950.

[25]Delone, N. B. and Krainov, V. P., *Atoms in Strong Light Fields*, Springer-Verlag, Berlin, FRG, 1985, p. 6.

[26]Ibid. p. 246

[27]Measures, R. M., *Laser Remote Sensing: Fundamentals and Applications*, Wiley-Interscience, New York, 1984.

[28]McKenzie, R. L. and Gross, K. P., *Applied Optics*, Vol. 20, No. 12, June 1981, pp. 2153-2165.

[29]Crosley, D. R., *Optical Engineering*, Vol. 20, No. 4, 1981, pp. 511–21.

[30]Measures, R. M., *Laser Remote Sensing: Fundamentals and Applications*, Wiley-Interscience, New York, 1984.

[31]Birge, R. R.,"One-Photon and Two-Photon Excitation Spectroscopy," in *Ultra-sensitive Laser Spectroscopy*, edited by D. S. Kliger, Academic Press, New York, 1983, pp. 109–174.

[32]Crosley, D. R. and Smith, G. P., *Optical Engineering*, Vol. 22, 1983, pp. 545–553.

[33]Lucht, R. P., Salmon, J. T., King, G. B., Sweeney, D. W., and Laurendeau, N. M., *Optics Letters*, Vol. 8, No. 7, July 1983, pp. 365–367.

[34]Bischel, W. K., Perry, B. F., and Crosley, D. R., *Applied Optics*, Vol. 21, 1982, pp. 1419–1429.

[35]Fairbank, W. M., Hänsch, T. W., and Schawlow, A. L., *Journal of the Optical Society of America*, Vol. 65, Feb. 1975, pp. 199–204.

[36]Lucht, R. P., Laurendeau, N. M., and Sweeney, D. W., *Applied Optics*, Vol. 21, No. 20, Oct. 1982, pp. 3729–3735.

[37]Bogen, P. and Lie, Y. T., *Applied Physics*, Vol. 16, No. 2, June 1978, pp. 139–145.

[38]Van Veen, N., Brewer, P., Das, P., and Bersohn, R., *Journal of Chemical Physics*, Vol. 77, 1982, pp. 4326–4329.

[39]Massey, G. A. and Lemon, C. J., *IEEE Journal of Quantum Electronics*, QE-20, No. 5, May 1984, pp. 454-457.

[40]Alden, M., Wallin, S., and Wendt, W., *Applied Physics B*, Vol. B33, No. 4, Apr. 1984, pp. 205–212.

[41]Placzek, G., "Rayleigh and Raman Scattering." Translated from Akademische Verlagsgesellschaft G.m.b.H., Leipzig, *Handbuch der Radiologie*, Heft 6, Teil 2, 1934, pp. 209–374.

[42]Sushchinskii, M. M., *Raman Spectra of Molecules and Crystals*, Israel Program for Scientific Translations, New York, 1972.

[43]Anderson, A., (ed.), *The Raman Effect, Vol. 2: Applications*, Marcel Dekker, New York, 1973.

[44]Bass, F. and van den Hout, K. D., *Physica A*, Vol. 95A, No. 3, March 1979, pp 597–601.

[45]Lewis, J.W.L., Williams, W. D., and Powell, H. M., "Laser Diagnostics of a Condensing Binary Mixture Expansion Flow Field," *Rarefied Gas Dynamics, Vol. II*, edited by M. Becker and M. Fiebig, DFVLR, Porz-Wahn, FRG., 1974, pp. F7.1–F7.8.

[46]Wegener, P. P. and Stein, G. D., "Light Scattering Experiments and Theory of Homogeneous Nucleation in Condensing Supersonic Flow," Combustion Institute, International Symposium on Combustion, University of Poiters, Poiters, France, 1968, pp. 1183–1190.

[47]Daum, F. L. and Gyarmathy, G., "Condensation of Air and Nitrogen in Hypersonic Wind Tunnels," *AIAA Journal*, Vol. 6, Mar. 1968, pp. 457–465.

[48]Clumpner, J. A., "Light Scattering from Ethyl Alcohol Droplets Formed by Homogeneous Nucleation," *The Journal of Chemical Physics*, Vol. 55, No. 10, Nov. 1971, pp. 5042–5045.

[49]Beylich, A. E., "Condensation in Carbon Dioxide Jet Plumes," *AIAA Journal*, Vol. 8, No. 5, May 1970, pp. 965–967.

[50]Lewis, J.W.L. and Williams, W. D., "Raman and Rayleigh Scattering Diagnostics of a Two-Phase Hypersonic N_2 Flowfield," *AIAA Journal*, Vol. 13, 1975, pp. 709–710.

[51]Williams, W. D. and Lewis, J.W.L., "Experimental Study of the Reservoir Temperature Scaling of Condensation in a Conical Nozzle Flow Field," *Progress in Astronautics and Aeronautics, Vol. 51: Part II*, Ed. J. L. Potter, 1976, pp. 1137–1151.

[52]Lewis, J.W.L. and Williams, W. D., "Profile of an anisentropic nitrogen nozzle expansion," *The Physics of Fluid*, Vol. 19, No. 7, July 1976, pp. 951–959.

[53]Silvera, I. F., Tommasini, F., and Winjgaardeu, R. J., *Progress in Astronautics and Aeronautics*, Vol. 51(II), 1977, pp. 1295–1304.

[54]Barrett, J. J. and Myers, S. A., "A New Interferometric Method for Studying Periodic Spectra Using a Fabry-Perot Interferometer," *Journal of the Optical Society of America*, Vol. 61, No. 9, 1971, pp. 1246–1251.

[55]Barrett, J. J., "The Use of a Fabry-Perot Interferometer for Studying Rotational Raman Spectra of Gases," *Laser Raman Gas Diagnostics*, Plenum Press, New York, 1974, pp. 63–85.

[56]Barrett, J. J. and Harvey, A. B., "Vibrational and Rotational-Translational Temperatures in N_2 by Interferometric Measurements of the Pure Rotational Raman Effect," *Journal of the Optical Society of America*, Vol. 65, No. 4, Apr. 1975, pp. 392-398.

[57]Lewis, J.W.L. and Weaver, D. P., "An Investigation of the Use of the Fabry-Perot Interferometer for Flowfield Diagnostics Using Raman Scattering," *AEDC-TR-76-61*, June 1976.

Electron Beam Flourescence Measurements of Nitric Oxide

Robert J. Cattolica*
Sandia National Laboratories, Livermore, California

Abstract

Electron beam-excited fluorescence from nitric oxide (5%) in a mixture with argon at a total pressure of 0.100 torr and ambient temperature (293° K) was studied. Using a 700-element, imaged-intensified, linear diode array detector with a 0.32-m spectrometer (0.03 nm/diode), we recorded simultaneously spectrally resolved fluorescence from the A $^2\Sigma^+ \rightarrow$ X $^2\Pi$ electronic transition, the gamma band, of nitric oxide from 229 to 250 nm. The strongest features observed in this portion of the fluorescence spectrum are the (0,1) and (0,2) vibrational bands. The fluorescence spectrum over this section of the gamma band was recorded as a function of electron beam accelerating voltage from 8 to 20 keV, with beam current ranging from 0.040 to 0.135 mA. The spectrally integrated fluorescence signal from the (0,1) band, normalized by the electron beam current, was used to determine a normalized fluorescence excitation rate. The normalized fluorescence excitation rate decreased by a factor of 5 as the electron beam accelerating voltage was increased from 8 to 20 keV. The absence of a strong (1,3) band indicates that excitation by primary electrons with cross sections proportional to Franck-Condon factors is not sufficient to explain the fluorescence spectrum. Additional excitation by secondary electrons and cascading from the higher-lying C $^2\Pi$ and D $^2\Sigma^+$ electronic states both can contribute to the relatively higher population in the v'=0 level of the A $^2\Sigma^+$ state observed in the electron beam fluorescence spectrum of nitric oxide.

Introduction

The renewed emphasis on hypersonic research has stimulated a resurgence of interest in experimental methods for the study of high-

---

*Member Technical Staff, Combustion Research Facility.

speed flows. Improvement in the physical and chemical models used in computational fluid dynamic simulation of hypersonic flows requires a detailed experimental data base. Optical diagnostics provide the capability to make nonintrusive measurements of density, temperature, velocity, and species concentration in hypersonic flows. The short test time available in hypersonic wind tunnels or flight experiments necessitates spectroscopic methods capable of producing high signal levels. Fluorescence methods based on laser or electron beam excitation satisfy this requirement. For flight experiments, electron beam excitation offers a number of advantages over laser excitation, such as small device size, high electrical efficiency, and multiple-state and multiple-species excitation. In addition, a general foundation for the application of the electron beam fluorescence (EBF) technique to hypersonic flows is well established.[1]

Measurement of nitric oxide in hypersonic flows is of interest for the operation of combustion- and shock-driven wind tunnels and in the high-temperature air chemistry produced by high-Mach-number shock waves during hypersonic flight. The initial analysis of electron-beam fluorescence of nitric oxide identified major emission features.[2] Subsequent high-temperature wind-tunnel studies[3] of NO concentration and temperature using EBF illustrated the need for a fundamental understanding of the role of secondary electron excitation, vibrational energy transfer, and quenching in the EBF spectrum. Modern electro-optic instrumentation now makes it possible to examine the influence of these phenomena on the EBF spectrum from nitric oxide.

Experiment

In the experiments that are reported, the electron beam fluorescence spectrum from nitric oxide (5%) in argon background gas was studied as a function of electron beam energy. The experimental apparatus used in this study is shown in Fig. 1. A 2-liter chamber was maintained under a vacuum with a low-speed mechanical pump (2 l/s). Gas was introduced into the chamber through a flow-control valve that operated with feedback control to maintain a constant pressure of 0.100 torr. The electron beam entered the experimental chamber through a differentially pumped, 1-mm-diam orifice. The electron gun (Kimball Physics model EMG-12) was maintained at a pressure of 2 x 10^{-4} torr with a turbomolecular pump (50 l/s). After traversing the experimental chamber, the electron beam was collected with a water-cooled Faraday cup, and the current was measured.

Optical access to the chamber was provided through 80-mm-diam ultraviolet-quartz (uv) windows. The fluorescence light was collected and focused onto the entrance slit of a 0.32-m spectrometer with a pair of uv camera lenses (105 mm, f/4.5). A 2400-groove/mm holographic grating provided a dispersion of 1.2 nm/mm at the exit focal plane of the spectrometer. An image-intensified (microchannel plate) photodiode-array detector with 700 diodes (2 mm x 0.023 mm) on 0.025-mm centers was used to record simultaneously 21 nm of the spectrum. The diode array was operated by a controller (Princeton Instruments model ST120) interfaced to an IBM-AT computer. The 0.025-mm entrance slit gave a detection bandwidth of 0.03 nm for each diode. The spatial resolution in the experiment was determined by the spectrometer slit width and the diode-array pixel height (0.025 x 2.0 mm). The electron beam was less than 2.0 mm in diameter; thus, the entire cross section of the electron beam was observed with the spectrometer.

Results and Discussion

An example of a typical electron-beam-fluorescence spectrum of nitric oxide is presented in Fig. 2. This spectrum is from a NO(5%)/Ar gas mixture at a pressure of 0.100 torr that was excited by a 0.150-mA, 15-keV electron beam. The spectrum was obtained by integrating the fluorescence on the photodiode-array detector for 2 s and accumulating 30 measurement cycles for a total integration time of 60 s. The spectrum was corrected for the fixed pattern noise of the detector accumulated for the same period of time. The spectrum was then normalized for the detector spectral sensitivity, determined by measuring the spectrum of a calibrated deuterium lamp, and wavelength calibrated with the spectrum of a mercury-discharge lamp. The beam current was measured on each readout cycle of the diode array with an analog-to-digital converter and recorded along with the diode array signal for use in normalizing the fluorescence spectrum for variation in beam current during the experiment.

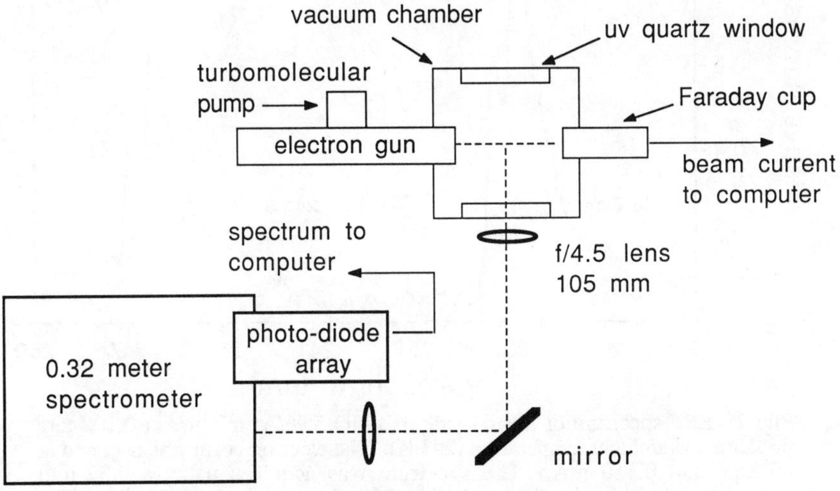

Fig. 1 Electron beam fluorescence (EBF) experiment.

The spectral features in Fig. 2 are from various vibrational bands of the A $^2\Sigma^+ \rightarrow$ X $^2\Pi$ electronic transition, the gamma bands. The prominent features in this spectrum are the $\gamma(0,1)$ and $\gamma(0,2)$ vibrational bands. The $\gamma(0,1)$ band is stronger than the $\gamma(0,2)$ band due to its higher transition probability.[4] Also indicated in Fig. 2 are the wavelength locations of the $\gamma(1,2)$ and $\gamma(1,3)$ bands. The presence of the $\gamma(1,3)$ band can be observed in the tail of the $\gamma(0,2)$ band. The $\gamma(1,2)$ band is almost two orders of magnitude weaker than the $\gamma(1,3)$ band because of an unfavorable Franck-Condon factor[4] and is not discernible in the spectrum.

If primary electron excitation was the only mechanism populating the v'=0 and 1 levels in the A $^2\Sigma^+$ upper state, the $\gamma(1,3)$ band would be expected to be approximately two-thirds as strong as the $\gamma(0,2)$ band. The spectrum in Fig. 2 indicates a higher relative population in the v'=0 level. Secondary electron excitation and radiative transfer to the A $^2\Sigma^+$ state from the C $^2\Pi$ and D $^2\Sigma^+$ electronic states can contribute to this effect. The higher-lying vibrational levels (v'>0) in the C and D states predissociate. Because of the predissocation, these electronic states radiate to the A $^2\Sigma^+$ state preferentially from the v'=0 level.[5] The radiative lifetimes for the C $^2\Pi$ (32 ns) and D $^2\Sigma^+$ (24 ns) states are

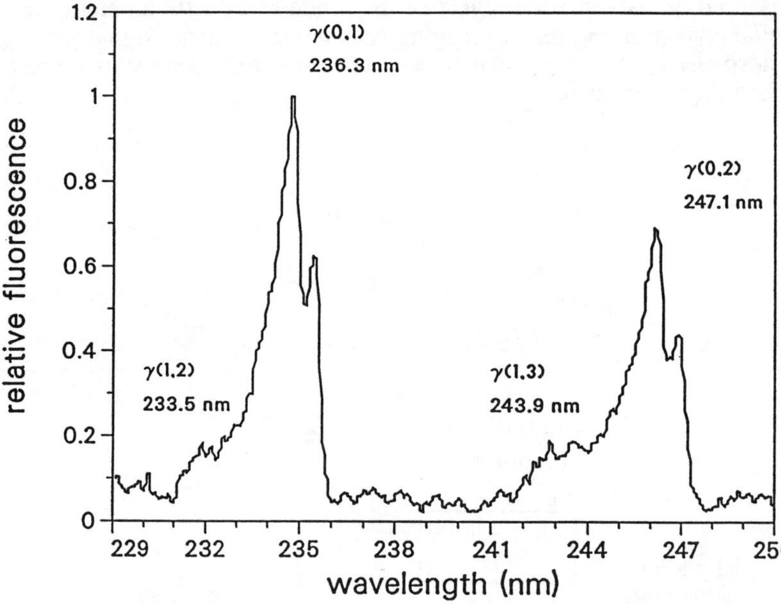

Fig. 2 EBF spectrum of nitric oxide in a NO(5%)/Ar mixture at 0.100 torr pressure and ambient temperature (293 K). The electron beam was operated at 15 keV and 0.150 mA. The spectrum was acquired using a 0.32-mm spectrometer with an image-intensified, 700-element photodiode array. The signal integration time was 60 s.

much faster than A $^2\Sigma^+$ (205 ns) state[5] and therefore provide a rapid, indirect excitation mechanism that can significantly alter the vibrational population distribution in the A state. Electronic quenching and vibrational energy transfer can also play a role in the preferential fluorescence from the v'=0 progression in the gamma bands. However, at the pressure at which this spectrum was taken, both electronic quenching and vibrational energy transfer rates are negligible compared to the spontaneous emission rate from the v'=0 level.

For nitric oxide concentration measurements at or below ambient temperature (293° K), the γ(0,1) band, which is also the strongest band in the v'=0 progression, would be the least affected by the overlap of adjacent spectral bands. To understand the relationship between the γ(0,1) band signal strength and electron-beam characteristics, the behavior of this feature as a function of electron beam accelerating voltage was studied.

The EBF spectrum from nitric oxide from 230 to 248 nm over a range of accelerating voltage from 8 to 20 keV is presented in Fig. 3. The spectra were corrected for detector sensitivity and normalized by the beam current as in Fig. 2. The fluorescence signal in Fig. 3 has been plotted relative to the peak of the γ(0,1) band at an accelerating voltage of 8 keV. To improve the signal-to-noise ratio, these spectra were obtained by integrating on the diode-array detector for 2 s and accumulating 300 scans for a total integration time of 600 s. As in Fig. 2, the primary features in these spectra are the γ(0,1) and γ(0,2) bands.

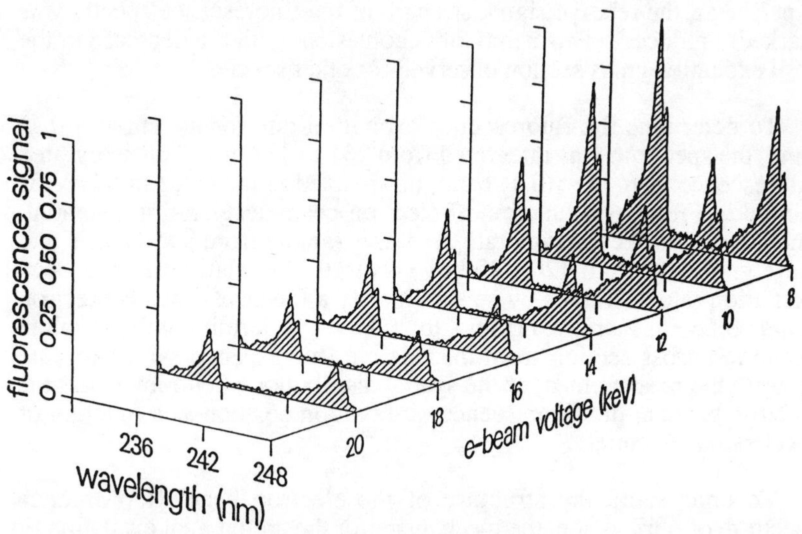

Fig. 3 Normalized electron beam-excited fluorescence spectra from the gamma band of nitric oxide from 230 to 248 nm, plotted as a function of electron beam accelerating voltage from 8 to 20 keV.

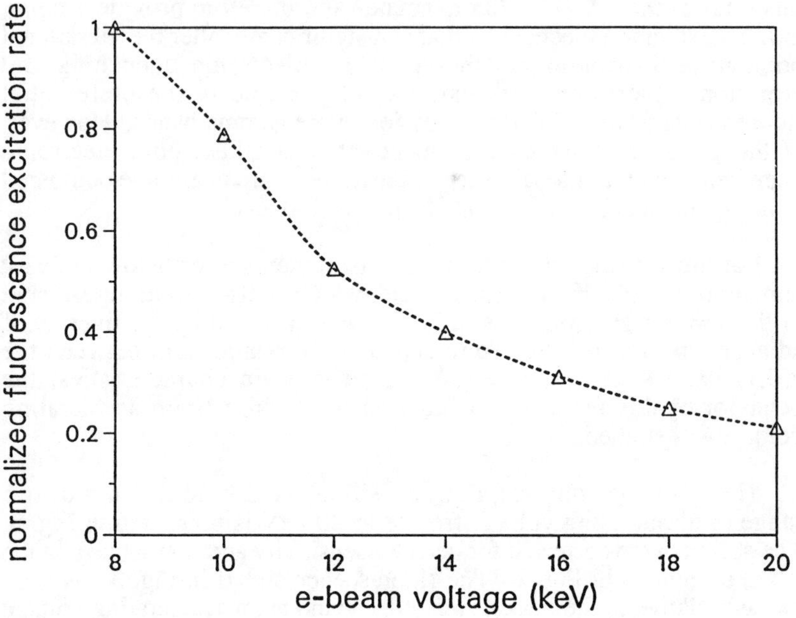

Fig. 4 Normalized fluorescence excitation rate of the γ(0,1) band of nitric oxide from 8 to 20 keV.

As the electron beam accelerating voltage was increased in the experiment, the relative signal strength of the fluorescence spectra was markedly reduced. This behavior is consistent with the decrease in the EBF excitation cross section observed for other species.[1]

To determine the fluorescence excitation rate for the entire γ(0,1) band, the spectrum was integrated from 231 to 237 nm. This integrated fluorescence from the γ(0,1) band, normalized to the value at 8 keV, is plotted in Fig. 4 as a function of electron beam accelerating potential. The fluorescence excitation rate decreases rapidly from 8 to 12 keV and more gradually from 12 to 20 keV. Overall, the relative fluorescence excitation rate decreases by approximately a factor of 5 as the electron beam voltage is increased from 8 to 20 keV. Determining the absolute excitation cross section from the relative fluorescence excitation rate requires the measurement of the spatial distribution of current across the electron beam at the fluorescence observation position as a function of accelerating potential.

To understand the structure of the electron beam fluorescence spectrum of nitric oxide, the mechanism for the preferential excitation of the v'=0 level in the A $^2\Sigma^+$ state must be determined. The absence of a strong (1,3) band indicates that primary electron excitation is not sufficient to describe the observed fluorescence spectrum. Additional excitation by secondary electrons and radiative transfer from the higher-lying C $^2\Pi$ and D $^2\Sigma^+$ electronic levels both could contribute to the relatively higher population in the v'=0 level of the A $^2\Sigma^+$ state. To determine how the secondary electron population affects the observed fluorescence spectrum, the background gas can be varied. The cascading phenomenon can be followed directly by observing direct emission from the C and D states or by laser probing of these levels with subsequent fluorescence detection. At higher pressure, consideration of selective de-excitation processes becomes important. Both electronic quenching and vibrational energy transfer must also be understood to model properly the electron-beam fluorescence spectrum of nitric oxide for higher density applications.

Acknowledgment

This work was supported by the U. S. Department of Energy.

References

1. Muntz, E. P., "The Electron Fluorescence Technique," AGARDograph Vol. 132, Dec. 1968.

2. Muntz, E. P. and Marsden, D. J., *Rarefied Gas Dynamics.*, edited by J. H.Deleeuw, Academic, New York, 1968,Vol. II pp. 1257-1286.

3. Petrie, S. L. and Komar, J. J., "Electron Beam Analysis of the Properties of Molecular Nitrogen and Nitric Oxide in the AFFGL 2-Foot Electrogasdynamics Facility," AFFDL Rept. AD-779 087, Feb. 1974.

4. Piper, L. G. and Cowles, L. M.,"Einstein Coefficients and Transition Moment Variation for NO ($A^2\Sigma - X^2\Pi$) Transition Journal of Chemical Physics Vol. 85, 1986, pp. 2419-2422.

5. Brzozowski, J., Erman, P., and Lyyra, M., Predissociation Rates and Perturbations of the A, B, B', C, D, and F States in NO Studied Using Time Resolved Spectroscopy, Physica Scripta Vol. 14, 1976, pp. 290-297.

Measurements of Freejet Densities by Laser Beam Deviation

J. C. Mombo-Caristan,* L. C. Philippe,* C. Chidiac,†
M. Y. Perrin,‡ and J. P. Martin§
Laboratoire d'Energetique Moleculaire et Macroscopic Combustion du Centre Nationale de la Recherche Scientifique, Chatenay Malabry, France
and
Ecole Centrale Paris, Chatenay Malabry, France

Abstract

We report accurate local density measurements in axisymmetric supersonic freejets using a laser beam deviation technique associated with an Abel inversion calculation. The influence of the nozzle geometry in the early stages of the expansion is investigated. Measured densities are compared with calculations obtained by the method of characteristics.

Introduction

The knowledge of the freejet local density near the nozzle exit is essential for comprehension of the subsequent physical phenomena such as rotational relaxation, condensation, etc.[1] Different methods of density probing have been developed, among which Pitot tube measurements play a major role. However, this method can introduce disturbances in the probe region. Nonintrusive investigations by laser interferometry,[2] striation slope technique,[3] spontaneous Raman diffusion,[4] laser-induced fluorescence,[5] or electron-induced fluorescence[6] can also provide the density gradients in gas flows. In this paper, we present a laser beam deviation technique[7,8] that is

---

* Ph.D. student.
† Assistant Professor.
‡ C.N.R.S. Senior Scientist.
§ E.C.P. Research Director.

simple both in conception and implementation. For axisymmetric flows, the Abel inversion[9,10] permits the reconstruction of the density field in the different regions of the supersonic freejet.

Theory and Abel Inversion for an Axisymmetric Freejet

A light beam crossing a medium where the refractive index n is not constant is deflected toward the stronger n values regions (see Fig. 1) according to the law[11]

$$\overrightarrow{du} = \vec{\nabla}_\perp \bar{n} \, ds \tag{1}$$

where $\bar{n}$ = Ln (n) and ds is the infinitesimal path length.

Then the total deflection is given by

$$\vec{\theta} = \int_{s_1}^{s_2} \frac{\overrightarrow{du}}{ds} \, ds \tag{2}$$

Since the deviations are small compared with the jet dimensions, and n is constant outside of the jet, Eq. (2) can be written as

$$\vec{\theta} = \int_{-\infty}^{+\infty} \vec{\nabla}_\perp \bar{n}(x,y,z) \, dy \tag{3}$$

The $\vec{\theta}$ components θ_x and θ_z are

$$\theta_x(x,z) = \int_{-\infty}^{+\infty} \frac{\partial \bar{n}}{\partial x}(x,y,z) \, dy \tag{4}$$

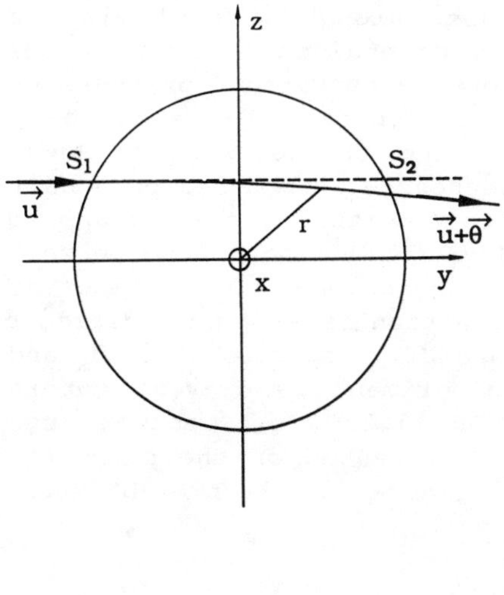

Fig. 1 Deflection light beam crossing a nonhomogeneous axisymmetric medium.

$$\theta_z(x,z) = \int_{-\infty}^{+\infty} \frac{\partial \bar{n}}{\partial z}(x,y,z)\, dy \tag{5}$$

The refractive index n is related to the density ρ by[11]

$$n - 1 = \beta \left(\rho/\rho_s\right) \tag{6}$$

where β is the Gladstone-Dale constant, which depends on the gas nature and the wavelength of the light beam,[12] and ρ_s is the gas density for the standard temperature and pressure conditions. Changing the integrals (4) and (5) into cylindrical coordinates in the (y,z) plane, and considering the axisymmetry of the freejet, Eq.(6) and the Abel inversion lead to the density and the density gradient along the jet axis:

$$\frac{\partial \rho^*}{\partial x}(x,r) = -\frac{\rho_s}{\beta \rho_0}\frac{1}{\pi}\int_r^{+\infty} \frac{\partial \theta_x}{\partial z} \frac{1}{\sqrt{z^2 - r^2}}\, dz \tag{7}$$

$$\rho^*(x,r) = \frac{\rho_s}{\beta \rho_0} \int_r^{+\infty} \frac{u}{\pi}\, du \int_u^{+\infty} \frac{\partial}{\partial z}\left(\frac{\theta_z}{z}\right) \frac{1}{\sqrt{z^2 - u^2}}\, dz \tag{8}$$

with

$$\rho^*(x,r) = \left(\rho(x,r) - \rho_1\right)/\rho_0 \tag{9}$$

where ρ_0 is the stagnation gas density and ρ_1 is the background gas density.

Experimental Setup

The supersonic freejet is produced by the expansion of pure nitrogen or pure argon through a nozzle into a vacuum chamber. The vacuum is maintained by Roots pumps (Edwards 2600), which condition the background pressure P_1 for each stagnation pressure P_0. Typically, P_1 = 1.5 bar, which sets P_0/P_1 = 750. The vacuum chamber and the He-Ne laser beam are kept fixed. The nozzle is mounted on a pair of perpendicular Micro-Controle translational stepping motors (1 μm accuracy). A 200-mm focal length lens between the laser and the supersonic freejet leads to a beam waist of 200 μm in the jet axis. A position-sensing photodetector delivers up two voltages, U_x and U_z, proportional to θ_x and θ_z respectively. The whole experiment is driven (motors control, acquisition) by a PDP 11/23 microcomputer; the Abel inversion code is also implemented on the computer. The three nozzle geometries considered in the present study

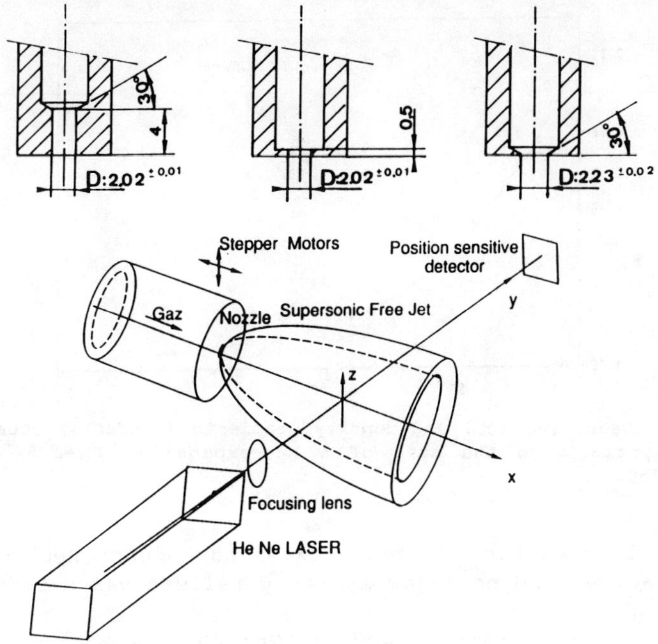

Fig. 2 Experimental setup and nozzle geometries. Lenghts are expressed in mm.

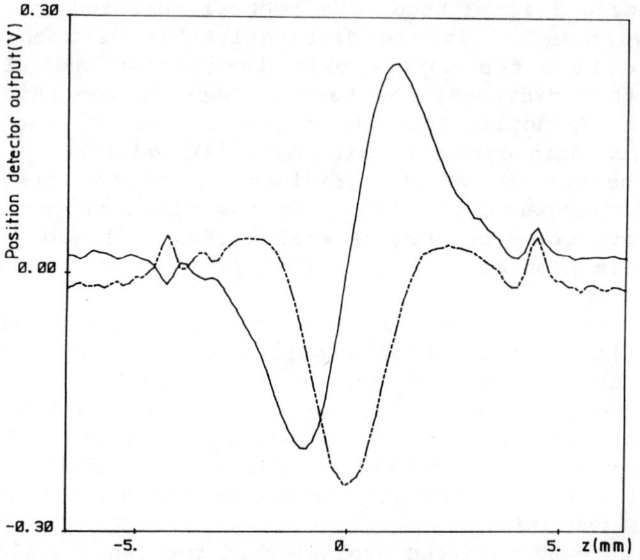

Fig. 3 Axial $U_x(z)$ (-- —) and radial $U_z(z)$ (——) deviograms in a N_2 supersonic freejet out of nozzle 2. P_0=1.5 bar; P_1=2.0 mbar; x=2.0 mm.

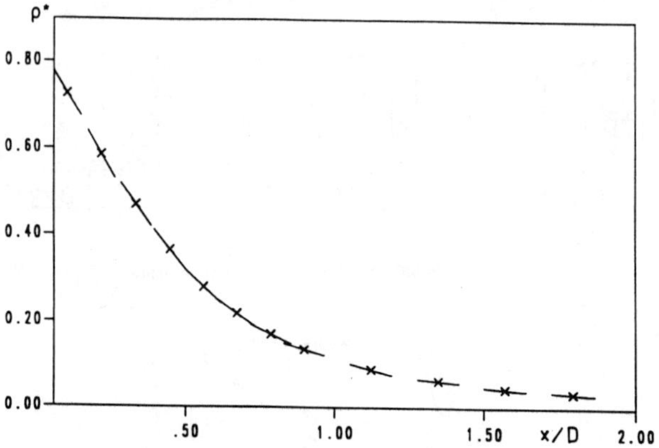

Fig. 4 Densities (X) and density gradients (tangents) obtained with nozzle 2 on the axis of a N_2 expansion. P_0=1.5 bar; P_0/P_1=750.

are detailed on Fig. 2. The nozzles have been checked with a microscope, and no major symmetry failure was detected.

Results and Discussion

On Fig. 3, an example of raw radial and axial deviograms in a nitrogen supersonic freejet exhausting from nozzle 2 is plotted. The central core and the barrel shock surrounding it are distinguishable as soon as the distance from the nozzle exit is greater than 1.5 mm. Beyond this distance, the barrel shock is too thin to be detected. We notice that the character even or odd of the deviogram is in agreement with Eqs. (4) and (5).

Density and density gradient along the axis can be deduced independently from the experimental radial and axial deviograms by Abel inversion [Eqs. (7) and (8)]. An example is plotted on Fig. 4 that proves the coherence of the results.

On Fig. 5, the density profiles in a nitrogen supersonic freejet out of the different nozzles for x = 2 mm are shown. The influence of the nozzle geometry is illustrated. Nozzles 0 and 1 generate wide freejets, with a sharp density peak; however, this character is less marked for freejet out of nozzle 2. The lateral shock and the boundary layer are responsible for of the density increase far from the axis.

Figure 6 shows the evolution of the density along the jet axis. Close to the nozzle exit, the ρ^* evolution depends strongly on the nozzle geometry. In particular, nozzle 1 gives axial densities much lower than those

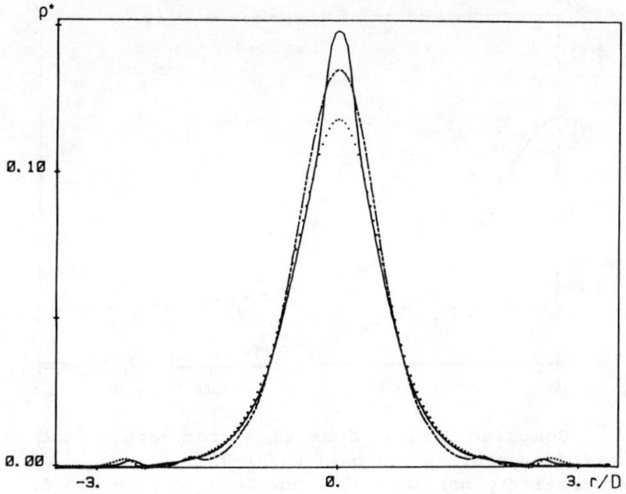

Fig. 5 Influence of the geometry of the nozzle on the density profiles of a N_2 supersonic freejet. P_0=1.5 bar; P_0/P_1=750; x=2.0 mm. (------) Nozzle 0; (——) nozzle 1; (-·-·-) nozzle 2.

issuing from the other nozzles. On the other hand, far from the nozzle, these differences vanish, allowing us to compare measurements with densities predicted by the Ashkenas and Sherman[13] formula. Because of viscosity effects in the nozzle, an effective diameter D^*, lower than the geometrical diameter D, must be introduced. A summary of the D^*/D ratios that give the best agreement with the Ashkenas and Sherman formula for the nozzles 0 and 2 is given in Table 1. No D^*/D ratio is given for nozzle 1 because of the anomalous behavior mentionned earlier. A numerical code using the method of characteristics as described in Ref. 14 has been developed[15] in order to complete our study for small x/D values. The results obtained with the sonic surface in the vertical plane of the nozzle exit are plotted on Fig. 6. The agreement with the measurements obtained through nozzle 0 is quite good.

Table 1 Values of D^*/D for argon and nitrogen with P_0=1.5 bar and T_0=300 K.

Nozzle \ Gas	N2	Ar
0	1	-
2	0.95	0.98

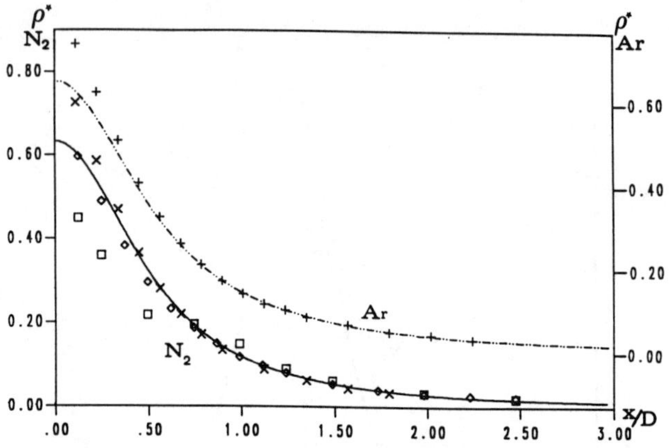

Fig. 6 Compared axial densities for argon and nitrogen supersonic freejets. P_0=1.5 bar; P_0/P_1=750.
N_2 experiments: ◇, nozzle 0; □, nozzle 1; ×, nozzle 2.
N_2 theoretical: (——), method of characteristics.
Ar experiments: ,nozzle 2.
Ar theoretical: (--——), method of characteristics.

Nozzle 2 densities are much greater at short distances. Our explanation follows Murphy and Miller's[1] conclusion. The M = 1 surface might not be flat but thrown outward from the nozzle. Assuming an isentropic expansion, our density measurements lead to M = 1 on the axis at x/D = 0.2. Our numerical code is not presently able to handle such curved sonic surface shapes.

In the course of this work, we have increased the stagnation pressure, keeping P_0/P_1 approximatively constant. Spatial stationary oscillations appear on the deviograms for $P_0 > 2$ atm, preventing analysis of the results by means of the Abel inversion. Nevertheless, a possible explanation of this phenomenon is the presence of acoustic waves in the chamber or the destruction of the jet symmetry with an increasing P_0.

Conclusion

We have determined accurate densities and density gradients in axisymmetric supersonic freejets near the nozzle exit using the laser deviation technique associated with Abel inversion. Three different nozzle have been investigated. At short distances they yield different results. Beyond 1.5-diam. distance, these differences become negligible. The experimental method has also permitted to point out an anomalous behavior of one of the nozzles. The oscillation phenomenon limits the stagnation pressure to moderate values.

References

[1]Murphy, H. R. and Miller, D. R., "Effects of Nozzle Geometry on Kinetics in Free-Jets Expansions," Journal of Physical Chemistry, Vol.88, 1984, pp. 4474-4478.

[2]Kobayashi, H., Nakagawa, T. and Nishida, M., "Density Measurements in Free-Jets by Laser Interferometry," Proceedings of the 14th Rarefied Gas Dynamics," edited by H. Oguchi, Univ. of Tokyo Press, Tokyo, Japan, 1984, Vol. 1, pp. 501-508.

[3]Schinn, G. W. and Measures, R. M., "STROPE: A New Species Density Gradient Measurement Technique," Applied Optics, Vol. 23, April 1984, pp. 1258-1266.

[4]Godfried, H. P., Silvera, I. F. and Van Straaten, J., "Rotational Temperatures and Densities in H_2 and D_2 Freejet," Proceedings of the 12th Rarefied Gas Dynamics," edited by S.S. Fisher, AIAA, New York, 1980, Vol. 2, pp. 772-784.

[5]McDaniel, J. C., Baganoff, D. and Byer, R. L., "Density Measurement in Compressible Flow Using Off-resonnant Laser-induced Fluorescence," Physics of Fluids, Vol. 25, July 1982, pp. 1105-1107.

[6]Lengrand, J. C., Allègre, J. and Raffin, M., "Experimental Investigation of Underexpanded Exhaust Plumes," AIAA Journal, Vol. 14, May 1976, pp. 692-694.

[7]Simpson, C. J. S. M., Chandler, T. R. D. and Strawson, A. C., "Vibrational Relaxation in CO_2 and CO_2-Ar Mixtures Studied Using a Shock Tube and a Laser Schlieren Technique," Journal of Chemical Physics, Vol. 51, Sept 1969, pp. 2214-2219.

[8]Jackson, W. B., Amer, N. M., Boccara, A. C. and Fournier, D., "Photothermal Deflection Spectroscopy and Detection," Applied Optics, Vol. 20., April 1981, pp. 1333-1344.

[9]Azinhera, J., "Caractéristiques locales dans des écoulements réactifs. Détermination des masses volumiques et populations rotationnelles par des techniques Laser 1986-1," Ph.D. Thesis, Ecole Centrale des Arts et Manufactures, Chatenay-Malabry, France, 1986.

[10]Noll, R., Haas, Werki, B., C. R. and Herziger, G., "Computer Simulation of Schlieren Images of Rotationally Symmetric Plasma Systems: a Simple Method," Applied Optics, Vol. 25, March 1986, pp. 769-774.

[11]Born, M. and Wolf, E., Principles of Optics, Pergammon, Oxford, U.K., 1975.

[12]L'Air Liquide, Encyclopédie des gaz, Elsevier, Amsterdam, 1976.

[13]Ashkenas, H. and Sherman, F. S., "The Structure and Utilization of Supersonic Free Jets in Low Density Wind Tunnels," Proceedings of the 5th Rarefied Gas Dynamics," edited by J. H. de Leeuw, Academic, New York, 1965, Vol. 2, pp. 84-105.

[14]Zucrow, M. J. and Hoffman, J. D., Gas Dynamics, Wiley, New York, 1976, Vol. 2.

[15]Goffe, D., "Modélisation numérique par la méthode des caractéristiques, et étude expérimentale de jets supersoniques libres," Ph.D. Thesis, Ecole Centrale des Arts et Manufactures, Chatenay-Malabry, France, 1984.

Turbulence Measurement of a Low-Density Supersonic Jet with a Laser-Induced Fluorescence Method

M. Masuda,* H. Nakamuta,† Y. Matsumoto,†
K. Matsuo,‡ and M. Akazaki‡
Kyushu University, Fukuoka, Japan

Abstract

The transition from laminar to turbulent flow in a low-density underexpanded supersonic jet may become important for the steady operation of neutral beam injectors for the fusion plasma heating. Fluctuation of flow parameters in a jet is generated in the shear layer of the jet boundary as the backpressure is increased. To investigate fluctuation, it seems desirable to perform experiments with nonintrusive diagnostic methods. In the present experiments, the laser-induced fluorescence method was used to measure the fluctuation of a low-density axisymmetric argon jet with high spatial and temporal resolution. With this technique, the critical Reynolds number above which fluctuation was generated in the shear layer was obtained, and the fluctuation distribution of the fluorescence intensity in the jet was also clarified.

Nomenclature

I_f = time-averaged intensity of laser-induced fluorescence, arbitrary unit
p_o = stagnation pressure, Pa
p_w = backpressure of the orifice, Pa
Re = Reynolds number based on the orifice diameter and the orifice exit condition

---

*Associate Professor, Department of Energy Conversion Engineering.
†Graduate Student, Department of Energy Conversion Engineering.
‡Professor, Department of Energy Conversion Engineering.

r = radial distance from jet centerline, mm
r_e = orifice radius, mm
z = axial distance from orifice exit, mm
z_M = Mach disk location, mm
Δ_{rms} = root-mean-square of the fluctuation amplitude of the laser-induced fluorescence intensity divided by its time-averaged value, %

Introduction

The transition from laminar to turbulent flow of a low-density underexpanded supersonic jet may become important for the steady operation of neutral beam injectors for the fusion plasma heating.[1,2] Also, the turbulence of the jet has important relations with the generation of the jet noise. When the backpressure of an orifice is kept to a high vacuum, the entire flowfield remains laminar. As the backpressure is increased, the fluctuation of flow parameters is generated in the shear layer of the jet boundary, and this fluctuation has been suggested to cause a jet noise.[3] Fluctuation or instability of a freejet in moderate as well as high Reynolds-number range was studied both experimentally[4]

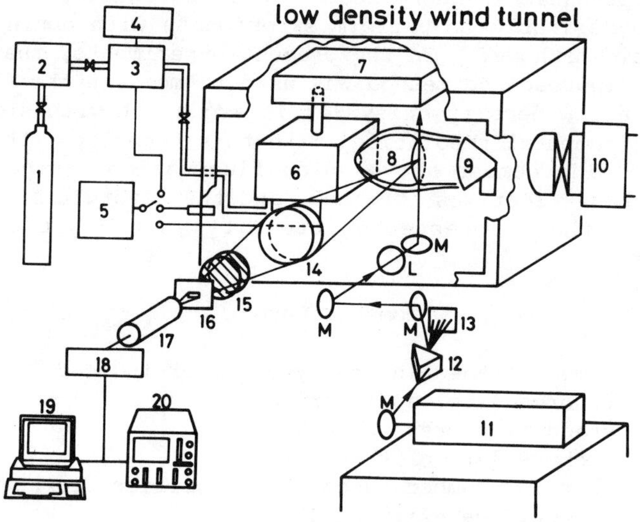

Fig. 1 Low-density wind tunnel with laser-induced fluorescence diagnostic system. 1:Gas reservoir; 2:rotameter; 3:iodine cell; 4:temperature controller; 5:Baratron gage; 6:plenum chamber; 7:traversing device; 8:jet; 9:LN_2 trap; 10:vacuum pumps; 11:A^+ laser; 12:prism; 13:beam dump; 14:Fresnel lens; 15:filter; 16:slit; 17:photomultiplier; 18:amplifier; 19:microcomputer; 20:oscilloscope.

and theoretically,[5] and the characteristics, including the growth rate of instability waves in a jet, were obtained. Also, in the fully turbulent regime, the numerical analysis based on the Navier-Stokes model was developed,[6] and, recently, the density fluctuation in a jet was measured with the Rayleigh scattering technique.[7] However, the information on the transition from laminar to turbulent flow in a jet is still limited.

To investigate fluctuation, it is desirable to perform experiments with nonintrusive diagnostic methods. Fluctuation is expected to occur at the ratio of the stagnation to backpressure of about 80 and a backpressure of about 50 Pa. In this pressure regime, however, the gas density is too high to be measured with the electron beam excitation method and also too low with the conventional optical techniques such as the schlieren, shadow graph, and interferometric methods. In the present experiments, therefore, the laser-induced fluorescence (LIF) method,[8,9] which had high spatial and temporal resolution, was used to measure the fluctuation of the low-density axisymmetric argon jet. With this technique, the critical Reynolds number above which fluctuation was generated in the shear layer was obtained, and the fluctuation distribution of the LIF intensity in the jet was also clarified.

Experimental Equipment

Figure 1 shows a schematic diagram of the low-density supersonic wind tunnel[1] and the LIF diagnostic system. To apply the LIF method to rarefied gas flows, it is necessary to add to jet a small amount of seed material, which is easily excited with the laser and emits fluorescence. Seed material used in the present experiments is iodine. Argon in a high-pressure reservoir flows through the rotameter into the iodine cell. This cell is kept to the temperature of 300 K and contains crystalline iodine, which is sublimated into the argon gas as seed. The amount of seed is determined with the saturated vapor pressure of iodine and is about 4 % by weight. The gas with seed molecules then flows into the stagnation chamber attached to the traversing device and issues into the wind tunnel through the orifice with 4 mm in diameter as an underexpanded supersonic jet. Iodine in the jet is recovered with the copper frustum cooled by liquid nitrogen and set in the downstream of the jet. The gas is then evacuated with vacuum pumps.

The laser used in the present experiments is an argon ion laser (Spectra-Physics model 168-07) with the maximum power output of 3 W. Since this laser is operated without an etalon (wavelength selector) in the cavity, the laser beam consists of various wavelengths. Therefore, the line with a wavelength of 514.5 nm is selected with the prism, because the cross section for the combined electronic, rotational, and vibrational excitation of an iodine molecule is expected to be great with this wavelength.[9] The laser beam is focused onto the jet centerline with lenses and mirrors. The diameter of the laser beam waist is 0.5 mm. The LIF is collected with two Fresnel lenses, each with a diameter of 256 mm and a focal length of 318 mm, to maximize the solid angle of the light collection. The fluorescence is then transmitted through two optical filters with the lower cutoff wavelength of 530 and 540, nm, respectively, to block the background stray light caused with the reflection and scattering of the laser light on the wind-tunnel window and walls. The fluorescence intensity is measured with the photomultiplier through a slit and analyzed with a microcomputer. The spatial resolution is determined both with the diameter of the laser beam waist and with the slit width, and it is estimated to be 0.5×0.5 mm; the temporal resolution depends on the speed of the analog-to-digital conversion of the microcomputer and is 20 μs.

The wind tunnel and the collection optics are placed on the rubber, which absorbs the vibration caused with vacuum pumps.

Results

Figures 2a and 2b show the axial and radial distributions of the time-averaged LIF intensity in the jet. It was pointed out that the LIF intensity was not directly proportional to the gas density and was also dependent on the gas temperature.[10] This figure, therefore, is not regarded as being indicative of the density distribution in the freejet. Nevertheless, the typical features of the underexpanded supersonic freejet, such as the rapid decrease in density along the jet centerline or the structure of the barrel shock wave, are clearly seen.

Figures 3a-3d show the radial distributions of the time-averaged and the fluctuation LIF intensity for various z locations. As described previously, the fluctuation of the LIF intensity is regarded as being composed of both density and temperature fluctuation. As shown in Fig. 3, fluctuation is not large for small z; especially, almost no fluctuation is observed in the isentropic core region of the jet. However, fluctuation is seen to be generated in the shear layer of the jet boundary, and its magnitude increases gradually with the increase in z. Under this jet condition, the Mach

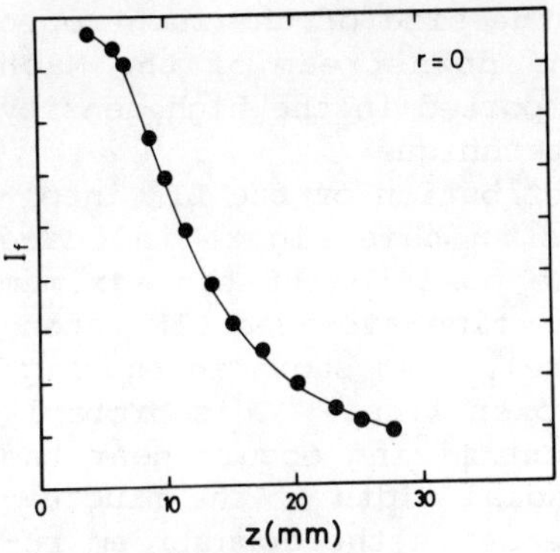

a) axial distribution,

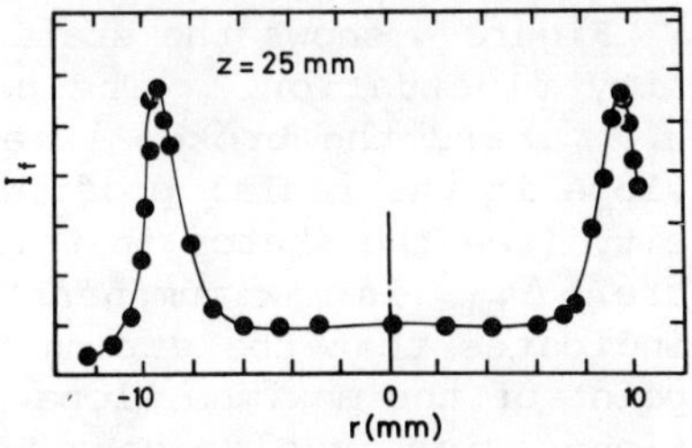

b) radial distribution.

Fig. 2 Distribution of time-averaged LIF intensity (p_o/p_w=96, p_o=7330 Pa).

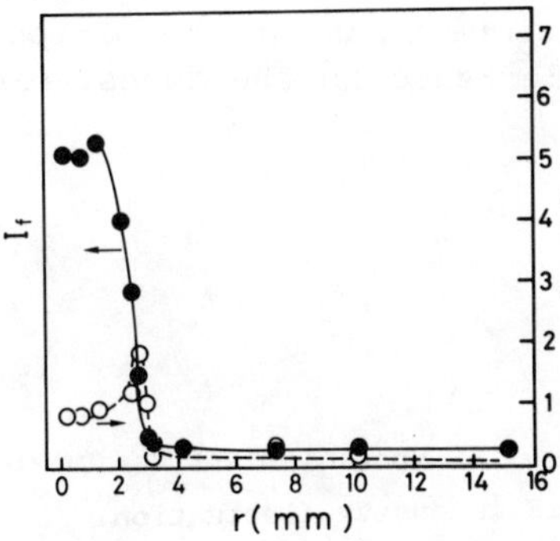

a) z=1.5 mm,

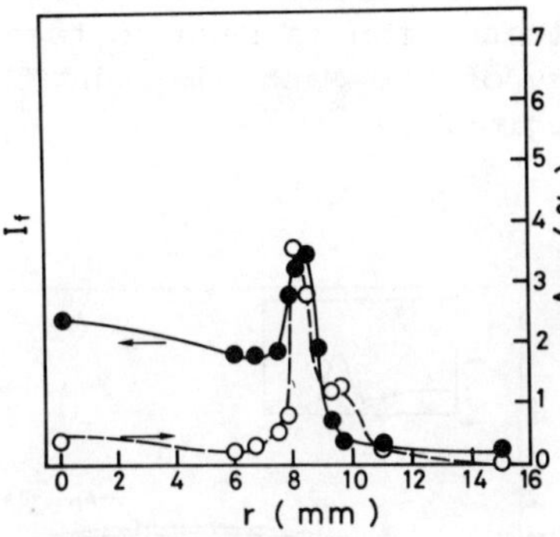

b) z=15 mm,

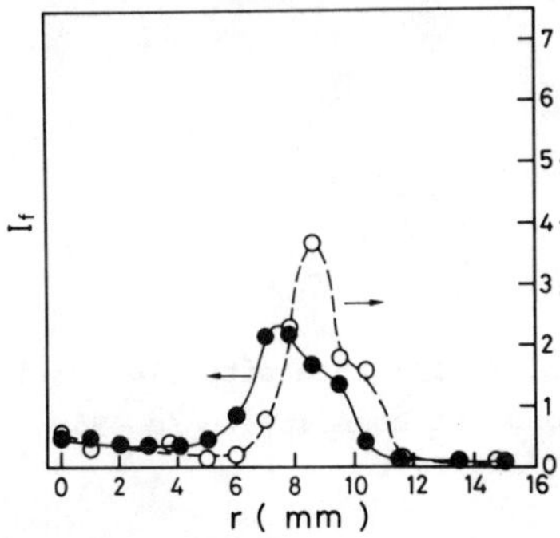

c) z=27 mm,

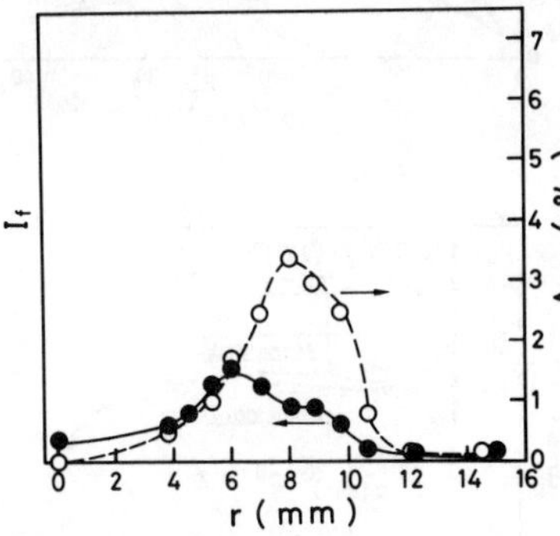

d) z=35 mm.

Fig. 3 Radial distributions of time-averaged and fluctuation LIF intensity (p_o/p_w=96, p_o=7330 Pa):

disk is located at $z \simeq 26$ mm, and the size of the isentropic core is shown to decrease in the downstream of the Mach disk. Similar results were also reported in the high density jet with the Rayleigh scattering technique.[7]

Figure 4 shows the spatial distribution of the LIF intensity fluctuation. The contours in this figure indicate Δ_{rms}, and the broken line is the position of the maximum slope in the radial profile of the time-averaged LIF intensity (see the sketch in this figure). As shown in the figure, Δ_{rms} is maximum near the broken line. This probably indicates that the strong turbulent mixing occurs near the point of the maximum slope. The radial width of the fluctuation region is also seen to increase in the downstream region.

Figure 5 shows Δ_{rms} in the jet boundary as a function of the point number indicated in the figure. These points correspond to the locations where the radial distribution of Δ_{rms} takes the maximum value. This figure clearly shows that the fluctuation grows along the jet boundary. Also, the fluctuation is seen to take the maximum value at the periphery of the Mach disk and then to decrease in the downstream region.

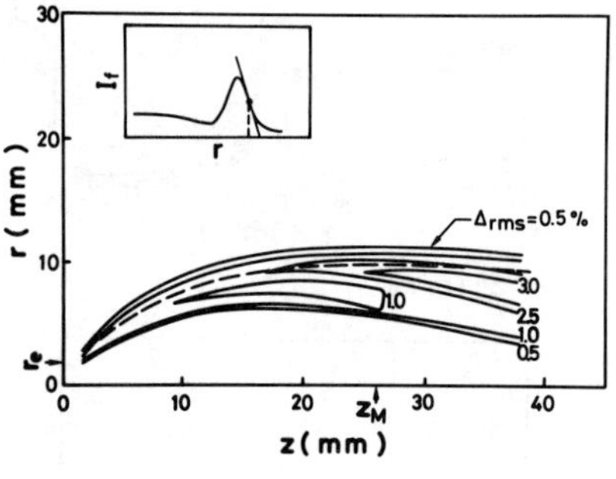

Fig. 4 Spatial distribution of LIF intensity fluctuation (p_o/p_w=96, p_o=7330 Pa).

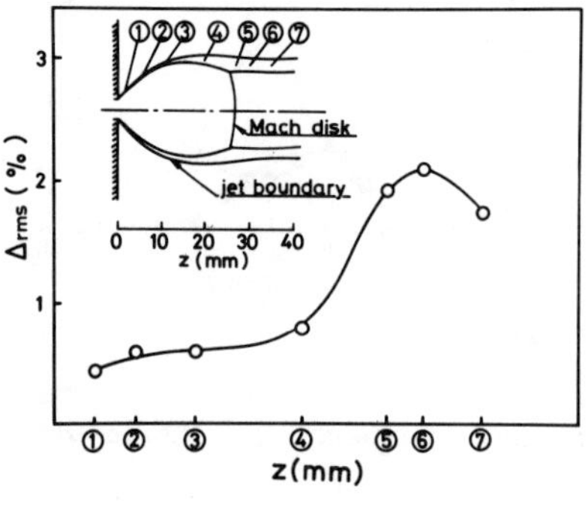

Fig. 5 LIF intensity fluctuation along jet boundary (p_o/p_w=96, p_o=7330 Pa).

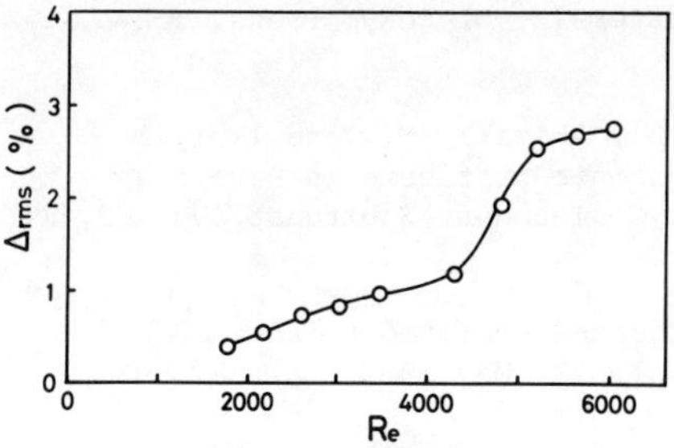

Fig. 6 LIF intensity fluctuation vs Reynolds number (p_o/p_w=83).

Figure 6 shows Δ_{rms} as a function of the Reynolds number based on the flow condition at the orifice exit. Fluctuation is taken at the point where this takes the maximum value through the entire flowfield (point 6 in Fig. 5). Because of the definition of the Reynolds number, this does not include the effect of the backpressure and may not be an appropriate parameter for correlating the fluctuation of jets. Even so, it can be seen that fluctuation suddenly increases near Re=4000, and this phenomenon seems similar to the usual transition from laminar to turbulent flow.

Conclusions

To investigate the transition from laminar to turbulent flow in a supersonic underexpanded jet, the LIF method was employed. This method was nonintrusive and had the advantage of detecting the critical Reynolds number above which fluctuation was generated in the shear layer of the jet boundary. The present experiments showed the detailed distribution of the LIF intensity fluctuation in the jet, and this fluctuation was shown to be generated in the jet boundary for the jet with the backpressure above 50 Pa and the ratio of the stagnation to backpressure of over 80. The critical Reynolds number above which fluctuation was generated in the shear layer was about 4000.

Acknowledgments

The authors are grateful to H. Yano, T. Shimakawa, and S. Kinoshita for their assistance in the experimental work. This work was partly supported by a Grant in Aid for Scientific Research, Ministry of Education, Science and Culture, Japan.

References

[1]Masuda, M., Yano, H., Akazaki, M., and Ikui, T., "Structures of Low Density Supersonic Jets for the Neutralization of Fast Ion

Beams," Rarefied Gas Dynamics, University of Tokyo Press, Tokyo, Vol. 2, 1984, pp. 1031-1038.

[2]Masuda, M., Nakamuta, H., Yano, H., Akazaki, M., and Ikui, T., "Mass Recovery Efficiency of Supersonic Diffusers in Rarefied Flows," Rarefied Gas Dynamics, B. G. Teubner, Stuttgart, Vol. 2, 1986, pp. 585-594.

[3]Pao, S. P. and Seiner, J. M., "Shock-Associated Noise in Supersonic Jets," AIAA Journal, Vol. 21, May 1983, pp. 687-693.

[4]Morrison, G. L. and McLaughlin, D. K., "Instability Process in Low Reynolds Number Supersonic Jets," AIAA Journal, Vol. 18, July 1980, pp. 793-800.

[5]Tam, C. K. W. and Burton, D. E., "Sound Generated by Instability Waves of Supersonic Flows," Journal of Fluid Mechanics, Vol. 138, Jan. 1984, pp. 249-271.

[6]Dash, S. M., Wolf, D. E., and Seiner, J. M., "Analysis of Turbulent Underexpanded Jets, Part I: Parabolized Navier-Stokes Model, SCIPVIS," AIAA Journal, Vol. 23, April 1985, pp. 505-514.

[7]Novopashin, S. A., Perepyolkin, A. L., and Yarygin, V. N., "The Use of Pulse Lasers for Flow Visualization and Local Density Measurements in Free Jets," Rarefied Gas Dynamics, B. G. Teubner, Stuttgart, Vol. 2, 1986, pp. 623-632.

[8]Teshima, K. and Nakatsuji, H., "Structures of Freejets from Slit Orifices," Rarefied Gas Dynamics, B. G. Teubner, Stuttgart, Vol. 2, 1986, pp. 595-604.

[9]McDaniel, J. C., Jr., "Investigation of Laser-Induced Iodine Fluorescence for the Measurement of Density in Compressible Flows," Stanford Univ. Department of Aeronautics and Astronautics Report No. 532, Stanford Univ., Stanford, CA, 1982.

[10]Cheng, S., Zimmermann, M., and Miles, R. B., "Separation of Time-Averaged Turbulence Components by Laser-Induced Fluorescence," Physics of Fluids, Vol. 26, April 1983, pp. 874-877.

Measurement of Aerodynamic Heat Rates by Infrared Thermographic Technique at Rarefied Flow Conditions

Jean Allègre,* X. Hériard Dubreuilh,* and M. Raffin*
Société d'Etudes et de Services pour Souffleries et Installations Aérothermodynamiques (SESSIA), Meudon, France

Abstract

The use of commercially available infrared (IR) scanning system for measurement of convective heat rates is described. Preliminary measurements are performed on a flat plate located within the test section of a low-density continuous-flow facility. Flow conditions are characterized by a Mach number of 20.2 and a Reynolds number of 2280. Convective heat rates as low as 0.5 kW/m^2 are measured at the model wall. Basic system, data reduction, and result comparisons are briefly presented. The data accuracy is discussed and limitation of IR scanning system is pointed out for large optical incident angles of the scanning camera axis over the investigated surface.

Introduction

For conventional wind tunnels, characterized by rather high flow densities, several instruments and techniques are available for aerodynamic heat rate measurements.[1] One can mention thin-film gages, coaxial surface thermocouples, thin-skin techniques, infrared thermography, phase-change paint technique, liquid crystals, and thermographic phosphor technique.

Under rarefied flow conditions, heat rate measurements require particular care, and the choice between available techniques is limited. In the low-density wind tunnel SR3 of Meudon, heat rate measurements have to be performed under hypersonic conditions. Stagnation pressures are ranging from a few tenths of a bar up to 120 bars, and stagnation

* Research Engineer, Laboratoire d'Aerothermique.

temperatures may reach 1600°K. Corresponding Reynolds numbers (Re) based on the model length range between 10^3 and 10^5. The thin-skin technique and phase-change paint technique were used satisfactorily at the highest values of Re. The thin-skin technique is considered the most accurate and reliable method, but limitations arise when analyses have to be realized on complicated shape models. For such configurations, the phase-change paint technique or liquid crystals allows a continuous thermal mapping over the coated surface to be obtained but becomes quite inaccurate for convective heat rates lower than 6 kW/m^2. For such considerations, the infrared (IR) thermographic technique was implemented. This method is more sensitive than the phase-change paint technique and may provide thermal mapping over complex models.

Infrared Scanning System

Commercially available IR systems are of two basic types : the scanning mirror infrared camera with single detector or detectors arrays, and the vidicon-type infrared camera with electronic scanning and pyroelectric detectors.

Systems under development use matrix of detectors that will reduce the noise and problems associated with the mechanical scanning. Useful informations on basic characteristic and potentialities of thermographic systems may be found in recent papers.[2,3]

Among the existing systems, the one which appears to be the most widely used in wind-tunnel testing is the AGA Thermovision System with horizontally and vertically rotating mirrors and a cryogenically cooled detector (Fig. 1).

For our test conditions, the chosen IR scanning system is an AGA Thermovision 782 SWB camera (Fig. 2). It is associated with an on-line IBM PC-AT microcomputer and a Sharp JX720 color printer for real-time color-coded analysis of surface temperature distribution and hardcopy of instant thermograms. The main components of the IR data system are presented in Fig. 3.

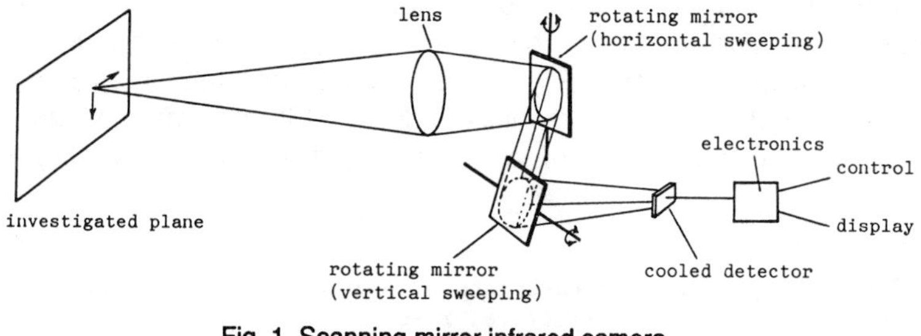

Fig. 1 Scanning mirror infrared camera.

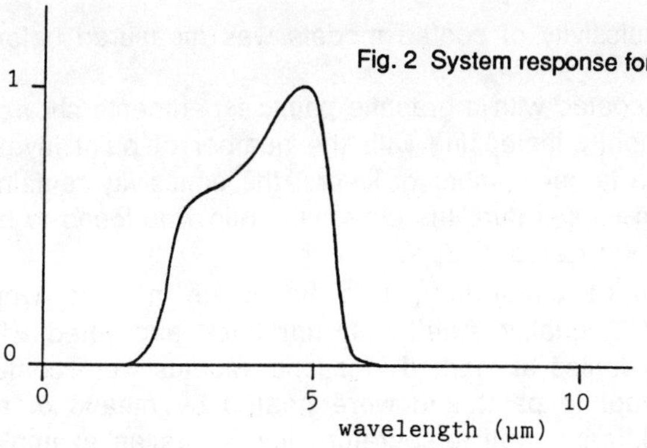

Fig. 2 System response for the AGA 782 SWB scanner.

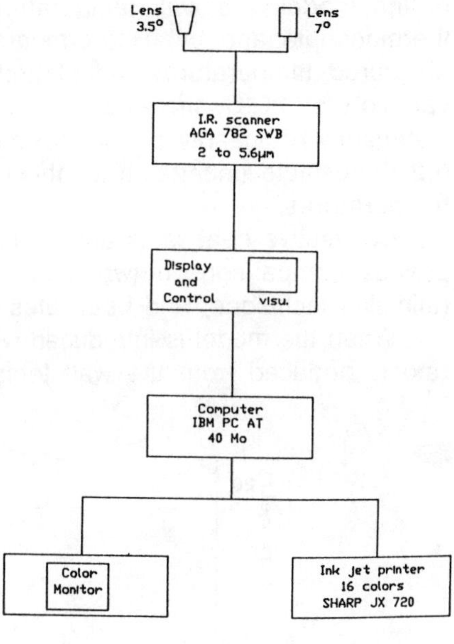

Fig. 3 Infrared data system diagram.

The CATS (Computer Aided Thermography Software) program is implemented on the IBM PC-AT Computer. The various menus are used for data initialization regarding color tables and measurement parameters, for handling thermal images with storage onto fixed disk or retrieval of images from it, and for analysis of thermal images.

Infrared System Calibration and Measurement Procedure

The main difficulty of temperature measurement by IR thermography is quantifying with accuracy the model surface emissivity. For our present experiments, emissivity of coated models was measured before the wind-tunnel tests.

For a steel surface coated with a graphite paint, experiments showed that the emissivity is slightly increasing with the number of paint layers up to three layers. For a larger number of layers, the emissivity remains almost constant. At room temperature the emissivity ratio was found to be close to 0.87 for most of the tested models.

For scanning camera calibration, thin flat plate models were manufactured in ARMCO quality steel, 0.4 mm thick and fitted with chromel alumel thermocouples at marked locations. Models were coated with three layers of graphite paint and were heated by means of an electrical air dryer at different wall temperature levels. As an example, starting from an initial temperature value T_0 measured at a marked point of the model, the wall temperature increase was measured both by thermocouple and by IR thermography when the model was heated. Measured temperatures are plotted in Fig. 4. For a wall temperature variation of 20°C, the error is less than 0.4°C, corresponding to an uncertainty of linearity of 2% maximum. This takes also into account the 0.2°C absolute uncertainty of the IR system in the range of measured temperatures.

Convective heat rates are measured on thin skin models in order to provide comparisons between heat rates measured by thermocouples (thin-skin technique) and heat rates measured by IR thermography.

When the model is introduced within the heated flow, convective heat rate is deduced from the wall temperature increase vs time. The wall

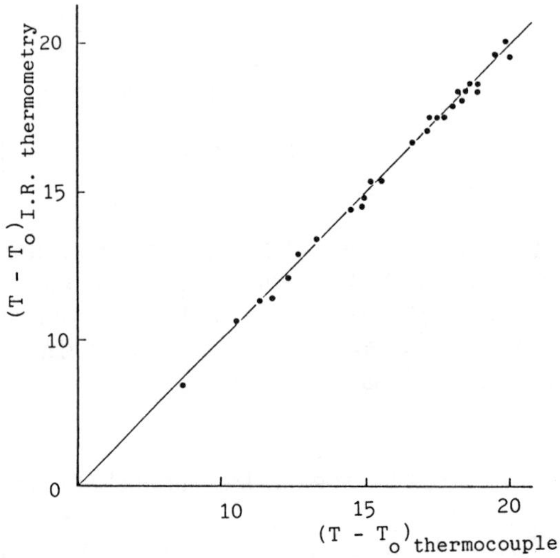

Fig. 4 Wall temperature variation measured by thermocouple and by IR thermography.

temperature is measured either by thermocouples for marked points or by IR thermography for the whole investigated model surface.

The principle of the measurement is identical for IR thermography and for the thin-skin technique. At the time of heat flux measurement, successive thermograms are recorded with a frequency up to six thermograms per second during 3 -15 s. The analysis of thermograms makes it possible to quantify the heat rate value at each point of the investigated wall surface for a known thin-skin material.

The thin-skin technique for heat rate measurement in wind tunnels has been used for many years and remains one of the most accurate, reliable methods available. The reduction of thin-skin temperatures T to heat rate quantity q involves the calorimetric heat balance for the thin-skin, which can be written as

$$q = \rho\, c\, b \frac{dT}{dt}$$

where ρ and c are the model material density and specific heat, respectively and b is the model skin thickness.

Thermal radiation and heat conduction effects on the thin-skin element are neglected in the preceding relationship, and the skin temperature response is assumed to be due to convective heating only. In our test conditions in a wind tunnel, wall temperature variations during the heat rate measurements are quite limited and do not exceed 5°C. In such conditions, thermal radiation and heat conduction effects caused by temperature gradients along the skin are indeed small and are appropriately neglected.

Experimental Set-up and Model

The SR3 facility is an open-jet wind tunnel (Fig. 5). The test section centerline is 1.5 m distant from the scanning camera. According to the model size, the scanning camera is mounted with two different lens. One 7-deg lens is used for investigation of models up to 15 cm long. Another 3.5-deg lens is used to perform measurements over smaller model areas. A cylindrical test chamber of 2 m diam surrounds the open-jet section, and the scanning camera optical path goes through a 0.15-m-diam disk transparent to the IR and transparent in the visible region of the spectrum. This disk is mounted on the chamber wall and included to the IR system when the scanning camera is calibrated.

Experiments in a wind tunnel are performed on a thin-skin model located within the low-density hypersonic flow of the SR3 facility. The model described in Fig. 6 is a flat plate 60 mm wide, 80 mm long, and 10 mm thick. The truncated leading edge is 2 mm thick. Five thermocouples are inserted along the flat plate at distances of 20, 30, 40, 50, and 60

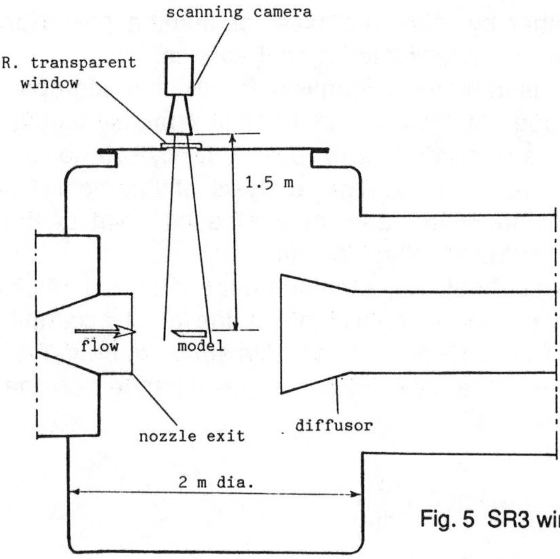

Fig. 5 SR3 wind tunnel, mounting arrangement.

mm from the leading edge. Thermocouples are inserted in a flat sheet 0.4 mm thick made of ARMCO steel material.

A movable sting support actuated by a pneumatic jack allows the injection of the model within the flow, which has been previously started and stabilized. A mechanical switch mounted on the movable sting support activates the begining of thermograms recording as soon as the model is injected through the test section. Maximum speed of data recording corresponds to six thermograms per second. During the same time period, temperatures delivered by the five thermocouples are entered to a graphic recording device.

Flow Conditions and Preliminary Heat rate Measurements

The hypersonic nitrogen flow is characterized by a Mach number of 20.2 and a Reynolds number of 2280 based on the flat plate length. Stagnation pressure and stagnation temperature are 3.5 bars and 1100°K, respectively.

Before the heat rate measurements, the model wall temperature is close to 290°K (room temperature).

For the present tests two angles of attack of 0 and 21.6 deg are considered.

The flat plate is injected through the flow and wall temperature variations vs time are measured by IR thermography. For the five marked points on the plate corresponding to the thermocouple locations, recorded wall temperatures are plotted in Fig. 7 and 8 for respective flat plate

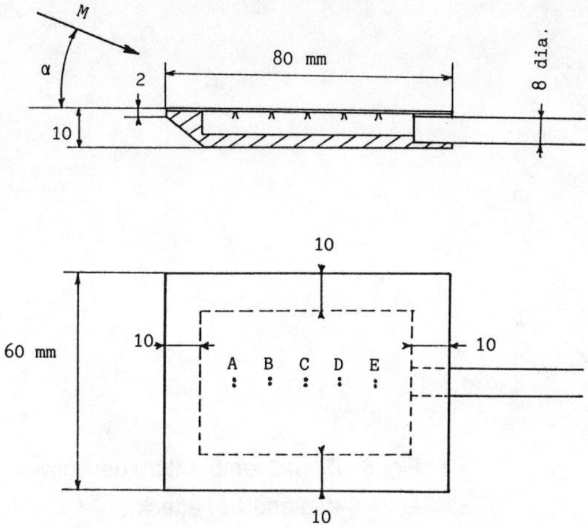

Fig. 6 Flat plate model.

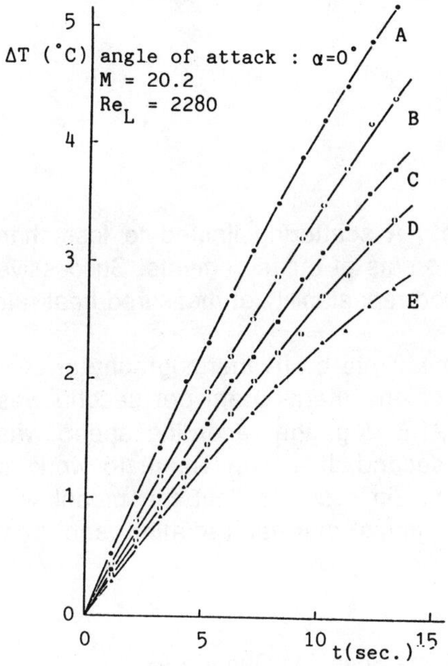

Fig. 7 IR wall temperature derivatives at 0-deg angle of attack.

angles of attack of 0 and 21.6 deg. Simultaneously, wall temperature variations on thermocouples are also recorded. From measured temperature derivatives and knowing the skin material thermal characteristics, heat rates are experimentally deduced from IR thermography and from thermocouple measurements.

Convective heat rate values are plotted in Fig. 9. A reasonable cross-checking appears between heat rate values measured by IR

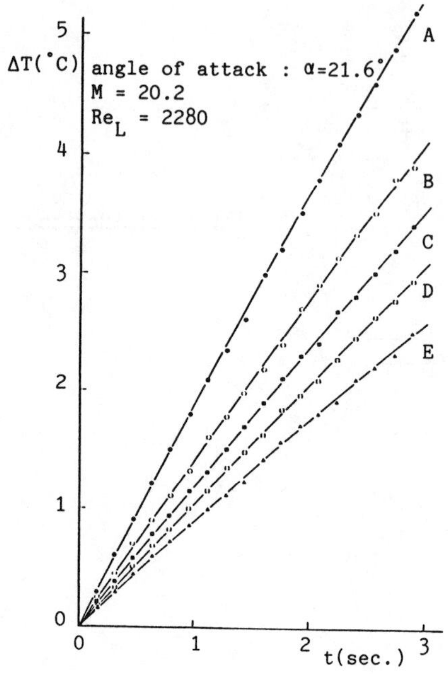

Fig. 8 IR wall temperature derivatives at 21.6-deg angle of attack.

thermography and by thermocouples. A scattering limited to less than 10% may be noticed between the two sets of measurements. Successive measurements have also shown a good repeatability of measured heat rate data for the same flow conditions.

When recording the convective heat rate by IR thermography at zero angle of attack, a speed recording of one thermogram per second was chosen. At an angle of attack of 21.6 deg, the recording speed was increased to six thermograms per second. It is convenient to work at recording speeds as high as possible in order to limit the model wall heating during the test and to minimize thermal radiation and heat conduction through the wall.

Effect of Tilt Angle Between the Scanning Camera Axis and The Investigated Model Surface

An experimental device was built to measure apparent local wall temperatures and apparent local wall emissivities when the axis of the scanning camera was tilted over the model surface.

A cylindrical container, mounted vertically, 6.7 cm in diameter and 10.5 cm high, is made of thin sheet of aluminum. It is open in its upper side and filled up with water at various known temperatures. A stirring

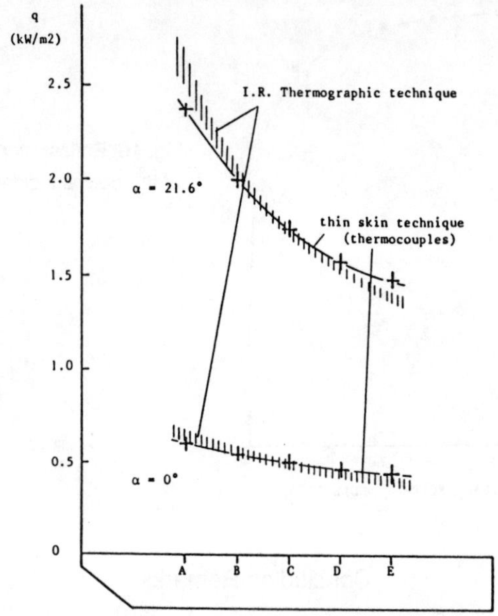

Fig. 9 Convective heat rates measured at rarefied hypersonic flow conditions.

rod is used to get a uniform temperature inside the whole container. Results are showing that, even if the container is not isotherm on its whole height, the effective wall temperature is constant around the central section of the container at half the distance between the upper side and the bottom.

Because of the low heat capacity of the thin wall, the surface temperature around the central region of the container is uniform and equal to the water temperature. The container is previously coated with three layers of graphite paint. A thermogram of the visualized container allows to get the apparent temperature distribution around the container.

At a given tilt angle θ, the effective temperature measured by thermocouple and the apparent wall temperature measured by IR thermometry allows the emissivities vs the optical incident angle of the scanning camera to be obtained (θ = 0 deg corresponds to an incident angle normal to the model surface). For various tilt angles up to 80 deg, apparent emissivities are related to the emissivity at 0 deg tilt angle. Emissivity ratio is plotted vs the optical incident angle of the scanning camera in Fig.10. For tilt angles up to 60 deg, the apparent emissivity does not reach values less than 95% of the normal emissivity. For tilt angles more than 60 deg, recorded values show that the apparent emissivity decreases quite drastically and has to be taken into account for experimental wall temperature measurement by IR thermography.

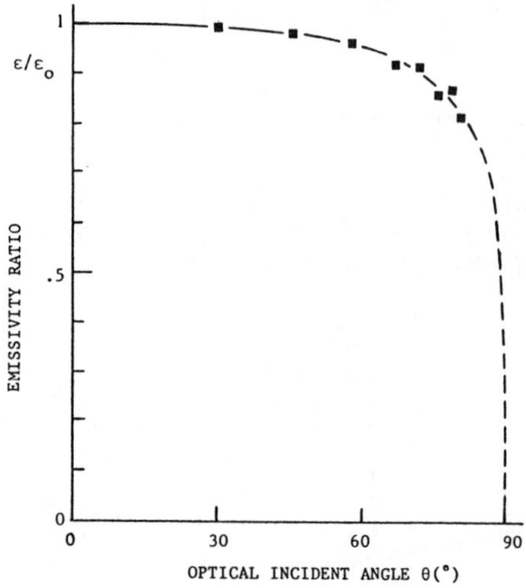

Fig. 10 Emissivity ratio as a function of the optical incident angle.

Concluding Remarks

Preliminary measurements of heat rates have been performed by IR thermography on a flat plate model under hypersonic rarefied flow conditions. Comparisons between IR heat rate values and heat rates measured by the thin-skin technique have shown a reasonable agreement between the two sets of experimental data. The scattering is limited to less than 10% for the present tests.

Before wall temperature and heat rate measurements, local wall emissivity over coated models is experimentally quantified. With three layers of graphite paint and a scanning camera axis directed normally to the investigated surface, a mean emissivity of 0.87 is measured.

The emissivity distribution is also obtained at different tilt angles of the scanning camera axis with respect to the model surface. For tilt angles exceeding 60 deg, the apparent emissivity decreases markedly. The apparent emissivity data as a function of the tilt angle has to be accounted for when wall temperatures and heat rates over complex models have to be quantified.

First measurements by IR thermography have been made on thin-wall models in order to validate the system by reference to thin-skin technique results. Further investigations are planned to get heat rate values over solid models made of low thermal diffusivity material.

Acknowledgments

The present IR thermography system has been developed with financial support of Aérospatiale les Mureaux and Direction des Engins.

References

[1]Matthews, R. K., "Developments in Aerothermal Test Techniques at the AEDC Supersonic/Hypersonic Wind Tunnels", GAMNI-SMAI Meeting (Groupe pour l'Avancement des Méthodes Numériques de l'Ingénieur-Société de Mathématiques Appliquées et Industrielles) Paris, Dec. 7-11, 1987.

[2]Monti, R., "Flow-Visualization and Digital Image Processing", Von Karman Institute Lecture Series 1986-09, Rhode St. Genese, Belgium, June 9-13, 1986.

[3]Carlomagno, G. and de Luca, L., "Heat Transfer Measurements by Means of Infrared Thermography", Fourth International Symposium on Flow Visualization, Paris, Aug. 26-29, 1986.

Experimental Investigation of CO_2 and N_2O Jets Using Intracavity Laser Scattering

Roland G. Schabram,* Alfred E. Beylich,† and
Evgenij M. Kudriavtsev‡
Stosswellenlabor, Technische Hochschule, Aachen,
Federal Republic of Germany

Abstract

A laser-intracavity spectrometer has been developed to measure local gas properties in free-jets produced by orifices with diameters D = 1 - 10 mm. Rotational and vibrational temperatures were measured on the axis of jets of CO_2 and N_2O. Considerable degrees of condensation were found above certain limiting values of p_0D (p_0 being the stagnation pressure) producing a deviation of the temperature from the expected isentropic expansion case as well as a deviation from the molecular Rayleigh light scattering. When condensation sets in, the usual freezing of vibrational temperatures is reduced. Apparently a very efficient energy exchange between molecules and clusters causes a further depopulation of vibrational levels. Condensation effects can also be noticed by a broadening, for instance, of lines in the rotational Raman spectrum.

---

*Graduate Student.

†Professor, Stosswellenlaboratorium.

‡Visiting Scientist, on leave from Lebedev Physical Institute, Moscow.

Introduction

The process of energy partitioning between the components of gas mixtures and among the degrees of freedom of the individual species components during the expansion in a supersonic flow is of fundamental interest for several applications. If the redistribution among the energy levels of the degrees of freedom is fast compared to the exchange of energy between the degrees of freedom, for instance, among the levels of rotation, among the levels of a vibrational mode, or even among the vibrational modes of a molecule, as compared to translational-vibrational (t-v), translational-rotational (t-r), or rotational-vibrational (r-v) transfer, then the problem is considerably simplified. It can be assumed that the state of the individual degrees of freedom can be described by an individual Boltzmann distribution or a temperature T_t, T_r, T_v, which will eventually tend to relax toward an equilibrium state described by a single temperature T. Especially for a freejet expansion, as it is studied here, such an approach seems to be adequate, as long as the gas is not too much rarefied.

With respect to the application of rapid gasdynamic expansions, the process of laser separation should be mentioned. To prepare the gas for the first step of selective excitation, it has to be cooled, which is only possible in a supersonic flow expansion. Another application is the gasdynamic laser, where the mechanisms that maintain and optimize the inversion must be known; detailed information can be obtained from measurement of local temperatures. In connection with aerospace applications, the study of nonequilibrium effects in supersonic flows is of basic interest, and the freejet itself is one of the most important devices to investigate, for instance, gas-wall interactions.

Experimental investigations that yield information on the local state variables are therefore of great importance. In this paper, we report on local measurements of spectra of linear molecules (CO_2 and N_2O) in the freejet from which we have obtained vibrational and rotational temperatures. Since condensation is a phenomenon that appears in most freejets when the stagnation state (p_0, T_0) is above a certain threshhold value, being of importance in most applications, we have also studied condensation and its influence on the gas state.

Experimental Setup and Diagnostic Techniques

A laser-Raman-intracavity spectrometer [1-3] was designed to perform local measurements in a freejet with typical shock barrel dimensions of 100-140 mm diam. To obtain high-intensity signals, the folded cavity was extended by 1400 mm; this made a careful design necessary with a vibrational decoupling from the laboratory ground and the low-density wind tunnel. The Brewster window of the laser plasma tube (Coherent Innova 12 argon-ion-Laser) was at the same time the vacuum seal to the wind tunnel (see Fig. 1). A concave mirror, M3, with radius r = 200 mm being opposite to the collecting lens L1, improves the intensity by a factor of 2. A Jarrel-Ash monochromator (type 78-493 with f = 750 mm) was modified by installing a stepper motor, which allowed complete automatization of the experiment by using a minicomputer. Photon counting was used, and the photomultiplier, type EMI 9863 B/350 with S20 cathode, was cooled to $-20^{\circ}C$, reducing the dark current to about 5 pulses/s.

Local rotational temperatures were obtained from the pure rotational Raman spectra. In Fig. 2, two typical examples are shown: upper spectra are measurements with the strong Rayleigh line at the

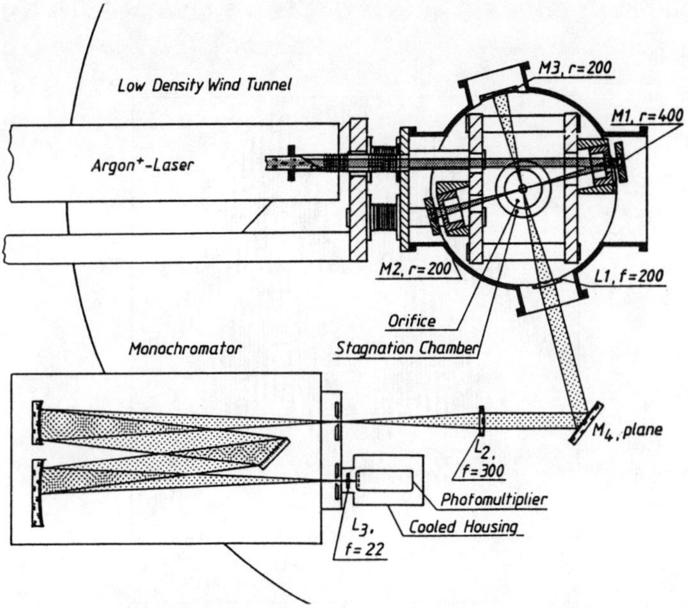

Fig. 1 Laser-Raman-Intracavity Spectrometer. Argon ionlaser. Coherent Innova 12. Monochromator: Jarrel-Ash 78-493 with f = 750 mm, 102 x 102 mm^2 grating with 1180 lines mm blazed at 4000 A.

center, and the lower spectra are numerically produced. Note the missing of every second line in the CO_2 spectrum and the alternating intensity in the N_2 spectrum due to spin $I_{CO_2} = 0$ and $I_{N_2} = 1$, respectively. Rotational temperatures may be obtained from the relation

$$I_{J\rightarrow J'} = C/Q_{rot}(\nu_0 + \Delta\nu_{J\rightarrow J'})^4 \, g_J S_J \exp[-BJ(J+1)hc/kT_r] \quad (1)$$

where $g_J = (2J+1)$ is the statistical weight and S_J the Hönl-London factors. For a Raman transition they are

$$S_J = \frac{3(J+1)(J+2)}{2(2J+3)} \quad \text{for} \quad \Delta J = +2 \quad (2a)$$

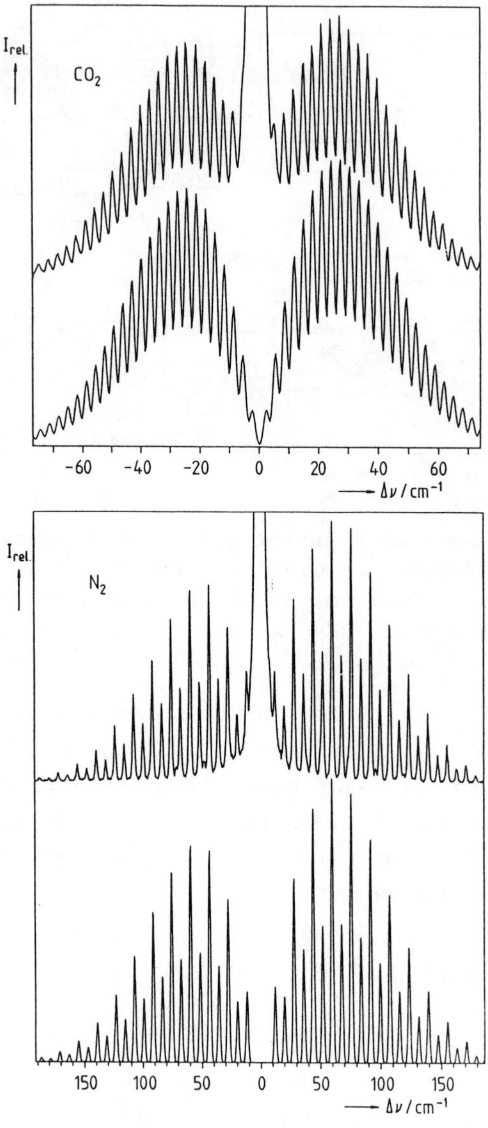

Fig. 2 Experimental (top) and calculated (bottom) pure rotational Raman spectra for CO_2 (top diagram) and N_2O (bottom diagram). T = 297° K.

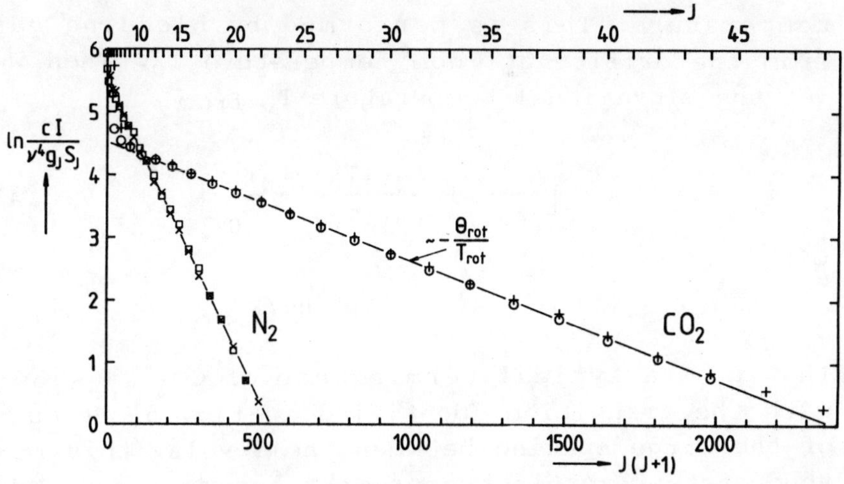

Fig. 3 Boltzmann plot for experimental data (CO_2 and N_2O). O, ◘, Stokes lines; +, × anti-Stokes lines. Deviation from linear behavior for low J due to Rayleigh line background.

$$S_J = \frac{3(J-1)J}{2(2J-1)} \qquad \text{for} \quad \Delta J = -2 \qquad (2b)$$

where C is a constant, Q_{rot} is the state sum for rotation, and ν_0 is the central wave number. A typical Boltzmann plot from which T_r can be obtained is shown in Fig. 3. Since all data are available in digital form, the reduction is performed by a computer program.

Vibrational temperatures can be obtained from the vibrational Raman transitions. For a Boltzmann-type population of the levels, we have at level v

$$n_v = (n/Q_{vib}) d_i \exp[- Q_v\, hc/kT_v] \qquad (3)$$

where n is the number density Q_{vib} the state sum for vibration, d_i a weight and G_v the term value of vibration. For determination of T_v from relative intensities belonging to two transitions (labeled 1 and 2), we need to calibrate at a known temperature. This is performed by bleeding gas from the orifice at room temperature T_0. Then we get the vibrational temperature T_v from

$$T_v = \left[\frac{1}{\theta_{21}} \ln\frac{I_{21}(T_0)}{I_{21}(T_v)} + \frac{1}{T_0} \right]^{-1} \quad (4)$$

with

$$\theta_{21} = (G_2 - G_1)\, hc/k$$

In Fig. 4 a typical term scheme of CO_2 is shown with the transition identified in Fig. 5. Because of the large spacing between the levels, this method of temperature measurement becomes very difficult because of low population of upper levels, if the temperatures are low.

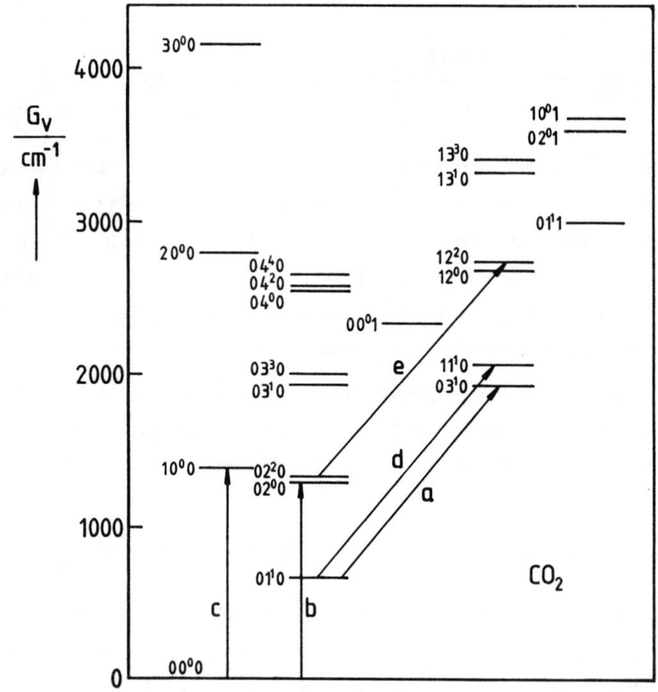

Fig. 4 Term scheme for vibrational modes of CO_2. Measured transition indicted by a - e.

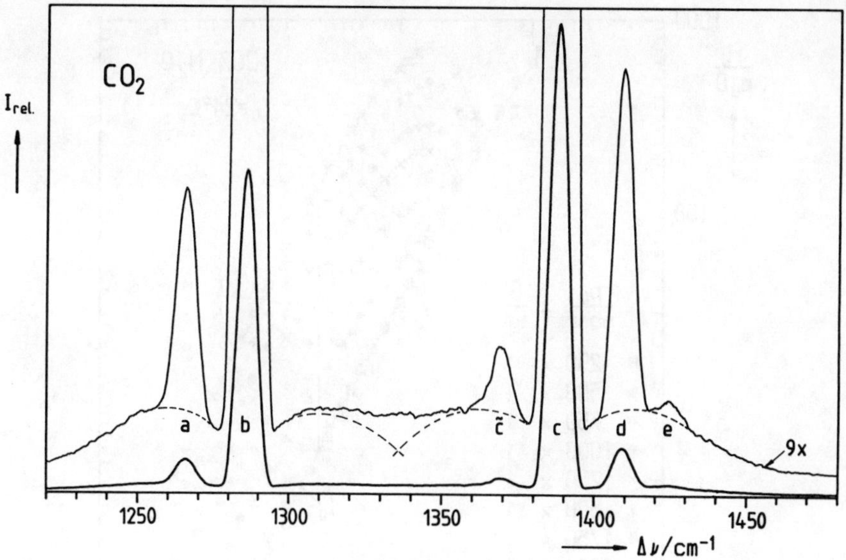

Fig. 5 Vibration-rotation bands of Raman spectrum for CO_2 at T = 297° K. Upper curve magnified 9 times. Letters correspond to transitions in Fig. 4.

Under the condition of clean, dust free gases, the intensity of the elastically scattered light (Rayleigh line) can be used for local measurements of total density. However, when condensation starts in the flow, we expect an increase of intensity due to the fact that dipole scattering is proportional g^2n_g (where g is the number of molecules in a cluster and n_g is the number density of cluster group g). We can assume-- and this is verified by the study of the polarization of the scattered light-- that the cluster diameter is always much smaller than the wavelength of the incident light beam. If we denote the light being scattered by the molecules as single particles by I_0, then the total intensity I due to the actual molecular and cluster scattering, related to I_0, can be written as

$$\frac{I}{I_0} = \frac{n_1\sigma_{12} + n_2 + \sum_{g=2}^{\infty} g^2 n_g}{n_1\sigma_{12} + n_{20}} \tag{5}$$

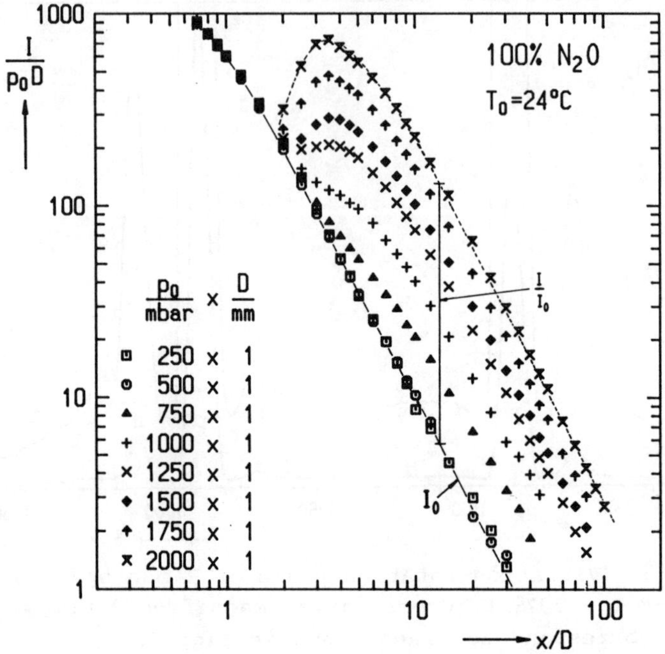

Fig. 6 Measurement of Rayleigh scattering on freejet axis for different p_0D shows increase of condensation with growing p_0. I_0, = pure molecular scattering; I_f, intensity for frozen condensation.

with

$$n_{20} = n_2 + \sum_{g=2}^{\infty} g\, n_g \tag{6}$$

where n_1 is the number density of a noncondensable gas component, the subscript 2 is for the condensable gas, and $\sigma_{12} = \sigma_1/\sigma_2$ is the ratio of the scattering cross sections of the species that can be easily obtained from experiment. In Fig. 6 a typical experimental result of the change of intensity on the jet axis x is plotted. For low p_0D we obtain the quantity I_0, which is, further downstream, proportional to x^{-2}. As soon as condensation starts, we notice a deviation with $I/I_0 > 1$. For evaluation of data it is more convenient to plot I/I_0 as function of x/D; this has been done in Fig. 7, where the definition of the onset point x_c and the frozen intensity I_f are also given.

Experimental Results and Comparison with Theory

The insensitivity of the Pitot pressure along the jet axis with respect to changes in the ratio of specific heats[4] allows a stream tube to be introduced for the flow on the axis which can be related to the orifice diameter. For the general case of such a one-dimensional flow with condensation and with vibrational and rotational relaxation, we have the following set of governing equations:[5]

$$\frac{d}{dx}\,[Au\{\varrho_1 + \varrho_2/(1-\chi_c)\}] = 0 \qquad (7a)$$

with the degree of condensation

$$\chi_c = \sum_{g=2}^{\infty} g n_g / n_{20}$$

$$\varrho u \,\frac{d}{dx}u + \frac{d}{dx}p_t = 0 \qquad (7b)$$

with total pressure

$$p_t = (n_1 + n_2)\,kT + \sum_{g=2}^{\infty} n_g\, kT_g \simeq p$$

and

$$\frac{d}{dx}\left(\frac{u^2}{2} + h\right) = -\frac{m_2}{\varrho}\left\{\Xi\left(\frac{1}{u}\frac{du}{dx} + \frac{1}{A}\frac{dA}{dx}\right) + \frac{d\Xi}{dx}\right\} \qquad (7c)$$

with

$$\Xi = \sum_{g=2}^{\infty} n_g (h_g - g h_2)$$

and

$$h = \left[\frac{5}{2} X_1 + \left(\frac{5}{2} \chi_c \right) X_2 \right] \frac{p}{\varrho (1 - X_2 \chi_c)} + X_2 (e_v + e_{rot})$$

and

$$\frac{de_v}{dx} = \frac{e_v^* - e_v}{u \tau_v} \tag{7d}$$

$$\frac{de_r}{dx} = \frac{e_r^* - e_r}{u \tau_r} \tag{7e}$$

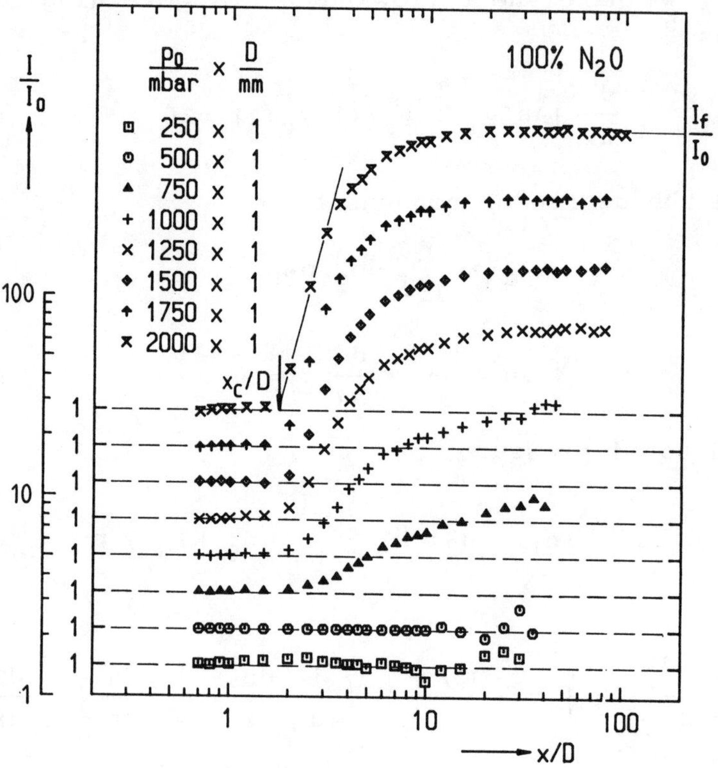

Fig. 7 Scattering intensity I related to molecular scattering I_0 on jet axis for N_2O and different p_0D. For better evaluation of onset point x_c and frozen intensity, I_f curves are shifted upward.

where X_i are number density concentrations of species i, e_v and e_r are the internal energies of vibration and rotation, respectively, which are functions of T_v and T_r, and the asterix indicates energies at T_t. For the present study we performed only calculations without condensation and used collision numbers z_v and z_r to fit the calculation to the experimental results.

In Fig. 8 results are shown for pure CO_2. In the upper diagram the temperatures of translation T_t, rotation T_r, and vibration T_v are plotted as functions of distance x/D. The dashed lines are calculated for $z_v = 20.000$ and $z_r = 10$ for p_0D = 500 mbar mm (these numbers are in fact close to what has been reported before,[6] except for z_r, which seems to be for the case of freejet expansion for N_2O and CO_2 closer to 10 than to 2.5). Note the rotational freezing for this value and the influence of condensation for larger p_0D. Marks indicate where Rayleigh scattering shows onset of condensation. For the same p_0D, the temperature rises for smaller D to higher values, indicating stronger condensation; this is in agreement with the light scattering data. A very remarkable effect is shown in the lower diagram, where T_v is on a linear scale. Whereas under normal conditions T_v freezes when going downstream (see $p_0D = 250 \times 2$ in Fig. 8), in the case of condensation there seems to be an effective mechanism that helps to depopulate vibrational levels further. For the same p_0D, the smaller D produces not only more condensation but also more efficient vibrational depopulation.

For N_2O (Fig. 9), we see similar phenomena, however, condensation is much less, the vibrational depopulation is more effective, and the difference of T_v for the same p_0D is more pronounced. Whereas the overall light intensity I* in Fig. 10 shows no difference for D = 1 and 2 mm (which is

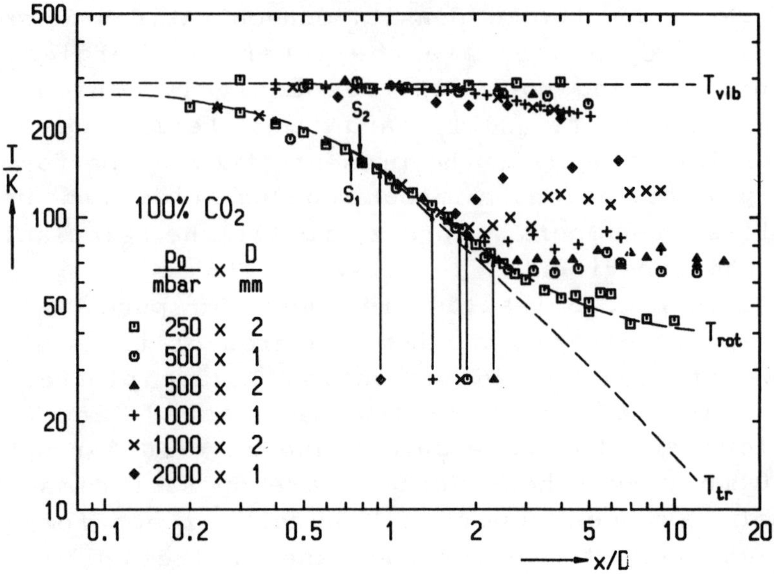

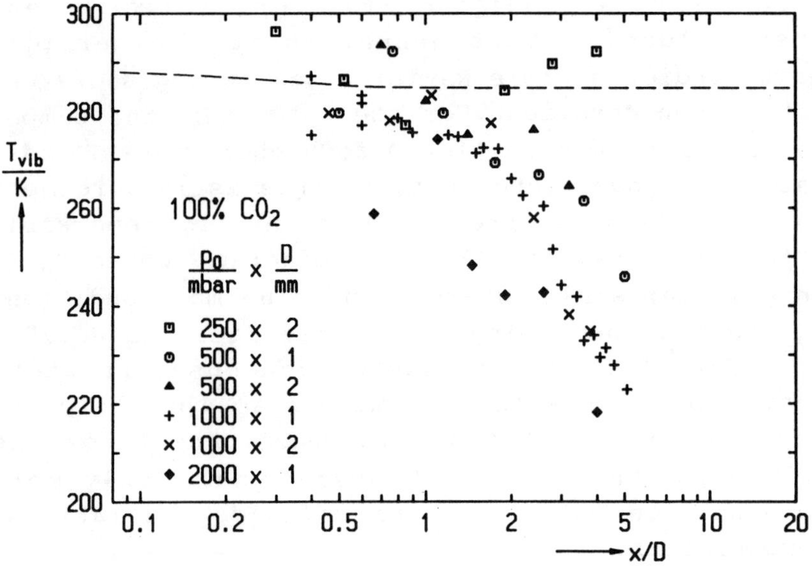

Fig. 8 Temperatures on jet axis of CO_2. Dashed lines, calculated results for p_0D = 500 mbar mm; z_v = 20.000, z_r = 10. Top diagram, vibrational and rotational temperatures obtained from experiments for different p_0D. Bottom diagram, T_v on linear scale.

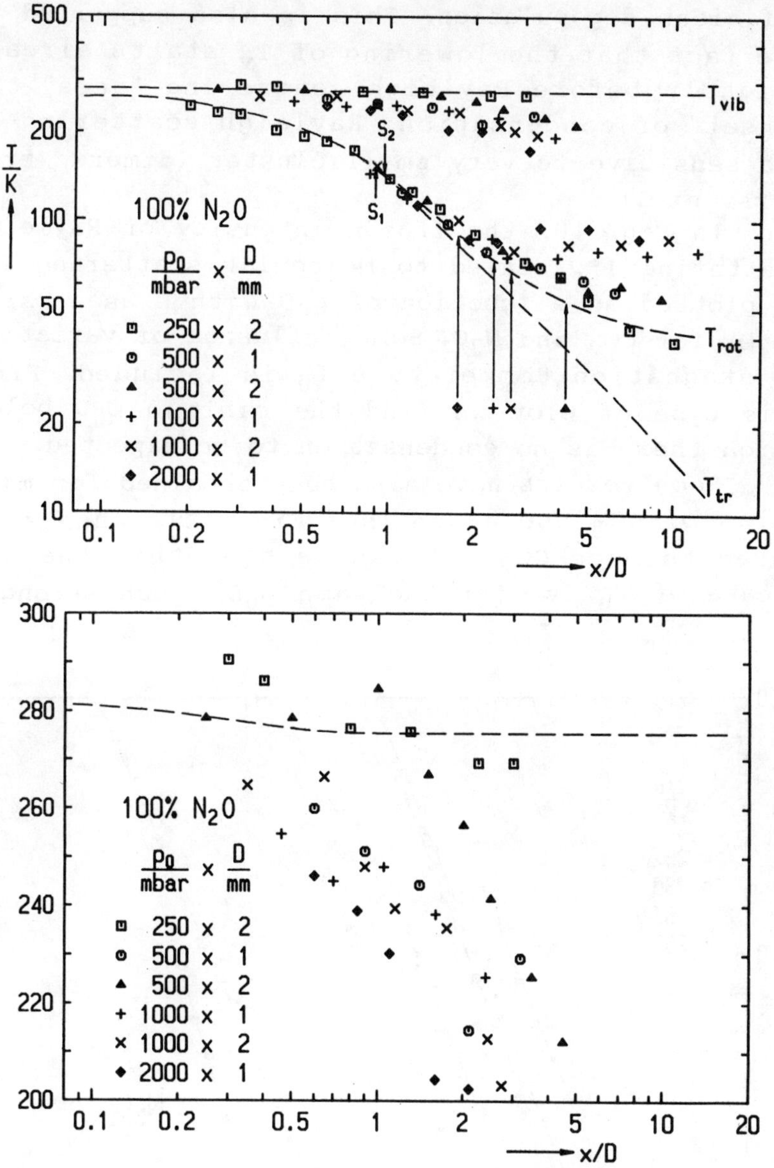

Fig. 9 Results similar to Fig. 8 but for N_2O.

unusual and has not been found before for other gases); here p_0D = 2000 x 1 has lower T_v than p_0D = 1000 x 2. One explanation could be that for p_0D = 2000 x 1 more smaller size clusters cause a more efficient depopulation. This is also supported by the fact that the lowering of T_v starts already slightly before Rayleigh scattering "sees" the "onset" of condensation; Rayleigh scattering is not sensitive to very small cluster (dimers, trimers, etc.).

In Fig. 10, the frozen intensity of Rayleigh scattering I*, scaled to molecular scattering I_0, is plotted as a function of p_0D with D as a parameter for CO_2 and N_2O. Some influence of variation of stagnation temperature T_0 is included. From this type of plot we find the minimum p_0D below which there is no condensation to be expected.

Some results have also been obtained for mixtures of CO_2 and argon (see Fig. 11). When compared to pure CO_2, it can be seen that the decrease of T_v is, for the same p_0D, much stronger

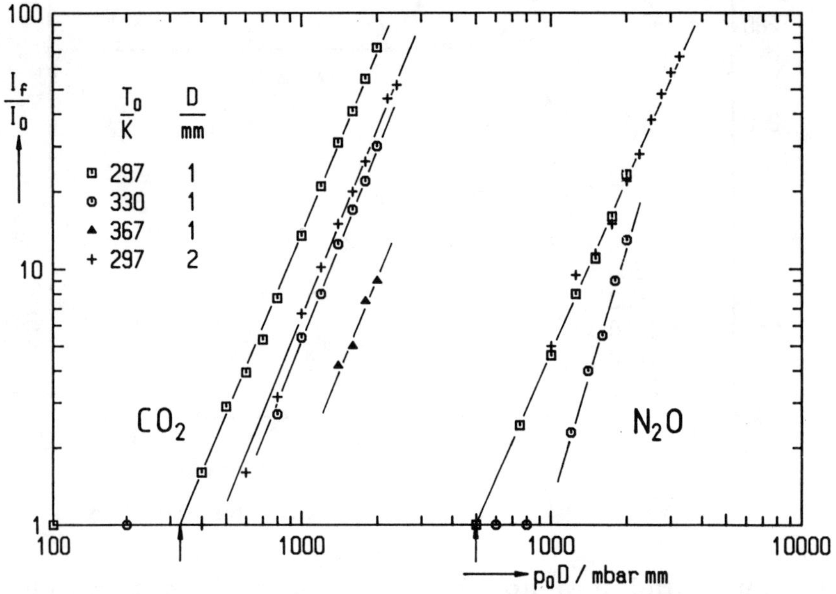

Fig. 10 Frozen intensity I_f/I_0 as a function of p_0D and D for CO_2 and N_2O. Arrows indicate onset points for condensation.

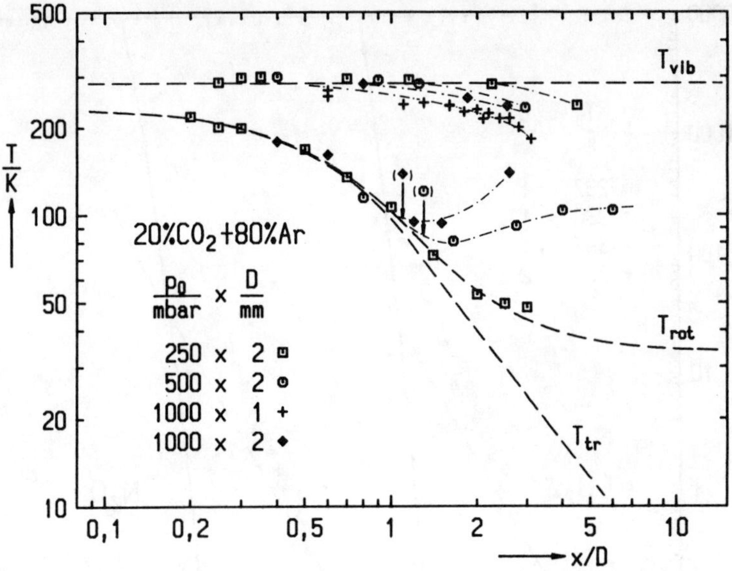

Fig. 11 Results similar to Fig. 8 but for 20% CO_2 + 80% Ar.

in this case; also, condensation is stronger, which is due to the fact that argon with $\gamma = 5/3$ helps to cool down the mixture more quickly.

From the Rayleigh scattering data and the temperature measurements in connection with the numerical results, it is possible to plot the line of onset of condensation in the p-T diagram (see Fig. 12). Supersaturation increases with lower density, and there exists a minimum expansion line (dashed curve) below which no condensation is to be expected. With respect to their onset behavior, CO_2 and N_2O are quite similar.

Discussion

With the laser-intracavity spectrometer it was possible to obtain space-resolved state variables of the expanded gas, especially the T_r and the T_v. Condensation has an important influence on the state of the expanding gas. With respect to T_r

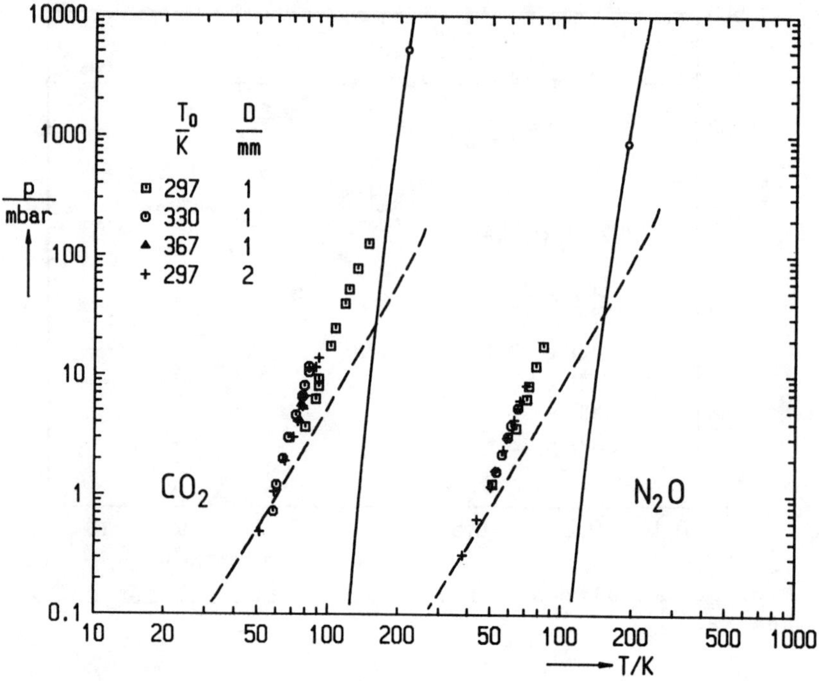

Fig. 12 Onset of condensation line in p-T diagram for CO_2 and N_2O. Full curve is saturation line; open symbol on saturation line is triple point. Dashed curve indicates expansion line below which there is no further condensation.

it causes a rise of temperature, that generally does not scale with p_0D; for smaller D and constant p_0D the released heat -- and in a sense also the degree of condensation -- is larger. Two effects may explain this. For smaller D and the same p_0D the saturation line is crossed at a higher density more upstream, and the ejection of molecules from cluster is an effect that does not scale with p_0D but is dependent on cluster temperature, shape, and size.

The onset points (i.e., the first noticeable deviation from isentropic behavior) obtained from temperature measurements (T_r) and from Rayleigh scattering are in good agreement. These scattering data, however, are difficult to interpret without further information on cluster number density n_c and average cluster size $\langle g^2 \rangle$, since $I \sim n_c \langle g^2 \rangle$, but from the temperature data it can be speculated that both quantities will grow with increase of $p_0 D$. There exists an important effect on T_v by condensation. Even before condensation is seen by Rayleigh scattering or the rise of T_r, T_v stops to freeze and decreases further; this effect is more pronounced for larger degree of condensation. The interaction with clusters seems to be a very efficient depopulation mechanism. This effect is more pronounced in N_2O than in CO_2, however, the classical collision number z_v is also about three times smaller for N_2O than for CO_2. When argon is added to CO_2, this depopulation becomes even stronger.

Condensation also has an influence, for instance, on the width of the rotational lines in the pure rotational Raman spectrum. With the onset of condensation, the ratio of line maxima and minima decreases considerably, and the lines become more fuzzy, although there should be enough intensity. It is difficult to say whether this is caused by superposition of the overall rotational spectrum of the cluster or by the molecules being bound -- but still partly rotating -- in the clusters, or by other unidentified effects.

Acknowledgment

This research was partly financially supported by the Deutsche Forschungsgemeinschaft.

References

[1]Silvera, I. F. and Tommasini, F., "Intracavity Raman Scattering from Molecular Beams: Direct Determination of Local Properties in an Expanding Jet Beam," Physical Review Letters, Vol. 37, July

1976, pp. 136-140.

[2]Luijks, G., Timmerman, J., Stolte, S., and Reuss, J., "Raman Analysis of SF_6 Molecular Beams excited by a CW CO_2 Laser," Chemical Physics, Vol. 77, May 1983, pp. 169-184.

[3]Schabram, R.G. and Beylich, A.E., "Temperature Measurement in Free Jet Expanded Gas Mixtures," Proceedings of the 15th International Symposium on Rarefied Gas Dynamics, edited by V. Boffi and C. Cercignani, Teubner, Stuttgart, FRG, 1986, Vol. 2, pp. 504-513.

[4]Ashkenas, H. and Sherman, F., "The Structure and Utilization of Supersonic Free Jets in Low Density Wind Tunnels," Proceedings of the 4th International Symposium on Rarefied Gas Dynamics, edited by J. H. de Leeuw, Academic, New York, 1966, Vol. II, pp. 84-105.

[5]Beylich, A.E. and Richarz, H.P., "Expansion of Gas Mixtures in Free Jets," Proceedings of the 14th International Symposium on Rarefied Gas Dynamics, edited by H. Oguchi, Univ. of Tokyo Press, Tokyo, 1984, Vol. I, pp. 457-466.

[6]Lambert, J.D., Vibrational and Rotational Relaxation in Gases, Clarendon, Oxford, UK, 1977.

High-Speed-Ratio Helium Beams: Improving Time-of-Flight Calibration and Resolution

R. B. Doak and D. B. Nguyen
AT&T Bell Laboratories, Murray Hill, New Jersey

Abstract

To obtain very high precision absolute calibration of our helium beam time-of-flight (TOF) measurements, we have developed a novel technique based on in situ variation of the flight path. We describe this technique briefly and demonstrate that the calibration can utilize TOF signatures other than that of a constant-temperature beam, taking as an example selective adsorption signatures in helium scattering from LiF(001). The baseline TOF resolution of the apparatus is considered next. We show that variable flight-path measurements allow the TOF spread caused by the beam velocity distribution to be separated from that caused by nonnegligible detector length (shutter function contributions are evaluated by varying the chopper speed). In addition, we discuss and document other factors that limit the TOF resolution and/or calibration accuracy, notably the radial (transverse) velocity distribution, real-gas effects, and clustering.

Introduction

Inelastic neutral helium atom scattering, by virtue of its strict surface sensitivity, its capacity to exchange large amounts of momentum at thermal energies, and the feasibility of generating helium beams of very high angular and energy resolution, is finding increasing utility as a probe of surface lattice dynamics and adsorbate vibrations.[1] Figure 1 illustrates schematically how such a measurement is made. A high-speed-ratio beam[2-4] is produced by skimming the core of a supersonic freejet expansion. This beam is pulsed with a mechanical chopper and scattered from a crystal surface. Creation or annihilation of surface phonons (Stokes and anti-Stokes transitions, respectively) will decrease or increase the energy of beam atoms correspondingly, altering the time of arrival of the inelastically scattered atoms at the detector. With a judicious choice of experimental conditions (varying beam energy, target temperature, and scattering angles), single-phonon interactions will dominate; it is a straightforward matter to extract the energy and wave vector of the participating phonon from the measured flight time and the known scattering angles. A set of measured time-of-flight (TOF) spectra is shown in Figure 2 for an arsenic-terminated Si(111) surface.[5] Both creation and annihilation events are present in this instance. As the angle of incidence is varied, different portions of the surface Brillouin zone are probed, shifting the single-phonon peaks and mapping out the surface phonon dispersion curves. In this set, the peaks at the extremes (880 and 1058 μs in the uppermost trace) are caused by annihilation and creation, respectively, of near-zone boundary phonons at 16.7 meV. The two annihilation peaks are well resolved although separated only by 4 meV. Note that the target has been rotated only by about 12 min of arc between the spectra. Although not obvious from these traces, the broad central peak in the uppermost trace actually comprises *four* inelastic peaks (creation events at 1.5, 1.8, 2.9, and 4.5 meV) as well as the barely resolved incoherent elastic peak. This is an *extreme* test of resolution since we are attempting to resolve a double acoustic mode (with phonons both parallel and antiparallel to the beam), near the zone center, almost under kinematical focusing[6] conditions (which produce the broad creation peak at 24.93 deg) and with a room-temperature beam. Nonetheless, it is clear that better TOF resolution could be useful.

For phonon measurements, beam resolution tends to be the dominant issue. Knowledge of the incident beam energy is also necessary to extract the phonon energy and wave vector, but an accuracy of a few percent

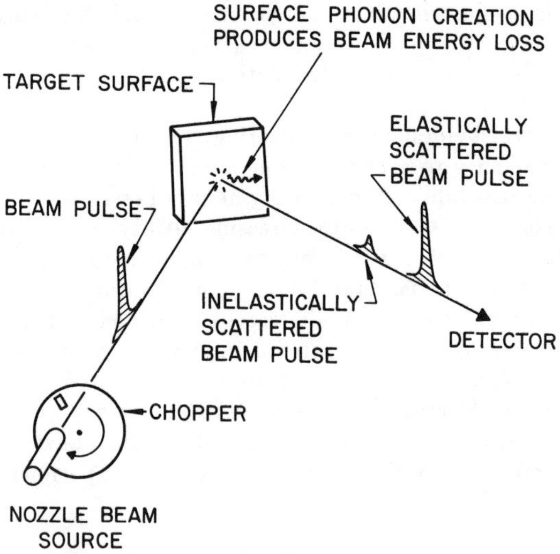

Fig. 1 Schematic diagram of inelastic surface scattering using a neutral helium beam. Creation (annihilation) of single surface phonons will increase (decrease) the target-to-detector flight time, allowing the phonon energy to be determined.

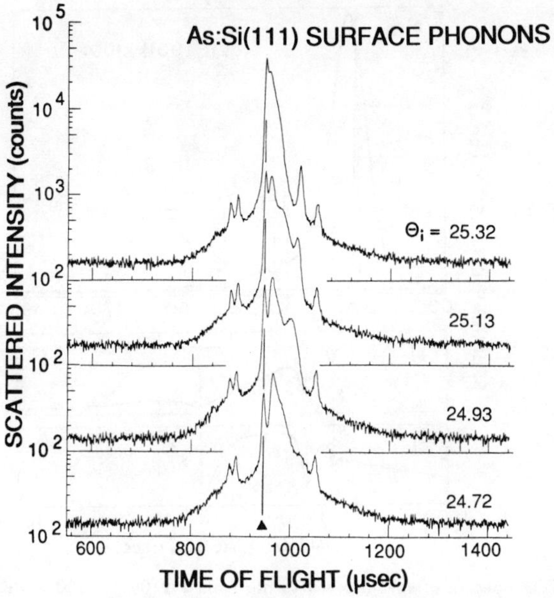

Fig. 2 Example of TOF spectra for a room-temperature helium beam scattered from a room-temperature As:Si(111) (1 x 1) surface along the <112> azimuth at a total scattering angle θ_d = 76.14 deg. (In this surface, the As atoms replace the outermost Si atoms in the top double layer, saturating all Si dangling bonds to remove the (7 x 7) reconstruction and produce an electrically and chemically inert surface.)

generally suffices.[7] In other instances, a precise knowledge of the beam energy may be the overriding concern. An example is given in Figure 3. The upper trace shows two TOF spectra of specular reflection from LiF(001), one with our usual high-resolution beam and a second with a vastly degraded resolution (much lower nozzle pressure). Expanding the vertical scale shows small hole-burning features in the tail of the broad distribution (middle curve). These are caused by the resonant transit of the helium atoms at particular incident energies through bound states of the surface potential well,[8] so-called selective adsorption. Such features have an intrinsic energy width that clearly can be much broader than the energy width of the high-resolution beam. For the purpose of many scattering experiments, it is of crucial importance to know exactly these bound state energies. Such information may be extracted from the hole burning, but it is much faster to sweep the high-resolution beam through the pertinent range of energies by varying the source temperature, recording TOF snapshots. Such a drift spectrum is shown in the bottom curve; the nozzle temperature is literally allowed to drift from liquid nitrogen to room temperature.[9] Each point represents a TOF spectrum of a few seconds duration, from which the instantaneous intensity and elastic TOF may be recovered. Here the dominant concern is then one of calibration, establishing the precision to

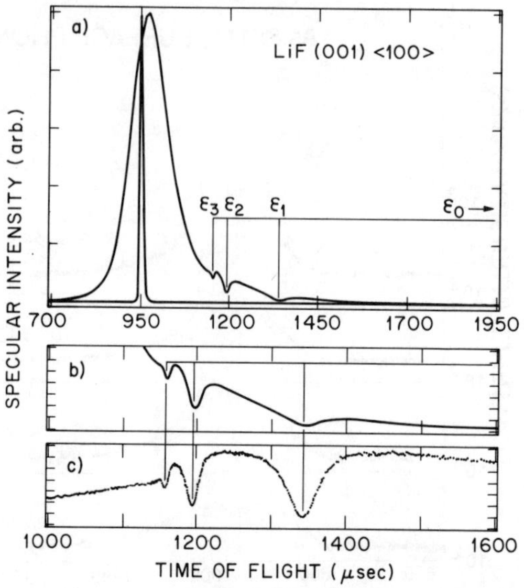

Fig. 3 TOF spectra of specular scattering from LiF(001) <100> with θ_d = 110 deg. a) Comparison of high-pressure, high-resolution beam (120 bar, 20-s integration time) with low-pressure, low-resolution beam (4 bar, 6-h integration time), both at room temperature. b) Expanded view of hole burning in the tail of a broad specular TOF peak due to selective adsorption. c) Drift spectrum of selective adsorption features. Each point represents one quick (3 s) TOF snapshot made as the nozzle temperature is drifted during roughly 15 min from liquid-nitrogen temperature to room temperature.

which the beam energy is known, given the particular measured flight time for each adsorption minimum.

These issues of calibration and resolution have not attracted much attention to date, perhaps because helium scattering is greatly superior in these respects to alternative methods of measuring surface phonon dispersion curves.[1] This superiority notwithstanding, we have found it worthwhile to devote some time to investigating such questions, with the general objective of advancing the state of the art. Those measurements form the crux of this paper, with particular emphasis on employing variable flight-path measurements to greatly improve calibration and to characterize explicitly the present limitations on TOF resolution with continuous-flow (non-pulsed) supersonic helium beams.

Apparatus

To investigate questions of TOF calibration and resolution, we have incorporated into our apparatus the capability of varying the target-to-detector distance in situ. An outline plan view is shown in Figure 4. The entire detector section is mounted on rails and may be translated about 0.71 m, varying the chopper-to-detector distance from 1.57 to 2.28 m. The beam-defining apertures track along a straight line to within 0.23 mm (rms) over this range. We note in passing that the total scattering angle θ_d (source-target-detector angle) is continuously variable from 50 to 180 deg (at the latter angle, the detector views the incident beam) by means of a flexible re-entrant bellows connecting the target chamber and detector section and five beam ports on the target chamber (both target chamber and source section rotate). Angular resolution (e.g., specular width) is roughly 0.2 deg. Pertinent dimensions are chopper-to-target distance, 0.479 m; skimmer-to-chopper distances, 0.125 m; and source-to-skimmer distance, 0.015-0.025 m, typically. Numerous differential pumping sections drop the ambient partial pressure of helium from about 5×10^3 Torr in the source chamber to around 10^{-14} Torr in the detector chamber. The continuous-flow nozzle (5-20 μ in diameter) is operated at temperatures ranging from 77 to 500 K, yielding incident beam energies of roughly 17 to 100 meV, respectively. The detector is a high-resolution quadrupole mass spectrometer. We employ very high-speed (200-MHz) digital counting, and the TOF spectra are stored in a custom-made multichannel scaler (MCS), 50-ns minimum channel width, operated under computer control via a CAMAC interface. The beam is chopped with a conventional slotted rotating disk chopper (2 slits, 0.53 mm wide, at a radius of 50 mm) operating at speeds of up to 1000 disk rotations/s. The apparatus is largely automated. We are able to change scattering angles; monitor system pressures and temperatures; start, stop, and store TOF spectra; measure angular distributions of scattered intensity; and so forth, in real time under computer control.

TOF Calibration

To establish the incident beam velocity, it is necessary to measure the flight time t_{cd} from chopper to detector over the flight path x_{cd}, whence

$$v = x_{cd} / t_{cd} \tag{1}$$

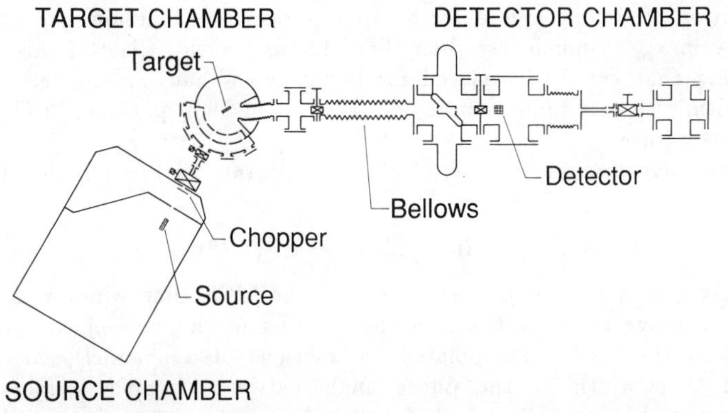

Fig. 4 Schematic plan view of experimental apparatus. The entire detector section translates along the scattered beam axis (flexing the connecting bellows) thus allowing the target-to-detector distance to be varied under ultra high vacuum.

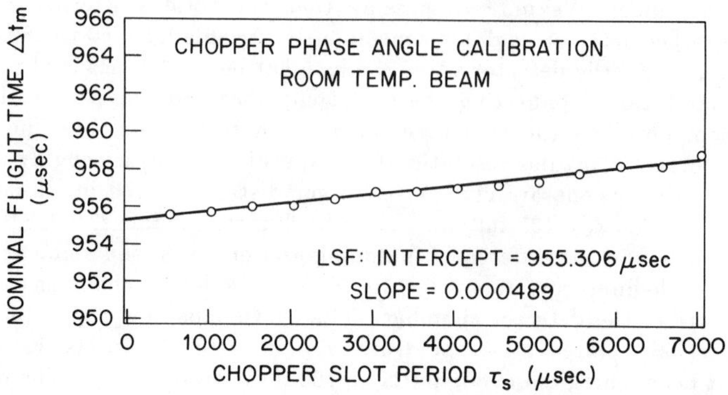

Fig. 5 Measurement of nominal TOF (specular scattering, LiF(001)) as a function of chopper speed to determine the chopper phase angle, which has been set very nearly equal to zero (giving near-zero slope and, in large part, removing any TOF dependence on chopper speed) by in situ positioning of the chopper relative to the beam axis.

TOF calibration then consists principally of measuring x_{cd} and establishing the zero flight time, i.e., determining exactly the time at which the chopper shutter opens, and removing any extraneous time shifts due to electronic signal propagation, detector drawout, nonnegligible response times resulting from electronic bandwidth limitations, etc. For many purposes,[10] a mechanical measurement of flight path yields a sufficiently accurate value of x_{cd}, and the time zero may be determined precisely enough by reversing the chopper rotation (to measure the chopper phase angle) and/or shining a laser beam through the chopper and into the detector ($c >> v_{He}$ giving essentially zero transit time). We are seeking much higher TOF calibration accuracy, however. Since it is the peak-centering accuracy that is of concern, it makes sense (for a reasonably behaved chopper shutter function, velocity distribution function, and detector spatial response function) to seek an accuracy in x_{cd} to much less than the detector length. Clearly, this is not something that can be measured mechanically. What can be determined mechanically to very high accuracy are *changes* in flight path, and this is the technique we use.

As we have discussed in detail elsewhere,[11] the corrected flight time is given by

$$t_{cd} = \Delta t_{dly} + (n+\tfrac{1}{2})\Delta t_{chn} - \alpha N_s \tau_s / 2\pi + \Delta t_p \qquad (2)$$

where Δt_{dly} is a time delay set to center the MCS time window about a particular range of flight times, n the location in channels of the peak of interest in the MCS (extrapolated to fractions of a channel), Δt_{chn} the channel time width, α the phase angle between chopper trigger pulse generation and beam pulse release (center of the shutter function), τ_s the slit period, and Δt_p the sum of electronic propagation delays, including detector drawout delays.

The chopper phase angle α is determined quickly by measuring the elastic flight time as a function of chopper slit period for the incident or specularly scattered beam as in Figure 5, the slope giving $-\alpha N_s/2\pi$ (the $\tau_s=0$ intercept gives the TOF with the phase angle removed). We have set $\alpha \simeq 0$ by positioning the photodiode pair that generates the chopper trigger pulse diametrically opposite the beam axis: fine-tuning is then possible by moving the chopper transversely with respect to the beam axis. This enables us to make the phase angle correction arbitrarily small, less than 0.5 μs between the value at $\nu_s = 1000$ Hz (a typical operating speed) and the extrapolated value at $\nu_s = \infty$.

We then measure the elastic TOF as a function of chopper-to-detector flight path, measuring accurately the increment Δx_{cd} relative to an arbitrary reference point near maximum extension of the connecting bellows. The changes in both flight time and flight path can be determined to very high accuracy (50 ns and 50 μm, respectively). A plot of the measured flight time vs Δx_{cd} yields a straight line, such as the upper line of Figure 6 for a liquid-nitrogen-cooled beam. Note that the slope of this line determines the beam velocity to very high precision (5×10^{-4}). A second measurement at a different beam velocity yields a different line (the lower line of Figure 6, for a room-temperature beam). The point of intersection of these two lines corresponds to zero flight path and zero flight time; the intersection coordinates thus give the flight path x^o_{cd} (from chopper to reference point) and electronic delay time Δt_p. Since only linear least-squares fits are involved, a detailed error analysis is straightforward.[11] This calibration yielded (using very conservative 95% confidence intervals)

$$x^o_{cd} = 2131.5 \pm 2.8 \text{ mm} \tag{3a}$$

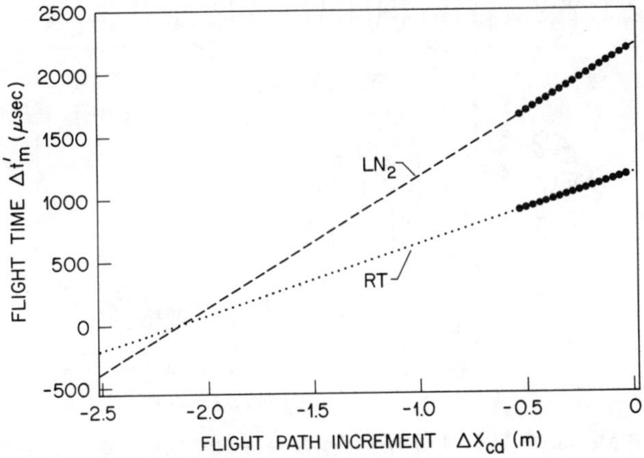

Fig. 6 TOF calibration curves. Elastic TOF (corrected for chopper phase angle) of specular scattering from LiF(001) as a function of change in the flight path relative to an arbitrary but well-defined reference near maximum extension. The curves for two different beam velocities intersect at zero flight path and zero flight time, determining both the flight distance to the reference point and any electronic/detector delays.

$$\Delta t_p = -5.0 \pm 2.4\ \mu\text{sec} \tag{3b}$$

To appreciate these error limits, we note that the detector ionization region could be as long as 14 mm (the physical length of the ion repeller cage); transit times across this distance would be 7.8 and 14.5 μs for the room-temperature and liquid-nitrogen-temperature beams, respectively.

The variable flight-path calibration technique is quite accurate and reasonably fast. Recording a full set of TOF spectra at a given beam temperature takes only a few minutes, although it may be necessary to let the nozzle temperature equilibrate for several hours upon changing from the first beam temperature to the second. It appears to be small (10 mK), short-term temperature variations of the liquid-nitrogen-cooled nozzle that limit the overall accuracy at present. This is a partial answer to an often-posed question; namely, why it is not possible to calibrate solely by very careful measurements of the nozzle temperature: the TOF measurement is simply a much better thermometer! More importantly, real-gas effects introduce a source pressure dependence into the terminal beam velocity (Figure 7). Such effects may be extracted (Figure 8) by measurements at different pressures, but the resulting complications mitigate any advantages

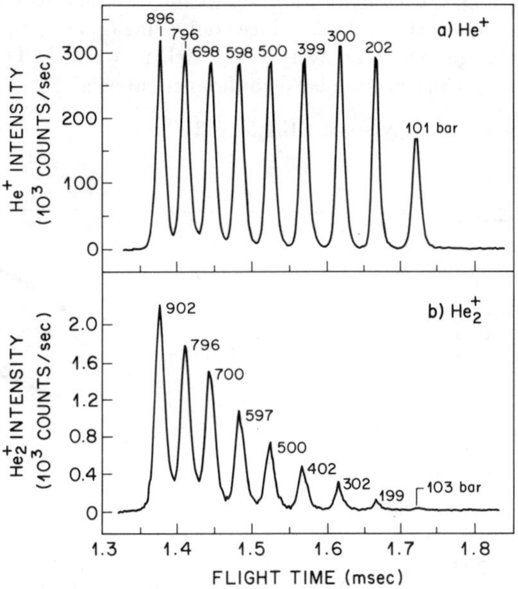

Fig. 7 a) Measured TOF of incident beam as a function of source pressure (3.8-μ-diam nozzle at liquid-nitrogen temperature, detector emission 0.010 mA). The shift in beam energy with source pressure indicates the presence of substantial real-gas effects in the expansion. b) As above but monitoring cluster formation as measured by the He_2^+ peak in the mass spectrum. The clustering rises exponentially above its onset at about 100 bar; this appears to be correlated with a leveling off of peak amplitude and peak width in the He^+ spectra of a).

over a variable flight-path measurement. Incidentally, measurements at different nozzle temperatures, both heated and cooled with respect to the room-temperature ambient, show that the shifts of Figure 7 *are* caused by real-gas effects and not just variation in temperature accommodation of of the gas when passing through the nozzle.

It is interesting to consider alternatives to two beam temperatures as velocity markers for the TOF calibration. Since we are concerned with surface scattering, one possibility that immediately comes to mind is to make use of velocity-dependent scattering signatures; diffraction is one obvious choice. A conventional measurement of diffraction angles, however, shifts the calibration basis from determination of increments in the flight path to increments in the scattering angle. We can measure the latter only to about 0.02 deg, or roughly a factor of 5 worse — relative to a reasonable range of variation — than a corresponding flight-path variation. Therefore, an alternative is to position on the specular and record drift spectra of the selective adsorption signatures, exactly as in Figure 3, at various flight paths. Note that this same scheme could be pursued at off-specular angles, using drift spectra to sweep diffraction peaks past the detector at characteristic beam energies. This would produce considerably sharper signatures — although in general, fewer — than may be produced with selective adsorption. We discuss only the use of selective adsorption features here.

The features of Figure 3 are mediated by out-of-plane reciprocal lattice vectors.[9] Thus, they may be split by rotating in azimuth a few degrees off of the <100> symmetry axis, as in Figure 9. At 3 deg off-axis, these three features have split to yield six distinct, well-separated features (lower trace, Figure 9), doubling the number of velocity signatures available for our calibration. In this geometry, we then record a series of drift spectra at different flight paths. The flight times corresponding to the minima were picked off by hand (no fitting at all) and are plotted in Figure 10, along with the direct TOF measurements using the room-temperature and liquid-

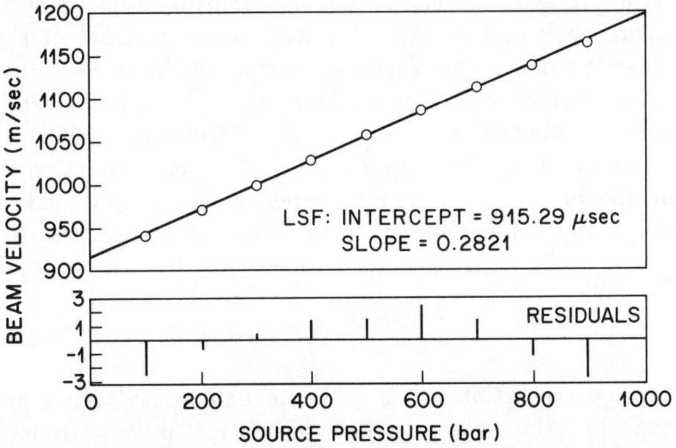

Fig. 8 Plot of measured beam velocity as a function of source pressure, extracted from the data of Fig. 7. An extrapolation to zero pressure gives a value for comparison with the ideal-gas relation $E_i = \gamma kT/(\gamma-1)$.

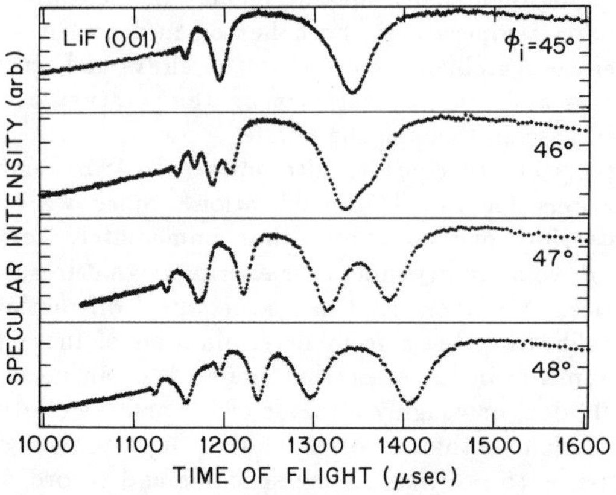

Fig. 9 Drift spectra for LiF(001) <100> at a total scattering angle of 110 deg. Rotating off the symmetry axis splits these bound state minima that are mediated by out-of-plane reciprocal lattice vectors. The six well-defined minima at 3 deg off-axis (bottom trace) are used for TOF calibration.

nitrogen-temperature beams. The accuracy clearly does not rival that of the direct TOF calibration, but undoubtedly could be improved vastly by use of a proper fitting routine, something we are pursuing at present. We note only that even in this rudimentary form, the TOF calibration accuracy using the selective adsorption features equals or exceeds that of most conventional calibration schemes.

We remark in passing that, as with the direct TOF calibration, the slope of each line is determined to very high accuracy, providing a precise measurement of the beam wave vector associated with each selective adsorption event.[9] In turn, this allows the individual bound state energies to be determined precisely, the limiting uncertainty being that of the scattering angles. The largest uncertainty, that in azimuthal angle, may be removed by extrapolating back to the symmetry axis using the data of Figure 9. Since we are positioned on the specular beam, we have also fixed $\theta_i = \theta_f$ by symmetry, leaving only the total scattering angle to be determined. This, in fact, can be verified/measured from the diffraction angles once the direct TOF calibration has been completed.[11] The end effect is the ability to determine the bound state energy corresponding to the peak center of each signature (or any other characteristic point) to a few tens of nano-electron volts.

TOF Resolution

The energy resolution of an inelastic helium-scattering measurement is determined by the time spread of signal pulses from the detector corresponding to a given scattering event of specified energy and momentum exchange. Therefore, it ultimately must include convolution of the scattering geometry and surface properties with the properties of the

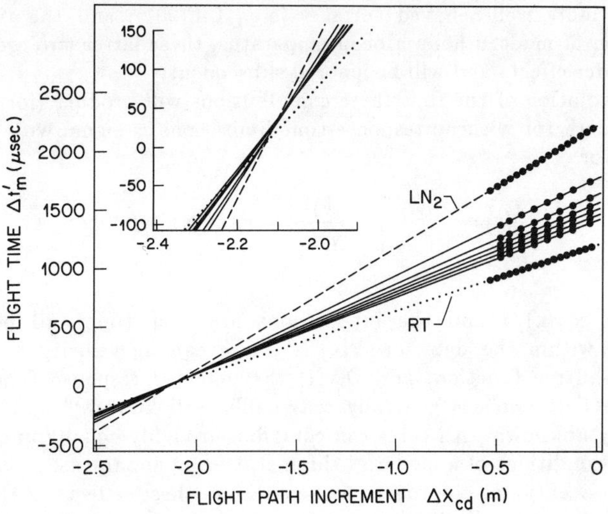

Fig. 10 TOF calibration curves using drift spectra signatures due to selective adsorption at different flight paths (solid lines). Curves showing the direct TOF measurements with two different beam velocities are also shown (dashed, dotted lines). Inset: expansion around the region of intersection at + 5.0 μs, - 2.1315 m.

incident beam. For example, creation of a surface phonon of very nearly the incident beam energy results in a very long flight time from target to detector and correspondingly good energy resolution. Annihilation of a high-energy phonon has the opposite effect, resulting in a short flight time and poor resolution. In general, both the angular resolution and the energy resolution will depend strongly not only on the apparatus but also on the crystal surface and on the momentum and energy exchanged.[12] For the sake of a baseline characterization, however, it is proper to consider alone the incident beam or specularly reflected beam, bearing in mind in the latter case that surface properties may still contribute (e.g., through zero-point energy or domain-size effects on beam coherence). Diffraction, unlike specular scattering, disperses the beam in angle according to its velocity distribution and may lead to monochromatization of the beam, depending on the specific collimation geometry (this is true for any nonzero parallel momentum exchange, including inelastic scattering). This has advantages and disadvantages for calibration purposes. Here we will consider only incident or specularly scattered beams. The primary contributions to the time spread at the detector are then the following:

1) Velocity distribution of the beam
2) Shutter function of the beam chopper
3) Nonzero detector length
4) Multiple paths from chopper to detector
5) Surface properties

For reasonably well-behaved surfaces (e.g., LiF(001)) and the tight beam collimation of modern helium beam apparatus, these latter two are generally second-order effects and will be ignored subsequently.

A convolution of the first three contributions will produce (for a density-sensitive detector with no response-time limitations) a signal with time from the detector of

$$I(t) = \int_{-L_d/2}^{L_d/2} \int_0^t \frac{F(v)}{v} S(t-\tau) D(v,x) d(v(\tau)) dx \tag{4}$$

where $v = x_{cd}(x)/\tau$ and the integrations are over time and position of ionization within the detector. F(v) is the streaming velocity distribution, S(t) the shutter function, and D(v,x) the detector response function. Of these, the first two are generally reasonably well known,[10] and the latter completely unknown. All three can contribute roughly equally under typical operating conditions of a modern helium-scattering apparatus.

The conventional approach to separating these effects is to assume functional forms for F(v), S(t), and D(v,x), then compute I(t) and compare with the measured TOF spectrum, adjusting the parameters as necessary to produce a satisfactory fit. Apart from the fact that D(v,x) is unknown, this method suffers from the problem that S(t) is influenced by the beam geometry and collimation and may be changed by, for example, an unrecognized beam clipping at a misaligned collimator. Alternatively, these various contributions may be separated experimentally by measuring the TOF peak shape as a function of chopper speed and flight path (this is readily apparent by considering the limit $\nu_s \to \infty$, $x_{cd} \to \infty$ making S and D delta functions with respect to F; these are not viable conditions for inelastic scattering, unfortunately!). Indeed, the pertinent data are exactly those of the preceding TOF calibration, but now considering peak shape/width as opposed to peak position.

To illustrate this, we adopt a simplified approach: the extension to a more rigorous treatment will be readily apparent. We treat the TOF broadening caused by shutter function, detector length, and velocity distribution as Gaussian in shape if expressed as isolated functions of time and convolute them by adding the squares of their widths, i.e.,

$$\Delta^2 = \Delta_F^2 + \Delta_S^2 + \Delta_D^2 \tag{5}$$

where Δ is thus the expected TOF width of the function I(t) given time widths of Δ_F, Δ_S, and Δ_D for the isolated effects of velocity distribution (over the flight path x_{cd}), shutter function, and detector length. We will be able to evaluate this Gaussian assumption subsequently. All time widths are much smaller than the chopper-to-detector flight time.

Proceeding in this vein, it is then possible to express the explicit dependencies of the various contributions; namely,

$$\Delta_F = x_{cd} \Delta v / v^2 \tag{6}$$

$$\Delta_D = L_d / v \tag{7}$$

$$\Delta_S = w_s N_s \tau_s / (2\pi r) \tag{8}$$

where v is the beam velocity, Δv the width of the velocity distribution ($\Delta v << v$), and L_d the detector length. The effective chopper slot width w_s may be determined by either the beam width or slot width as shown in Figure 11, depending on which is larger. We chop our circular cross-sectional beam with a slot about twice as wide ($w_b / w_s = 0.5$) since the effective beam width at the chopper varies slightly as we vary x_{cd}, and we wish this not to change the measured time width (FWHM) (see Figure 11).

The procedure then becomes obvious. Exactly as in the TOF calibration, it is possible to exploit the different dependencies to isolate the various effects. We use the FWHM of the elastic (specular) TOF peak as our characterization of peak width. As a function of chopper period, this takes the form of Figure 12. Clearly, at chopper periods above about 1 ms (500 rotations/s, disk speed) the shutter function dominates. At faster chopping speeds (shorter shutter functions), the curve levels off as velocity distribution and/or detector length effects take precedence. Least-squares fits [Eqs. (5) and (8)] to the data are shown in Figure 12 and tabulated in Table 1. The coefficient of τ_s^2 yields an effective shutter function width of 0.70 and 0.62 mm, respectively, for the liquid-nitrogen and room-temperature beams, which is in reasonable agreement with the measured slit width of 0.53 mm. At the risk of overinterpreting an admittedly simple model, we note that finite source size effects will indeed broaden the effective shutter width and more so for the colder beam, improving agreement.

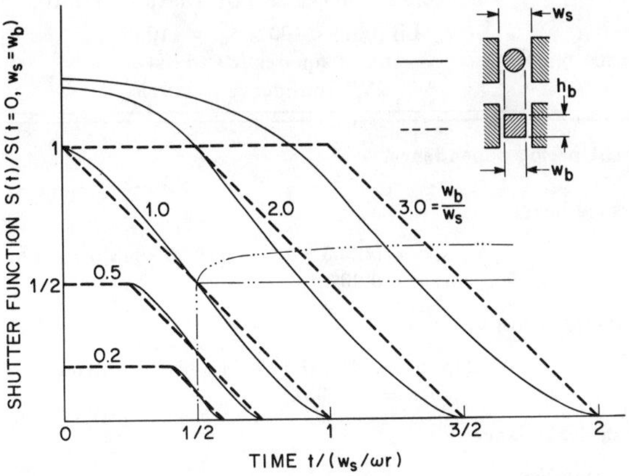

Fig. 11 Shutter functions for a rectangular slit chopping a beam of circular cross section (solid lines) and rectangular cross section (dashed lines); the abscissa is normalized with respect to slit width and disk tangential velocity ωr at the chopping radius. The locus of the FWHM points of this family of curves is shown as dot-dash lines. In our measurements (circular beam cross section), we have set $w_b / w_s \simeq 0.5$, making the peak width (FWHM) independent of the small changes in w_b / w_s as the flight path is varied.

Having determined the shutter function dependence, at least within the simple modeling of Eq. (5), we then measure the specular TOF width as a function of x_{cd} to separate out detector length and velocity distribution contributions. The results are shown in Figure 13. Least-squares fits of the

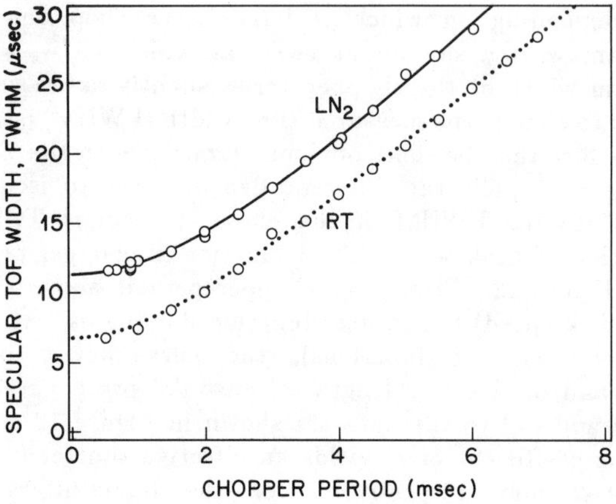

Fig. 12 Measured TOF peak width Δ (specular scattering from LiF(001)) as a function of chopper speed to extract shutter function. Data for both room temperature (RT) and liquid-nitrogen-cooled (LN_2) source are shown. The room temperature widths are from the same data set for which the flight times are plotted in Fig. 5.

Table 1 Specular TOF width (FWHM)
LiF(001) <100> θ_d = 110 deg;
Least squares fits to data;
95% confidence intervals

Chopper slit period dependence:

1) LN_2-cooled beam

$$[\Delta(\mu s)]^2 = (129.5 \pm 6.7) + (19.90 \pm .45)[\tau_s(ms)]^2$$
$$\text{corr.} = 0.99901$$

2) Room temperature beam

$$[\Delta(\mu s)]^2 = (45.8 \pm 7.1) + (15.49 \pm .30)[\tau_s(ms)]^2$$
$$\text{corr.} = 0.99954$$

Flight path dependence:

1) LN_2– cooled beam

$$[\Delta(\mu s)]^2 = (78.0 \pm 13.9) + (45.4 \pm 4.0)[x_{cd}\ (m)]^2$$
$$\text{corr.} = 0.9875$$

2) Room temperature beam

$$[\Delta(\mu s)]^2 = (133.90 \pm 4.3) + (9.34 \pm 1.24)[x_{cd}\ (m)]^2$$
$$\text{corr.} = 0.9721$$

form Eqs. (5) and (6) are drawn through the data and listed in Table 1. From these and Eqs. (5-8), we extract an effective detector length of 7.4 ±1.2 and 7.7 ±1.3 mm, respectively, for the liquid-nitrogen and room-temperature results. To check modeling sensitivity, additional lines have been plotted on Figure 13 by fixing the detector length at the limits given by the preceding error bars on detector length and adjusting the other contributions (Eq. (5)) to pass a line through the least-squares-fit value at x_{cd} = 1.692 m (our standard setting for TOF spectra). These lines do indeed lie visibly outside of an acceptable range, demonstrating that the modeling is, in fact, sensitive to parameter changes.

Other Oddities

Finite source size effects should show up in a measurement of the radial (transverse) velocity distribution of the beam. 13 This can be measured easily with the present apparatus by mounting a small aperture in the target manipulator and rotating the source to θ_d = 180 deg to send the incident beam directly through the aperture and into the detector. Moving the aperture transversely to the beam then sweeps the pattern of the apertured beam across the detector. We first used this to map out the transverse extent of the detector response function. The detector was then collimated down to limit detection to the flat central portion of this response function, producing essentially a step function in detection probability when sweeping the apertured beam across the detector collimator edges. We then varied the source pressure and temperature and aperture size, observing the changes in spreading of the beam to gage changes in radial velocity distribution. A typical set of data is shown in Figure 14; note that the angular spread increases with increasing source pressure. Similar increases were seen with decreasing source temperature. For a room-temperature beam, 3.8-μ-diam nozzle, and 200-bar source pressure, we estimated $\Delta v_r/v$

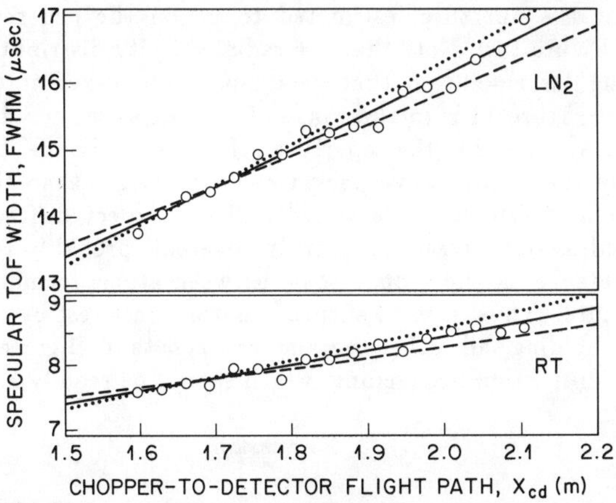

Fig. 13 Measured TOF peak width Δ (specular scattering from LiF(001)) as a function of flight path. Same data set as in Fig. 6.

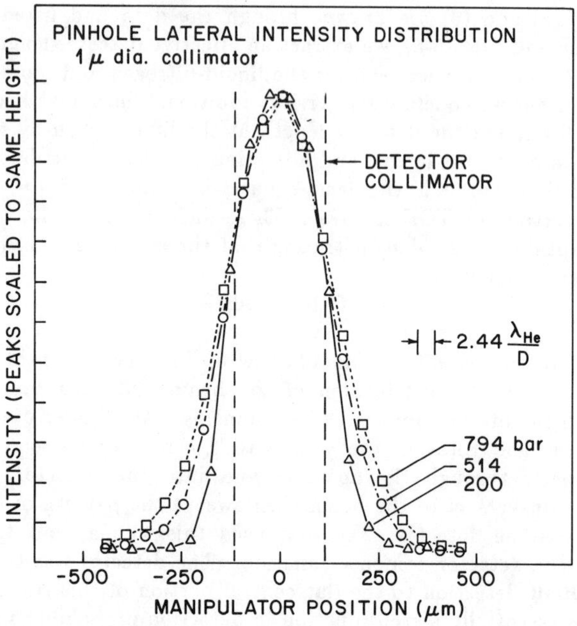

Fig. 14 Lateral profile of apertured incident beam as a function of nozzle pressure. A 1-μ-diam aperture is mounted in the target manipulator and moved transversely to the beam to sweep the apertured beam across the detector, itself collimated to yield a flat response function within the region shown (dashed vertical lines). The increase in spreading at higher source pressures indicates an increasing radial velocity distribution/transverse source extent.

(FWHM) $\simeq$ 1.7%. (This is a very rough estimate; no attempt was made to remove, for example, the smearing caused by background laboratory vibration at the aperture, estimated to be on the order of 10 μ on the abscissa of Figure 14.) Note that the radial velocity distribution represents a limit on angular resolution that generally is not taken into account. The spreading of Figure 14 is in the range of that expected for diffraction of the helium plane wave by the aperture. It is conceivable that the lateral coherence of the helium wave packet could be larger than the aperture size (1 and 4 μ in diameter). We would still not expect diffraction spreading, however, unless the interaction is truly coherent, probably requiring a strict spatial coherence on the atomic scale between atoms forming opposite sides of the aperture. Thus, it is no surprise that a direct test by varying aperture size while holding all other parameters constant (Figure 15) shows no evidence of diffraction broadening, which should be readily apparent.

Discussion

The resolution measurements are summarized in Table 2. The values obtained for both the shutter function and the detector length agree very well for the independent calculations for room-temperature and liquid-

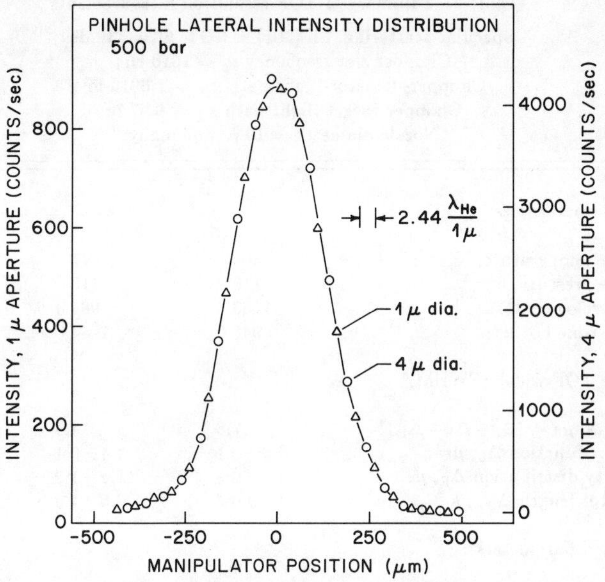

Fig. 15 As in Fig. 14 but for two different aperture sizes at a fixed pressure. A characteristic defraction spread for the 1-μ aperture is shown; this should scale with aperture size. The good agreement of 1-μ and 4-μ aperture results indicates that no diffraction of the helium wave packet is seen.

nitrogen-temperature beams. This is strong evidence that our simplistic modeling does lead to viable conclusions and that the assumption of Gaussian lineshapes is valid. In all events, these values would be the logical starting point for a refinement of the results by introducing a chosen detector response function D(v,x) into Eq. (5) and turning the crank.

To improve resolution, the time widths listed in Table 2 make the course of action clear. For a room-temperature beam, the shutter function, velocity distribution, and detector length are all limiting and must be reduced. For the liquid-nitrogen-cooled beam, a factor of 2 improvement could be attained by reducing the velocity spread to a value commensurate with the detector contribution. One intrinsic limitation in this regard is demonstrated in Figure 7 for a liquid-nitrogen-cooled nozzle. As the pressure is raised, the onset of clustering eventually is reached, beyond which further increases in pressure do not appear to improve either resolution or peak amplitude. This is also seen with a room-temperature beam for which the clustering temperature is about 1000 bar (3.8-μ-diam nozzle). The answer may be to reduce clustering by reducing source pressure and increasing source diameter (while maintaining the product constant to achieve the same resolution). This, of course, increases source throughput. Since continuous-flow beam machines are already at or near reasonable source pumping limitations, the next generation of helium beam machines may well be based on pulsed nozzle technology or multiply skimmed sources.

Table 2 Summary of TOF resolution measurements
Specular scattering, LiF(001) <100> θ_d = 110 deg;
Chopper slot frequency ν_s = 1010 Hz;
Chopper-detector flight path x_{cd} = 1.6916 m;
Chopper-target flight path x_{ct} = 0.47 ln
Nozzle diameter = 10 μ, nominally

Beam characteristics:		
Nozzle temperature, K	300	77
Nozzle pressure, bar	116	118
Beam velocity, m/s	1783.3	964.4
Flight time t_{cd}, μs	948.6	1754.0
Specular TOF width (FWHM)[a]:		
Total Δ, $\mu s = (\Delta_S^2 + \Delta_F^2 + \Delta_D^2)^{1/2}$	7.8	14.4
Shutter function Δ_S, μs	3.9 $\pm$0.1	4.4 $\pm$0.1
Velocity distribution Δ_F, μs	5.2 $\pm$0.8	11.4 $\pm$1.2
Detector length Δ_D, μs	4.3 $\pm$0.7	7.6 $\pm$1.3
Calculated parameters[a]:		
Effective shutter width, FWHM, mm	0.70 $\pm$0.01	0.62 $\pm$0.01
Detector length, FWHM, mm	7.7 $\pm$1.3	7.4 $\pm$1.2
Resolution:		
TOF, $\Delta t/t$, FWHM	0.0082	0.0082
Velocity, $\Delta v/v$ FWHM	0.0055	0.0065
Difference, %[b]	49	26
Energy, true, FWHM/x_{cd}, meV	0.72	0.25
Energy, effective, FWHM/x_{td}, meV	1.51	0.44

[a] Error bars are 95% confidence intervals to least-squares fits and *do not* include systematic errors due to modeling!

[b] $(\Delta t/t - \Delta v/v)/(\Delta v/v)$

We emphasize the obvious fact that the velocity distribution $\Delta v/v$ is equal to the TOF distribution $\Delta t/t$ only in the limit that all contributions other than that of the velocity spread are negligible. It is quite valid to take $2(\Delta t/t)$ as an effective energy resolution of the apparatus but not, for example, to use $\Delta t/t$ to compute a speed ratio or analyze off-specular angular resolution. Doing so can introduce substantial errors (Table 2).

Finally, we note that the variable flight-path techniques described here are quite general and equally applicable to neutron scattering, ion scattering, or any other TOF-analyzed measurements. Moreover, it is a straightforward extension of the preceding treatment to consider inelastic scattering (varying the target-to-detector distance, as we indeed do with our present arrangement). This affords the possibility of, for example, explicitly providing an absolute calibration of energy exchange or explicitly measuring energy widths of phonon signatures.

References

1 Toennies, J. P., "The Study of the Forces between Atoms of Single Crystal Surfaces from Experimental Phonon Dispersion Curves," Solvay Conference on Surface Science, Austin, TX, Dec. 1987 to be published in the Springer Series in Surface Science.

2 Campargue, R., Lebehot, A., and Lemonnier, J. C., "Nozzle Beam Speed Ratios Above 300 Skimmed in a Zone of Silence of He Freejets," Progress in Astronautics and Aeronautics: Rarefied Gas Dynamics, Vol. 51, Pt. II, edited by J. Leith Potter, AIAA, New York, 1977, pp. 1033-1045.

3 Brusdeylins, G., Meyer, H.-D., Toennies, J. P., and Winkelmann, K., "Production of Helium Nozzle Beams with Very High Speed Ratios," Progress in Astronautics and Aeronautics: Rarefied Gas Dynamics, Vol. 51, Pt. II, edited by J. Leith Potter, AIAA, New York, 1977, pp. 1047-1059.

4 Toennies, J. P. and Winkelmann, K., "Theoretical Studies of Highly Expanded Free Jets: Influence of Quantum Effects and a Realistic Intermolecular Potential," Journal of Chemical Physics, Vol. 66, May 1977, pp. 3965-3979.

5 Doak, R. B. and Nguyen, D. B., "Phonons on a Surface with a Mass Defect: As:Si(111)(1x1)," submitted for publication.

6 Benedek, G., Brusdeylins, G., Toennies, J. P., and Doak, R. B., "Experimental Evidence for Kinematical Focusing in the Inelastic Scattering of Helium from the NaF(001) Surface," Physical Review B: Solid State, Vol. 27, Feb. 1983, pp. 2488-2493.

7 Brusdeylins, G., Doak, R. B., and Toennies, J. P., "High-Resolution Time-of-Flight Studies of Rayleigh Surface Phonon Dispersion Curves of LiF, NaF, and KCl," Physical Review B: Solid State, Vol. 27, March 1983, pp. 3662-3685.

8 Brusdeylins, G., Doak, R. B., and Toennies, J. P., "Observation of Selective Desorption of One-Phonon Inelastically Scattered He Atoms from a LiF Crystal Surface," Journal of Chemical Physics, Vol. 75, Aug. 1981, pp. 1784-1793.

9 G. Derry, D. Wesner, S. V. Krishnaswamy, and D. R. Frankl, "Selective Adsorption of ^{3}He and ^{4}He on Clean Surfaces of NaF and LiF," Surface Science, Vol. 74, May 1978, pp. 245-258.

10 Auerbach, D. J., "Velocity Measurements by Time-of-Flight Methods," Atomic and Molecular Beam Methods, Vol. 2, edited by G. Scoles, Oxford Univ. Press, New York, to be published.

11 Doak, R. B. and Nguyen, D. B., "Absolute Calibration of an Atomic Helium Beam Time-of-Flight Apparatus by Flight Path Variation," Review of Scientific Instruments, Vol. 59, Sept. 1988, pp. 1957-1964.

12 Doak, R. B., "Measurement of Surface Phonon Dispersion Relations for LiF, NaF, and KCl Through Energy-Analyzed Inelastic Scattering of a Helium Atomic Beam," Ph.D. Thesis, Massachusetts Institute of Technology, Cambridge, MA, Sept. 1981.

13 Abuaf, N., Anderson, J. B., Andres, R. P., Fenn, J. B., and Miller, D. R., "Studies of Low Density Supersonic Jets," Rarefied Gas Dynamics, Academic, New York, 1967, pp. 1317-1337.

Velocity Distribution Function in Nozzle Beams

O. F. Hagena*
Kernforschungszentrum Karlsruhe, Karlsruhe,
Federal Republic of Germany

Abstract

The present theoretical models for the velocity distribution in nozzle beams are discussed. It is shown that the differential intensity changes from $V^3\exp[-(V-S)^2]$ to $V^2\exp[-(V-S)^2]$ when the divergence of the flowfield that produces the beam must be taken into account. Neglect of this divergence may introduce errors in the flowfield properties obtained from the molecular-beam velocity distribution, with flow temperature being too high and flow velocity being too low. For higher speed ratios $S > 5$, the deviations are less than 5 %. The distribution function valid for the divergent flow satisfies the enthalpy balance.

Introduction

Analysis of the velocity distribution function of molecular beams obtained from nozzle sources is a standard procedure to determine the flow velocity u and translational temperature T of the beam, for example. These data can then be used to characterize the supersonic flowfield from which the nozzle beam was extracted, e.g., source temperature T_0, speed ratio S, terminal translational temperature T_∞. The standard experimental technique for velocity analysis is the time-of-flight (TOF) method: The beam is chopped by a rotating chopper disk, and the velocity distribution is derived from the arrival times as registered by a detector placed at a distance L downstream of the chopper.

To extract the flow velocity u and temperature T (or other moments of the distribution function), the usual practice is based on a curve fit of the measured signal to a theoretical signal, with, for example, u and T as free parameters. Of course, corrections due to the finite length of the chopper pulse and due to the nonideal detector response, including its associated electronics, must be applied if necessary. Problems arise, however, with the correct form of the theoretical velocity distribution. The respective dispute

*Senior Research Scientist, Institut für Mikrostrukturtechnik.

can be traced as far back as 1920, when Albert Einstein corrected the oven-beam distribution function used by Otto Stern, the famous pioneer of molecular-beam research, changing the pre-exponential factor from v^2 to v^3.[1] The different exponent was caused by the fact that the detector was sensitive to the molecular flux -- number of molecules arriving per unit time--in contrast to detectors sensitive to the particle number density--number of molecules per unit volume. In the nozzle-beam community, a similar dispute as to the proper exponent n of the pre-exponential faktor v^n has a different basis: The value for n depends on the type of flowfield from which the nozzle beam is extracted, changing from n = 3 to n = 2 if the divergent nature of the flowfield has to be taken into account.[2] Although this analysis has been accepted, confirmed, and used by other groups,[3-6] at present there are an increasing number of papers that--unaware of this divergence problem--use a velocity distribution function that does not apply to the respective experimental conditions. The present paper summarizes the conditions for which n = 2 or n = 3 is correct and gives information as to the systematic errors when using the incorrect exponent. In addition, since u and T depend on the respective exponent n, it is demonstrated how the enthalpy balance can be used to check the correctness of either v^2 or v^3.

Velocity Distribution in a Parallel Flow

For a gas at rest, the velocity distribution function, or differential density, is given by the Maxwell distribution f(V):

$$f(V) \equiv (1/N)\,dN/dV = (4/\sqrt{\pi})\,V^2 \exp(-V^2) \qquad (1)$$

N is the total number of particles in the respective volume, and dN is the fraction of N with speeds in the interval dV centered on V. For simplicity, we use nondimensional velocities:

$$V = v/\beta \qquad (2)$$

where $\beta = \sqrt{2kT/m}$ is the most probable thermal speed. The mean kinetic energy associated with the distribution function (Eq.(1)) equals (3/2)kT.

Another basic result of gas-kinetic theory (see, for example, Ref.7) is the distribution function for the particle flux g(V), or differential intensity:

$$g(V) \equiv (1/J)\,dJ/dV = 2V^3 \exp(-V^2) \qquad (3)$$

J is the total number of particles impinging on a surface element per unit time (particle flux), and dJ is the fraction of J with speeds in the interval dV centered on V. The mean kinetic energy associated with the distribution function (Eq.(3)) equals 2kT. Note that g(V) is valid for the differential intensity of the so-called oven beam. Both the differential density f(V) and the differential intensity g(V) are independent of the direction of the particle speeds.

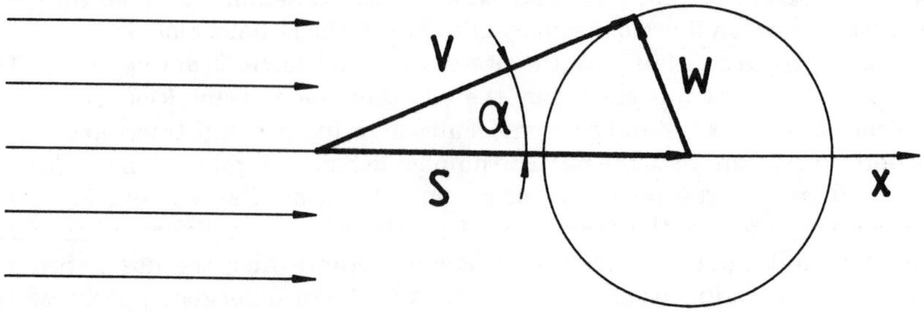

Fig. 1 Parallel flowfield. S = flow velocity, V = particle velocity, W = thermal velocity. All velocities in units of the most probable thermal speed, ß, see Eq.(2).

Changing from the system at rest to a system moving with the nondimensional velocity S (called speed ratio), see Fig. 1,

$$S = u/ß \tag{4}$$

which is oriented along the x axis (parallel flowfield), the differential density is now represented by a distribution function that is, of course, no longer isotropic but depends on the angle between the direction of S (x axis) and the direction of the particle velocity V; the same holds for the differential intensity. The result for the latter is

$$g(V,S,\alpha) \equiv (1/I)\, dI/dV = K_1 V^3 \exp(-(V - S\cos\alpha)^2) \tag{5}$$

where K_1 is the normalization constant; I, in contrast to J, the particle flux, or intensity, per unit solid angle in the direction α; and dI the fraction of I with speeds in the interval dV centered on V. If α = 0, Eq.(5) reduces to

$$g(V,S,\alpha = 0) = K_2 V^3 \exp[-(V-S)^2] \tag{6}$$

For S = 0, Eqs. (5) and (6) yield the oven-beam result of Eq. (3).

The variation of I with angle is given by[8]

$$I(\alpha)/I(\alpha = 0) = [(1.5 + S^2\cos^2\alpha)/(1.5 + S^2)]\cos^2\alpha \exp(-S^2\sin^2\alpha) \tag{7}$$

On a qualitative basis, comparison of Eqs. (5) and (6) shows that the distribution function for particles moving under an angle α changes in the sense that S is replaced by Scos, i.e., the "effective" speed ratio decreases with increasing α. Consequently, both the average velocity as well as the mean kinetic energy are highest for particles moving parallel to S, i.e., for α = 0. and decrease with increasing α. The contribution of these "slower" particles (α > 0) to the total energy flux decreases due to the factor $\exp(-S^2 \sin^2\alpha)$ in Eq. (7) so that flows with low values of S exhibit a wider angular range of particle trajectories (thermal divergence).

If a parallel flow is used to produce a molecular beam, by extracting a small portion with a skimmer, a detector placed far downstream of the skimmer registers only particles with speeds parallel to S, $\alpha = 0$, and the differential intensity is described by Eq. (6).

Velocity Distribution in a Divergent Flow

As shown in the schematic of Fig. 2, the flowfield used to produce the so-called nozzle beam is similar to that from a point source. For this divergent flow the particles seen by a detector far downstream of the skimmer have again velocities V parallel to the beam axis. With respect to the S vector characterizing the orientation of their respective streamtube, however, they are inclined by an angle which varies from zero to a maximum value α which is determined by the source-skimmer geometry, see Fig. 2:

$$\sin \alpha = r/x \tag{8}$$

Here r is the radius of the skimmer opening and x the skimmer distance. To obtain the intensity at the detector and the differential intensity, one has to sum up the contributions of the individual streamtubes in the skimmer plane. The result for I, correcting a typographical error in Ref.2, was

$$I = I_{is}\,[1 - \cos^3\alpha \exp(-S^2 \sin^2\alpha)] \tag{9}$$

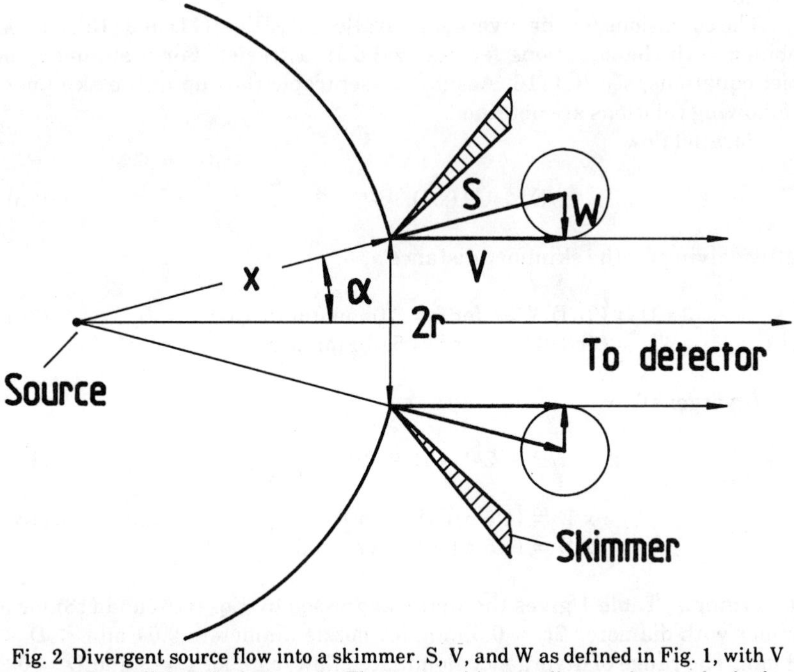

Fig. 2 Divergent source flow into a skimmer. S, V, and W as defined in Fig. 1, with V arranged parallel to the beam axis. The source-skimmer distance is x, and α is the angle toward the skimmer lip.

I_{is}, the isentropic intensity,[2] is the upper limit for the intensity on the beam axis.

For the differential intensity, we give only the (simpler) formulas for the two limiting cases: If the beam is divergence-dominated, the result is

$$g(V) = K_3 V^2 \exp[-(V-S)^2] \quad (10)$$

which is valid for

$$\sin \alpha \geq 2/S \quad (11)$$

In this case, the intensity is very close to the maximum value: From Eqs. (9) and (11), one gets

$$I \geq 0.98 I_{is} \quad (12)$$

The parallel case, for which Eq. (6) is valid, applies for

$$\sin \alpha \leq 1/(3S) \quad (13)$$

The intensity is now given by

$$I = I_{is} \sin^2\alpha (S^2 + 3/2) \leq I_{is}/9 \quad (14)$$

These results for I, Eqs. (12) and (14), show that I approaches its limit only, if the flow becomes divergence-dominated and thus all streamtubes that could contribute to the axial intensity are allowed to do so--thanks to a sufficiently big skimmer.

The conditions for the divergent/parallel case, Eqs. (11) and (13), can be combined with the equations for S(x) valid in a freejet. For a summary of freejet equations, see Ref. 16. Assuming isentropic flow up to the skimmer, the following relations are obtained:

Parallel flow:

$$r/D \leq 1/9\, (x/D)^{(f-2)/f} \quad (15)$$

or, after solving for the skimmer distance x,

$$x/D = (9r/D)^3 \quad \text{for } f = 3 \text{ (monatomic gas)} \quad (16)$$
$$x/D = (9r/D)^{5/3} \quad \text{for } f = 5 \text{ (diatomic gas)}$$

Divergent flow:

$$r/D \geq 2/3\, (x/D)^{(f-2)/f} \quad (17)$$

$$x/D = (1.5\, r/D)^3 \quad \text{for } f = 3 \quad (18)$$
$$x/D = (1.5\, r/D)^{5/3} \quad \text{for } f = 5$$

As an example, Table 1 gives the limits expressed by Eqs. (16) and (18) for a skimmer with diameter $2r = 0.5$ mm, for nozzle diameters $0.04 \text{ mm} < D < 0.28$ mm. The range of divergence-dominated flow, e.g., for $f = 3$ and $D = 0.2$ mm, ends at a distance of 1.32 mm, whereas the parallel flow begins at 284

Table 1 Limits for nozzle–skimmer distance[a]

Nozzle diameter, D/mm	Minimum distance, parallel flow, x/mm		Maximum distance, divergent flow, x/mm	
	f = 3	f = 5	f = 3	f = 5
0.04	7,119.14	33.03	32.96	1.67
0.08	1,779.79	20.81	8.24	1.05
0.12	791.02	15.88	3.66	0.80
0.16	449.95	13.11	2.06	0.66
0.20	284.77	11.30	1.32	0.57
0.24	197.75	10.00	0.92	0.50
0.28	145.29	9.03	0.67	0.46

[a]Calculated from Eqs. (16) and (18) for a skimmer diameter 2r = 0.5 mm. If the perpendicular temperature is higher than the respective isentropic value at the skimmer, the effective minimum distance beyond which the flow can be treated as parallel is longer than the value given in the table. Likewise, the maximum distance up to which the flow is divergence-dominated is shifted downstream, too. The parameter f = 3 and 5 refers to monatomic and diatomic gas, respectively.

mm. This rather long transition zone (factor $6^3 = 216$) reflects the factor of 6 difference in the term ($S\sin\alpha$) required--somewhat arbitrarily, see Eqs. (11) and (13)--to distinguish between the two types of flowfields.

The results of the present analysis need to be consistent with energy conservation: If the intensity on the axis is equal to the isentropic value (divergence-dominated), the energy balance must be that of an isentropic flow, too. As discussed earlier, the parallel flow has a higher on-axis velocity, and the energy conservation does not apply. The mean kinetic energy associated with the particle flux, whose differential intensity is g(V), is given by

$$\langle E_{kin}\rangle = 0.5m\langle v^2\rangle = 0.5m\beta^2\langle V^2\rangle \tag{19}$$

where $\beta^2 = 2kT/m$, as in Eq.(2), and

$$\langle V^2\rangle = \int_0^\infty V^2 g(V)dV \Big/ \int_0^\infty g(V)dV \tag{20}$$

The somewhat lengthy solutions of Eq.(20), including terms $\exp(-S^2)$ and $[1 + \mathrm{erf}(S)]$,[9] reduce for $S > 2$ to the following simple result for the mean kinetic energy $\langle E_{kin}\rangle$ valid for distribution functions g(V) with different pre-exponential factors V^n

$$\begin{aligned} \langle E_{kin}\rangle &= 0.5mu^2 + 1.5kT && \text{for } n = 1 \\ \langle E_{kin}\rangle &= 0.5mu^2 + 2.5kT && \text{for } n = 2 \\ \langle E_{kin}\rangle &= 0.5mu^2 + 3.5kt && \text{for } n = 3 \end{aligned} \tag{21}$$

where $g(V) = V^n \exp[-(V-S)^2]$. For an isentropic expansion, the flow velocity u must satisfy the enthalpy balance:

$$0.5mu^2 = 2.5\,(kT_0-kT) \quad \text{(monatomic gas)} \tag{22}$$

Comparison with Eq.(21) shows that only the distribution g(V) of Eq. (10), with n = 2, is consistent with this balance for an isentropic flow; the forms with V^1 and V^3 give lower and higher flux-averaged kinetic energy, respectively.

The problem of divergence has been discussed in Ref.10 in terms of the "virtual source model". For the divergence-dominated case, which they define by $S\sin\alpha > 1$ as compared with our more conservative > 2, see Eq. (11), they arrive at a flux distribution function $V^1\exp(-(V-S)^2)$, whereas for the parallel flow case, they also use $V^3\exp[-(V-S)^2]$. This result is inconsistent with the energy balance for an isentropic flow.

Consequences for the TOF-Analysis of Nozzle Beams

Figure 3 shows a TOF signal, obtained either as a real- time picture on the CRT of an oscilloscope or after processing with signal-averaging equipment. The TOF technique introduced for nozzle beams by Becker and Henkes[11] transforms the differential intensity into a corresponding distribution function of arrival times, and the appropriate transformations for the various types of detectors are well known.[11-13] For the usual density-sensitive through-flow ionization detector the ideal TOF signal is

$$U^0(t) = \text{Konst.}\; t^{-(n+1)} \exp\{-[L/(t\beta)-S]^2\} \tag{23}$$

L is the length of the flight-path chopper-detector, n the exponent in the V factor of g(V), and S and ß as defined in Eqs.(2) and (4). The two unknowns in Eq.(23), ß and S, or u and T, are fully determined by knowing two points of the

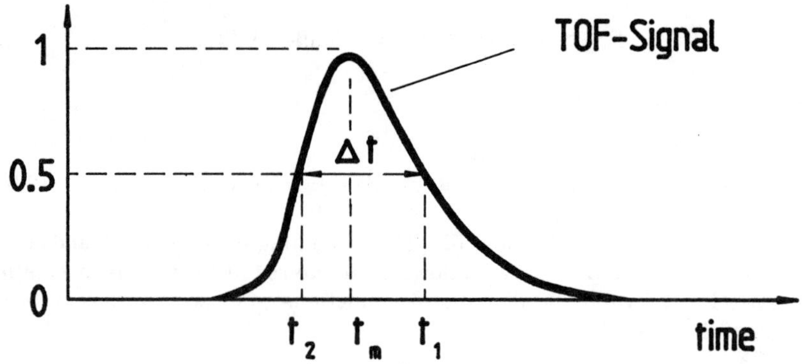

Fig. 3 Time-of-flight detector signal, showing the time variation of the signal due to the time-of-arrival distribution, including the characteristic times used in the signal analysis.

Table 2 Data for TOF analysis (n = 2 and 3 refers to the divergent and parallel flux distribution function and a density-sensitive detector)

	n = 2		n = 3	
S	t_m/t_1	t_2/t_1	t_m/t_1	t_2/t_1
1.50	0.67849	0.50646	0.70622	0.53889
1.75	0.69719	0.52948	0.72114	0.55810
2.00	0.71488	0.55168	0.73547	0.57680
2.25	0.73144	0.57285	0.74910	0.59484
2.50	0.74682	0.59289	0.76194	0.61211
2.75	0.76102	0.61173	0.77399	0.62853
3.00	0.77409	0.62939	0.78523	0.64408
3.25	0.78610	0.64588	0.79569	0.65876
3.50	0.79713	0.66127	0.80541	0.67258
3.75	0.80724	0.67561	0.81442	0.68556
4.00	0.81654	0.68897	0.82279	0.69775
4.25	0.82509	0.70142	0.83055	0.70920
4.50	0.83297	0.71304	0.83775	0.71995
4.75	0.84024	0.72389	0.84445	0.73004
5.00	0.84695	0.73402	0.85068	0.73952
5.50	0.85895	0.75239	0.86189	0.75682
6.00	0.86932	0.76855	0.87167	0.77217
6.50	0.87834	0.78285	0.88025	0.78584
7.00	0.88626	0.79558	0.88783	0.79806
7.50	0.89325	0.80696	0.89455	0.80904
8.00	0.89946	0.81718	0.90055	0.81895
8.50	0.90501	0.82641	0.90593	0.82792
9.00	0.91000	0.83479	0.91078	0.83608
9.50	0.91450	0.84241	0.91517	0.84353
10.00	0.91859	0.84937	0.91916	0.85035
12.00	0.93169	0.87210	0.93204	0.87269
14.00	0.94121	0.88893	0.94143	0.88932
16.00	0.94842	0.90189	0.94857	0.90216
18.00	0.95407	0.91217	0.95417	0.91235
20.00	0.95861	0.92050	0.95868	0.92064

signal curve. As shown in Fig. 3, the usual choice is t_m and t_1 and/or t_2, the times for the maximum and the half-intensity. The result is[12]

$$u = h(a)\,(L/t_m) \tag{24}$$

$$\beta^2 = 2kT/m = g(a)\,(L/t_m)^2 \tag{25}$$

$$S = u/\beta = h(a) / \sqrt{(g(a))} \tag{26}$$

Table 3 Effect of exponent n (divergent/parallel type distribution) on the speed distribution characteristics S, g, h, and t_m/t_o

		n = 2				n = 3			relative change		
1	2	3	4	5	6	7	8	9	10	11	12
t_m/t_1	S	g	h	t_m/t_o	S	g	h	t_m/t_o	dg/g	dh/h	dt/t
0.68	1.52	0.2064	0.6904	0.9963	1.07	0.2379	0.5243	0.9325	−0.153	0.241	0.064
0.70	1.79	0.1720	0.7419	0.9902	1.40	0.1929	0.6141	0.9271	−0.122	0.172	0.064
0.72	2.08	0.1432	0.7853	0.9872	1.73	0.1571	0.6859	0.9290	−0.097	0.127	0.059
0.74	2.39	0.1186	0.8221	0.9861	2.08	0.1279	0.7443	0.9347	−0.078	0.095	0.052
0.76	2.73	0.0977	0.8535	0.9862	2.46	0.1037	0.7926	0.9421	−0.062	0.071	0.045
0.78	3.12	0.0796	0.8805	0.9871	2.88	0.0835	0.8329	0.9501	−0.049	0.054	0.038
0.80	3.57	0.0641	0.9038	0.9885	3.36	0.0666	0.8668	0.9580	−0.039	0.041	0.031
0.82	4.10	0.0508	0.9238	0.9902	3.91	0.0523	0.8954	0.9656	−0.030	0.031	0.025
0.84	4.74	0.0394	0.9409	0.9919	4.58	0.0403	0.9195	0.9727	−0.023	0.023	0.019
0.86	5.55	0.0297	0.9555	0.9936	5.41	0.0302	0.9397	0.9790	−0.017	0.017	0.015
0.88	6.60	0.0215	0.9677	0.9951	6.48	0.0218	0.9565	0.9845	−0.012	0.012	0.011
0.90	8.05	0.0148	0.9778	0.9965	7.95	0.0149	0.9702	0.9892	−0.008	0.008	0.007
0.92	10.19	0.0094	0.9859	0.9978	10.11	0.0094	0.9812	0.9931	−0.005	0.005	0.005

where $a = t_m/t_1$ or t_m/t_2. The functions g and h are given with n as a parameter:

$$g(a) = (a-1)^2 / [\ln 2 + (n+1)(1-a+\ln a)] \quad (27a)$$

$$h(a) = 0.5\,[2-(n+1)g(a)] \quad (27b)$$

It is often more convenient to use only the two half-intensity points for data analysis.[14] This requires a numerical solution of Eq.(26), which gives the pair of a values for a given S and thus the value for t_2/t_1. Table 2 lists the data obtained for $n = 2$ and $n = 3$. For a given S, going from $n = 2$ to 3 gives higher values for both time ratios, with the differences decreasing with increasing speed ratio S.

For isentropic expansions, u and T are not independent but correlated through the energy balance with the source temperature T_o: From Eq. (22) and with the definition of the limiting velocity

$$u_o = L/t_o = \sqrt{(2.5kT_o/m)} \quad (28)$$

the combination of Eqs. (24), (25), and (27) gives an expression for the variation of the ratio t_m/t_o with a:

$$(t_m/t_o)^2 = h(a)^2 + 2.5g(a) \quad (29)$$

Table 3 lists the results for S, g, h, and t_m/t_o as functions of the measured time ratio t_m/t_1. Data are given for both $n = 2$ and 3, and the last three columns (10-12) give the relative changes,$[g(2)-g(3)]/g(2)$. The data demonstrate the importance of the proper choice for the exponent n: For $S > 10$ the effects are below 0.5% but increase to around 3% for $S \approx 3$ up to as much as some 10% for $S < 2$. Of special interest is the comparison of columns 5 and 9: For $n = 2$, t_m is essentially constant, being at most 1.4% lower than t_o; for $n = 3$, however, it varies by 7%.

For nozzle beams with high speed ratios $S > 10$, the preceding effects are negligible. Flow velocity is obtained from t_m, and S is represented accurately by[14]

$$S = [2\sqrt{(\ln 2)}] \cdot t_m / (t_1 - t_2) = 1.665\, t_m / \Delta t \quad (30)$$

Concluding Remarks

Thus far, the present analysis has not considered a number of complicating issues that may change the expected distribution function. First to mention are the effects in the transition from continuum, collision-dominated flow to collisionless molecular flow. As discussed earlier,[2] it is the perpendicular velocity component that determines the thermal spread, and the variation of $T_\perp$ has to be known. Certain features of the simplified theory can be readily checked by experiment, i.e., the effect of increasing skimmer diameter 2r: For parallel flow, intensity should increase in proportion to skimmer area, see Eq. (14). For divergent flow conditions, enlarging the skimmer should enlarge the beam profile, not its intensity--assuming otherwise ideal skimmer performance, i.e., neglible distortions due to scattering losses up- and downstream of the skimmer.

Fortunately, the success of most experiments using nozzle beams is not endangered because of uncertainties about the proper velocity distribution function: The beam is simply used as a source of an intense, highly collimated molecular beam with little spread in particle velocities. If, however, one starts to extract values for u, T, or S from a measured TOF signal and wants to establish energy balances for the beam,[15] one has to make a careful analysis as to whether the divergent or parallel flow distribution function has to be used.

References

[1]Stern, O., "Eine direkte Messung der thermischen Molekulargeschwindigkeit," Zeitschrift für Physik, Vol. 3, 1920, pp. 417-421.

[2]Hagena, O. F. and Morton, H. S., "Analysis of Intensity and Speed Distribution of a Molecular Beam from a Nozzle Source," Rarefied Gas Dynamics, edited by C. L. Brundin, Academic Press, New York, 1967, pp. 1369-1384.

[3]Alcalay, J. A. and Knuth, E. L., "Molecular-Beam TOF Spectroscopy," Review of Scientific Instruments, Vol. 40, 1969, pp. 438- 447.

[4]LeRoy, R. L. and Govers, T. R., "Ideal Intensities of Supersonic Molecular Beams," Canadian Journal of Chemistry, Vol. 48, 1970, pp. 1743-1747.

[5]Andersen, J. B., "Molecular Beams from Nozzle Sources," Molecular Beams and Low Density Gasdynamics, edited by P. P. Wegener, Marcel Dekker, New York, 1974, pp. 1-91.

[6]Scott, J. E., "Some Aspects of Atomic and Molecular Beams Produced by Aerodynamic Acceleration," Entropie, Vol. 30, 1969, 1-11.

[7]Ramsey, N. F., Molecular Beams, Clarendon Press, Oxford, 1956.

[8]Morton, H.S., private communication, 1965

[9]Wilmoth, R. G. and Hagena, O. F., "Scattering of Argon and Nitrogen off Polycrystalline Nickel," Report AEEP-4038-105-67U, Univ. of Virginia, RLES, Charlottesville, VA, August 1967.

[10]Beijerinck, H. C. W., Menger, P., and Verster, N. F., "Shape Analysis of Non-Maxwell Boltzmann Parallel Velocity Distributions in Supersonic Expansions of Noble Gases," Rarefied Gas Dynamics, edited by R. Campargue, CEA Press, Paris, 1979, pp. 871-884.

[11]Becker, E. W. and Henkes, W., "Geschwindigkeitsanalyse von Lavalstrahlen," Zeitschrift für Physik, Vol. 146, 1956, pp. 320- 332.

[12]Hagena, O. F. and Henkes, W.,"Thermische Relaxation bei Düsenströmungen," Zeitschrift für Naturforschung, Vol. 15a, 1960, pp. 851-858.

[13]Hagena, O. F. and Varma, A. K., "TOF Analysis of Atomic and Molecular Beams," Review of Scientific Instruments, Vol. 39, 1968, pp. 47-52.

[14]Hagena, O. F., Scott, J. E., and Varma, A. K., "Design and Performance of an Aerodynamic Molecular Beam and Beam Detection System," Rept. AST-4038-103-67U, Univ. of Virginia, RLES, Charlottesville, VA, June 1967.

[15]Haberland, H., Buck, U., and Tolle, M., "Velocity Distribution of Supersonic Nozzle Beams," Review of Scientific Instruments, Vol. 56, 1985, pp. 1712-1716.

[16]Hagena, O. F., "Nucleation and Growth of Clusters in Expanding Nozzle Flows," Surface Science, Vol. 106, 1981, pp. 101-116.

Cryogenic Pumping Speed for a Freejet in the Scattering Regime

J.-Th. Meyer*
DFVLR, Göttingen, Federal Republic of Germany

Abstract

Pumping of a free jet with a cryogenic pump in the presence of a background gas is investigated. The theoretically constant pumping speed of the cryopump is observed only in a small range of stagnation pressures. The loss in pumping speed can be explained by a scattering of the free jet, which is characterized by the length r_{bbs} of the free molecular part of the jet and a parameter Kn_{bbs} describing the background gas penetration into the jet.

Nomenclature

A	= area, m^2
$b\,(\Theta)$	= mass flow distribution, see Eq. (6)
$\bar{c}$	= mean thermal speed, m/s
d^*	= sonic orifice diameter, m
Kn_{bbs}	= background penetration parameter, see Eq. (5)
$\dot{m}_{tot}$	= total mass flux, kg/s
$\dot{m}_{th}$	= see Eq. (8), kg/s
$\dot{N}$	= particle flux, 1/s
$\dot{N}_{ev}$	= evaporated particle flux from the cryopump, 1/s
n	= particle density, 1/s
$n_{j,ref}$	= particle density of the jet at a reference point, 1/m^3

*Research Scientist, Institute for Experimental Fluid Mechanics.

p = pressure, Pa

R = gas constant, J/kg · K

r = distance measured from the source point along a streamline, m

r_{bbs} = see Eq. (4), m

r_c = radius of the cryopump, m

$\overline{S}_p$ = mean pumping speed, m^3/s ; see Eq. (7)

$S_{p,th}$ = thermal pumping speed, m^3/s

T = temperature, K

$u_{j,\infty}$ = limiting value of free-jet macroscopic velocity, m/s

w = transmission probability

α = sticking coefficient

Θ = angle from the centerline of the free-jet, rad

Θ_c = angle from the centerline of the free jet to the cryopump, rad (see Fig. 4)

λ = mean free path, m

κ = ratio of specific heats

Subcripts

b = background

bj = background disturbed by the jet

c = cryopump

j = jet

jb = jet disturbed by the background

Introduction

With the use of cryopumps, it is possible to obtain a very low final pressure in a vacuum chamber. The pumping speed is a function of the surface area and the velocity of the particles hitting the cryopump surface. To increase pumping efficiency, instead of pumping at background conditions with a low random thermal particle speed, a configuration is desirable where flow particles directly hit the surface.

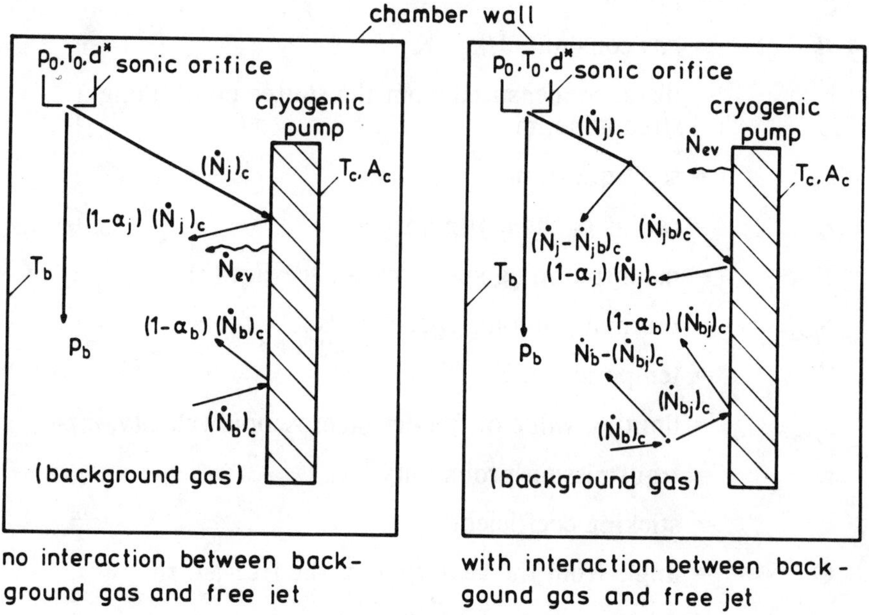

Fig. 1 Particle fluxes to the cryopump surface.

Cryogenic Pumping Speed

In Fig. 1a, the particle fluxes to the cryopump are shown. The jet particle flux $\dot{N}_j$ expands from a sonic orifice into the vacuum chamber. A portion $(\dot{N}_j)_c$ condenses directly on the cryopump surface with the sticking coefficient α_j. Furthermore, a particle flux $(\dot{N}_b)_c$ of background gas molecules condenses on the cryopump with a sticking coefficient α_b. Taking into account the evaporated particle flux $\dot{N}_{ev}$, the total condensed particle flux becomes

$$\dot{N}_c = (\alpha_j \dot{N}_j + \alpha_b \dot{N}_b)_c - \dot{N}_{ev} \quad [1/s] \tag{1}$$

To obtain the cryogenic pumping speed, Eq. (1) is divided by the background number density

$$S_p = \dot{N}_c / n_b \quad [m^3/s] \tag{2}$$

Equation (1) is valid under two assumptions:

1) Background gas particles and jet particles reach the cryopump surface without any interaction.

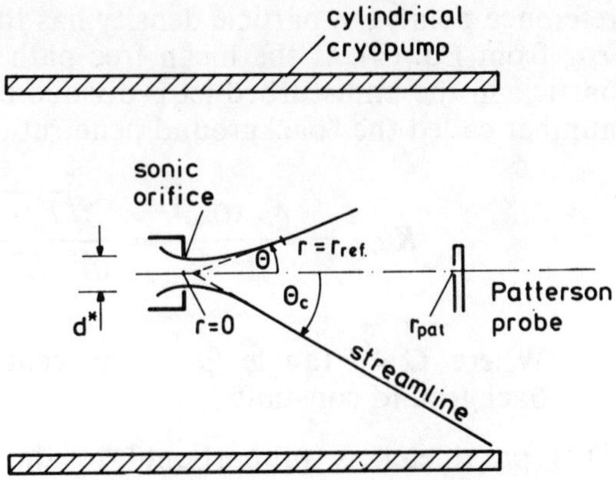

Fig. 2 Coordinate system.

2) Background gas does not penetrate into the free jet. In this case only free jet particles condense on the cryopump surface, $(\dot{N}_b)_c = 0$.

If the first assumption is violated, the particle flux is reduced due to interaction (see Fig. 1b)

$$\dot{N}_c = (\alpha_j \dot{N}_{jb} + \alpha_b \dot{N}_{bj})_c - \dot{N}_{ev} \quad [1/s] \tag{3}$$

$\dot{N}_{jb}$ is the attenuated flux of free-jet particles at the cryopump surface resulting from collisions with background gas. $\dot{N}_{bj}$ is the particle flux onto the pump from the disturbed background.

Definition of the Scattering Region

To get an understanding of the physical parameters responsible for the attenuation of the jet and the background particle flux, their interaction was treated theoretically with a simple model proposed by Muntz et al [1]. This model assumes an exponential decay of the background particle density upstream along a streamline. The distance where the frequency of mutual background particle collisions equals the collision frequency of background particles with free-jet particles is calculated for each streamline (r_{crit} in Ref.1) as:

$$r_{bbs} = r_{ref} \sqrt{\frac{n_{j,ref}}{n_b} \ \frac{1}{\sqrt{2}} \ \left(\frac{u_{j,\infty}}{\bar{c}_b} + 1 \right)} \quad [\mathrm{m}] \tag{4}$$

r_{ref} is the distance of a reference point from the virtual source point on a streamline where the jet is undisturbed (see Fig. 2). At this reference point, the particle density has to be calculated to compute r_{bbs} from Eq. (4). If the mean free path λ_{bj} of a background gas particle in the undisturbed jet is divided by r_{bbs}, a type of Knudsen number called the "background penetration parameter" is obtained:

$$Kn_{bbs} = \frac{\lambda_{bj}(r_{bbs})}{r_{bbs}} = \frac{\sqrt{T_o \cdot T_b}}{d^* \sqrt{p_o \cdot p_b}} \; C \qquad (5)$$

Where $C = f$ (angle Θ to the centerline, free- jet, and background constants).

This parameter is proportional to the reciprocal value of the Muntz-Hamel-Maguire-rarefaction parameter in Ref. 1. For the experiments, C has a value of

$$C(\Theta = 0 \text{ deg}) = 4.498 \text{ x } 10^{-5} \qquad [\text{Pa} \cdot \text{m/K}]$$

on the centerline. Kn_{bbs} also can be expressed as the ratio r_{bbs}/r_p; r_p is the distance where the background particles are attenuated to 1/e by the jet $(n_{bj} = n_b \cdot e^{-r_p/r})$.

With the help of r_{bbs}, r_p, and Kn_{bbs}, it is possible to distinguish three background-jet interaction regions in a first approach (see Fig. 3):

1) $Kn_{bbs} > 1$: Background gas particles travel very far into the free jet. The undisturbed free jet is small.

2) $Kn_{bbs} = 1$: Only a small portion of background particles reaches the region inside r_{bbs}.

3) $Kn_{bbs} < 1$: The region inside r_{bbs} becomes free of background gas particles.

Obviously the best working conditions of the cryopump should be achieved for $Kn_{bbs} < 1$ and the distance r_{bbs} larger than the length of the cryopump cylinder.

Experiments

The experimental set up is shown in Fig. 4. A cylindrical cryogenic pump was attached to two cold heads, which supply 20 W at a temperature of 20 K. A free jet from a sonic orifice was positioned

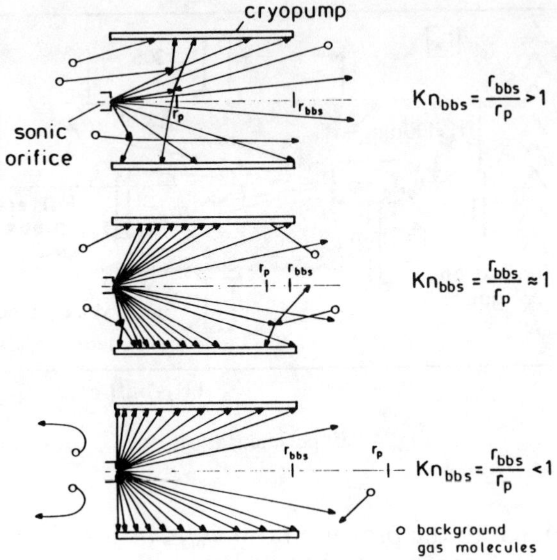

Fig. 3 Scattering regions in a cryopumped free jet.

on the centerline of the cryopump. A defined part

$$b(\Theta) = \int_{0}^{\Theta_c} \dot{m}(\Theta)\, d\Theta \,/\, \dot{m}_{tot} \tag{6}$$

of the free-jet mass flux $\dot{m}_{tot}$ reaches the background gas directly. The angle Θ_c is the cone half- angle of streamlines from the sonic orifice to the edge of the cylinder. During the experiments, it is possible to vary $b(\Theta)$ by shifting the position of the sonic orifice.

As a test gas only N_2 was used with a stagnation temperature of $T_o = 293$ K. The stagnation pressure was varied between 100 and 10,000 Pa.

The background gas pressure, p_b, was measured by an ionization gage at different locations. The wall temperature, T_b, of the liquid nitrogen (LN_2) - cooled surrounding baffle was measured with thermocouples. The temperature of the cryopump was determined by PT-100 resistances and an H_2 vapor pressure manometer.

From p_o, T_o, p_b, and T_b, the mean pumping speed is calculated:

$$\overline{S}_p = \dot{m}_{tot}\; R\; T_b \,/\, p_b \sim p_o \,/\, p_b \quad [\mathrm{m}^3/s] \tag{7}$$

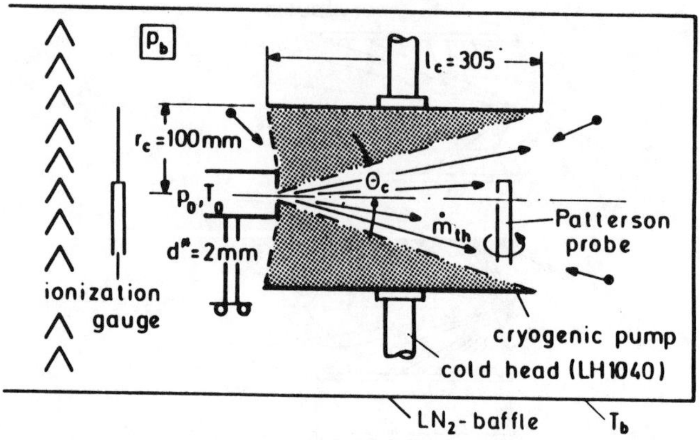

Fig. 4 Experimental set-up.

It is proportional to the pressure ratio p_0/p_b, which is the measured quantity. With the thermally pumped mass flow

$$\dot{m}_{th} = b(\Theta)_{ex}\ \dot{m}_{tot} \quad [\text{kg/s}] \tag{8}$$

which is the portion of the total mass flow entering the background directly, the angular mass flux distribution is measured by

$$b(\Theta)_{ex} = S_{p,th} \,/\, \overline{S}_p = \dot{m}_{th} \,/\, \dot{m}_{tot} \tag{9}$$

To determine $b(\Theta)_{ex}$, the knowledge of $S_{p,th}$ is necessary by which $\dot{m}_{th}$ can be calculated. $S_{p,th}$ was measured using a different gas inlet, so that the total mass flux was pumped thermally as background gas. To obtain the thermal pumping speed of the inner and outer surfaces of the cryopump separately, the outer or inner surface of the cylinder was covered. The background pressure was varied by an artificial leakage. Pressures in the free jet were measured with a free molecular pressure probe (Patterson probe) turned into the flow direction and perpendicular to it.

Results and Discussion

In Fig. 5, the theoretical value of the necessary cooling power vs the stagnation pressure is shown for thermal and direct pumping. As a result of the higher velocity of the direct flow, the supplied cooling power is consumed at lower mass fluxes; i.e., lower stagnation pressures than in the case of a thermal gas inlet. The experimental points will be explained subsequently.

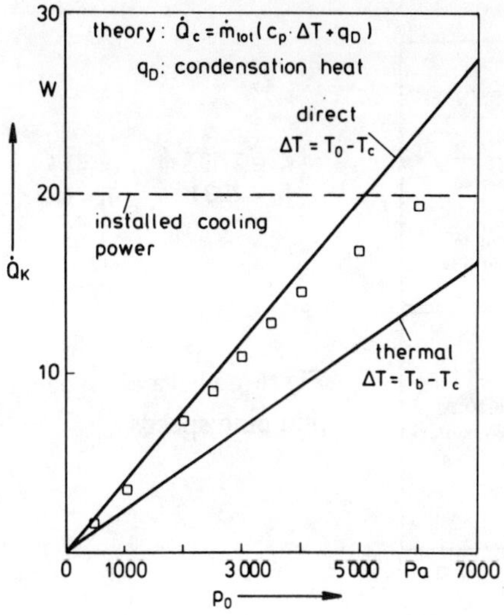

Fig. 5 Cooling power of the cryopump and heat release.

The measurements of the thermal pumping speed are shown in Fig. 6. The gas inlet was positioned such that all particles reached the surface with the LN_2-baffle temperature. Three different configurations were tested: all surfaces of the cryopump were accessible to background particles, the inner surface was covered, and, finally, the outer surface was covered. As a result, the measured pumping speed depends on the background pressure. The reason seems to be that the accessibility to the cylinder surfaces for the particles depends on their mean free path. In Fig. 6, a theoretical free molecular value is shown:

$$S_{p,th} = w \; [\, \bar{c}\,(T_b) \,/\, 4 \,] \; 2 \; \pi \; r_c^2 \quad [\mathrm{m}^3/\mathrm{s}] \tag{10}$$

The transmission probability w accounts for those particles traversing the cylinder without hitting the wall. It has the value $w = 0.6$ after Clausing.[2] From the equation for continuum mass flow into vacuum, an upper limit for a cylinder of infinite length is obtained as

$$S_{p,th} = \sqrt{R \; T_b} \;\; 2 \; \pi \; r_c^2 \cdot \Gamma(\kappa) \quad [\mathrm{m}^3/\mathrm{s}] \tag{11}$$

where $\Gamma(\kappa) = 0.685$ for N_2.

The existence of two limits is also detected if the inner surface is covered. This is caused by the asymmetric installation of the cylinder in the vacuum chamber (see Fig. 6).

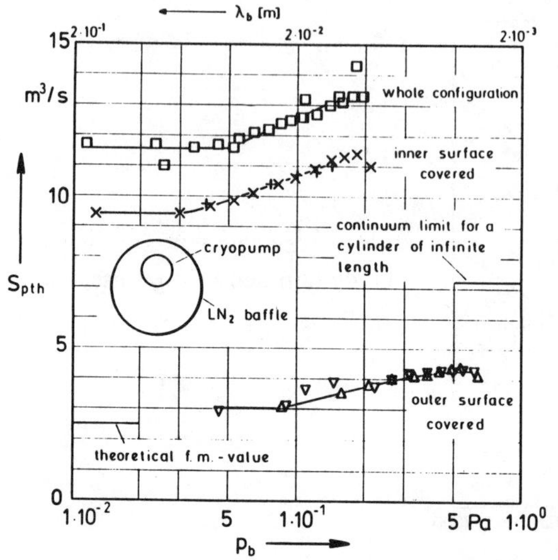

$d^* = 2 \cdot 10^{-3}$ m , $\varkappa = 1.4$
$T_0 = 293$ K , $T_b = 120$ K

Fig. 6 Thermal cryogenic pumping speeds.

Direct pumping measurements were performed in two different vacuum chambers. For evaluation of these experiments, the measured values, p_b and p_o, are plotted, because the mean pumping speed after Eq. (7) includes direct and thermal pumping. The ratio p_b/p_o is proportional to the reciprocal of the pumping speed. The aim of the experiments is to accomplish a low background pressure.

The dimensionless background pressure vs the stagnation pressure is shown in Fig. 7. The constant theoretical value of the pressure ratio (at $S_{p,th} = 15.4$ m^3/s , $b(\Theta) = 0.125$) and the values of r_{bbs} and Kn_{bbs} at three selected points are added. In the lower stagnation pressure region ($p_o < 1000$ Pa), the value of Kn_{bbs} is greater than 1. This shows that the background gas can penetrate very far into the free jet. Furthermore, the distance r_{bbs} is 80% of the cylinder length. Hence, the jet is disturbed before reaching the cryopump surface, and the pumping speed is reduced. In the range of 2000 Pa $< p_o <$ 4000 Pa , the expected constant pumping speed was observed. In this region, Kn_{bbs} is nearly 1 and r_{bbs} is 120% of the cylinder length. This means that the background gas cannot penetrate the jet, and the jet reaches the cryopump surface nearly undisturbed. At stagnation pressures $p_o >$ 4000 Pa, the temperature of the cryopump begins to rise as a result of limited cooling power (see Fig. 5). This causes an increase in background pressure as a result of increased evaporation and reduced condensation. Therefore, although Kn_{bbs} is less than 1 as desired, r_{bbs} reduces to 75% of the cylinder length. That

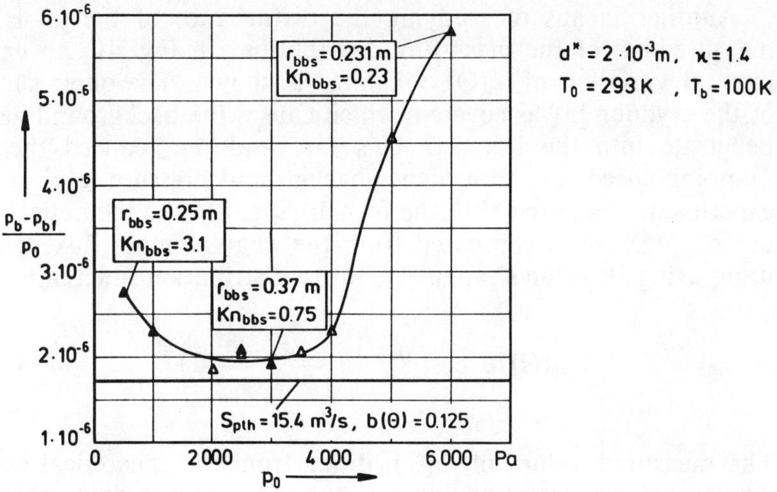

Fig. 7 Pressure ratio p_b/p_o versus stagnation pressure.

means the free-jet molecules cannot reach the cryopump surface directly from the source.

In Fig. 8, results of experiments are shown performed under three different final packground pressures p_{bf}, which are set through an artificial leackage before starting the jet. This causes different behavior in the scattering regime. At a stagnation pressure of approximately 3000 Pa, a sudden drop in the background pressure ratio occurs. A comparison of the Mach disk location[3] with the distance of the freezing surface[4] reveals that a shock cannot cause this behavior. As this sudden drop of p_b/p_o has been observed under different environmental conditions, $Kn_{bbs} \approx 1$, the explanation is still pending. From this point on a constant increase in p_b/p_o or a decrease in the pumping speed is observed as a result of the mechanisms described earlier.

A first step for the investigation of the flow inside the cryopump was done using a freemolecular pressure probe. The experimental and theoretical pressures of the Patterson probe results are shown in Fig. 9. The pressure probe is at the same position in the cryopump for both measurements, but the orifice was moved 20 mm. The measurements do not show any dramatic effect, as in Fig. 7. In the lower stagnation pressure regime, the measured relative mass flux to the probe is larger, which is caused by scattering and the fact that background molecules can reach the pressure probe directly through the rarefied jet. For higher stagnation pressures the measurements approach a constant value.

Another means of studying the disturbance of the jet is the measurement of the mass flux distribution. In Fig. 10, an experimental variation of $b(\Theta) = \dot{m}_{th}/\dot{m}_{tot}$ is shown. The outer surface of the cryopump was covered, which causes the background gas to penetrate into the free jet. This fact and the reduced thermal pumping speed lead to a higher background pressure level in this experiment compared with the former (Fig. 8). The theoretical values of $b(\Theta)$ were computed from the angular mass flux distribution using Boynton's[5] angular density distribution function:

$$\rho(\Theta) \approx \cos^{2/(\kappa-1)}\left(\frac{\pi}{2} \cdot \frac{\Theta}{\Theta_{lim}}\right) \qquad (12)$$

The measured values of $b(\Theta)$ differ from the theoretical curve. Analysis of the corresponding r_{bbs} and Kn_{bbs} values shows that the extension of the jets is the same for all stagnation pressures, but the ability of the background gas to penetrate into the jet is smallest for $p_o = 3000$ Pa. Hence the deviation from the theoretical curve is smallest in this case.

To investigate the behavior of the cryogenic pumping speed in a higher background pressure region, an experiment with the same configuration as described in Fig. 10 was performed. The orifice was kept at a fixed position, and the stagnation pressure was varied between 100 and 8000 Pa. The results are shown in Fig. 11. As

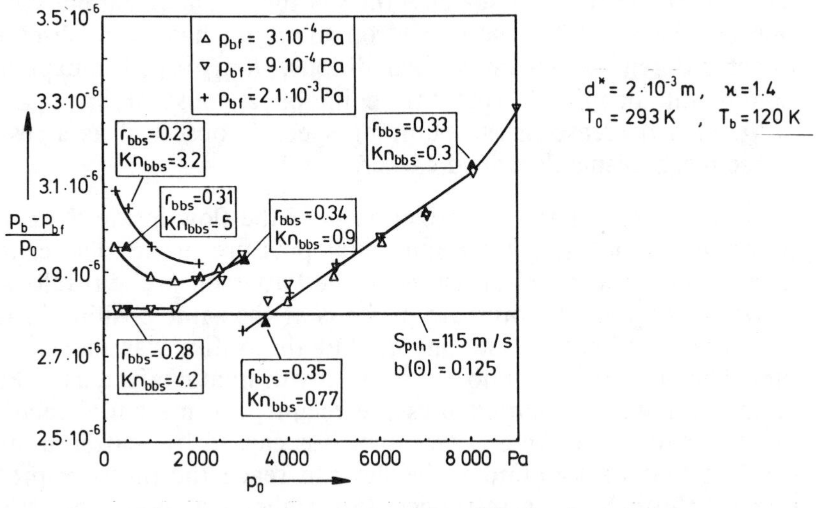

Fig. 8 Direct cryogenic pumping with different final pressures.

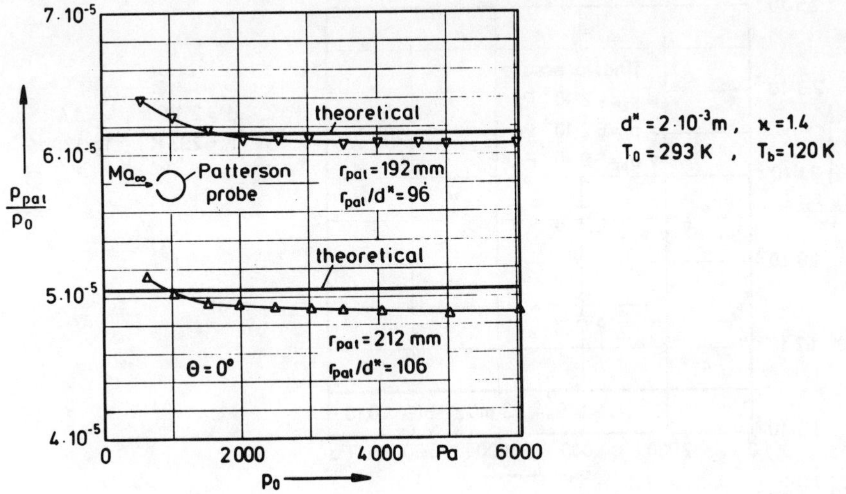

Fig. 9 Patterson probe measurements inside the cryopump.

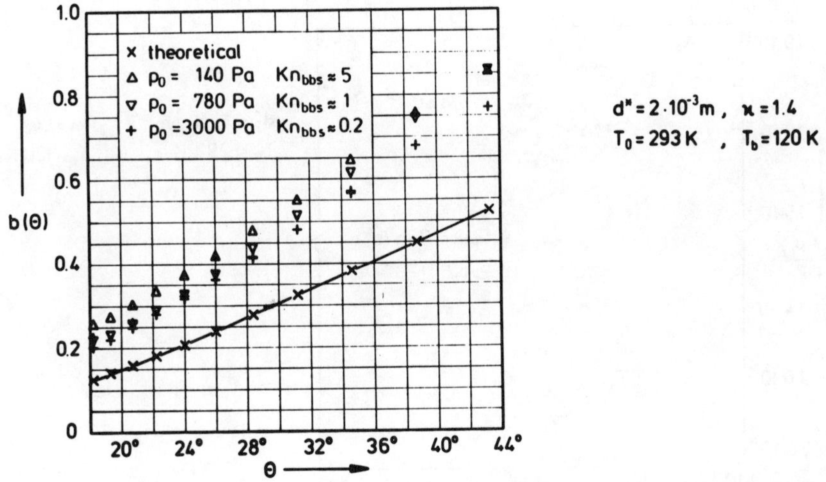

Fig. 10 Variation of $b(\Theta) = \dot{m}_{th}/\dot{m}_{tot}$ with Θ for different stagnation pressures.

starting conditions, two different final pressures p_{bf} were used. In the case of the higher final pressure, the theoretical value of the pressure ratio is not reached because the jet dimension (i.e., r_{bbs}) is too small. In the case of the lower final pressure, scattering occurs at low stagnation pressures. The jet dimension is greater than in the first case; however, as a result of the high Kn_{bbs}, the jet does not reach the surface undisturbed. At a stagnation pressure of 1000

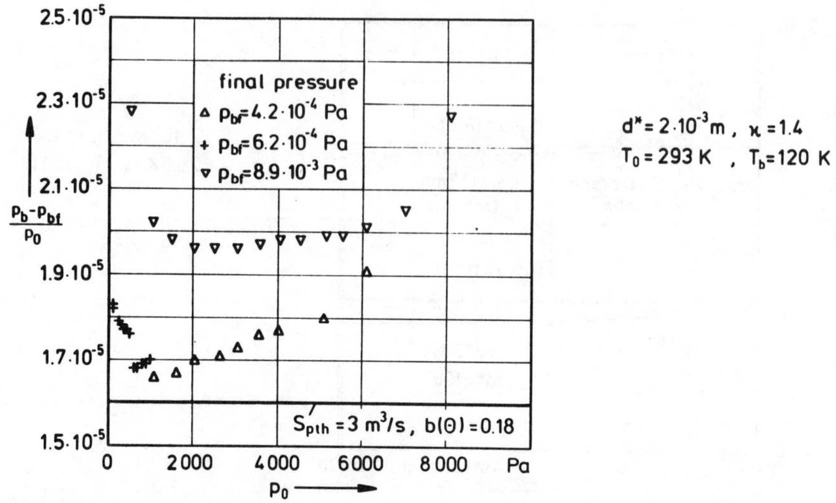

Fig. 11 Measurement of $(p_b - p_{bf})/p_o$ as function of p_o at low and high final pressures p_{bf}.

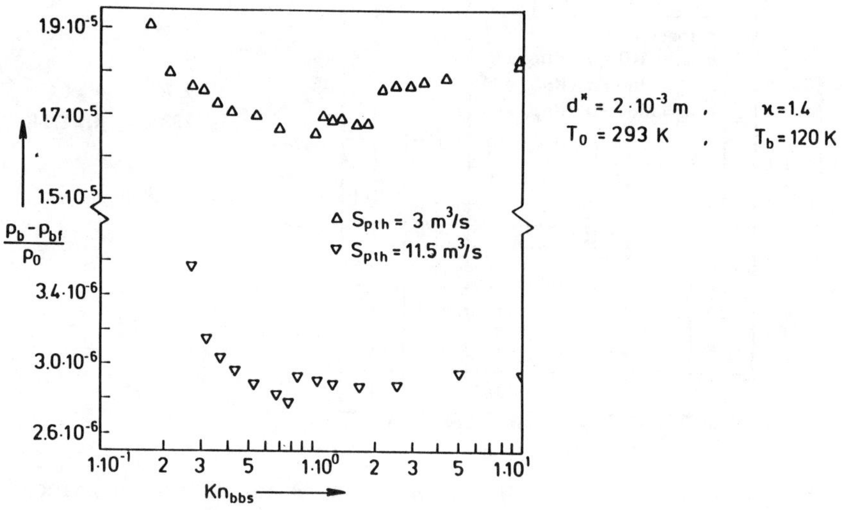

Fig. 12 Correlation of $(p_b - p_{bf})/p_o$ with Kn_{bbs} at different environmental conditions and thermal pumping speeds.

Pa, the lowest value of p_b/p_o is reached. Kn_{bbs} is nearly 1 and r_{bbs} is 65% of the cylinder length. For higher stagnation pressures, Kn_{bbs} becomes lower and r_{bbs} decreases as a result of higher background pressure, which is caused by the reduced thermal pumping speed.

In Fig. 12 $(p_b - p_{bf})/p_o$ is plotted vs the background penetration parameter Kn_{bbs}. Figure 12 is derived from the lower curve in Fig. 11 and an experiment of Fig. 8. Despite the different levels of background pressure, which are caused by different thermal pumping speeds, the optimal work conditions are achieved for $Kn_{bbs} \approx 1$, confirming that Kn_{bbs} is a useful correlation parameter.

Conclusion

The experiments have shown that the background gas pressure affects the achieved pumping speed of a cylindrical cryogenic pump, which surrounds a free jet. In the region of $p_o < 4000$ Pa , with $d^* = 2$ mm, and $T_o = 293$ K , the cooling power of the cryopump was sufficient to maintain a temperature of 20 K. In this region, the actual loss in the pumping speed is caused by flow phenomena. The thermal pumping speed of a cylinder is dependent on the mean free path in the background gas.

The loss in the direct pumping speed for a free jet is caused by the penetration of background molecules into the jet. Collisions lead to a deflection of the jet particles and to a decrease of the jet particle velocity. On the other hand, background molecules colliding with jet molecules can reach the cryopump surface with a greater probability due to the increase in directed velocity onto the cryosurface. The experiments show that the effect of jet disturbance by background molecules dominates.

The influence of the background gas can be analyzed by evaluating the distance r_{bbs} and the background penetration number Kn_{bbs}. With the help of these quantities, it is possible to identify the operating conditions where scattering occurs.

For a closer look at the jet disturbance caused by background gas penetration and its influence on direct cryogenic pumping, the use of different gas species and a higher cooling power are necessary. It seems possible to extend the current model to support the planning of these advanced investigations.

References

[1]Muntz, E. P., Hamel, B. B. and Maguire, B. L., "Some Characteristics of Exhaust Plume Rarefaction", AIAA Journal Vol. 8, Sept. 1970, pp. 1651-1658.

[2]Clausing, P., "Über die Strömung sehr verdünnter Gase durch Röhren von beliebiger Länge", Annalen der Physik , Vol. 12, 5.Folge, 1932, p. 961.

[3]Ashkenas, H. and Sherman, F. S., "The Structure and Utilization of Supersonic Free Jets in Low Density Wind Tunnels", Rarefied Gas Dynamics , Fourth Symposium, Supplement 3 , Vol. II, edited by J. H. de Leeuw, Academic Press, New York, 1966, p. 84.

[4]Bird, G. A., "Breakdown of Continuum Flow in Free-Jets and Rocket Plumes", Progress in Astronautics and Aeronautics : Rarefied Gas Dynamics , Vol. 74, p. 2, edited by S. S. Fisher, AIAA, New York, 1981, pp. 681-694.

[5]Boynton, F. P., "Highly Underexpanded Jet Structure. Exact and Approximate Calculations". AIAA Journal Vol. 5, Sept. 1967, pp. 1703-1705.

Effectiveness of a Parallel Plate Arrangement as a Cryogenic Pumping Device

K. Nanbu,* Y. Watanabe,† and S. Igarashi†
Tohoku University, Sendai, Japan
and
G. Dettleff‡ and G. Koppenwallner§
German Aerospace Research Establishment, Göttingen, Federal Republic of Germany

Abstract

Effectiveness of a parallel plate arrangement as a cryogenic pumping device is examined by use of the direct simulation Monte Carlo method. The arrangement consists of two parallel plates of 80 and 20°K and an end plate. A free molecular flow is first incident on the 80°K plate cooled by liquid nitrogen. Because of the reflection on the surface of this plate, the molecules lose most of their kinetic energy and slowly leave for the 20°K plate cooled by liquid helium. After undergoing some molecular collisions in the gap between the two plates, most molecules finally arrive at the surface of the 20°K plate and are condensed. This paper examines the effects on the amount of noncondensing flux of the condensation coefficient for the 20°K plate, the density of the incoming stream, and the geometry of the arrangement.

I. Introduction

The characteristics of apparatus such as attitude control thruster used in space is measured in vacuum chamber with nonzero background gas pressure. The background gas often has an appreciable effect on the measured data. One of the classical techniques for reducing the background

---

*Professor, Institute of High Speed Mechanics.
†Research Associate, Institute of High Speed Mechanics.
‡Research Scientist, Institute for Experimental Fluid Mechanics.
§Division Chief, Institute for Experimental Fluid Mechanics.

gas pressure is the cold trap; molecules incident on a cold surface are condensed on it. In this paper we propose an arrangement with two cold plates shown in Fig. 1 and examine its effectiveness as a cryogenic pumping device. Our future goal is to cover the whole inner surface of a vacuum chamber with an assembly of such devices and to attain a negligibly low background pressure.

In Fig. 1 a free molecular flow of constant velocity is first incident on the upper plate 1 (80°K) cooled by liquid nitrogen. When molecules with high speed are diffusely reflected on this plate, they lose most of their kinetic energy and leave with the velocity corresponding to 80°K for plate 2 (20°K) cooled by liquid helium. After undergoing some intermolecular collisions in the gap between the two plates, a few molecules may escape from the gap, but most of the molecules arrive at the surface of the 20°K plate and are condensed there. The incoming stream is not directed directly to the 20°K plate so that the amount of the expensive liquid helium can be saved; since the kinetic energy of the molecules incident on the 20°K plate is very low after the reflection at the 80°K plate, the heat released on the 20°K plate is small and the liquid helium is thus saved. The purpose of the present study is to examine the effects of various factors on the amount of noncondensing flux and to obtain a design data for the parallel plate arrangement.

II. Outline of Calculation

As shown in Fig. 1, the arrangement is in a freemolecular flow with uniform velocity U_∞ and density ρ_∞. The trajectories of all molecules are parallel straight lines in the far upstream region. We consider the case when the gradient $\tan\beta$ of the trajectory is given by

$$\tan\beta = b/\ell_1 \qquad (1)$$

Fig. 1 Parallel plate arrangement for cryopumping.

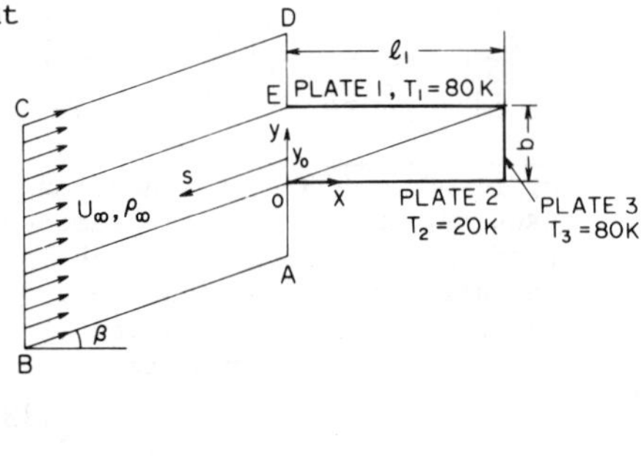

where b is the gap size and ℓ_1 is the length of plate 1. In this case all molecules directed to the arrangement would be incident on plate 1 if there were no molecular collision. Actually, each molecule collides with others not only in the inside of the arrangement but also in the outside region near to it.

The domain of calculation is the inside and the region ABCD, where AO = OE = ED = b and BC is at $x = -\ell_1$ (see Fig. 1). The calculation has been done by use of the direct simulation Monte Carlo (DSMC) method.[1] The inside is divided into I × J cells and the outside is divided into I × (3J) cells; all cells have the same volume. The collision stage of the DSMC method is treated by the modified Nanbu sheme.[2,3] The molecular model is the hard sphere. The condition at the surface of the plates is as follows. On plate 1 every molecule is reflected. However, it is condensed on plates 2 and 3 with the probability α_2 and α_3, respectively. The reflection is always the diffuse one.

The total mass flux $\dot{m}_i$ of the incoming stream is given by

$$\dot{m}_i = \rho_\infty U_\infty b \cos\beta$$

Let $\dot{m}_c$ be the total mass flux condensed on the inner surfaces of the arrangement. The ratio f of noncondensed flux to the incoming flux is

$$f = 1 - (\dot{m}_c/\dot{m}_i) \tag{2}$$

In designing the arrangement we need the condition for which f takes the smallest values. To examine the effect on f of various factors, we chose the following condition as the standard:

$$U_\infty = 780 \text{ m/s}, \quad \rho_\infty = 0.2 \times 10^{-6} \text{ kg/m}^3$$

$$\ell_1 = \ell_2 = 0.3 \text{ m}, \quad b = 0.1 \text{ m}$$

$$\alpha_2 = 1, \quad \alpha_3 = 0$$

$$T_1 = 80°\text{K}, \quad T_2 = 20°\text{K}, \quad T_3 = 80°\text{K} \tag{3}$$

We are planning to perform an experimental study for the condition in Eq. (3). The test gas is nitrogen. The total crosssection of a nitrogen molecule is

$$\sigma_T = 4.42 \times 10^{-19} \text{ m}^2$$

In the DSMC calculation we introduce nondimensional variables by dividing velocity by U_∞, length by ℓ_1, and time by ℓ_1/U_∞. (We denote the nondimensional variables by a caret placed over the symbol.) After this it is immediately seen that the present problem is governed by the two parameters b/ℓ_1 and

$$\frac{\ell_1 \sigma_T \rho_\infty}{m} = \frac{\ell_1}{2^{1/2}\,\lambda_\infty} \quad (\equiv Kn^{-1}) \tag{4}$$

where m is the mass of a molecule and λ_∞ is the mean free path in the incoming stream. The quantity in Eq. (4) is nothing but the inverse Knudsen number Kn^{-1}. It is 0.570 at the standard condition in Eq. (3). The ratio λ_∞/b is then 3.72. The flow is near the free molecular regime, but we should regard the flow in the arrangement as the transitional flow because, as we see later, a high-density region appears; hence, the local Knudsen number is much smaller than λ_∞/b.

In the modified Nanbu method the collision probability P_{ij} of the collision pair (i,j) plays a central role. It is given by

$$\begin{aligned} P_{ij} &= n\sigma_T g_{ij}\Delta t/N \\ &= \frac{Kn^{-1}}{N_\infty} \cdot \frac{V_\infty}{V}\, \hat{g}_{ij}\, \Delta\hat{t} \end{aligned} \tag{5}$$

where n is the number density, g_{ij} is the magnitude of the relative velocity between molecules i and j, Δt is the time step, N is the number of simulated molecules in a cell with volume V, V_∞ is the volume of the standard cell located in the far upstream region, and N_∞, which is the input data, is the number of simulated molecules in the volume V_∞. The time step should be chosen in such a way that the following two conditions are satisfied:

$$U_\infty \Delta t \lesssim \Delta x \text{ or } \Delta\hat{t} \lesssim \Delta\hat{x} \tag{6}$$

$$P_i \simeq NP_{ij} = n\sigma_T g_{ij}\Delta t << 1 \tag{7}$$

where Δx ($= \ell_1/I$) is the cell dimension in the x direction and P_i is the collision probability of molecule i. The first is called the Courant condition and the second is required by the principle of uncoupling.[1,2] The use of the modified Nanbu method requires the additional condition $P_{ij} < 1/N$. However, this is weaker than the condition in

Eq. (7). Our choice of the parameters is

$$N_\infty = 50,\ \Delta\hat{t} = 0.05$$

$$I = 20,\ J = 9$$

Since $\Delta\hat{x} = 0.05$, the condition in Eq. (6) is satisfied. The nondimensional form of the condition in Eq. (7) is

$$\Delta\hat{t} \ll [\ Kn^{-1}(n/n_\infty)\hat{g}_{ij}\ ]^{-1}$$

The value of the right-hand side is 0.58 for $Kn^{-1} = 0.570$, $n/n_\infty \sim 3$, and $\hat{g}_{ij} \sim 1$. (As shown later, the maximum of n/n_∞ is about 3.) We now see that the condition in Eq. (7) is satisfied for our choice of $\Delta\hat{t}$.

The flow properties are sampled after the disappearance of the initial transient. We have sampled the number of molecules condensed in the arrangement, the density ρ, and the x and y components $\bar{c}_x$, $\bar{c}_y$ of the flow velocity. When we use the arrangement, we have to know the flow region disturbed by its presence. Let us consider the attenuation of a molecular beam along the line s in Fig. 1 that is parallel to the undisturbed flow. The attenuation ξ is defined by

$$1 - \xi = \frac{\rho u}{\rho_\infty U_\infty} = \frac{\rho(\ \bar{c}_x \cos\beta + \bar{c}_y \sin\beta\ )}{\rho_\infty U_\infty} \tag{8}$$

where ρ is the density and u is the component of the flow velocity in the direction of the undisturbed flow. Of course, $\xi = 0$ in the undisturbed flow; ξ increases with decreasing s due to molecular collisions.

III. Results and Discussion

Our concern is in the effects of the following parameters on the f of the noncondensed gas, ξ, and the density field: α_2, α_3, ρ_∞, b, ℓ_2, and the presence or absence of plate 3. When we change one parameter, all others are fixed at the standard condition. The result for f is summarized in Table 1.

We first see that, as is expected, f increases as α_2 decreases. The effect of α_2 is small. Note that the number 0.1084 is corresponding to the standard condition. We have also tentatively calculated the value of f by disregarding molecular collisions. It is 0.170 at the standard condition. This shows that the molecular collisions

Table 1 Ratio f of noncondensed flux to incoming flux

Conditions	f
$\alpha_2 = 1.00$	0.1084
$\alpha_2 = 0.95$	0.1125
$\alpha_2 = 0.90$	0.1184
$\alpha_3 = 0$	0.1084
$\alpha_3 = 1$	0.0899
$\rho_\infty = 0.1 \times 10^{-6}$ kg/m³	0.1383
$\rho_\infty = 0.2 \times 10^{-6}$ kg/m³	0.1084
$\rho_\infty = 0.4 \times 10^{-6}$ kg/m³	0.0653
b = 0.08 m	0.0909
b = 0.10 m	0.1084
b = 0.12 m	0.1229
$\ell_2 = 0.3$ m	0.1084
$\ell_2 = 0.4$ m	0.0476
Plate 3 is present	0.1084
Plate 3 is absent	0.3487

play an important role in the present problem. If the temperature of plate 3 is 20°K, i.e., $\alpha_3 = 1$, the value of f somewhat decreases. However, since this effect is small, there is no need to cool plate 3 by helium. Most of the molecules reflected on plate 3 are condensed on plate 2, so that the cooling of plate 3 has only small effect on f. The effect of the freestream density is large, and the value of f decreases with increasing ρ_∞. As ρ_∞ increases, the density and hence the collision frequency increase inside the arrangement; a molecule has a greater chance of striking plate 2. The effect of ρ_∞ suggests that in plume experiments it is better to set up the arrangement at the position not too far from the plume core.

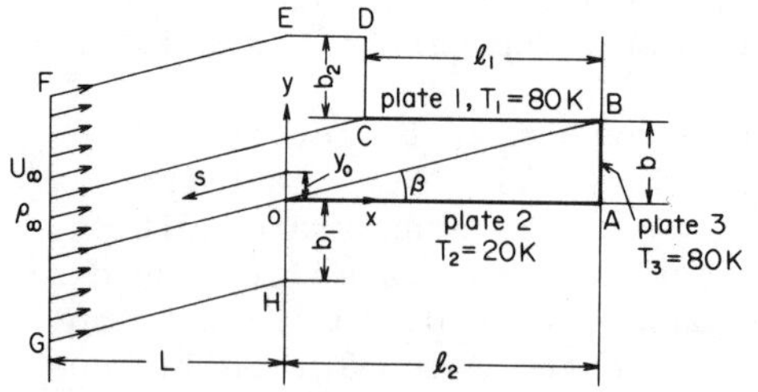

Fig. 2 Arrangement with elongated plate 2.

The value of f decreases with b. The smaller the gap, the more probable it is that a molecule reaches plate 2. Note, however, that the total number of the arrangements necessary to cover the inner wall of a vacuum chamber increases when b is small. Figure 2 shows the arrangement plate 2, which is elongated. In this case the gradient $\tan\beta$ of the incoming stream is b/ℓ_2, not Eq. (1). The domain of calculation is OABCDEFGHO. Table 1 shows that the effect of ℓ_2 is quite large. Roughly speaking, f is halved when ℓ_2 is elongated by 30%. This is because of the additional condensation on the elongated part of plate 2. Plate 3 is not of no use. In fact, Table 1 shows that the removal of plate 3 triples the value of f. Plate 3 keeps molecules in the arrangement, which promotes condensarion.

Figure 3 shows the density contours. We can see how far the region disturbed by the presence of the arrangement penetrates into the free stream and where condensation

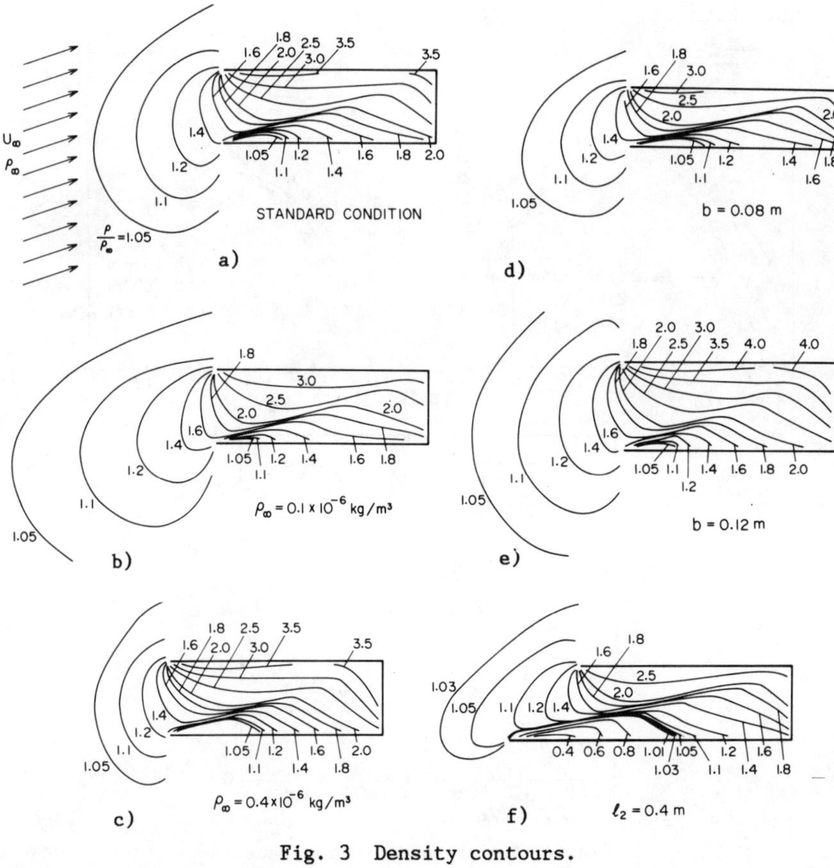

Fig. 3 Density contours.

actively occurs. Figure 3a is the result for the standard condition. Since the density along plate 2 increases with the distance from the leading edge, so does the amount of condensation. The effect of ρ_∞ on the density field can be seen from Figs. 3a - 3c. The lower the density ρ_∞, the wider the disturbed region. The effect of b can be seen from Figs. 3a, 3d, and 3e. In Table 1 the value of f increases with b. Consistent with this, the disturbed region becomes wider as b increases. The effect of the elongation of plate 2 is seen from Figs. 3a and 3f. For the longer plate 2 the disturbed region is narrow because the value of f drastically decreases.

It is important to see where a strong condensation occurs. Figures 4a - 4c show the distributions of the local condensation rate $\dot{m}(x')$ on plate 2, where $x' = \ell_2 - x$.

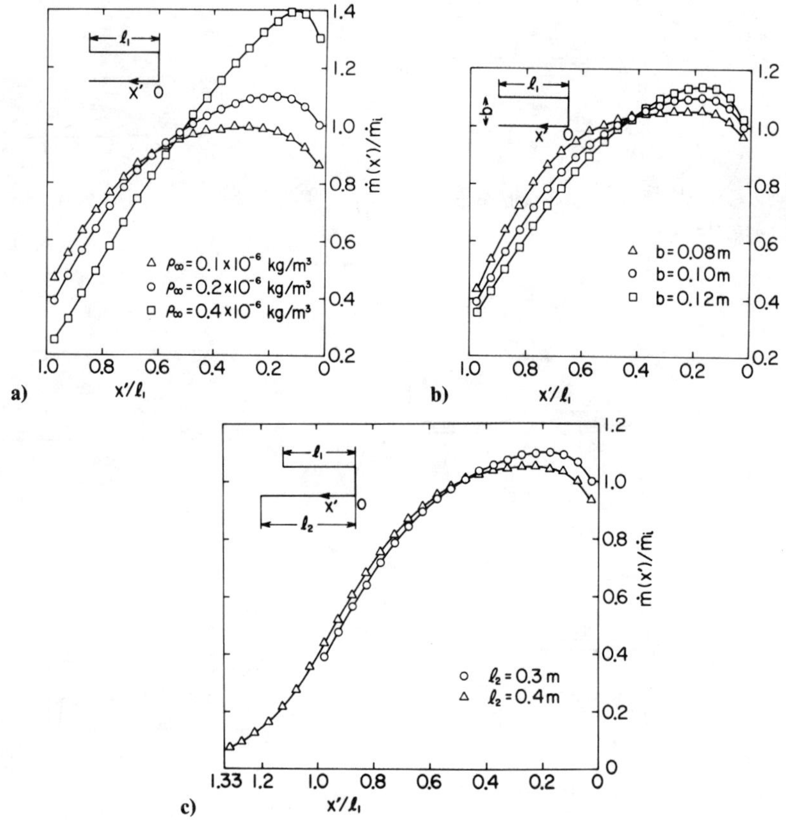

Fig. 4 Local condensation rate.

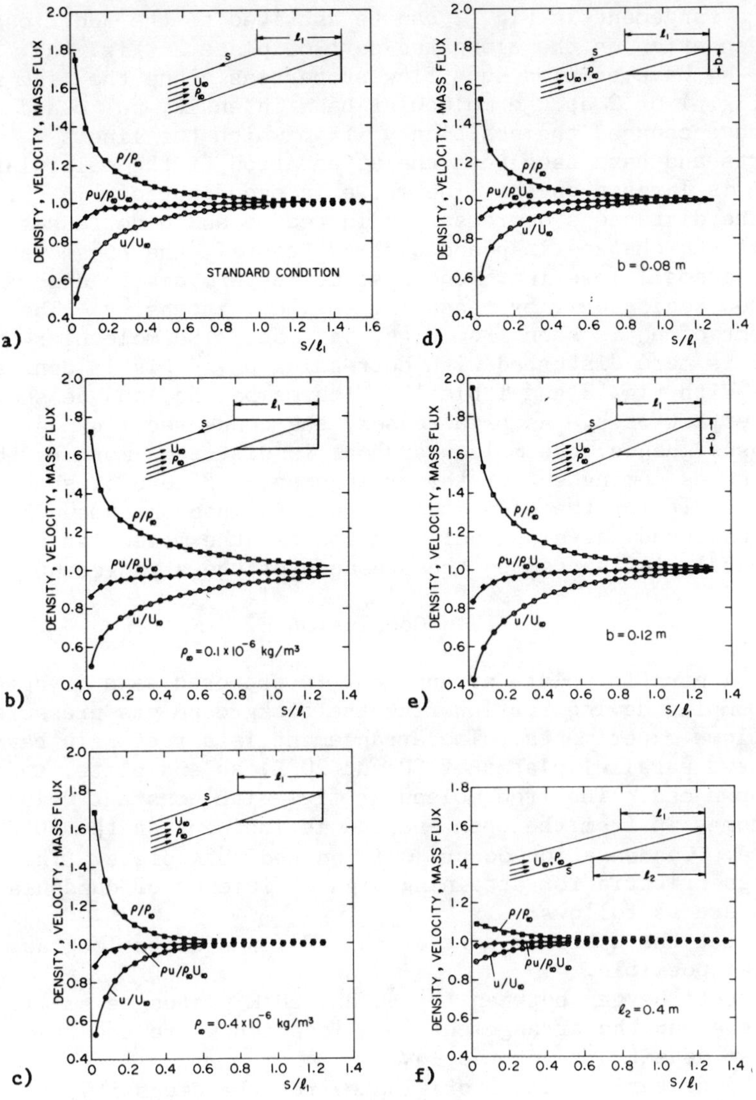

Fig. 5 Density ρ, velocity u, and beam intensity ρu.

Note that

$$\dot{m}_c = \int_0^{\ell_2} \dot{m}(x')\, dx'$$

As ρ_∞ increases, $\dot{m}(x')$ becomes large near the rear end of plate 2 (Fig. 4a). A wide region of high $\dot{m}(x')$ appears in case of small b (Fig. 4b). The drastic decrease of f for the arrangement in Fig. 2 can be ascribed to the additional condensation on the elongated part of plate 2 (Fig. 4c).

We have sampled such flow properties along the line s in Fig. 1 or 2 as the molecular beam intensity ρu, ρ and u. We have changed the position y_0 from which the line s starts and have seen that the attenuation of the molecular beam is largest on the line shown in each part of Fig. 5. As the distance s decreases, ρ increases and u decreases. Since the changes in ρ and u almost cancel, the ρu shows only a small lowering. One must not determine the disturbed region only by measuring the beam intensity. The effect of ρ_∞ is seen from Figs. 5a - 5c. The molecular beam is more disturbed with decreasing ρ_∞. This is consistent with Fig. 3 and Table 1. Figures 5a, 5d, and 5e show the effect of b. As b increases, the condensed flux decreases; hence, the molecular beam is disturbed more by the molecules coming out of the arrangement. Figure 5f shows the result for the elongated plate 2. Since the amount of noncondensed gas drastically decreases, the molecular beam is hardly influenced by the presence of the arrangement.

IV. Conclusion

A parallel plate arrangement is proposed as a cryogenic pumping device for lowering the background gas pressure in plume experiments. The arrangement is a rectangle having two parallel plates of 80 and 20°K, an end plate, and an open end. The free molecular flow with constant velocity comes in from the open end and is incident on the 80°K plate. Condensation occurs only on the 20°K plate. The design criteria for attaining high efficiency of condensation are as follows:

1) Let the upstream part of the 20°K plates be long as far as possible.
2) Let the gap between the 80 and 20°K plates be small.
3) Set up the arrangement at a location where the free-stream density is not too low.

The criteria 2 and 3 originate from the necessity to increase the collision frequency within the arrangement. The design criteria are probably applicable to three-dimensional arrangement with another two end plates. Because of the additional molecular reflection on these two plates, the amount of condensation is expected to be greater in the three-dimensional arrangement than in the two-dimensional one.

References

[1]Bird, G. A., Molecular Gas Dynamics, Clarendon, Oxford, U.K., 1976, pp. 118-132.

[2]Nanbu, K., "Theoretical Basis of the Direct Simulation Monte Carlo Method," Proceedings of the Fifteenth International Symposium on Rarefied Gas Dynamics, Vol. 1, Teubner, Stuttgart, FRG, 1986, pp. 369-383.

[3]Illner, R. and Neunzert, H. "On Simulation Methods for the Boltzmann Equation," Transport Theory and Statistical Physics, Vol. 16, Nos. 2 & 3, 1987, pp. 141-154.

Chapter 3. Particle and Mixture Flows

Aerodynamic Focusing of Particles and Molecules in Seeded Supersonic Jets

J. Fernández de la Mora,* J. Rosell-Llompart,† and P. Riesco-Chueca
Yale University, New Haven, Connecticut

I. Introduction

Following the observations of Becker et al[1] on substantial enrichment of the heavy component in the core of a supersonic H_2-Ar free jet, the subject of aerodynamic concentration and separation of species in gas mixtures has been permanently present in the rarefied gas dynamics symposia. Much of the research effort on aerodynamic separation was originally directed toward the enrichment of ^{235}U in UF_6-H_2 mixtures,[2] a problem to which decreasing attention has been devoted through the last decade. However, heavy molecule focusing in the core of seeded free jets of He or H_2 is of considerable current analytical importance (for instance in interfaces between gas or liquid chromatographs and mass spectrometers[3]) and is a key issue in the generation of molecular beams.[4,5] Aerodynamic enrichment of species remains therefore a subject worthy of attention, still filled with unsettled issues. This paper addresses the basic questions of how much, and how, can one concentrate the heavy gas in a seeded jet.

In the limit of extreme mass disparity, a partial answer to the questions formulated earlier has been available since 1967, when Israel and Friedlander found conditions when the whole stream of microscopic particles introduced by an air jet into a vacuum exhibited a vanishing angle of divergence far from the source.[6] Considering that the number density of air had droped by many orders of magnitude whereas that of the particles had remained practically unchanged through the expansion, the seed species must undoubtedly have undergone an extraordinary degree of enrichment.

Such a singular aerosol focusing behavior in supersonic jets has been observed repeatedly over the last two decades.[7-10] In particular, Dahneke and his colleages[8] have identified a single parameter β governing

Presented as an Invited Paper.

*Associate Professor, Mechanical Engineering Department.
†Graduate Student, Mechanical Engineering Department.

the phenomenon, which appeared in principle to control also the behavior of jets of mixed gases.[8] However, although the validity of their analysis was confirmed by experiments with submicron particles, Dahneke[11] has failed to observe similar focusing effects in He-Xe and H_2-Xe mixtures accelerated through conical nozzles with a semiangle of convergence of 15 deg. Unpublished work at Yale in 1986 using H_2-CBr_4 and H_2-Hg mixtures measured also rather modest molecule focusing. In conclusion, the evidence that existed when this investigation was initiated was that a stream of suspended aerosol particles could be easily focused at large distances from a free jet source within an area comparable to that of the nozzle throat. But the analogous effect with jets of mixed gases appeared to be far more elusive, even in mixtures with molecular mass ratios exceeding 165.

Because the problem under consideration involves rather complex non-equilibrium conditions, our study of it will begin by considering an extreme oversimplification,[12] slowly building some design intuition to guide experiments,[13] and subsequently incorporating essential missing physical features.[14] The extreme asymptotic behavior is the "deterministic" limit of infinite mass ratio, where Brownian motion is negligible for the seed species, which in addition is taken to be infinitely dilute.[12] As first pointed out by Reis and Fenn,[15] this idealization corresponds physically to the behavior of aerosol suspensions and will be considered in Secs. II and III. The corresponding deterministic analysis provides design insights that are subsequently tested in aerosol beam experiments.[13] In particular, it emerges that highly convergent nozzle geometries strongly favor the focusing phenomenon and are absolutely essential in order to focus single molecules. It is also found that aerosol beams can be concentrated practically without limit into a region of nearly vanishing dimensions,[12,13] far more sharply than previously observed.

Armed with such information, the problem of focusing molecules is considered again. Section IV examines experimentally the structure of $W(CO)_6$ and CBr_4 beams seeded in He and H_2 jets accelerating through the simplest among the promising highly converging geometries, a thin-plate orifice.[14] Under optimal circumstances, one then observes heavy gas beams narrower than the nozzle itself, even five nozzle diameters downstream from the source. However, although molecular and aerosol beam-widths are indistinguishable at pressures just a few times the critical focusing pressure, within the optimal focusing range, beams of heavy molecules are never smaller than some 35% of a nozzle diameter d_n, and fail to exhibit the sharp boundaries characteristic of their aerosol analogs. These important differences between the spacial distribution of aerosols and molecules in seeded jets are attributed to Brownian motion, a phenomenon still ignored in the analysis.

The fundamental questions of what is the ultimate limit toward which the focusing phenomenon may be pushed and what are the corresponding scaling laws are addressed theoretically in Sec. V by incorporating

Brownian motion within the picture. This task is undertaken within a hypersonic theory providing a finite set of hydrodynamic equations, whose systematic closure relies on the smallness of the ratio between the thermal speed of the heavy gas and its mean convective velocity.[16] These equations are reduced in Sec. VB through an approximate boundary-layer analysis (valid near the jet axis) which shows that the minimum beam width is linear with the square root of the the ratio of molecular masses m/m_p. The near-axis equations admit physically relevant Gaussian particular solutions that reduce the problem to the integration of several coupled first-order ordinary differential equations. From them, one may determine asymptotically the local structure of the focal region (Sec. VC). These equations are solved numerically in Sec. VD for the simple but realistic case of a two-dimensional (2D) potential jet. In agreement with experiment, the beam widths for this problem are found to be larger than $5\ d_n\sqrt{(m/m_p)}$.

II. Deterministic Limit (Aerosols)

Under conditions in which their Brownian movement may be neglected, the motion of small spherical aerosol particles suspended in a carrier gas may be described through Newton's equation

$$d\mathbf{u}_p/dt = -(\mathbf{u}_p - \mathbf{u})/\tau \tag{1}$$

where $\mathbf{u}_p$ is their velocity, $\mathbf{u}$ is the velocity field of the suspending gas, and the force coupling the two has been taken to be linear in the slip velocity $\mathbf{u}_p - \mathbf{u}$ through the proportionality constant τ, the particle relaxation time.[17] Although the range of validity of this linear drag law is limited, it can be extended systematically for gas mixtures of disparate masses interacting elastically, as discussed in Ref. 18. Such a generalization would introduce additional dependences on source temperature and intermolecular potential, but would not modify the formulation or qualitative features of the problem.

The particle trajectories may be determined straightforwardly once τ and $\mathbf{u}(\mathbf{x},t)$ are specified together with appropriate injection (initial) conditions for $\mathbf{u}_p$. The parameter τ, directly related to the friction coefficient,[17] is a function of the local thermodynamic variables, which will be made dimensionless with a characteristic value U of the velocity $\mathbf{u}(\mathbf{x},t)$ and a characteristic geometric length L of the problem

$$S = \tau U/L \tag{2a}$$

This group, often called the Stokes number, measures the degree in which $\mathbf{u}_p$ and $\mathbf{u}$ are uncoupled. It also arises naturally in the case of compressible flows, where the dependence of τ on temperature and pressure must be accounted for.[8] Using the fluid sound speed c for U and the nozzle diameter d_n for L, the corresponding Stokes number for sonic flows is related to the similarity parameter β first introduced by Dahneke and his colleagues[8] through

$$S = c_o \tau_o / d_n = \beta / 4.27 \qquad (2b)$$

where the subscript o refers to the stagnation conditions upstream from the nozzle.

Experimentally, the essential features of axisymmetric hypersonic aerosol beams have been derived from the diameter of the particle deposit collected on greased surfaces placed perpendicular to the jet axis far downstream from the nozzle. Typical results from such measurements may be seen in Fig. 1 (a re-elaboration of Fig. 5 from Ref. 8), which shows far-field aerosol beam widths in terms of the single variable β. All data correspond to particles in the free molecular regime, accelerated in air through a conical nozzle with a semiangle of 15 deg. They collapse closely into a single curve confirming the suitability of S (or β) as the fundamental parameter governing the process. The upper positive section of this figure is typical of the behavior observed in earlier aerosol beam experiments, where, as the source pressure is decreased (S is increased), the beam width first becomes narrower, reaches a minimum (the focused condition), and becomes wider again. Notice that the figure shows an outer and an inner diameter, corresponding to the fact that only an annular section of the jet was seeded with particles upstream from the nozzle. The aerosol deposit thus had a ring shape. Notice also that the mirror image of the figure in the negative side extends the positive part continuously, showing that the two particle trajectories bounding the annular beam actually cross the jet axis (Israel and Friedlander[6] have already pointed out that the portion of this curve beyond the minimum correponds to particles that have crossed the axis ahead of the collector).

One can infer from the preceding discussion that, by probing the particle beam at moderate distances from the source, around the crossing or focal point, one should expect to find substantially more concentrated regions of particles than those seen before. This hypothesis was first confirmed, both analytically and numerically for symmetric nozzles in Ref. 12, which we review here. An analysis of Eq. (1) near the axis of symmetry shows that, with great generality, the governing equations admit "moving" singular points that must be interpreted as foci.[12] All particle trajectories originating near the axis do focus at exactly the same location on the axis (Fig. 2). Foci are thus singular points of infinite particle concentration. Concrete illustrations of the structure and features of the focal region result from numerical solutions of Eq. (1) in 2D irrotational flows, whose simplicity[19] permits detailed and accurate calculations of the fluid velocity field and the particle trajectories.

Let us consider the case illustrated in Fig. 2, where an inviscid incompressible jet issues from a 2D symmetric nozzle whose straight walls form an angle 2α. Not surprisingly, particles may have sufficient inertia to retain some of their original speed of convergence and cross the

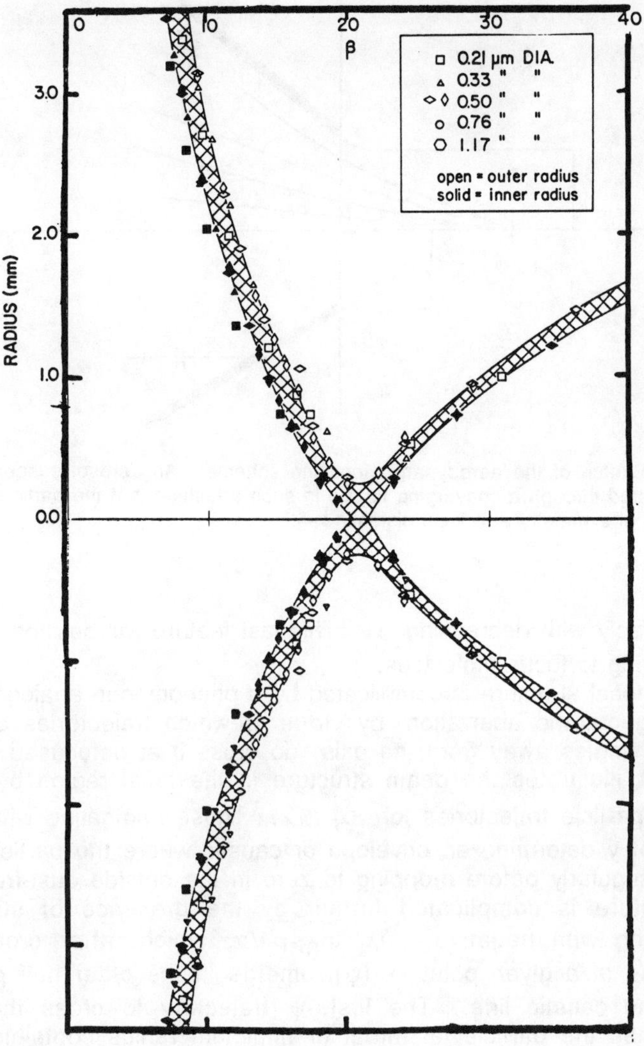

Fig. 1 Diameters from annular particle beams collected on a greased surface, taken from Fig. 5 of Ref. 8, and combined with their reflection on the horizontal axis. In terms of the variable β (β=4.27S), all of the data for particles in the free molecule limit collapse together.[8] Notice that the original two positive branches continue each other smoothly on the negative side after interchanging the inner and outer ring radii (d_n=0.2mm).

axis. Figure 3 shows the position x_{co} found numerically for the focus (the crossing point of near-axial particle trajectories) as a function of the Stokes number $S=\tau U/b$ for several values of the angle of convergence α. Notice the existence of a critical value S^* of S at which x_{co} tends to infinity and below which no focusing occurs. Observe also that S^* grows

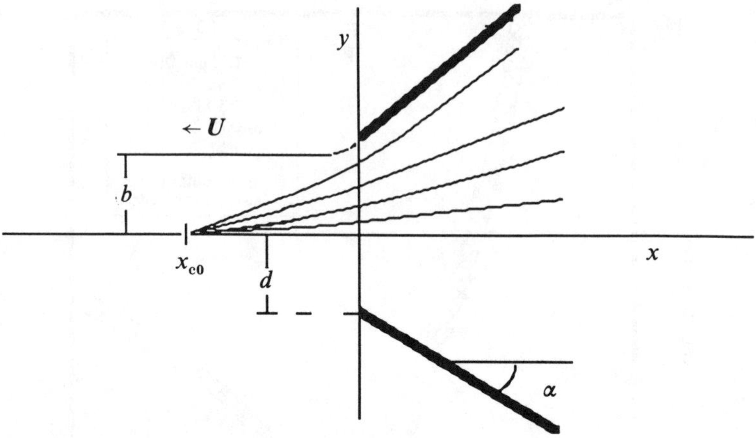

Fig. 2 Sketch of the aerodynamic focusing scheme. An aerosol suspension is accelerated through a converging nozzle in such a fashion that the particles cross the axis of symmetry at a focal distance x_{co}.

considerably with decreasing α. This last feature will be most important when trying to focus molecules.

The focal structure is complicated by a phenomenon analogous to the optical geometric aberration, by virtue of which trajectories originating at large angles away from the axis do cross it at defocused positions. Figure 4 illustrates the beam structure in the focal region by showing various particle trajectories for $\alpha=\pi/2$. Those originating with positive values of y determine an envelope or caustic where the particle density peaks singularly before dropping to zero in the outside dust-free region. The picture is complicated further by the presence of streamlines originating with negative $\psi=\theta_{initial}/\alpha$, which, after crossing the centerline at a given point $x_c(\psi)$, emerge in the other half plane and cross the caustic line. The first $-\psi$ trajectory to cross the caustic determines the particle jet throat or minimum radius containing all the particles.

Summing up, the following conclusions from Ref. 12 are most relevant for our purposes here: **1)** A focal point exists and is infinitely sharp in symmetric nozzles for the streamlines near the axis of symmetry. **2)** There is a "geometric aberration" that defocuses the streamlines originating far from the jet centerline. In spite of it, numerical examples show that the width of the focal region may be made over two orders of magnitude smaller than the nozzle diameter by restricting the region where particles are seeded to a moderate angle away from the axis. This angle may be higher than $\pi/4$ for the case of a jet exiting through a slit in a thin plate. **3)** Focusing occurs only for particles characterized by a value of the Stokes number greater than a

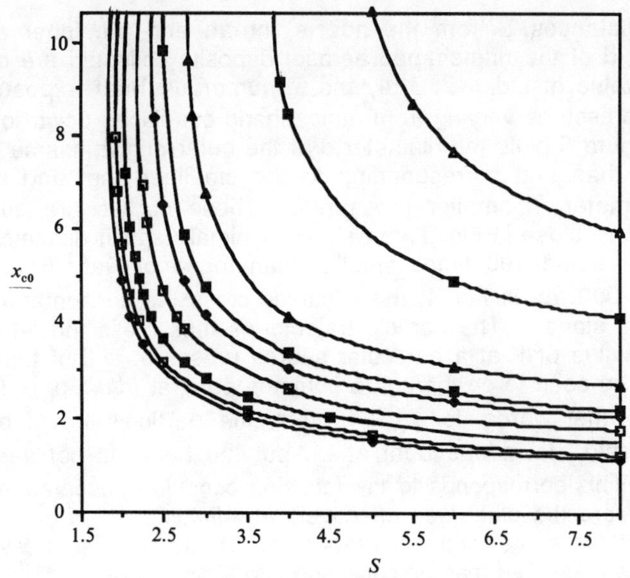

Fig. 3 Crossing point x_{co}/d_n for trajectories originating near the axis as a function of the Stokes number $S = U\tau/b$ for different wedge half-angle α in the two-dimensional flow of Fig. 2: $\alpha = \pi/8$ (Δ); $\alpha = \pi/6$ (■); $\alpha = \pi/4$ (▲); $\alpha = 3\pi/10$ (■); $\alpha = \pi/3$ (◆); $\alpha = 2\pi/5$ (◘); $\alpha = \pi/2$ (◘); $\alpha = 3\pi/5$ (◘); $\alpha = 2\pi/3$ (◆).

critical S^* (typically of order unity), whose value may be reduced considerably by using rapidly converging nozzles. **4)** The focus is rather sharp except perhaps near critical conditions, under which the focal point tends to infinity and geometric aberration effects appear to be singular. For that reason, high-resolution focusing might be easier to attain at finite distances from the source. This observation might explain why previous aerosol beam experiments (where the collecting surface was invariably hundreds of nozzle diameters dowstream from the source) only achieved a modest degree of focusing.

III. Experiments on High-resolution Aerosol Focusing

All the data reported in this section correspond to highly supersonic air jets with pressure ratios of several hundreds acros the nozzle orifice. The nozzle is a sharp round hole drilled on a thin plate with d_n=0.5 mm.

A. Annular deposits

For reasons of simplicity, we will first describe observations made with a particle-seeding system identical to that previously employed by Dahneke et al[8] (thoroughly described in Ref. 20), which seeds the particles symmetrically only in a relatively narrow annular region. The free jet is intercepted perpendicularly by a greased surface held at varying distances L from the nozzle throat, and the inner and outer diameters d of the ring-shaped aerosol deposits collected are measured. At each value of L/d_n= 2, 3, 4, and 5, numerous brief exposures were taken at pressures varying from atmospheric conditions down to some 30 torr. Figure 5 plots the diameter d of the outer ring at source pressures p_o larger than that corresponding to the smallest ring, and minus the inner diameter at smaller pressures. These curves are qualitatively analogous to those in Fig. 1, but show minimal deposit diameters around 3% of d_n, some 100 times smaller than those of Ref. 8, where L/d_n exceeded 300. As in Fig. 1, these curves cross the horizontal axis with a continuous slope. The particle trajectories thus cross the jet centerline for each value of L at a particular source pressure. All of these curves may also be seen to go through a common point at $(d/d_n, p_o)$= (0.35, 240 torr). In other words, for p_o=240 torr, the particles travel parallel to the axis (d/d_n is independent of L), but the beam is not very sharply focused. This corresponds to the focusing condition observed in previous studies where the plate was effectively at infinity.

Except for some slight discrepancies at the highest pressures, the four curves measured may be collapsed into a single one:

$$d/d_n = 0.35 + (0.224 + 0.337L/d_n)F(p_o) \qquad (3)$$

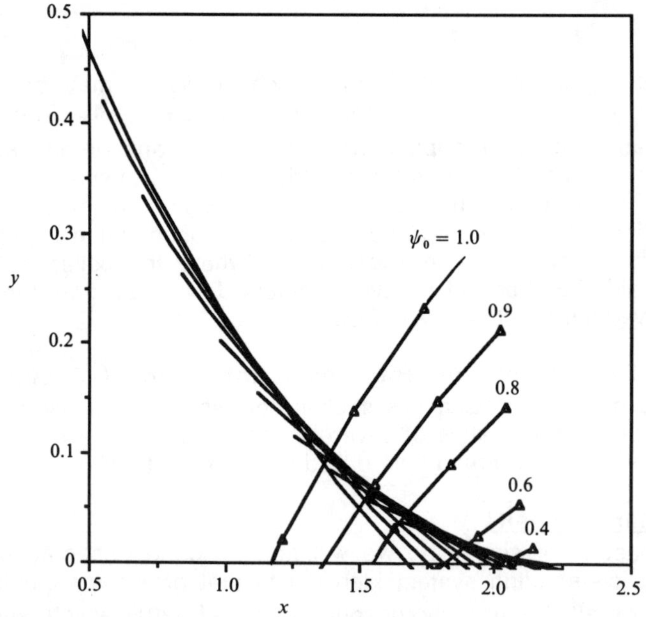

Fig. 4 Particle streamlines originating from the upper and lower half planes in the 2D flow of Fig. 2 with $\alpha = \pi/2$ and S = 4. Also represented is their envelope or caustic.

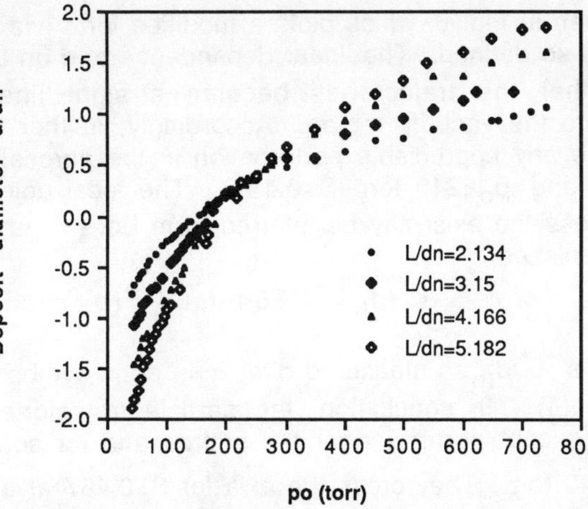

Fig. 5 Dimensionless diameter d/d_n of the annular particle deposits as a function of the source pressure for several nozzle to plate distances L/d_n. Positive and negative values of d correspond to the outer and inner ring diameters, respectively. Nominal particle diameter d_p=0.245 μm.

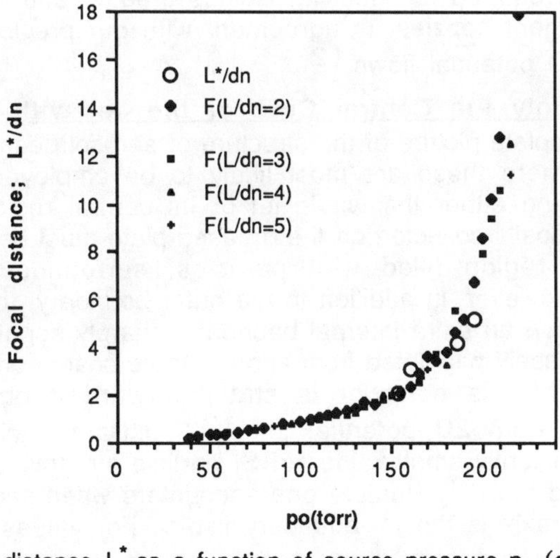

Fig. 6 Focal distance L^* as a function of source pressure p_o (d_p=0.245 μm). Open symbols are determined directly from the point where the curves of Fig. 5 cross the horizontal axis. Filled data points are extracted from the conditions in which d/d_n in Eq. (4) vanishes and are therefore an extrapolation into values of L for which no direct experimental information is available. The function $F(p_o)$ which collapses into a single curve the aerosol deposit diameters measured at four different distances L is defined in Eq. (3). Notice the vertical asymptote as $p_o \rightarrow$ 240 torr.

as may be seen in Fig. 6, which plots a modified form [via Eq. (4)] of the function $F(p_o)$ so defined. The linear dependence of d on L shown in Eq. (3) implies that the trajectories become straight lines soon after penetrating into the vacuum region. Accordingly, neither the jet nor the plate introduce any appreciable perturbation in the aerosol path, at least when $L>2d_n$ and $p_o<240$ torr ($S>0.457$). The focal point where these trajectories cross the axis may be inferred from Eq. (3) as the value of L for which d vanishes:

$$L^*/d_n=-0.664-1.04/F(p_o) \qquad (4)$$

Figure 6 shows L^*/d_n as measured directly in open symbols, and also as given by Eq. (4). In conclusion, the particle trajectories studied are straight lines except perhaps near the source and for source pressures larger than 240 torr. They cross the axis for $S>0.457$ at a distance from the nozzle that tends to zero as $S\to\infty$ and to infinity as $S \to 0.457$.

It is instructive to make some comparisons here with the results obtained by Dahneke et al.[8] at large distances from the source also with a conical nozzle. For a value of the half-angle $\alpha=15$ deg, these authors found that $S^*(15\text{ deg})=4.8$ ($\beta=20.5$), whereas we obtain $S^*(90\text{ deg})=0.46$, nearly an order of magnitude smaller. The ammount of particle inertia required for focusing is thus greatly reduced when using sharply convergent nozzles, in agreement with our previous predictions in Fig. **3** for 2D potential flows.[12]

B. Seeding only the Central Core of the Jet with Particles

A more complete picture of the structure of aerosol beams in the mode of operation where these are most likely to be employed results from uniformly seeding either the whole jet or its central region. Naturally, each of the deposits collected on the greased plate must be made up of a finite circular region filled with particles, surrounded by a clean background. However, in addition to the outer boundary, the deposits we collect often have an extra internal boundary, sharply separating an outer region more densely populated from another more scarce on particles. The interpretation of this behavior is straightforward if one recalls the earlier results from 2D potential jets,[12] illustrated schematically in Fig. 4. At sufficiently small values of S, particle streamlines never cross each other, and the only feature one encounters when sectioning the jet normally to its axis is the jet boundary itself. For values of S of order one, however, the outermost trajectories ($\Psi=\pm1$ in Fig. 4) cross inward through their neighbors, initiating the formation of a caustic (now the outer boundary). The trajectory labeled $\Psi=1$ thus ceases to mark the outer boundary but still creates an internal boundary. The region between the caustic and this internal boundary appears as brighter than the core region because each point in it is reached by more than one particle streamline. We shall reserve the symbol d for the diameter of the boundary corresponding to the Ψ=1 trajectory. The diameter of the caustic will be denoted by c.

Experimentally, one may seed only the core of the jet with particles by adding sheath air to the outer boundary of the gas, thus eliminating the particles from some of the trajectories around Ψ=1. The principal practical consequence of this operation is the narrowing of the beam waist marked by the intersection of the caustic and the outermost trajectory originating from the opposite side of the beam. Figure 7 represents the caustic c and its mirror image -c as filled symbols, superposed to d as open symbols for a fixed source pressure p_o=162.6 torr (S=1.48), and a varying ratio r of seeded air to total air flow. Several features of this figure deserve attention. First, in the range 1<r<0.3, the form of the caustic is independent from r. The small discrepancies observed may be attributed mostly to the slight inaccuracies in the measurement of L. Second, the inclination of the curve d(L) decreases substantially with decreasing r. Notice finally that the beam throat diameter t (intersection of the caustic with the outer boundary) takes the approximate values t/d_n=0.1, 0.1, 0.06, 0.03, and 0.02 for r=1, 0.85, 0.504, 0.388 and 0.292, corresponding roughly to an average enrichment of the particle beam ($rd^2{}_n/t^2$) of 100, 85, 140, 430, and 730, respectively. Evidently, this focusing effect is truly singular.

C. Particle Beam Structure in the Free Molecular Regime

The experimental data for the caustic shown in Fig. 8 correspond to the free molecular regime and should faithfully model the structure of unsheathed beams of heavy molecules in the limit of large mass disparity.

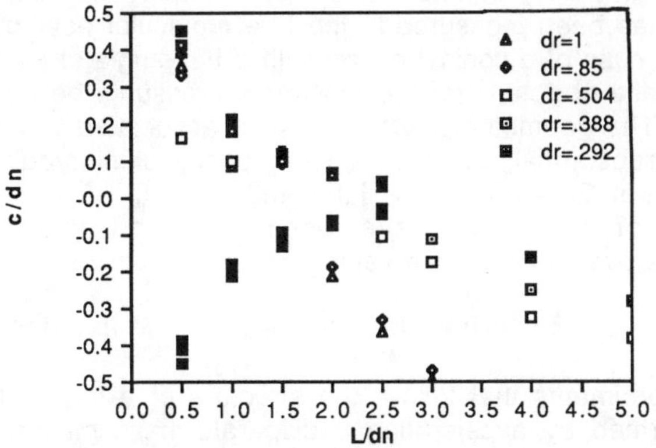

Fig. 7 Effect on the deposit structure of adding particle-free sheath air on the outer region of the jet (d_p=0.496 μm). We represent the width of the caustic c and its mirror image -c as filled symbols, superposed to d as open symbols. The source pressure takes the single value p_o=162.6 torr (S=1.48). The different data points represent conditions with different proportions 1-r of sheath air.

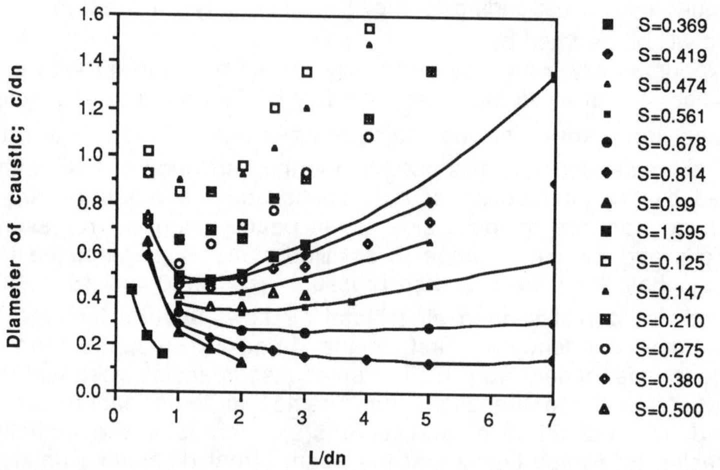

Fig. 8 Caustic c for particles in the free-molecular regime, with nominal diameters of 0.109 μm (open symbols) and 0.261 μm (filled symbols joined with a continuous line) as a function of the distance L.

The measurements were performed with polystyrene spheres supplied by Duke Scientific with nominal diameters of 0.261 and 0.109 μm. The critical condition for the caustic to become parallel to the axis corresponds to S=0.82.

In conclusion, most qualitative features of highly focused aerosol beams predicted for incompressible 2D jets have been observed in highly supersonic axisymmetric aerosol jets. The structure of the deposits collected on effectively infinite flat surfaces introduced perpendicularly to the beam has been measured in the free molecular asymptote and its vicinity for the case of a conical nozzle with a half-angle of 90 deg in a jet of air with a ratio of upstream to downstream pressures between 700 and 460 (Fig. 8). The geometric aberration effect associated with this nozzle geometry is substantial, but may be practically eliminated by seeding only the central 30-40% of the jet with aerosol. Average absolute enrichments of over 700 have been observed for the particle concentrations over their source value.

IV. Experiments on Focusing Molecules

In the experiments that follow, the structure of a highly dilute heavy gas beam formed by accelerating a disparate-mass mixture through a thin-plate orifice is determined by scanning the jet with a small detector differentially sensitive to the heavy gas.[14] The seed species is sufficiently dilute to produce only a slight perturbation in the carrier gas flow, whose structure is therefore well known.[21] For a particular nozzle geometry, the heavy gas beam structure depends exclusively on the nature of the carrier gas, on the ratio p_o/p_1 between the stagnation pressures p_o at the source, and the background pressure p_1 in the vacuum chamber downstream the nozzle orifice, and on the Reynolds number Re

$$Re = \gamma p_o d_n / \mu_o c_o \qquad (5)$$

where d_n is the nozzle throat diameter, and γ, μ_o, and c_o are the carrier gas specific heat ratio, viscosity coefficient and sound speed, $c=\sqrt{(\gamma kT/m)}$, respectively. The subscript o denotes stagnation conditions upstream from the nozzle, T is the absolute gas temperature, m its molecular mass, and k is Boltzmann's constant.

In the limit of very large mass disparity, the structure of the heavy gas beam depends on the variables γ, Re, and p_o/p_1, as well as on the inertia parameter S [Eq. (2b)] which, with the help of Einstein's law relating τ to the diffusivity of the mixture D [Eq. (16)], may be expressed as

$$S = Fe/Re \qquad (6)$$

where the Fenn number Fe is a weakly temperature dependent property of the gas mixture measuring the disparity of relaxation scales for the two species. It may be written in terms of the mixture viscosity-to-mass diffusivity ratio $Sc=\mu/\rho D$, and the molecular mass ratio m_p/m as

$$Fe = \gamma m_p/(mSc) \qquad (7)$$

The values of the parameters Sc, m_p/m, and Fe estimated in Ref. 14 for the mixtures used in this investigation are displayed in Table 1, where we use 1.4 and 5/3 for γ in H_2 and He, respectively.

A first fundamental difference between this and earlier sections on aerosol focusing stems from the much larger values of Fe typical of aerosol particles. Because focusing phenomena require values of S of at least order one (Fig. 8), Eq. (6) implies that the jet Reynolds numbers of inertial interest are at most of order Fe. As a result, aerosol focusing occurs typically in the large Reynolds number asymptotic region, where the jet structure is practically independent of Re, but this is no longer

Table 1 Estimated values for Sc, m_p/m, and Fe [14]

Mixture	m_p/m	Sc	Fe
He-$W(CO)_6$	87.92	4.46	32.86
H_2-$W(CO)_6$	174.57	3.77	64.83
He-CBr_4	82.86	4.16	33.20
H_2-CBr_4	164.51	3.51	65.62
H_2-C_2Cl_6	117.43	3.51	46.44

the case for gas mixtures, characterized by values of Fe typically smaller than 100. Two findings from the deterministic limit are most relevant here. First, viscosity (more precisely, rotationality) disfavors focusing phenomena[12]; second (Secs. II and III), focusing occurs only above a critical value S^* of S, which decreases as the angle of convergence of the nozzle increases. Earlier attempts at focusing molecules used moderately converging nozzles characterized by values of S^* near 5 (Fig. 1). For He-Xe mixtures, with Fe≈24, the corresponding maximum value of Re compatible with focusing would thus be around 5, a value at which the jet structure is dominated by viscosity. Not surprisingly, neither Dahneke nor the authors observed any important focusing of molecules. But the situation can be improved substantially with a nozzle geometry appropriately designed to reduce both S^* and viscous phenomena. Indeed, for a thin-plate orifice and a highly supersonic flow, S^* is below 1, as seen in Fig. 8. For that same geometry, it is well known that the viscous boundary layer near the nozzle walls remains moderately small due to the strong acceleration, even at Reynolds numbers as small as 10 (the importance of the viscous effects acting on a sonic thin-plate orifice can be derived from measurements of discharge coefficients, which take values close to the high Re asymptote even for Re=10)[22] Accordingly, in a H_2-CBr_4 mixture (Fe≈66) accelerating through a thin-plate orifice, the critical value of Re for focusing is around 70, where viscous effects are still fairly small. Under these conditions, molecular focusing should be easily observable.

A second most important difference between the present and the previous two sections results from Brownian motion. In the region of greatest interest, where S=0(1), the relative importance of thermal

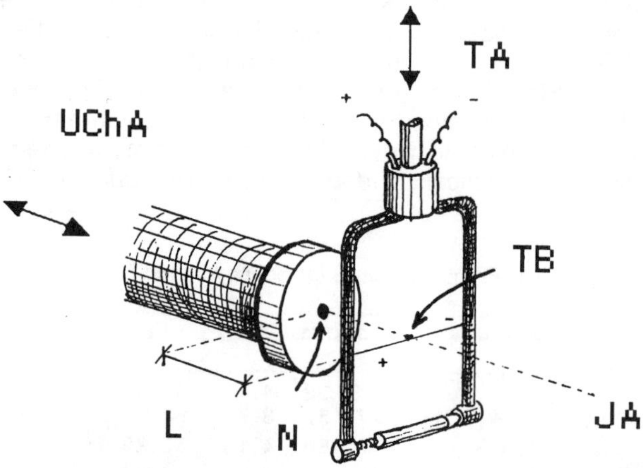

Fig. 9 Layout of the thermocouple probe inside the test chamber. Its transverse position r and its distance from the nozzle exit plane L can be independently adjusted.

agitation may be measured by the ratio of the convective to the thermal velocity of the heavy particles, a quantity that is of the order of $(m/m_p)^{1/2}$ for a sonic jet. Accordingly, denoting by Δ the width of the heavy gas stream [defined more precisely in Eq.(9)],

$$\Delta_{min}/d_n \sim (m/m_p)^{1/2} \tag{8}$$

This group is extremely small in aerosol suspensions, but takes nonnegligible values of around 0.1 for the mixtures considered here. We shall see that Brownian motion is in fact the fundamental mechanism limiting the sharpness with which molecules may be aerodynamically focused and that, unfortunately, Eq. (8) estimates the corresponding broadening effect by a factor several times smaller than what the data show.

A. Experimental Arrangement

Experimentally, the heavy gas beam structure is measured by translating a small detector sensitive to the heavy gas across a jet diameter (Fig. 9), at varying axial distances L from the nozzle and at different values of the source Stokes number. S is controlled through the carrier gas pressure. The detector is a small chromel-constantan thermocouple bead (supplied by Omega Inc. with an output of 1 mV for a temperature increment of 16.7 K), whose response is discussed in Sec. IVB. The small bead diameter (about 14% of d_n) ensures a relatively high spacial resolution in the measurement, as well as a small interference with the jet structure.

The central element of the experiment is a cross-shaped vacuum chamber with four circular openings, within which the free jet encounters the thermocouple probe. One of the windows houses the nozzle assembly and another, perpendicularly, the thermocouple assembly as shown in Fig. 10. Each of these two elements can be displaced inwardly and outwardly with their respective micrometers, which give control on the radial (r) and axial (L) coordinates of the thermocouple bead, respectively.

Depending on the position of two valves, the light gas may or may not be passed through a seeding cell upstream the nozzle, which contains crystals of the heavy species kept at room temperature. Vapors are continuously generated by sublimation of the crystals and entrained by the light gas flowing freely through the cell. A valve just ahead of the seeding cell controls the fraction of light gas flow going through it and thus the final mass fraction of seed gas in the mixture. This quantity is measured pneumatically[14] and maintained at around 1%.

B. Thermocouple Probe

Our detector is inspired by the work of Maise and Fenn,[23] who found anomalously high recovery temperatures in supersonic jets of mixed gases with disparate masses, where the heat-transfer coefficient is considerably larger for the heavy than for the light gas. A thermocouple

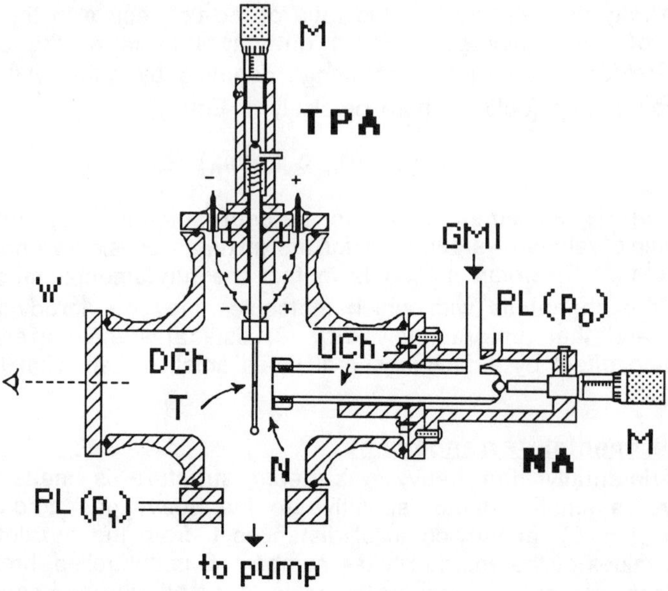

Fig. 10 Sketch of the test chamber. The mixture flows from the upstream chamber (uch) into the downstream chamber (dch), through an orifice nozzle (n). The attached assemblies can be displaced in and out with the help of their respective micrometers (M).

immersed in a H_2 jet seeded with CBr_4 molecules is overheated substantially by the energetic collisions with the heavy vapor, and is therefore capable of sensing it selectively.

The operation of this probe is illustrated in Fig. 11, which shows the voltages recorded while scanning the thermocouple radially across He-$W(CO)_6$ jets at a constant distance L/d_n=3.1. The narrower curves, shown with symbols, plot the increment in the signal resulting from seeding the heavy species on a given flow of light gas. This excess voltage shall subsequently be called DV, where the symbol V will be reserved for the voltage measured in the unseeded jet. Although the detailed interpretation of DV is not straightforward, one can show that it measures approximately the kinetic energy flux of the heavy gas, so that the radial width of the DV(r) curves in Fig. 11 is closely related to the beam width.[14] This conclusion provides a quantitative tool to study the physical dimension of the heavy gas jet, the fundamental parameter needed to characterize the process of aerodynamic focusing.

The continuous lines in figure 11 represent the thermocouple voltages corresponding to different source pressures in a jet of pure He. Because there are some small losses of heat to the electrodes through the thermocouple wires, V represents some sort of a nonadiabatic recovery temperature (the adiabatic recovery temperature, T_a is usually defined

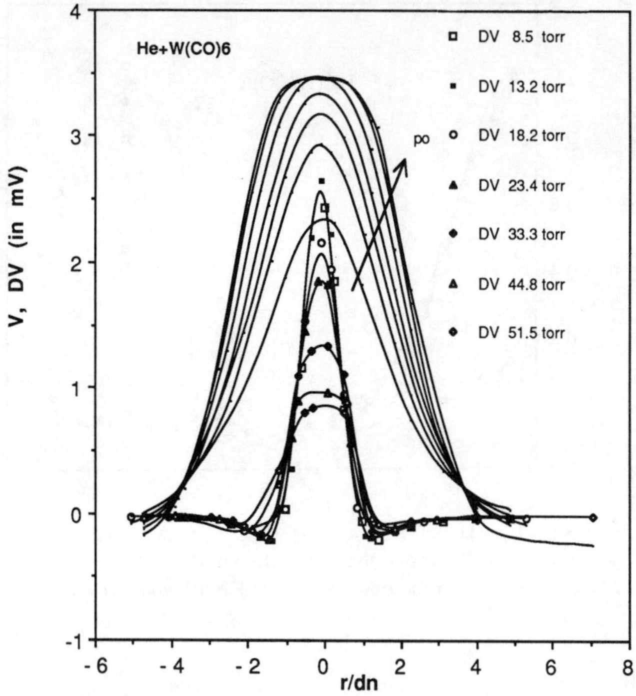

Fig. 11 Transverse voltage profiles collected at a fixed distance from the nozzle $L=3.1d_n$ (with $d_n=0.57$ mm) in a He-$W(CO)_6$ beam for the values of the source pressure: p_o=8.5, 13.2, 18.2, 23.4, 33.3, 44.8, and 51.5 torr. The voltage measured in the unseeded jet is plotted with solid lines without symbols. Lines with symbols show the excess voltage DV due to the heavy gas.

as the temperature reached by a thermally insulated body immersed in a fluid). In the continuum limit, T_a is smaller than the stagnation temperature T_o.[24] while the situation is reversed for a hypersonic flow in the free-molecule limit , where T_a/T_o tends to $2\gamma/(\gamma+1)$.[25]

C. Measured Heavy-Gas Beam Structures

In this subsection we report on sets of DV profiles (normalized with their maximum value) from jets of He and H_2 seeded with $W(CO)_6$ and CBr_4 in the limit of small heavy gas mas fraction $\varepsilon << 1$, at distances from the source L between 0.5 and 8 nozzle diameters.

In the range $0.4<S<2.2$, all of the data taken for $L/d_n>2$ collapse approximately into a single curve, provided the radial variable is normalized with a characteristic beam width. Defining the normalizing radius $\Delta/2$ as the radial position at which the signal DV is half of the

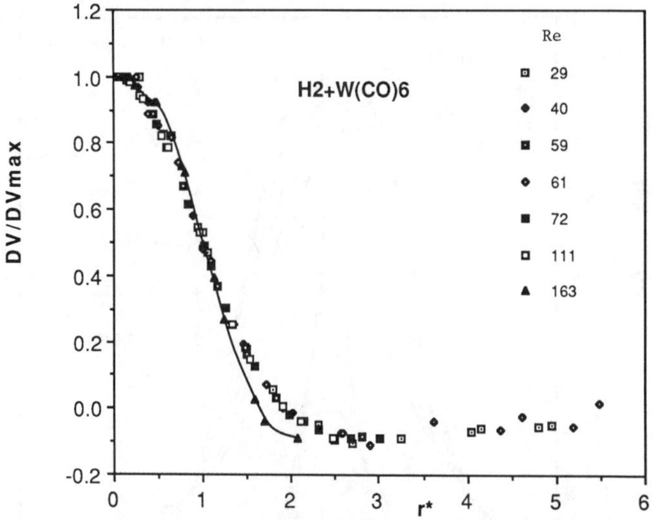

Fig. 12 Normalized profiles DV/DV_{max} vs r*, collected at distances from the nozzle larger than $L/d_n=2$, where they tend to an asymptotic curve. Symbols with a solid line correspond to the case of largest Re for which S = 0.39.

maximum

$$DV(\Delta/2) = DV_{max}/2 \tag{9}$$

and using r^* to denote the normalized radial coordinate

$$r^* = 2r/\Delta \tag{10}$$

one may confirm in Fig. 12 the universality of the DV/DV_{max} vs r^* curve in all of the region $L/d_n>2$; $0.4<S<2.2$. The same figure shows how the profile with the largest Re departs from the self-similar shape. Figure 13 shows normalized DV profiles for $S<0.5$, exhibiting a boundary with increasing sharpness as S decreases.

In conclusion, except for the highest values of the Reynolds numbers and the lowest values of L/d_n, the dependence of these curves on p_o and L enters only through the normalization factor $\Delta/d_n=F(S,L/d_n)$. The heavy gas beam may thus be fully characterized through the universal "Rosellian" curves of Fig. 12, together with experimental information on the beam width parameter Δ/d_n. In fact, even for the profiles that do not fully collapse into the Rosellian shape, their corresponding "width" Δ/d_n [defined in Eq. (9)] provides a most useful measure of the actual radial dimension of the beam. Such widths are shown in Fig. 14 as a function of L/d_n with S acting as a parameter. Each of these curves characterizes the shape of the beam, roughly as one would see it emerging from the nozzle with the help of some hypothetical visualization technique. Notice first that the smallest widths arise for H_2 mixtures at L/d_n around 1. Second,

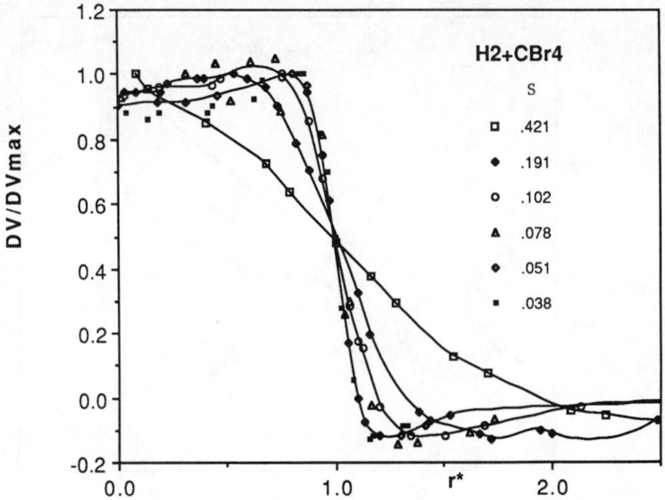

Fig. 13 DV/DV_{max} vs r* in the region of small S for a H_2-CBr_4 mixture and a distance to the nozzle plane L/d_n=3.2.

for a given value of L, Δ decreases first with increasing S, reaches a minimum, and increases subsequently with S. For each S, the axial position $L^*(S)$ at which the minimum width is reached plays a role analogous to a focal point. This behavior may be seen in Fig. 15, which cross plots data from Fig. 14 with S in the horizontal axis for a given value of $L=3.2d_n$.

Notice also that the smallest value of Δ in Fig. 14 is around $0.35d_n$, whereas aerosol beams could be concentrated far more sharply within one tenth of a nozzle diameter (Fig. 7). We shall have more to say on this, which we attribute to Brownian motion. Finally, it is worth remarking that some points in Fig. 14b show beam widths smaller than d_n even 5 nozzle diameters downstream from the source. Although the corresponding degree of concentration is far inferior to those attainable with aerosols, even this limited enrichment considerably exceeds anything previously observed in jets of mixed gases.

D. Beam Broadening and Brownian Motion

The sharpness of the boundary of the heavy gas beam at small values of S (Fig. 13) is reminiscent of the behavior of aerosol beams, where the absence of Brownian diffusion maintains the beam boundary as the same step function shape with which it is injected in the vacuum chamber at the nozzle exit. Naturally, as Re increases, diffusion phenomena decrease, and the heavy gas approaches the aerosol behavior. In fact, the S dependence in Fig. 13 resembles the time evolution of an initial concentration step under the effect of diffusion. Accordingly, because the diffusion process depends equally on the time as on the diffusion

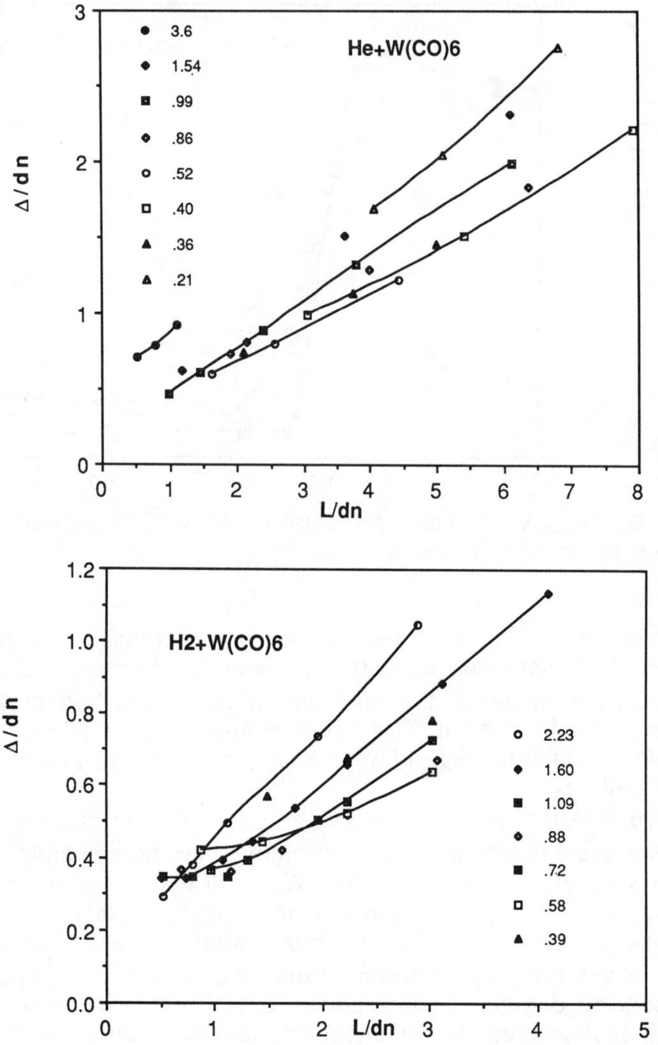

Fig. 14 Full heavy gas stream widths at half the maximum Δ/d_n vs distance to the nozzle plane L/d_n for He-W(CO)$_6$ (a), and H_2-W(CO)$_6$ (b) jets. The Stokes number S has been calculated with Eq. 6 using the values of Fe given in Table 1.

coefficient D, and because D varies linearly with S, interpreting the broadening of the heavy gas jet boundary as resulting from Brownian motion is practically unavoidable. With this perspective, it is interesting to look again at Fig. 15, comparing the previously mentioned results on heavy beam widths with those reported in Fig. 8 for aerosol jets in the free-molecule limit expanding also through a thin-plate orifice. Such a reference to the aerosol behavior should only be taken as

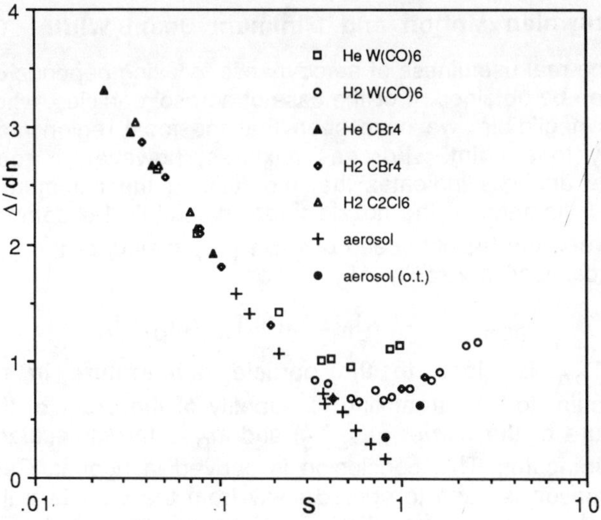

Fig. 15 Heavy molecule beam widths cross plotted from Fig. 14, jointly with the diameter of the aerosol caustic (filled circles) from Fig. 8 as a function of S for $L/d_n=3.2$.

semiquantitative because the ratios p_o/p_1 were higher there than they are here (however, this difference does not influence much the particle trajectories, as discussed in note 19 of Ref. 13) and also because some of the data shown in Fig. 15 use He as the carrier gas, and the light gas jet structure depends on the value of γ. Nonetheless, one sees in both figures that the curves corresponding to jets of mixed gases tend to the aerosol points as S diminishes. The analogy between particles and molecules therefore holds rather well regarding their dynamic behavior in jets, but this is true only at pressures large enough for diffusion to be unimportant. The much sharper minimum width seen in the aerosol curve is clear evidence of the fact that Brownian motion sets a definitive limit to the resolution with which a jet of mixed gases may be focused. This point is further confirmed by the differences shown in Fig. 15 between the mixtures using He and those using H_2 as the carrier gas. The minimum width observed is considerably smaller for the latter, as expected in a problem where the importance of Brownian motion is measured by the ratio between the thermal velocity of the heavy gas and the mean flow velocity, a quantity of the order of $(m/m_p)^{1/2}$ for a sonic flow. However, the rough estimate given in Eq. (8) for the minimum width is considerably smaller than what the previously mentioned experiments show. The incorporation of Brownian motion within our previous deterministic theoretical framework seems to be essential in the description of molecular focusing in mixtures where m/m_p is only of order 100.

V. Brownian Motion and Minimum Beam Width: Theory

The real usefulness of aerodynamic focusing depends on how sharp a focus can be obtained. For the case of aerosol particles, whose Brownian motion is negligible, we have seen that the focal region may be shrunk practically to a point. For gas mixtures, however, a rough order of magnitude analysis indicates that the ratio of the minimum beam width Δ_m to the diameter of the nozzle throat d_n should be comparable to the ratio between the fastest speed of wave propagation of the heavy gas and its mean convective velocity U_p

$$\Delta_m/d_n \sim \sqrt{(3kT_{nn}/m_p)}/U_p \tag{11}$$

where T_{nn} is close to the particle temperature in the direction perpendicular to the streamline (a quantity of the order of the stagnation temperature of the carrier gas, T_o) and m_p is the molecular mass of the heavy molecules. This conclusion is derived in rigor in Ref. 26, where, the information is seen to spread away from the particle trajectories by a small angle μ whose magnitude is given by the right-hand-side of Eq. (11). The gas emerging from a point source at the nozzle throat would thus acquire a width of order μd_n [as stated in Eq. (11)] after travelling a distance of order d_n. For a converging nozzle (Fig. 2), the maximum velocity U that the carrier gas attains at the throat is its sound speed, whereas U_p is at most of the order of U. The ratio Δ_m/d_n therefore becomes for a sonic jet

$$\Delta_m/d_n \sim \sqrt{(m/m_p)} \tag{12}$$

where m is the molecular mass of the carrier gas. According to Eq. (12), one would expect being able to concentrate a beam of gold atoms carried in a sonic H_2 jet within a region 10 times smaller in diameter than the nozzle throat diameter, reaching Au densities 100 times larger than at the source. Unfortunately, however, the previously mentioned experiments with H_2-$W(CO)_6$ mixtures have shown that such an estimate grossly underpredicts the focal width Δ_m by a factor larger than 5 (Sec. IV). Rather than a value of Δ_m/d_n of 7.5% as given by Eq. (12), measured minimal beam widths were near 35% of d_n. This result, if indicative of a fundamental limit, would reduce the potential concentrating power of the phenomenon by a factor larger than 25 below our earlier expectations. For that reason, the present section addresses theoretically the question of how much focal broadening Brownian motion introduces into a beam that would, on its absence, concentrate into a region of nearly vanishing dimensions.

A. Mathematical Formalism

The rigorous theoretical description of Brownian motion in the situation illustrated in Fig. 2 is in principle rather complex. The conditions required for decoupling the motion of the heavy and the light gas in order for the former to converge into a focus require a considerable departure from an equilibrium state, under which the ordinary diffusion equations of hydrodynamics fail.[27] This problem belongs fully to the kinetic theory of gases and must therefore be solved within its framework. For any nozzle geometry leading to aerodynamic focusing, an attempt at a numerical solution to the kinetic problem appears to be hopeless, even within the simplest picture ignoring particle-particle collisions and taking the light gas to be in equilibrium and unperturbed by the dilute heavy component. The same unpromising conclusion still holds after a systematic exploitation of the mass disparity is used to reduce the collision integrals of the Boltzmann equation into their far simpler Fokker-Planck differential form.[28]

A possible numerical attack to the Fokker-Planck equation could be based on solving the stochastic Langevin equation, as Gupta and Peters have done under similar circumstances.[29] However, for the present situation where the right hand side of Eq. (11) is a small number, a simpler description can be based on a well established hypersonic theory (Hypersonic closures of the hydrodynamic equations of a pure gas under non-equilibrium conditions exploit the fact that, although the pressure tensor appearing in the momentum equation is not known, it is small compared with the convective flux of momentum. The same holds with the unknown heat flux in the equation for the pressure tensor, and so forth for the highest order moments. A closed set of equations for the moments of order n or smaller may thus be obtained at any arbitrary order $n>2$ by neglecting the moments of order $n+1$ appearing in the equation of nth order[30-33]), recently addapted to problems of nonequilibrium Brownian motion[16,26,34]. This approach provides a rigorous closure to a finite set of hydrodynamic equations (the hypersonic equations), which retain the effects of thermal agitation, by keeping the pressure tensor $\mathbf{P}_p$ in the momentum conservation equation and neglecting the heat flux in the conservation equation for $\mathbf{P}_p$. In addition, the terms accounting for the momentum and energy transfer between the two species will be written in their simplest form, which holds in the free molecular limit only when the relative velocity between the two species is small compared with the speed of sound of the pure light gas. More general expressions for these terms derived systematically in an expansion in the mass ratio m/m_p may be found in Ref. 18. The light gas is treated as incompressible at the Euler level, with a known velocity field $\mathbf{U}(\mathbf{x})$, and a constant temperature T. The corresponding closed hypersonic equations for the heavy gas are[16]

$$D\lambda_p + \nabla\cdot\mathbf{U}_p = 0 \tag{13}$$

$$D\mathbf{U}_p + ((\boldsymbol{T}_p\cdot\nabla)\lambda_p + \nabla\cdot\boldsymbol{T}_p)k/m_p = (\mathbf{U} - \mathbf{U}_p)/\tau \tag{14}$$

$$D\boldsymbol{T}_p + (\boldsymbol{T}_p\cdot\nabla)\,\mathbf{U}_p + ((\boldsymbol{T}_p\cdot\nabla)\,\mathbf{U}_p)^T = 2(T\boldsymbol{I} - \boldsymbol{T}_p)/\tau \tag{15}$$

where $\mathbf{I}$ is the unit tensor, superscript T denotes the transposed tensor, and τ is the particle relaxation time related to the mixture diffusion coefficient D through Einstein's law:

$$\tau = Dm_p/kT \tag{16}$$

D is the standard convective derivative:

$$D \equiv \partial/\partial t + \mathbf{U}_p \cdot \nabla \tag{17}$$

and λ_p is the logarithm of an appropriately normalized particle density ρ_p:

$$\lambda_p = \log(\rho_p/\rho_{po}) \tag{18}$$

Instead of the heavy gas pressure tensor $\mathbf{P}_p$, we have used the temperature tensor $\boldsymbol{T}_p$:

$$\mathbf{P}_p = \rho_p k \boldsymbol{T}_p / m_p \tag{19}$$

Making Eqs. (13-15) dimensionless by introducing the characteristic velocity, temperature, and length scales U_R, T_R, and L_R, respectively,

$$\mathbf{U}_p \leftarrow \mathbf{U}_p/U_R,\quad \boldsymbol{T}_p \leftarrow \boldsymbol{T}_p/T_R,\quad \mathbf{U} \leftarrow \mathbf{U}/U_R,\quad T \leftarrow T/T_R,\quad \mathbf{x} \leftarrow \mathbf{x}/L_R \tag{20}$$

in terms of S and the small parameter ε (a sort of an inverse Mach number)

$$S = U_R\tau/L_R, \qquad \varepsilon^2 = kT_R/(m_p U_R^2) \tag{21}$$

Eqs. (13-15) become

$$D\lambda_p = -\nabla \cdot \mathbf{U}_p \tag{22}$$

$$D\mathbf{U}_p + \varepsilon^2 [(\boldsymbol{T}_p \cdot \nabla)\lambda_p + \nabla \cdot \boldsymbol{T}_p] = (\mathbf{U} - \mathbf{U}_p)/S \tag{23}$$

$$D\boldsymbol{T}_p + (\boldsymbol{T}_p \cdot \nabla)\, \mathbf{U}_p + ((\boldsymbol{T}_p \cdot \nabla)\, \mathbf{U}_p)^T = 2\,(T\boldsymbol{I} - \boldsymbol{T}_p)/S \tag{24}$$

The effect of Brownian motion appears now in Eq. (23) as proportional to ε^2, originating from the divergence of $\mathbf{P}_p$. Rather than by a near-equilibrium closure, this term is given by the tensorial energy Eq. (24), where an unknown heat flux term of order ε has been neglected.

Because the hypersonic equations (22-24) are fully hyperbolic within the region of parameter space where they are at all valid ($\varepsilon \ll 1$), their numerical solution for complex flows can be addressed by the method of characteristics.[26] Here, however, we shall only attempt an even simpler near-axis analysis.

B. Hypersonic Equations Near the Axis of Symmetry

Let us extend the deterministic near-axis results that showed the existence of focal points[12] to incorporate Brownian motion. Defining a parameter ϕ taking the value 1 for axisymmetric flows and 0 in two-dimensional situations, Eqs. (13-15) for the steady problem symmetric around the x axis become

$$D\lambda + u_x + v_y + \phi\, v/y = 0 \tag{25}$$

$$Du + \varepsilon^2(D_x \lambda + T_{xx,x} + T_{xy,y} + \phi T_{xy}/y) = (u'-u)/S \tag{26}$$

$$Dv + \varepsilon^2[D_y \lambda + T_{xy,x} + T_{yy,y} + \phi(T_{yy} - T_{zz})/y] = (v'-v)/S \tag{27}$$

$$DT_{xx} + 2D_x u = 2(T'-T_{xx})/S \tag{28}$$

$$\mathcal{D}T_{yy} + 2\mathcal{D}_y v = 2(T'-T_{yy})/S \tag{29}$$

$$\mathcal{D}T_{xy} + \mathcal{D}_x v + \mathcal{D}_y u = -2T_{xy}/S \tag{30}$$

$$\mathcal{D}T_{zz} + 2v\phi\, T_{zz}/y = 2(T'-T_{zz})/S \tag{31}$$

where the operator $\mathcal{D}$ was defined in Eq.(17), and

$$\mathcal{D}_x \equiv T_{xx}\,\partial_x + T_{xy}\,\partial_y\,; \quad \mathcal{D}_y \equiv T_{xy}\,\partial_x + T_{yy}\,\partial_y \tag{33}$$

The primed variables correspond to the light gas, and T_{zz} stands for the azimuthal temperature in the axisymetric case.

Because the hypersonic equations written earlier hold only in the limit $\varepsilon^2 \ll 1$, this feature can be exploited further by recognizing that a region of order ε around the axis can be considered a boundary layer characterized by the following internal variables:[16]

$$y = \varepsilon\eta, \quad v = \varepsilon w, \qquad T_{xy} = \varepsilon\theta_{xy} \tag{34}$$

There, with errors of order ε^2, u, Txx, Tyy and Tzz depend only on x. Simultaneously, also by symmetry, the carrier gas velocity and temperature fields can be expanded in powers of y,

$$u'(x,y) = u'_0(x) + y^2 u'_1(x) + \ldots = u'_0(x) + \varepsilon^2\eta^2 u'_1(x) + \ldots \tag{35}$$

$$v'(x,y) = y[w'_0(x) + y^2 w'_1(x) + \ldots] = \varepsilon\eta w'_0(x) + \varepsilon^3\eta^3 w'_1(x) + \ldots \tag{36}$$

$$T'(x,y) = T'_0(x) + y^2 T'_1(x) + \ldots = T'_0(x) + \varepsilon^2\eta^2 T'_1(x) + \ldots \tag{37}$$

To lowest order the boundary-layer form of Eqs.(25-31) is

$$\mathcal{D}_0\,\lambda + u_{0x} + w_\eta + \phi w/\eta = 0 \tag{38}$$

$$\mathcal{D}_0\,u_0 + (u_0 - u'_0)/S) = 0 \tag{39}$$

$$\mathcal{D}_0\,w + (w - w'_0)/S + T_2\lambda_\eta + T_{2\eta} + \phi(T_2 - T_3)/\eta = 0 \tag{40}$$

$$\mathcal{D}_0 T_1 + 2T_1 u_{0x} = 2\,(T_0' - T_1)/S \tag{41}$$

$$\mathcal{D}_0 T_3 + 2\phi w\, T_3/\eta = 2(T_0' - T_3)/S \tag{42}$$

$$\mathcal{D}_0 T_2 + 2T_2 w_\eta = 2(T_0' - T_2)/S \tag{43}$$

where the variables u_1 and θ_{xy} are uncoupled from the rest, and

$$\mathcal{D}_0 = u_0\partial_x + w\partial_\eta \tag{44}$$

The preceding equations are independent of ε, thus having absorbed the dependence of the problem on the mass ratio m/m_p. As anticipated in Eq. (11), because η is of order 1, the minimum beam width will be of order ε. Furthermore, the near-axis equations admit exact Gaussian solutions such that, if ρ_p is initially a Gaussian function of η, it will remain Gaussian farther downstream. Thus, when

$$\lambda(x=0,y) = \lambda_{00} + \lambda_{10}\eta^2 \tag{45}$$

then the problem admits the following exact solution:

$$\lambda = \lambda_0(x) + \eta^2\lambda_1(x), \qquad \theta_{xy} = \eta\beta(x) \tag{46}$$

$$w = \eta w_0(x), \qquad u_1 = p(x) + \eta^2 p(x) \tag{47}$$

where the functions of a single variable λ_0, $\lambda_1(x)$, $\beta(x)$, w_0, $p(x)$ and $p(x)$ obey the system of ordinary differential equations

$$u_0 d\lambda_0/dx + w_0(1+\phi) + du_0/dx = 0 \tag{48}$$

$$u_0 du_0/dx + (u_0 - u_0')/S = 0 \tag{49}$$

$$u_0 d\lambda_1/dx + 2w_0\lambda_1 = 0 \tag{50}$$

$$u_0 dw_0/dx + (w_0 - w_0')/S + w_0^2 + 2T_2\lambda_1 = 0 \tag{51}$$

$$u_0 dT_2/dx + 2[T_2 w_0 + (T_2 - T_0')/S] = 0 \tag{52}$$

$$u_0\, dT_1/dx + 2T_1 du_0/dx - 2(T_1 - T_0')/S = 0 \tag{53}$$

$$u_0 \frac{dp}{dx} + p\left(\frac{1}{S} + \frac{du_0}{dx}\right) + T_1 \frac{d\lambda_0}{dx} + \frac{dT_1}{dx} + \beta(1+\phi) = 0 \tag{54}$$

$$u_0 \frac{dq}{dx} + q\left(\frac{1}{S} + \frac{du_0}{dx}\right) - \frac{u_1'}{S} + 2qw_0 + T_1\frac{d\lambda_1}{dx} + 2\beta\lambda_1 = 0 \tag{55}$$

$$u_0 \frac{dT_3}{dx} + 2\frac{T_3 - T_0'}{S} = 0 \ \text{(for } \phi=0\text{)}; \quad T_3 = T_2 \ \text{(for } \phi=1\text{)} \tag{56}$$

This system of equations enjoys the considerable advantage over its predecessors of having a degree of complexity comparable to that of the deterministic equations (Sec. 3.1 of Ref. 12). Equation (49) is actually identical to its deterministic counterpart, whereas Eq. (51) involves the new "diffusive" term $2T_2\lambda_1$, whose determination requires the two additional Eqs. (50) and (52) for λ_1 and T_2. λ_0, T_1, T_3, p, and q do not affect the fundamental variables u_0, w_0, T_2, and λ_1. The deterministic limit is recovered exactly when $\lambda_1=0$ initially (ρ_p is then independent of η and there are no radial concentration gradients). Then, because Eq. (50) is homogeneous, λ_1 remains null. However, for any non vanishing initial value of λ_1, when $w_0>>1$ near the deterministic focus, λ_1 grows exponentially and opposes the term w_0^2 in Eq. (51) (notice that $\lambda_1<0$, $T_2>0$), which was responsible in the deterministic limit for the focal point singularity. Brownian motion thus allows for a minimal focal width of order ε but blocks the singularity proper. This may be seen in greater rigor through an analysis in the vicinity of the focus, which closely parallels that in Sec. 3.1 of Ref. 12 for the deterministic limit.

C. Structure of the Focal Region

If the beam has an initial width of order one, then $\lambda_1=0(\varepsilon^2)$, so that the behavior is nearly deterministic until near the focus. Accordingly, Brownian effects are confined to a small region in the vicinity of the focus where the linear terms inversely proportional to S are small with respect to the others in Eqs.(50-52) (exactly as in the deterministic limit). Accordingly, Eqs.(50-52) may be written as

$$u_0\frac{d\lambda_1}{dx}+2\lambda_1 w_0=0;\quad u_0\frac{dw_0}{dx}+w_0^2+2T_2\lambda_1=0;\quad u_0\frac{dT_2}{dx}+2T_2 w_0=0 \qquad (57)$$

The similarity between the first and the third of these equations leads to $\lambda_1=-\alpha^2 T_2$, where α is an unknown constant. This first integral makes it possible to rewrite the last two equations in phase space as a homogeneous equation that may be integrated analytically to yield

$$w_0=\frac{B}{\alpha 2^{1/2}}\,\frac{s}{1+s^2},\quad T_2=\frac{B}{2\alpha^2}\,\frac{1}{1+s^2},\quad s=\frac{B}{\alpha 2^{1/2}}\,\frac{x-x_0}{u_0} \qquad (58)$$

where B and x_0 are other unknown constants, and x_0 can be interpreted as the axial location of the focus. Notice that as $s \gg 1$, w_0 behaves as $1/s$ and T_2 as s^{-2}, as in the deterministic limit, whereas near s=0 these diverging tendencies are controlled by the Brownian movement. The magnitude of greatest interest is the beam width $\Delta\eta$,

$$\Delta\eta=\sqrt{(-\lambda_1)} \qquad (59)$$

which is linear with x with a slope $B^{1/2}/(\alpha u_0)$ as $s \gg 1$, and has the minimum value

$$\Delta\eta_{min}=\sqrt{(2/B)} \qquad (60)$$

independently from the value of the asymptotic slope $B^{1/2}/(\alpha u_0)$. Physically, this result implies that seeding the heavy gas only in the central region of the jet upstream from the nozzle would not reduce the minimum focal width. Indeed, the value of the constant B may be derived as the deterministic limit of T_2/w_0^2, which is independent of the initial value assigned to the beam width. We had observed this interesting result (ineffectiveness of sheathing) in a limited number of earlier experiments in H_2-$W(CO)_6$ mixtures impacting on a solid surface.

A final note of caution is due here regarding the use of these results for gas mixtures where m/m_p is only of order 0.01, so that ε is not sufficiently small to sharply separate the η and the y variables and the minimum focal width is nearly $0.35d_n$. For such cases it would be more accurate to solve the full near-axis equations numerically through the focal region.

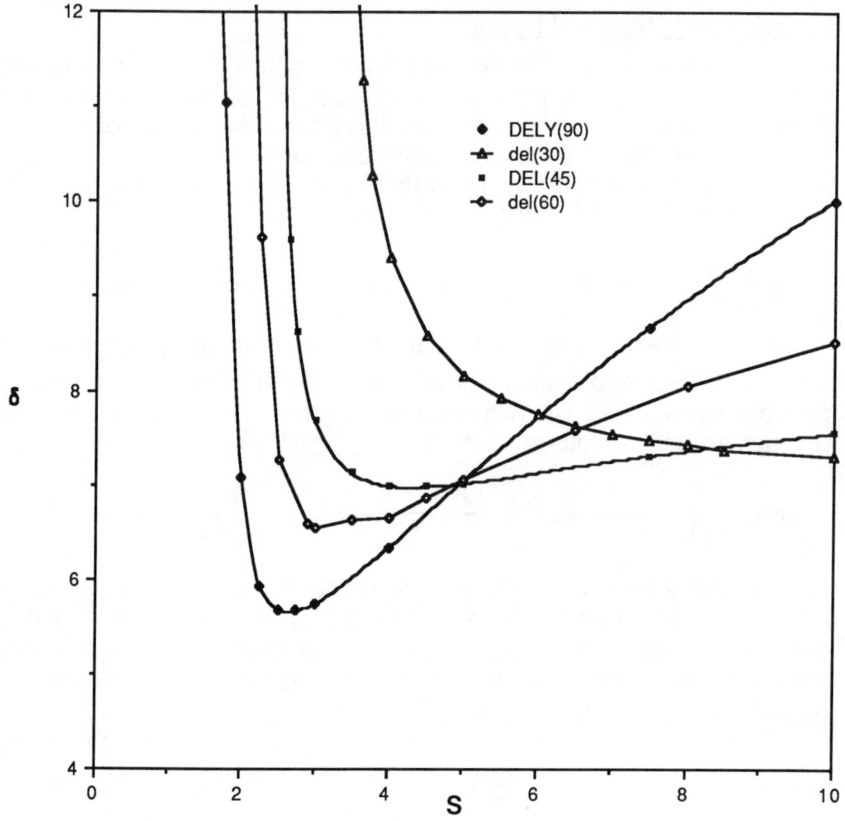

Fig. 16 Calculated minimal beam widths $\Delta\eta$ for the 2D nozzle of Fig. 2. The asymptote from left to right corresponds to angles α of 90, 60, 45, and 30 deg. $S = U\tau/b$.

D Computed Beam Widths for 2D Converging Nozzles

In this subsection we apply the near-axis equations to the potential problem of a 2D converging nozzle considered in Section II at the deterministic level. Several additional initial conditions are required for the integration. For the particle temperature tensor and velocity, we take their deterministic value at the injection section. The initial value for λ_1 is related to the starting beam width and does have an effect on the final result except for very small values of this parameter, corresponding to the limit $\varepsilon \ll 1$. In the following we report only on preliminary results where the integration was started at $x = -2$ with $\lambda_1 = 10^{-5}$. The results for the minimum beam width $\Delta\eta_{min}$ are shown in Fig. 16 as a function of $S = U\tau/b$ for various angles of convergence α. Notice that the smallest of the minima arises in the vicinity of the sharpest angle $\alpha = 90$ deg. Unfortunately, even in this near-optimal circumstance, $\Delta\eta$ takes values close to 6, in rough agreement with our experimental observations. Accordingly, although Δ_{min}/d_p is proportional to ε, the relatively large value of the constant relating these two quantities makes it unlikely that heavy species absolute concentration factors as large as 100 might be reached in gas mixtures with mass ratios of only 100.

VI. Conclusions

The problem of determining how much a beam of heavy molecules seeded in supersonic jets of He and H_2 may be concentrated has been considered. Using theoretical and experimental insights from the deterministic limit of aerosol beams, some of the most favorable focusing circumstances have been found to be associated with sharply converging nozzles. With the help of a thermocouple probe, heavy gas beam structures in mixtures with mass ratios of order 100 have been studied in hypersonic jets issuing from a thin-plate orifice. Experimentally one finds minimal beam widths of the order of 35% of a nozzle diameter d_n around one nozzle diameter downstream from the source. When focused, most of the seed gas may be found within one nozzle area, even five nozzle diameters away from the throat. These results are confirmed by a Gaussian approximation of the near-axis hypersonic theory, which shows that $\Delta_{min}/d_n \approx 6\varepsilon$.

Acknowledgments

We are in debt to J. B. Fenn, S.V. Hering, R. Fernandez-Feria, B. Dahneke, B. L. Halpern, S. Fuerstenau, S. Cohen, D. E. Rosner, S. Gomez, R. E Apfel, K Sreenivasan, and other colleages and students at Yale for their support in ideas, instruments, and cash, without which this paper would not have been written. This work has been financed by National Science Foundation Grant CBT-8612143, U.S. Department of Energy Grant DE-FG02-87ER13750, and Department of Defense Grant ARO, contract GAAL 03-87-K-0127D

References

[1]Becker, E. W., Bier, K. and Burghoff, H., "Die Trennduse. Ein neues Element zur Gas-und Isotopentrennung," Zeitschrift fur Naturforschung, Vol. **A10**, 1955, pp 565-572.

[2]Ehrfeld, W., Elements of Flow and Diffusion Processes in Separation Nozzles, Springer- Verlag, New York, 1983. See also Touryan, J. K., Muntz, E. P., Talbot, L. and Von Halle, E., "Proceedings of a Workshop on Gasdynamic Problems in Isotope Separation," Sandia Lab. Rept. SAND-75-0121, 1979.

[3]Ryhage, R., "Use of Mass Spectrometer as a Detector and Analyzer for Efluents Emerging from High Temperature Gas Liquid Chromatography Columns," Analytical Chemistry, Vol. 36, 1964, pp. 759-764.

[4]Anderson, J. B., Andres, R. P. and Fenn, J. B. , Advances in Chemical Physics, Vol. 10, edited by J. Ross, Interscience, New York, 1966.

[5]Herschbach, D. R., "Molecular Dynamics of Elementary Chemical Reactions," Les Prix Nobel en 1986, The Nobel Foundation, Stokholm, 1987.

[6]Israel, G. W. and Friedlander, S. K., " High Speed Beams of Small Particles," Journal of Colloid and Interface Science, Vol. **24**, 1967, pp. 330-337.

[7]Israel, G. W. and Whang, J. S, "Dynamical Properties of Aerosol Beams," University of Maryland, College Park, MD, TN- BN 709, 1971.

[8]Dahneke, B. E., Hoover, J. and Cheng, Y. S., "Similarity Theory of Aerosol Beams," Journal of Colloid and Interface Science, Vol. 87, 1982, pp. 167-179.

[9]Dahneke, B. E. and Cheng, Y. S., "Properties of continuum source particle beams," Journal of Aerosol Science, Vol. **10**, 1979, pp.257-274 .

[10]Dahneke, B. E. and Hoover, J., "Size Separation of Aerosol Beam Particles," Rarefied Gas Dynamics, Vol 1, edited by S. S. Fisher, AIAA, New York, 1982.

[11]Dahneke, B. E., private communication, 1987.

[12]Fernandez de la Mora, J. and Riesco-Chueca, P., "Aerodynamic Focusing of Particles in a Carrier Gas," Journal of Fluid Mechanics, Vol. 195, 1988, pp. 1-21.

[13]Fernandez de la Mora, J. and Dahneke, B. E., "Experiments on Aerodynamic Focusing of Particles in Supersonic Jets, " to be submitted to the Journal of Fluid Mechanics, 1989.

[14]Rosell-Llompart, J., Experiments on Molecule Focusing in Seeded Supersonic Jets," Master's Thesis presented at Universidad Autonoma de Barcelona, Physics Department, June 1988. Also, paper with the same title submitted to the Journal of Chemical Physics, 1988.

[15]Reis, V. H. and Fenn, J. B., "Separation of Gas Mixtures in Supersonic Jets", Journal of Chemical Physics, Vol. 39, 1963, pp. 3240-3250.

[16]Riesco-Chueca, P. and Fernandez de la Mora, J., "Brownian Motion far from Equilibrium, a Hypersonic Approach," to be published in Journal of Fluid Mechanics, 1989.

[17]Friedlander, S. K. , "Smoke,Dust and Haze", Wiley, New York, 1977, Sec. 4.6.

[18]Riesco-Chueca, P., Fernandez-Feria, R. and Fernandez de la Mora, J., "Interspecies Transfer of Momentum and Energy in Disparate-Mass Gas Mixtures," Physics of Fluids, Vol. **30**, 1987, pp. 45-55.

[19]Milne Thomson, L. M., "Theoretical Hydrodynamics," MacMillan, London, 1938.

[20]Dahneke, B. E., "Apparatus for Separation of Gas Borne Particles," U.S. Patent 4,358,302, Nov. 9, 1982 .

[21]Sherman, F. S., "A Survey of Experimental Results and Methods for the Transition Regime of Rarefied Gas Dynamics," Rarefied Gas Dynamics, edited by

J. A. Laurmann, Academic, New York, 1963, pp. 228-260; also Ashkenas, H. and Sherman, F. S., "The Structure and Utilization of Supersonic Free Jets in Low Density Wind Tunnels," Rarefied Gas Dynamics, edited by J. H. de Leeuw, Academic, New York, 1966, pp. 84-105.

[22]Liepmann, H., "Gaskinetics and Gasdynamics of Orifice Flow," Journal of Fluid Mechanics, Vol. 10, 1961, pp. 65-79. Also Smetana, F. O. , Sherrill, W. A., & Schort, D. R., "Measurements of the Discharge Characteristics of Sharp-edged and Round-edged Orifices in the Transition Regime", Rarefied Gas Dynamics, edited by C. L. Brundin, Academic, New York, 1967, pp. 1243-1256.

[23]Maise, G, and Fenn, J. B. "Thermal Recovery Factors in Supersonic Flows of Gas Mixtures," Journal of Heat Transfer (ASME), Vol. 94, 1972, pp. 29-35.

[24]Schlichting, H., Boundary Layer Theory, McGraw-Hill, New York, 1968, Chapter XIII.

[25] Oppenheim, A. K. , "Generalized Theory of Convective Heat Transfer in Free-Molecule Flow," Journal of Aeronautical Sciences, Vol. 20, 1953, pp. 49-58.

[26]Riesco-Chueca, P., Fernandez-Feria, R. and Fernandez de la Mora, J., "Method of characteristics description of Brownian motion far from equilibrium," published elsewhere in these volumes.

[27]Fernandez de la Mora, J., Halpern, B. L. and Wilson, J. A.,"Inertial impaction of heavy molecules," Journal of Fluid Mechanics, Vol. 149, pp. 217-233.

[28]Wang Chang, C. and Uhlenbeck, G. E., in Studies in Statistical Mechanics, Vol. V; pp. 89-92, edited by J. De Boer and G. E. Uhlenbeck, North Holland, Amsterdam, 1970. Also Fernandez de la Mora, J. and Fernandez Feria, R., "Kinetic theory of gas mixtures with disparate masses," Physics of Fluids, Vol. 30, 1987, pp. 740-751.

[29]Gupta, D. and Peters, D. H., "A Brownian Dynamics Simulation of Aerosol Deposition onto Spherical Collectors," Journal of Colloid and Interface Science, Vol. 104, 1985, pp. 375-389.

[30]Freeman, N. C., "Solution of the Boltzmann Equation for Expanding Flows," AIAA Journal, Vol. 5, 1967, pp. 1696-1698.

[31]Freeman, N. C. and Grundy, R. E., "On the Solution of the Boltzmann Equation for an Unsteady Cylindrically Symmetric Expansion of a Monoatomic Gas into a Vacuum," Journal of Fluid Mechanics, Vol. 31, 1968, pp. 723-736.

[32]Hamel, B. B. and Willis, D. R., "Kinetic Theory of Source Flow Expansion with Application to the Free Jet," Physics of Fluids, Vol. 9, 1966, pp. 829-841.

[33]Edwards, R. M. and Chen, H. K., "Steady Expansion of a Gas into a Vacuum," AIAA Journal, Vol. 4, 1966, pp.558-561.

[34]Fernández-Feria, R., "On the Gasdynamics of Binary Gas Mixtures with Large Mass Disparity," Ph.D. Thesis, Yale University, New Haven, CT, Dec. 1987.

Experimental Investigations of Aerodynamic Separation of Isotopes and Gases in a Separation Nozzle Cascade

P. Bley* and H. Hein†
Kernforschungszentrum Karlsruhe, Karlsruhe, Federal Republic of Germany
and
J. L. Campos,‡ R. V. Consiglio,§ and J. S. Coelho §
Centro de Desenvolvimento da Tecnologia Nuclear, Belo Horizonte, Brazil

Abstract

The operational behavior of a separation nozzle cascade for uranium enrichment was investigated experimentally in a 10-stage pilot plant erected in Belo Horizonte, Brazil. The experiments performed were supported by computer simulation. The pilot plant was equipped with the advanced double deflection systems, and it was demonstrated that it allowed the desired performance to be achieved more easily than with a cascade with single deflection systems. A control system for the external UF_6 flows of separation nozzle cascades, in which false UF_6 flows were transported to the bottom of the cascade, was developed and tested sucessfully. The control system guarantees the required stable and uniform UF_6 distribution in the cascade and keeps the ^{235}U isotopic gradient at its present level for maximum separation power. At the top of a separation nozzle cascade, UF_6 and the light auxiliary gas must be separated from each other. Cryogenic separation, performed in the past, was replaced by separation in a three-stage gas separation cascade in which the UF_6 concentration was reduced by more than four orders of magnitude. The collective operation of the two cascades for gas separation and isotope separation did not raise any difficulties.

Introduction

For the large-scale enrichment of ^{235}U the separation nozzle process has been developed at the Karlsruhe Nuclear Research Center [1] In the separation nozzle the separating centrifugal forces are generated by aerodynamic means: Gaseous uranium hexafluoride mixed with a light

---

*Section Head, Institut für Mikrostrukturtechnik.

† Senior Scientist, Institut für Mikrostrukturtechnik.

‡ Section Head, Divisão de Enrequicimento.

§ Scientist, Divisão de Enrequicimento.

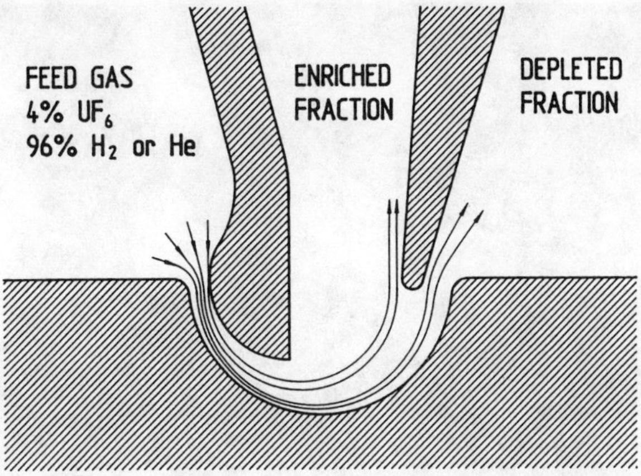

Fig. 1 Principle of the separation nozzle with single deflection system used in the first cascade.

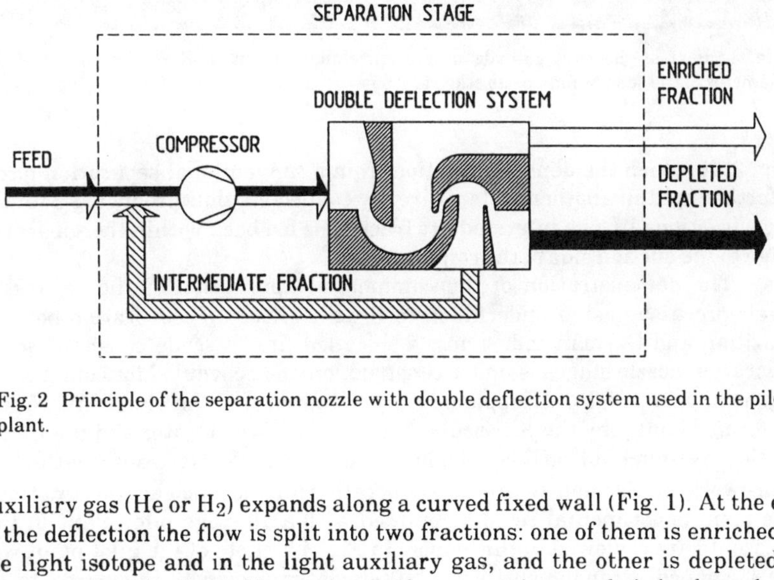

Fig. 2 Principle of the separation nozzle with double deflection system used in the pilot plant.

auxiliary gas (He or H_2) expands along a curved fixed wall (Fig. 1). At the end of the deflection the flow is split into two fractions: one of them is enriched in the light isotope and in the light auxiliary gas, and the other is depleted in these species. The Reynolds number in the separation nozzle is in the order of 100, and typical rarefaction phenomena are observed in the curved jet.

The experimental and theoretical investigations of these effects were presented at many symposia on rarefied gasdynamics.[2-7] Intensive planning of enrichment cascades was performed; e. g., the gasdynamics behavior of a cascade was investigated using a mathematical analysis of the gasdynamics network. The separation nozzle cascade is characterized by rarefied flow conditions in the separation nozzle and by continuum flow conditions in the other parts of the cascade (e. g., compressors, valves, etc.).[5] To make isotope separation more effective, the so-called double deflection system was proposed

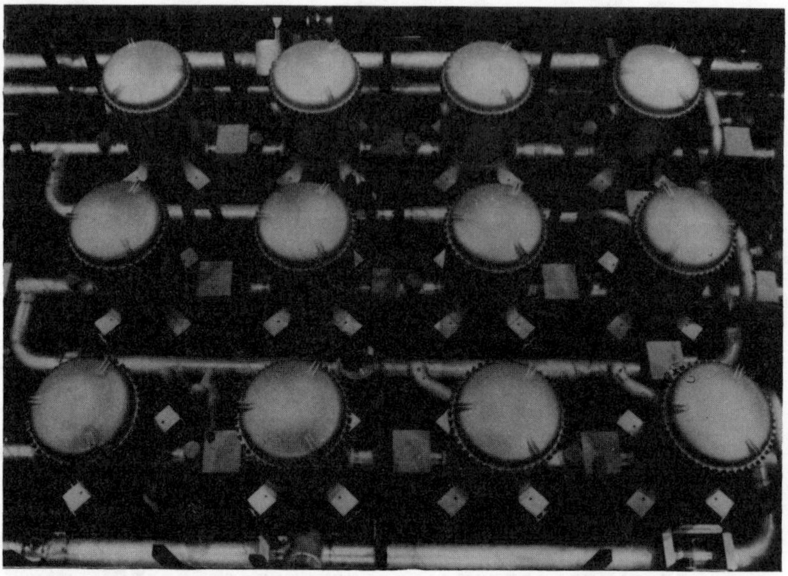

Fig. 3 Part of the first cascade nozzle enrichment plant in Resende, Brazil. The diameter of the separation nozzle stage is 1.5 m.

(Fig. 2) in which the depleted fraction from a conventional separation nozzle is further split in another system directly connected, thus producing a total of three fractions.[1,7] The intermediate fraction is fed back within the separation stage to the suction side of the compressor.

The demonstration of the commercial application of the separation nozzle process is being undertaken in Brazil, where, in cooperation between Brazilian and German industries, a so-called first cascade consisting of 24 separation nozzle stages is under construction and scheduled for commissioning next year (Fig. 3). Supplementary research and development work is performed jointly by the Karlsruhe Nuclear Research Center and the Centro de Desenvolvimento da Tecnologia Nuclear at Belo Horizonte within the framework of German-Brazilian cooperation in the nuclear energy sector. The main experimental results obtained with a 10-stage pilot plant in Belo Horizonte are reviewed in this paper. In Fig. 4 a photo of the pilot plant with the U-shaped arrangement of the 10 stages is shown.[8,9] Compared to the industrial stages of the first cascade in Resende, the stage throughput of the pilot plant is smaller by about four orders of magnitude.

Layout of the pilot plant

The pilot plant is equipped with 10 separation stages, and the main components of a stage are the tank with the double deflection system, the roots compressor, and the throttle valve to adjust the UF_6 cut of the stage to the nominal value (Fig. 5). The pilot plant is a square cascade with the

Fig. 4 View of the 10-stage separation nozzle pilot plant in Belo Horizonte, Brazil.

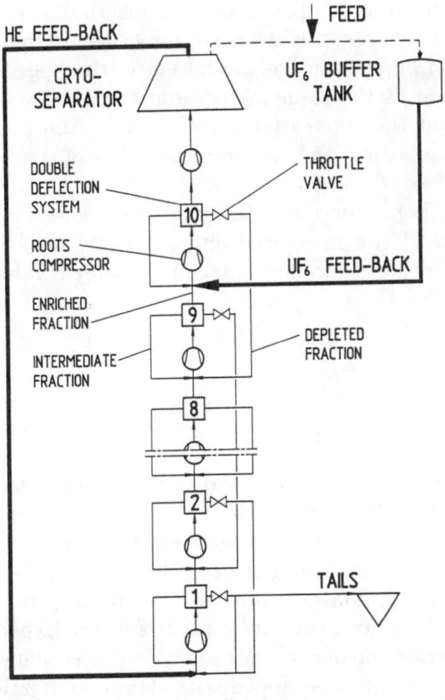

Fig. 5 Flow sheet of the 10-stage pilot plant with cryogenic He/UF_6 separation.

simplest interconnection: the enriched fraction from each stage is fed into the next upper stage, and the depleted fraction is fed into the stage immediately below. Therefore, the nominal UF_6 cut is 0.5.

To compensate in a separation nozzle cascade for the disproportionately large transport of light auxiliary gas to the top of the cascade, the bulk of the auxiliary gas stream is extracted at the top of the cascade with a gas separation unit and fed back to the bottom of the cascade, while UF_6 must be continuously fed back to the top of the cascade. In the first cascade, gas separation will be achieved by cryoseparators operating in a cyclic mode, and the UF_6 feedback will be achieved by a UF_6 buffer tank. For this reason, a cryogenic gas separation system had originally been installed in the pilot plant.[9]

To simplify the operation of the pilot plant, the feed is fed into the plant together with the UF_6 feedback; therefore, the pilot plant constitutes the stripping section of a cascade in which the ^{235}U concentration at the top is approximately constant, whereas the ^{235}U concentration at the bottom is reduced.

In all stages the pressures are measured in all fractions, and the UF_6 concentration of the feed gas N_0 is continuously recorded by α-ionization detectors.

The double deflection systems used in the pilot plant have a deflection groove with a radius of curvature of 0.75 mm for the first nozzle and of 0.4 mm for the second nozzle. Compared to the separation systems used in the first cascade and to technical separation systems that are being tested at the Nuclear Research Center in Karlsruhe, the critical dimensions are larger by factors of 7.5 and 30, respectively. Therefore, the operating pressure of the pilot plant is lower by the same factors and has a value of only 35 mbar. The expansion ratio at the separation nozzle is 2.5. The pilot plant is operated with an He/UF_6 mixture, and the nominal value of the UF_6 concentration of the feed gas is 4.2%.

The operating behavior of the pilot plant was simulated by computer models using all of the measured characteristic curves of the stage with double deflection system and the control algorithm developed.

Double Deflection System

The 10-stage pilot plant was equipped with double deflection systems. Experiments performed over a long period of time have proved the reliability of the cascade.

For example, in Fig. 6 the relative deviations in the operating pressure ($\Delta P_0/P_0$) and in the UF_6 concentration ($\Delta N_0/N_0$) of the feed gas are plotted over the stage number. In that experiment the throttle valve of stage 3 was adjusted to a false value. It can be seen that, within the limits of measuring accuracy, there are deviations only in the immediate neighborhood of the disturbed stage. The good correlation between the experimental results and the computer simulation of cascade operation is also shown in Fig. 6.

Adjusting cascades equipped with double deflection systems is much easier than adjusting cascades equipped with single deflection systems

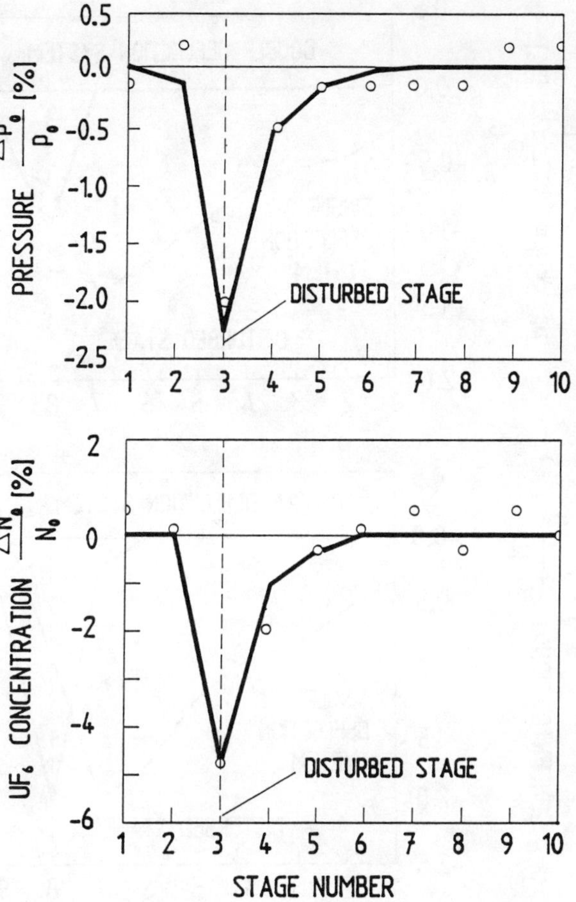

Fig. 6 Influence of false adjustment of the throttle valve in stage 3 on the operating pressure P_0 and UF_6 concentration N_0 of the feed gas. The curves drawn through the measured points were calculated with a cascade simulation computer code.

because the influence of one stage on another is less pronounced, as shown in Fig. 7. Therefore, the desired performance of a cascade with double deflection systems is more easily attained than that of a cascade with single deflection systems.[10]

Stabilization of the UF_6 Distribution and the Isotopic Gradient

The gasdynamic characteristic of an entire separation nozzle stage is determined mainly by the dependence of the UF_6 cut on the UF_6 content of the stage. This characteristic curve (Fig. 8) is the result of the combined influence of the process gas properties (pressure, UF_6 concentration, etc.) on

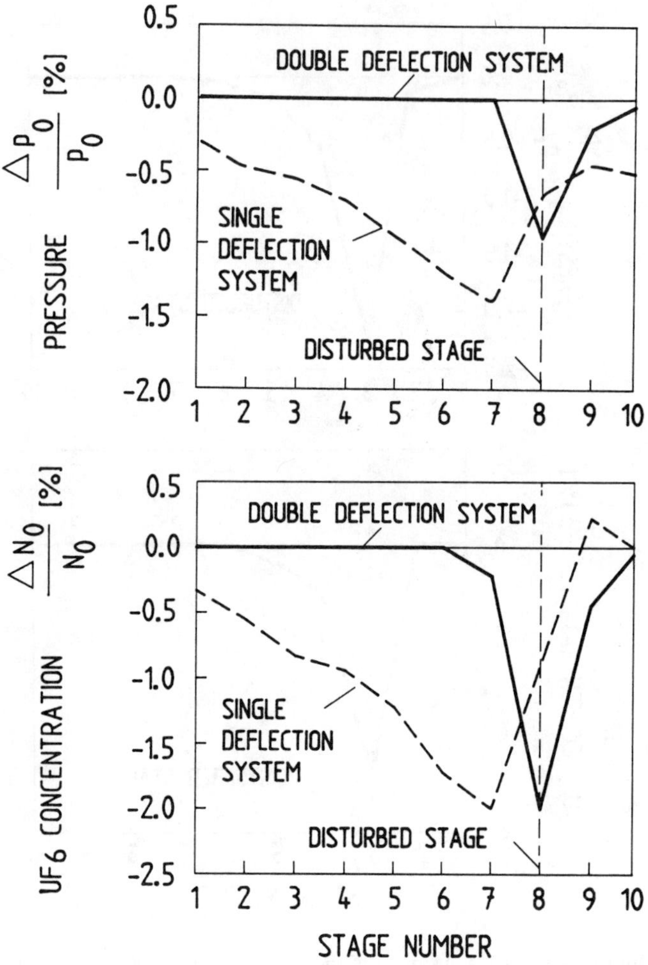

Fig. 7 Comparisation of the behavior of cascades with single deflection systems and cascades with double deflection systems. For both cascades a disturbance of the throttle valve in stage 8 was assumed, resulting in the same maximum disturbance in the UF_6 concentration N_0.

the compressor characteristic (mainly compression ratio) and on the separation characteristics of the separation system. If the cascade is operated in an area of the characteristic with a negative gradient (point A in Fig. 8), false flows of UF_6 are carried to the bottom of the cascade. For example, if the UF_6 content of a stage exceeds the nominal value, the UF_6 cut is diminished; i. e., the UF_6 flow in the depleted fraction, which is fed to the stage below, is increased. On the other hand, if the gradient of the characteristic curve is positive (point B in Fig. 8), false flows are carried to the top of the cascade where they are absorbed in the cryoseparator without any active regulation.

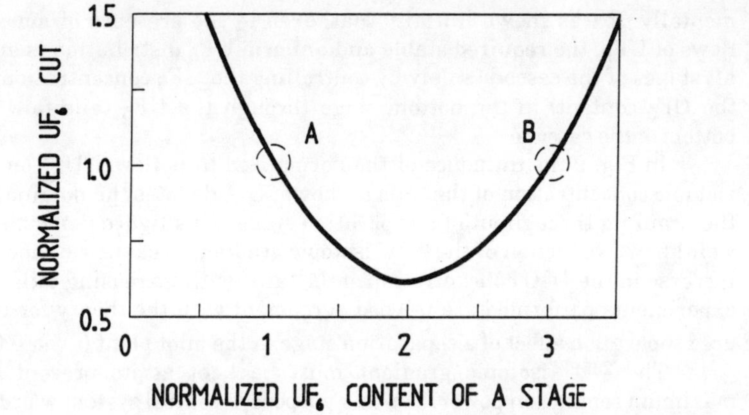

Fig. 8 Influence of the UF_6 content of a stage on the UF_6 cut of a stage. UF_6 cut equals UF_6 flow in the enriched fraction of a stage divided by UF_6 throughput of a stage. In the pilot plant, the nominal values (*) are UF_6 cut* = 0.5 and UF_6 content of a stage* = 0.4 g

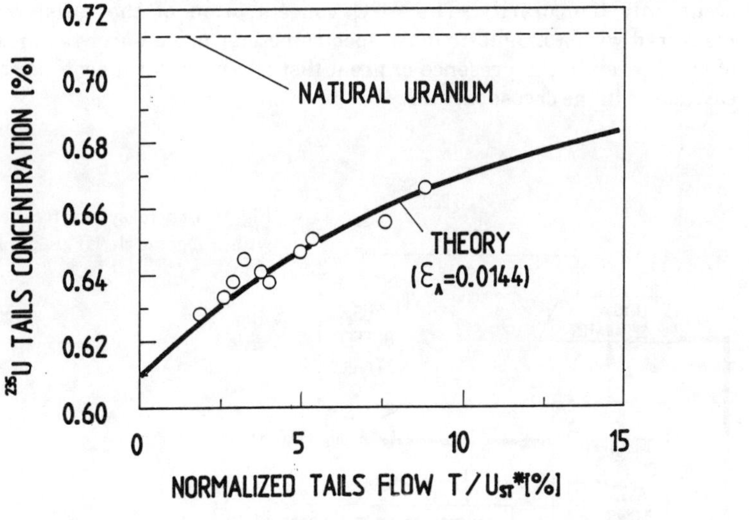

Fig. 9 Influence of the normalized tails flow T/U_{St}^* on the ^{235}U isotope concentration of the tails, where U_{St}^* is the nominal uranium throughput of a stage (180 g/h).

Under the operation conditions determined by economic viability assessments, the gradient of the characteristic curve is negative. In that case, the false flows of UF_6 must be dissipated quickly at the bottom of the cascade. False flows of UF_6 generated at the top of the cascade pass through the entire cascade and may result in a large reduction of the plant's separating power.

In the 10-stage pilot plant, the changes in UF_6 distribution and in the ^{235}U isotopic gradient caused by such false flows of UF_6 were studied experimentally. It was shown initially that, even in the presence of massive false flows of UF_6, the required stable and uniform UF_6 distribution is ensured at **all** stages of the cascade solely by controlling the UF_6 concentration $\mathbf{N_0}$ (i. e., the UF_6 content) in the bottom stage through the UF_6 tails flow **T** at the bottom of the cascade.

In Fig. 9 the influence of the normalized tails flow T/U_{St}^* on the ^{235}U isotopic concentration of the tails is shown; U_{St}^* denotes the nominal value of the uranium throughput of a separation stage. This figure demonstrates the well-known reduction of the ^{235}U isotopic gradient over the cascade, i. e., the increase in the ^{235}U concentration of the tails, with increasing tails flow. The experiments performed are in good agreement with the theory for the measured separation effect of a separation stage in the pilot plant ($\varepsilon_A = 0.0144$).

The ^{235}U isotopic gradient must be kept at its present level for maximum separation power. For this purpose, a control system was developed for the transport of UF_6 (Fig. 10); in this system the tails flow is used as a control variable, and the UF_6 feedback flow from the buffer tank at the top of the cascade is used as a manipulated variable.

The experiments performed demonstrated that combined control of the UF_6 content and of the UF_6 transport allow false flows of UF_6 to be dissipated quickly enough to ensure that minor deviations in the ^{235}U isotopic gradient occur only temporarily. The ^{235}U concentration of the tails, which was measured with a magnetic mass spectrometer, varied only within the limits of error, even in the presence of great disturbances in the isotope separation cascade or in the cryoseparator.

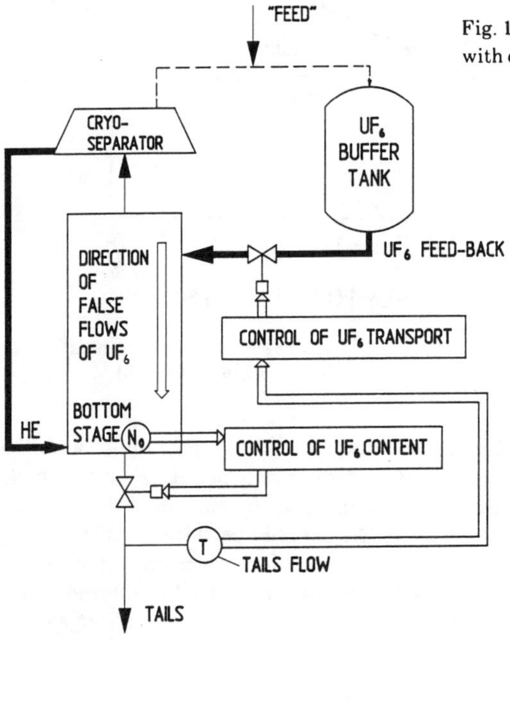

Fig. 10 Control system for the pilot plant with cryogenic He/UF_6 separation.

Gas Separation with Double Deflection Systems

The cryogenic separation of UF_6 and auxiliary gas (He or H_2) results in the handling of a large amount of UF_6 at the top of the cascade. An increase in the operating pressure of the cascade (which is inversely proportional to the characteristic dimensions of the separation system) results in extremely high expenditure for cryoseparation in the overall investment and operating costs for isotope separation. Therefore, it was proposed that the gas separation should be performed with double deflection systems, thus avoiding the cyclic mode of operation of the cryoseparators and the large UF_6 buffer at the top of the cascade. The gasdynamic separation of UF_6 and He also offers the advantages that the UF_6 holdup at the top of the cascade is eliminated, reducing the equilibrium time of the cascade drastically. Also, the control concept for the external UF_6 streams is simplified.

Since the time at which the detailed theory for gas separation cascades was proposed, the experiments described in this paper have been carried out.[6] To realize a gas separation cascade at the pilot plant in the last three stages, special double deflection systems for gas separation were mounted. In the pilot plant the connection of the gas separation stages is identical to the connection of the isotope separation stages. In Fig. 11 the new flow sheet of the pilot plant is shown. Three stages represent the gas separation cascade and seven stages the isotope separation cascade, which now consists of an enrichment section and a stripping section.

A special infrared process photometer equipped with a multipass reflection cell was used to measure very low UF_6 concentrations down to 1 ppm.[11]

It was predicted by computer simulation that the operation of the

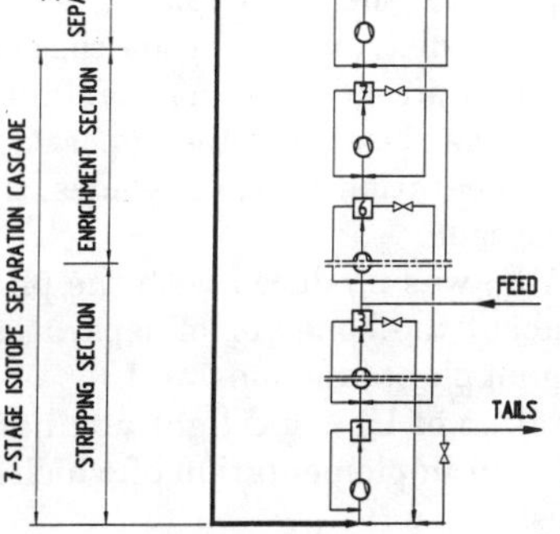

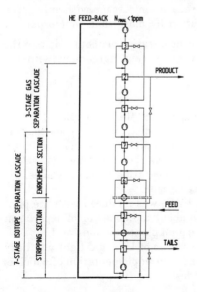

Fig. 11 Flow sheet of the pilot plant consisting of a three-stage gas separation cascade and a seven-stage isotope separation cascade.

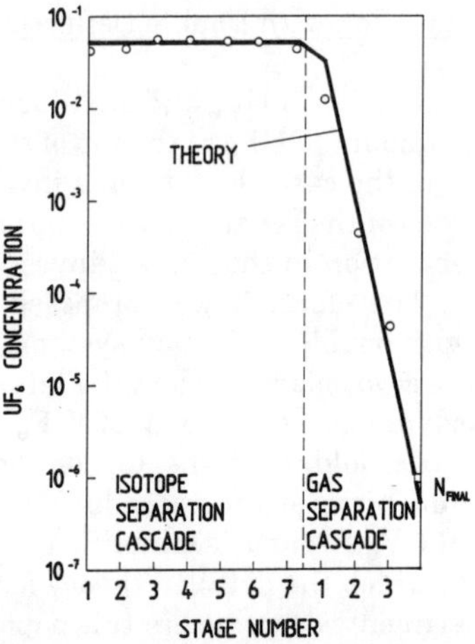

Fig. 12 Profile of the UF_6 concentration in the isotope separation cascade and in the gas separation cascade. The curve drawn through the measured points was calculated with a cascade simulation computer code. O, UF_6 concentration of the feed gas; □, final UF_6 concentration in the He feedback.

isotope separation cascade is influenced at a very small degree only by the positions of the throttle valves in the gas separation cascade. Therefore, a simple procedure was developed for setting up the combination of gas separation cascade and isotope separation cascade to give the required performance. The results of these theoretical considerations were fully verified by the experiments and the combined gas and isotope separation cascade operated without any problems.

The profile of the UF_6 concentrations measured in both cascades is shown in Fig. 12. It is evident that a gas separation cascade with only three stages is able to strip the UF_6 concentration by more than four orders of magnitude from about 1.5% to a final UF_6 concentration of about 1 ppm in the He feedback. The experimental results were in good agreement with the computer simulation.

Conclusions

A control system has been developed for the pilot plant, which guarantees the required stable and uniform UF_6 distribution in the cascade as well as keeping the ^{235}U isotopic gradient at its present level for maximum separation power. Because of the similarity between the operation characteristic of the pilot plant and technical separation nozzle cascades, the obtained results could be applied to the first cascade.

A small amount of slightly enriched UF_6 was produced with the pilot plant in order to supply more information about the behavior of separation nozzle cascades and to demonstrate that the pilot plant functions well.

At the pilot plant the advanced separation of UF_6 and light auxiliary gas was achieved by a gas separation cascade, an implementation of which is foreseen in further separation nozzle cascades.

References

[1]Becker, E. W., Bier, W., Bley, P., Ehrfeld, W., Schubert, K., and Seidel, D., "Uranium Enrichment by the Separation Nozzle Process," International Conference on Nuclear Power Experience, IAEA-CN-42/382, 1982, Nuclear Power Experience, Vienna 1983, Vol. 3, pp. 513-525.

[2]Becker, E. W., Bley, P., Ehrfeld, U. and Ehrfeld, W., "The Separation Nozzle, an Aerodynamic Device for Large-scale Enrichment of Uranium-235," 10th International Symposium of Rarefied Gas Dynamics, 1976, Progress in Astronautics and Aeronautics edited by J. L. Potter, published by AIAA 1977, Vol. 51, pp. 3-16.

[3]Bley, P., Ehrfeld, U. and Ehrfeld, W., "Enhancement of Nozzle Discharge Coefficients in Rarefied Flows of Disparate Mass Mixtures," 11th International Symposium on Rarefied Gas Dynamics, 1978, edited by Campargne, CEA Press, Paris 1979, pp. 241-252.

[4]Chatwani, A. U., Fiebig, M., Mitra, N. K., Schwan, W., Bley, P., Ehrfeld, W. and Fritz, W., "Tracer Monte-Carlo Simulation for an Isotope Separation Nozzle," 12th International Symposium on Rarefied Gas Dynamics, 1980, Progress in Astronautics and Aeronautics edited by S. S. Fischer, published by AIAA 1981, Vol. 74, pp. 517-540.

[5]Ehrfeld, W. and Fritz, W., "Analysis of Cooperative Behavior of Rarefied Flow and Continuum Flow Components in Separation Nozzle Cascades," 12th International Symposium on Rarefied Gas Dynamics, 1980, Progress in Astronautics and Aeronautics edited by S. S. Fischer, published by AIAA 1981, Vol. 74, pp. 642-655.

[6]Bley, P., Ehrfeld, W. and Schmidt, D., "Effective Separation of Disparate Mass Mixtures in a Single Stage by Double Deflection Separation Nozzles," 14th International Symposium on Rarefied Gas Dynamics, 1984, edited by H. Oguchi, University of Tokyo Press 1984, pp. 645-654.

[7]Ehrfeld, W. and Schelb, W., "Influence of Non-equilibrium Effects on Isotope Separation in Real Flow Fields of Separation Nozzles," 15th International Symposium on Rarefied Gas Dynamics, 1986, edited by V.Boffi and C. Cercignani, Teubner Stuttgart 1986, pp. 44 -55.

[8]Yadoya, R., Bley, P., Camara, A. S., Consiglio, R., Hein, H. and Linder, G., "Usina Piloto de 10 Estágios: sua Utilização na Pesquisa do Comportamento Dinâmica e Processo de Regulagem de uma Cascata de Jato Centrifugo Para Enriquecimento de Urânio," Ciência e Cultura (São Paulo), Vol. 38, No. 10, 1986, pp.1732-1739.

[9]Bley, P., Coelho, J. S., Hein, H. and Souza, A. S., "Stabilisierung der UF_6-Verteilung und des Isotopengradienten einer Trenndüsenkaskade zur Urananreicherung," Kernforschungszentrum Karlsruhe GmbH, KfK Report No. 4315, Feb. 1988.

[10]Fritz, W., Hoch, P., Linder, G., Schäfer, R. and Schütte, R., "Experimentelle Untersuchungen und Digitalrechner-Simulation des instationären Betriebsverhaltens von Trenndüsen-Kaskaden für die ^{235}U-Anreicherung," Chemie Ingenieur Technik, Vol. 45, No. 9-10, Sept. 1973, pp. 590-596.

[11]Berkhahn, W., Bley, P., Krieg, G. and Schmidt, D., "Test of the SPECTRAN Process Photometer for the Detection of Traces of Corrosive Gases," Technisches Messen, Vol. 51, No. 12, Dez. 1984, pp. 421-426.

General Principles of the Inertial Gas Mixture Separation

B. L. Paklin* and A. K. Rebrov†
Siberian Branch of the USSR Academy of Sciences
Novosibirsk, USSR

Abstract

Inertial separation may be considered both for the case of equilibrium diffusion in a flow with a pressure gradient and also for that of essentially nonequilibrium separation at the injection of a separable mixture into a gas flow. In the latter case, the separation effect is much more significant. In this paper the general analysis of separation for a scheme with injection normal to the background gas flow direction is given and the limiting possibilities of such a separation scheme are defined with due account of the influence of molecule size. It is shown that the choice of the angle of injection of the separable mixture into the gas flow plays a significant role in the separation processes. The optimal values of the enrichment factor considerably exceed those obtained in experiments. The optimization of local separation geometry defines the conditions for which the maximum efficiency of gas dynamic separation is attained.

Introduction

When a binary mixture of gases with disparate molecular masses is injected into a light gas, the heavy components penetrate deeper than the light ones and as a result the spatial separation of the components is formed. The separation scheme for a mixture injected into a flow of light gas normal to the carrier gas velocity (Fig. 1) has been investigated previously.[1,3]

This simple separation scheme is not optimal from the viewpoint of the most efficient separation. The development of theoretical concepts concerning the gas dynamic separation of gas mixtures and isotopes as well as the practical realization of separation effects require a more general consideration of separation processes and a theoretical exploration for maximum separation effects. The present paper is devoted to this problem and begins with analysis of this simple scheme.

---

*Institute of Thermophysics.

†Professor, Laboratory Chairman, Institute of Thermophysics.

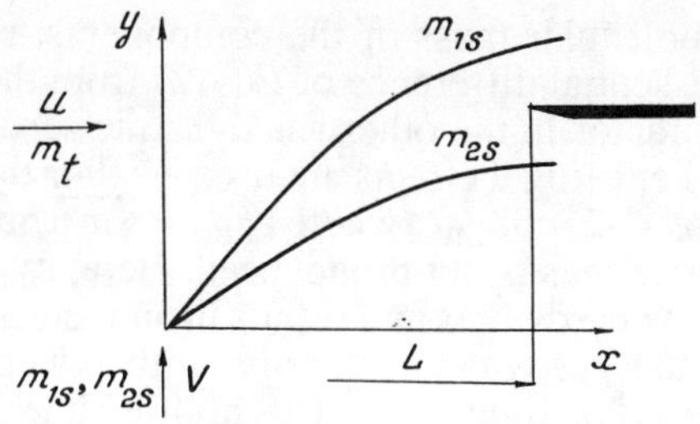

Fig. 1 Separation sceme with the mixture injection at right angle.

Basic Analysis of Separation Scheme

The main physical parameters defining the effect of inertial component separation have been presented in previous studies[2,3] and are presented schematically in Fig. 1. These parameters are as follows: the velocity ratios for the carrier gas ($S = u/u_*$) and injected mixture ($S_{in} = v/v_*$),[2] the ratio of mean mass velocities ($K = v/u$),[3] and a dimensionless complex of the molecular mass values for the injected mixture and carrier gas to be discussed later. Here u_* and v_* are the most probable velocities for interacting flow. The condition necessary to obtain the maximum possible component separation for the scheme under consideration was determined as a result of numerical modeling by a Monte-Carlo simulation for the test particles (Fig. 1).[2] The limiting situation of the component separation corresponds to maximum reachable gas velocity ratios and a maximum reachable value of their mean velocity ratios. As a result of the numerical calculation, with the assumption of a Maxwellian distribution in the velocities of carrier gas molecules and injected gas mixture prior to interaction with the flow of light gas, the dependence of the separation factor β determined with respect to the initial mixture composition f_o ($\beta = f(1-f_o)/f_o(1-f)$) has been obtained as

$$\beta \approx 1 + \varepsilon_{max}\,(1-\exp(-K))(1-\exp(-S_{in}))(1-\exp(-S)) \qquad (1)$$

where ε_{max} is the limiting enrichment factor. Practical limitation of the separation of the components is attained for values of the parameters K, S_{in}, and S of the order of 4 and result from the exponential character of the separation factor dependence [Eq. (1)] on the value of these parameters.

The value of the limiting enrichment factor for the separation scheme presented in Fig. 1 was determined both from analytical solutions[3,4] and also from modeling by Monte-Carlo simulation,[2] and was found to be

$$\varepsilon_{max} \approx -2\ln\theta\,\Delta m/\sqrt{m_t m_s} \qquad (2)$$

where θ is the component cut, $\Delta m = m_{1s} - m_{2s}$ the mass difference of the mixture molecules, and m_t and m_s the masses of the carrier gas molecules and an averaged molecular mass of the components, respectively. One should notice the essential difference of Eq. (2) from the known relations for the enrichment factor in the other gas dynamic separation methods in which diffusion or centrifugal effects are used.[5,6] For the case of a curved separation nozzle, $\varepsilon \approx -2\ln\theta\varepsilon_{f.m.}$, where $\varepsilon_{f.m.} = \sqrt{m_{1s}/m_{2s}}-1$, and thus the influence of the carrier gas is not pronounced. Here, ε_{max} depends on the molecular mass of the carrier gas and results in an increase in the value of ε by $2\sqrt{m_s/m_t}$ for the gasdynamic scheme with injection of fast heavy molecules into the cold, light gas. The high efficiency of the inertial separation method, particularly for heavy molecules, is attributed to this fact.

To substantiate the structural dependence of the enrichment factor ε on the molecular masses of the components and carrier-gas, one should perform an analysis of the motion of the heavy particle in the light gas.[3-7] It has been found[2,3] that the limiting separation of a mixture is realized at its injection into the light gas with very large velocity as compared to the velocity of carrier-gas. We now restrict the analysis to conditions where the mixture gas is injected into the quiescent gas with zero temperature. For Maxwellian molecules,[3] the relation follows the simple form

$$dv/dt = -(m_t/(m_s + m_t))\, v/\tau \tag{3}$$

where $1/\tau$ (= constant) is the collision frequency for the separating and background gas molecules.

One result from a previous study[8] is the conclusion that separation effects do not depend on the specific molecule's interaction potential. This result has been confirmed by numerical modeling. Therefore, consideration of the relaxation for Maxwellian molecules does not restrict the generality of considerations. The results of numerical modeling by a Monte-Carlo simulation of mean mass trajectories for molecule ensembles confirm the equivalence to the solution of Eq. (3) that can be considered as the governing relaxation equation for translational motion of molecules in the background gas.

The solution of Eq. (3) is of the specific relaxation form

$$v_y = v_{y_0} \exp(-t/\tau_*) \tag{4}$$

where $v_{y_0} = v_y$ at $t = 0$ and $\tau_* = \tau\,(m_s + m_t)/m_t$ is the characteristic time of translational relaxation. The expression for τ_* permits evaluation of the effective number of collisions necessary for molecule stagnation,

$$N_* \sim (m_s + m_t)/m_t$$

Integrating Eq. (4) with respect to time for a large period of time obtains the depth of heavy molecule penetration into the background gas $y_* \sim v\,\tau_*$. Taking into account the presence of the two components of the gas mixture, we shall obtain the difference in the penetration depth Δy_* for different components of the mixture with background molecules $\Delta y_* \sim$

$v\tau\Delta m/m_t$. One can assume that the molecular spatial distribution of each component with respect to the mean mass trajectory in the y-direction has a mean-square broadening $\delta \sim v\,\tau\,\sqrt{N_*}$. Let us expand the enrichment factor ε into a Taylor series in terms of the small parameter $\Delta y_*/\delta$ first introduced in Ref. 3, and restrict the expansion to the first term. From this result we obtain the following estimate

$$\varepsilon \sim \Delta m/(\sqrt{m_t}\sqrt{m_s + m_t})$$

As is readily seen, this result is in good agreement with Eq. (2) for the case where $m_s >> m_t$. The form

$$\varepsilon \sim (\Delta m/m_s)\sqrt{m_s/m_t}$$

is more convenient for the subsequent consideration.

Effects of Different Cross Sections

If a gas mixture injected into the light gas consists of molecules with different cross sections for interaction with the background molecules, this gives rise to an additional effect that can intensify or attenuate the separation effect associated with the difference in molecular masses of the components. For Maxwellian molecules one can readily carry out estimates analogous to the ones above, and obtain the following relation

$$\varepsilon \sim \Delta\,(m/\sigma)/(m_s/\bar{\sigma})\sqrt{m_s/m_t} \qquad (5)$$

where $\bar{\sigma}$ is the mean collisional cross section of the mixture and background molecules and $\Delta(m/\sigma) = m_{1s}/\sigma_{1s} - m_{2s}/\sigma_{2s}$. It is clear that the separation effect increases if $m_{1s} > m_{2s}$ or $\sigma_{1s} < \sigma_{2s}$ (and visa versa). The influence of the variation of cross section or size of the molecules to be separated was stated qualitatively in Ref. 9 by consideration of the barodiffusion in the framework of the Chapman-Enskog approximation for equilibrium diffusion discussed in Ref. 9 as "size diffusion." The size effect role for inertial separation of molecular flow has been pointed out in Ref. 10.

For molecules to be separated with similar molecular masses and cross sections, the expression for ε can be readily reduced to

$$\varepsilon \sim (\Delta\, m/m_s + \Delta\, \sigma/\bar{\sigma})\sqrt{m_s/m_t} \qquad (6)$$

Hence, it follows that the relative difference in the cross sections of interaction between the gas mixture and background molecules yields an equal contribution to the value of the enrichment factor as does the relative difference in molecular masses.

Numerical modeling of gas or isotope mixture separation has been carried out for a solid sphere model and confirms completely the dependence of Eq. (6). An illustration of the large values of enrichment

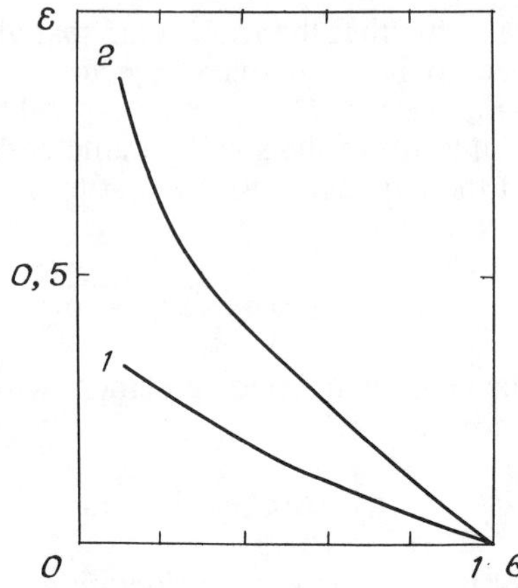

Fig. 2 $\varepsilon(\theta)$ at the limiting separation conditions for the mixture with $m_{1s} = 352$ amu, $m_{2s} = 349$ amu, $m_t = 2$ amu: 1, $\sigma_{2s} = \sigma_{1s}$; 2, $\sigma_{2s} = 1.01\sigma_{1s}$.

factor ε is given in Fig. 2 and shows the dependence for the case of zero temperature and conditions for limiting separation of mixture with $m_{1s} = 352$ a.m.u., $m_{2s} = 349$ a.m.u. at injection into the gas with $m_t = 2$ a.m.u.

Angular Variation Effects

Several schemes for the combined inertial and coherent radiation excitation-induced isotopic separation effects were suggested in Ref. 11. In the work of Ref. 10, the efficiency of gasdynamic separation schemes combined with the selective enlargement of the cross section of the light molecules by resonance excitation was emphasized. A consideration of the statement of the separation problem for the influence of the initial flow geometry demonstrates that the most interesting gasdynamic schemes are those for which separation has maximum efficiency at any point in space. In this context, the investigation of the influence of angular variation between the velocity vectors of separable and background gas is of interest. As a first step, we consider the effect of the angle for the mixture injection into the background gas in the separation process.

Let a mixture of different component molecules with the velocity v and having constant absolute value be injected at some angle into the uniform carrier gas flow moving with the velocity u (Fig. 3). We want to determine the optimal angle ($\tan\varphi_{opt} = v_y/v_x$) of the mixture injection for which the highest efficiency of the component separation is attained under the condition $C = v/u = $constant using the same assumptions as in previous investigations,[1-3] i.e., neglecting the interaction between the component molecules and their effect on the flow of the carrier gas. Let us also consider that the enrichment mixture was cut by a plate, as indicated previously, which is parallel to the direction of carrier gas motion.

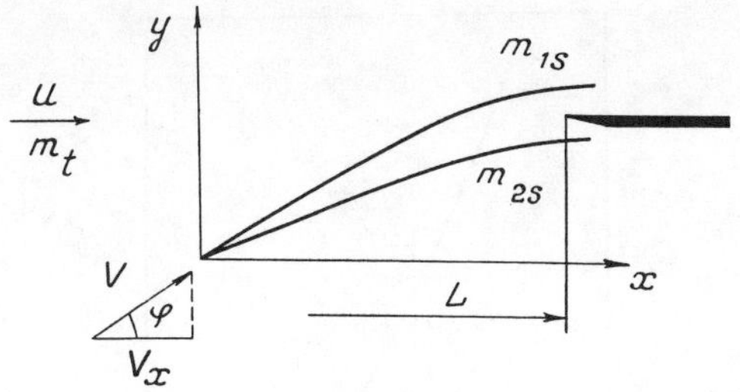

Fig. 3 **Separation scheme with the mixture injection at arbitrary angle.**

As a result of numerical modeling by a Monte-Carlo simulation for the scheme of the mixture injection normal to the direction of carrier gas motion, it follows that the separation efficiency increases with an increase of $K = v/u$. This result can be interpreted for the separation scheme with arbitrary injection angle. Let us transform to the inertial coordinate system moving with the velocity v_x. Separation effects should not change under Galilean transformation. In the new system of coordinates the velocity of the light gas is equal to $u - v_x$ and the velocity of the mixture injection is normal to the gas velocity and equals v_y. For the scheme of Fig. 1, the value $f = v_y/(u - v_x)$ (which is basically the function of one variable) serves as parameter K, since $v_y = \sqrt{(v^2 - v_x^2)}$ and $v = \text{const}$ according to the condition of the stated problem. While investigating the function $f(v_x)$ for a maximum, one can determine the injection velocity components (v_x, v_y), i.e., the optimal injection angle at which the maximum separation of the gas mixture components is attained.

One should consider two ranges of the C value variation: $C > 1$ and $C < 1$. It is obvious for $C > 1$ that $f \to \infty$ at $v_x \to u$ and $\tan\varphi_{opt} = v_y/v_x = \sqrt{(C^2 - 1)}$. Limiting separation will be attained under these conditions. Numerical modeling by Monte-Carlo simulation confirms this conclusion. The value of v_y, however, should not be very small and the earlier conclusion is valid when v_y is of the same order of magnitude as u.

For the case $C < 1$ we shall find the maximum value of the function f by the usual procedure:

$$f' = (v^2 - uv_x)/((u - v_x)^2\sqrt{v^2 - v_x^2}) = 0$$

At $v_x = v^2/u = Cv$ $f' = 0$ and $f'' = -1/uv(1 - C^2)^{5/2} < 0$, i.e., the function f actually attains a maximum at $v_x = Cv$. Then $v_y = v\sqrt{(1 - C^2)}$ and $\tan\varphi_{opt} = \sqrt{((1/C)^2 - 1)}$ and the maximum value of the function f is equal to $C/\sqrt{(1 - C^2)}$. The efficiency of the component separation depends on the value f_{max}. At $C \to 1$ $(C < 1)$, $f_{max} \to \infty$, and this provides the effect of possible limiting separation. Thus, a turn of the injection velocity vector allows us to obtain the same separation effect as at the mixture injection normal to the flow of light gas when the parameter K increases by $1/\sqrt{(1 - C^2)}$ times.

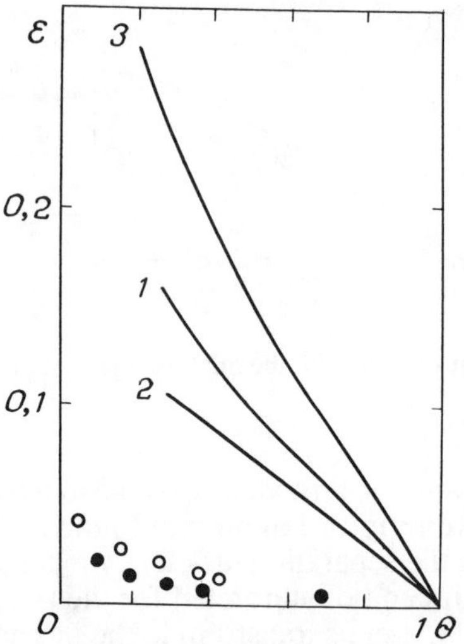

Fig. 4 $\varepsilon(\theta)$ **for the gas mixture with** $m_{1s} = 352$ **amu,** $m_{2s} = 349$ **amu,** $m_t = 2$ **amu for the case** $T_t = 0$**: 1,** $C = 0.8$**,** $\tan\varphi_{opt} = 0.75$**; 2,** $v_x = 0$**,** $v_y = 0.8u$**; 3, limiting separation conditions. Experimental data for separated nozzle: o 1.6%** UF_6 **+ 98.4%** H_2**;[12] • 5%** UF_6 **+ 95%** H_2**.[13]**

To verify the above relations and conclusions, the following separation situation was modeled: a mixture of gases with $m_{1s} = 352$ a.m.u. and $m_{2s} = 349$ a.m.u. and identical collisional cross section was injected into hydrogen gas where the value $C = 0.8$ was chosen, then $\tan\varphi_{opt} = 0.75$, and this is valid at $v_x = 0.64$ u and $v_y = 0.48$ u. Figure 4 shows the separation efficiency in cross section remote from the injection point, at distance $L = 10^4/n\sigma$ (n is the number density of light gas molecules, σ is the collisional cross section of the mixture and background gas molecules): 1 - curve $\varepsilon(\theta)$ for $C = 0.8$, $\tan\varphi_{opt} = 0.75$; 2 - $\varepsilon(\theta)$ for $v_x = 0$, $v_y = 0.8u$; 3 - $\varepsilon_{max}(\theta)$ is the limiting separation in the considered situation.

For the first case, the value $\varepsilon(\theta)$ coincides with the corresponding values of ε at $v_x = 0$, $v_y = 4/3$ u (as $f_{max} = 4/3$ at $C = 0.8$) that proves the above considerations. For the described case the same efficiency of the mixture separation is attained with the injection at optimal angle when the velocity is less by 5/3 times that at the injection normal to the flow. In addition, Fig. 4 shows the value of the enrichment factor attained in the experiments with the separated nozzles.[12,13] It follows from a comparison of these data with calculated ones that the possibilities of the optimization of the separation gasdynamic schemes are not yet exhausted.

Conclusion

A new dependence for the efficiency of gasdynamic separation as compared to the other known methods was found to follow from a consideration of the general relaxation equation for the motion of heavy particles in a light gas. The increase in the efficiency of gasdynamic separation was related to the increase of the flow velocity ratios and to optimization of the flow geometry. The results of the investigation of optimal angle offers a way for optimization of the field parameters in interacting flows. Together with the analysis of the limiting value of inertial separation effects a new point of view on the maximum possibilities of the economical inertial separation system for gas mixtures has been presented.

References

[1]Gallagher, R. J. and Anderson, J. B., "Isotope Separation in Crossed-Jet Systems," Proceedings of the Eleventh Symposium (International) on Rarefied Gas Dynamics, Vol. 1, CEA, Paris, 1979, pp. 629-637.

[2]Kusner, Yu.S., Paklin, B.L., and Rebrov, A.K., "On Limiting Situations of Gas Dynamics Separation," Proceedings of the Thirteenth Symposium (International) on Rarefied Gas Dynamics, Vol. 2, Plenum, New York, 1985, pp. 1313-1318.

[3]Chekmarev, S. F., "Gas or Isotopic Separation by Injection into Light Gas Flow," Proceeding of the Thirteenth Symposium (International) on Rarefied Gas Dynamics, Vol. 2, Plenum, New York, 1985, pp. 1297-1304.

[4]Kusner, Uy.S., "To the Theory of Gasdynamic Separation," Journal Doklady AN SSSR, Vol. 295, No. 2, 1981, pp. 359-361 in Russian.

[5]Muntz, E. P. and Hamel, B.B., "Rarefaction Phenomena in Gas and Isotope Separations," Proceedings of the Ninth Symposium (International) on Rarefied Gas Dynamics, Vol. 1, Gottingen, 1974, pp. B.1.1-10.

[6]Becker, E.W. Separation Nozzle: Uranium Enrichment, Topics in Applied Physics, Vol. 35, edited by Villani, S., Springer-Verlag, Berlin, 1979, pp. 245-268.

[7]Anderson, J. B., "Low Energy Particle Range," Journal of Chemical Physics, Vol. 63, No. 4, 1975, pp. 1504-1512.

[8]Ermolaeva, N.V., Ivanov, M.S., Kusner, Uy.S., and Nikolaev, V.I., "Statistical Theory of Gasdynamic Separation," Journal Zhurnal Tekhnicheskoi Fiziki, Vol. 56, No. 10, 1986, pp. 1873-1881 (in Russian).

[9]Waterman, P.C. and Stern, S.A., "Separation of Gas Mixtures in a Supersonic Jets," Journal of Chemical Physics, Vol. 31, No. 2, 1959, pp. 405-419.

[10]Dudnikov, V.G. and Chekmarev, S.F., "On Extracting of the Selectively Excited Particles in Crossed Jets," Journal Pisma v Zhurnal Tekhnicheskoi Fiziki, Vol. 7, No. 9, 1981, pp. 1174-1177 (in Russian).

[11]Winterberg, F., "Combined Laser Centrifugal Isotopic Separation Technique", Atomkernenergie (ATKE) Bd.30, Lfd.1, 1977, pp. 65-66.

[12]Becker, E.W., Bley, P., Ehrfeld, U., and Ehrfeld, W., "The Separation Nozzle - an Aerodynamic Device for Large-Scale Enrichment of Uranium-235," Proceedings of the Tenth Symposium (International) on Rarefied Gas Dynamics, Vol. 51, Part I, AIAA, New York, 1977, pp. 3016.

[13]Becker, E.W., Bier, W., Ehrfeld, W., and Eisenbeiss, G., "Physikalischen Grundlagen der Uran^{235} - Anreicherung nach dem Trenndusenverfahren," Z. Naturforch, Vol. 26a, No. 9, 1971, pp. 1377-1384.

Motion of a Knudsen Particle Through a Shock Wave

M. M. R. Williams*
University of Michigan, Ann Arbor, Michigan

Abstract

The motion of small particles, i.e. those lying in the Knudsen regime, are studied as they pass through a plane shock wave. The smallness of the particles enables their effect on the gas to be ignored and so the Mott-Smith method can be applied to calculate the gas atom velocity distribution function. It is then possible to calculate the force and rate of heat transfer on the particles subject to specified gas-surface interaction laws. These quantities then permit equations of motion and heat balance to be formulated from which the velocity and temperature of the particles is obtained as the shock is traversed.Characteristic lengths for particle velocity and temperature relaxation are determined and the influence of the shock thickness on the motion is studied. We note that the specific heat of the particle is an important parameter. The variation of the force and heat transfer rate between gas and particle is obtained across the shock.Results are presented for water droplets of Knudsen number 6.4 in air with a Mach number of 7.

I. Introduction

The motion of small particles in gas flows has long been a subject of interest in many diverse fields of science and engineering. Generally, the gas flow has been smooth and subsonic. However, there are situations such as the motion of solid particles in rocket motion exhausts, jet engines, and in the manifolds of internal combustion engines, where the gas undergoes severe temperature, density and velocity changes over a relatively small distance. Such conditions are akin to shock waves and therefore special methods are required to deal with them.

If the particles are large compared with a mean free path in the gas, then the standard equations of fluid mechanics, suitably modified to account for the presence of particles, can be employed. In this way, Marble[1] has calculated the temperature and velocity changes of the particle as it passes through a plane shock wave. The shock itself is considered to be infinitesimally thin, and

---

*Professor, Nuclear Engineering Department.

downstream from the shock-front there is shown to exist an equilibration region in which the particle temperature rises and the velocity decreases. The thickness of the equilibration region is found to depend on the density, size and thermal conductivity of the particles as well as on the viscosity of the gas.

Marble's theory begins to break down when the particle size becomes comparable to a mean free path, say about 0.1 m. Although it could be improved by means of slip boundary conditions, it is clear that for very fine particles the method would fail. Whilst the general problem of calculating the force on a particle in a shock wave for all Knudsen numbers would be a formidable one, it is possible to consider the free molecule limit in which it is assumed that the presence of the particle does not affect the molecular distribution function. In such a case, exact expressions for the force, rate of heat transfer and indeed any other integral property can readily be calculated. It is the purpose of this paper, therefore, to exploit this technique and to obtain the force acting on, and temperature of, a single fixed Knudsen particle at an arbitrary point in a shock wave. The particle is then allowed to move under these forces and its trajectory is calculated. The free molecule technique described above, relies upon the fact that the molecular distribution function for the problem is known. For a shock wave, considerable effort has been put into the study of infinite plane shocks and investigation of the shock region which is generally a few mean free paths thick. The most convenient procedure from the point of view of the present study is to make use of the Mott-Smith technique[2]. This assumes that the molecules in the shock wave can be represented by the sum of two Maxwell-Boltzmann distributions with different temperatures and mean velocities but with arbitrary densities. This bi-modal distribution in then substituted into the Boltzmann equation and an appropriate velocity moment is taken which subsequently leads to a simple, non-linear differential equation for the density as a function of position across the shock. In a later paper, Rosen[3] has used a variational principle to improve the accuracy of the method but the principle remains the same. Armed with this bi-modal distribution, which clearly changes from a pure Maxwellian with the upstream properties at $x=-\infty$ to a pure Maxwellian with the downstream properties at $x=+\infty$, we can obtain the desired statistical averages which ultimately lead to the force and heat transfer rate.

II. Theory

2.1 The force and heat transfer rate

The force on a particle moving with velocity V in a gas may be calculated, by definition, from the net rate of exchange momentum due to collisions with gas atoms. Thus,

$$\underline{F} = m \int d\mathbf{v}(\mathbf{v} - \mathbf{V})\left(\frac{\partial f}{\partial t}\right)_{\mathrm{coll}} \tag{1}$$

where m is the mass of a gas atom and $(\partial f/\partial t)$coll is the rate of change of the distribution function f(v,r) due to collisions, i.e., it is as though the particle is a component molecule of the gas. We have shown in a previous paper[4] that

$$\left(\frac{\partial f}{\partial t}\right)_{coll} = \int d\mathbf{v}'\, \sigma(\mathbf{v}' - \mathbf{V} \rightarrow \mathbf{v} - \mathbf{V})\, f(\mathbf{v}', \mathbf{r}) - |\mathbf{v} - \mathbf{V}|\sigma f(\mathbf{v}, \mathbf{r}) \tag{2}$$

where $\sigma(v' \rightarrow v)$ is the global scatterng kernel of the particle and

$$v\sigma_p = \int d\mathbf{v}'\, \sigma(\mathbf{v} \rightarrow \mathbf{v}') \tag{3}$$

where $\sigma_p = \pi R^2$ is the geometric cross section of the particle of radius R. It should be noted that s(v'v) depends on the particle surface temperature.

If the gas has an overall mean flow velocity of **U**, then we should write eqn (1) as

$$\mathbf{F} = m \int d\mathbf{v} (\mathbf{v} - \mathbf{V} - \mathbf{U}) \left(\frac{\partial f}{\partial t}\right)_{coll} \tag{4}$$

but because

$$\int d\underline{v} \left(\frac{\partial f}{\partial t}\right)_{coll} = 0 \tag{5}$$

we regain eqn (1). However, we are also interested in the rate of energy exchange between gas and particle, which is written as

$$\frac{dE}{dt} = \frac{1}{2} m \int d\mathbf{v} (\mathbf{v} - \mathbf{V} - \mathbf{U})^2 \left(\frac{df}{dt}\right)_{coll} \tag{6}$$

$$= \frac{1}{2} m \int d\mathbf{v} (\mathbf{v} - \mathbf{V})^2 \left(\frac{\partial f}{\partial t}\right)_{coll} - \mathbf{U}.\mathbf{F} \tag{7}$$

Using eqn (2) in eqn (1) and (7) we find

$$\mathbf{F} = m \int d\mathbf{v}\, \mathbf{v} \int d\mathbf{v}'\, \sigma(\mathbf{v}' \rightarrow \mathbf{v})\, f(\mathbf{v}' + \mathbf{V}, \mathbf{r}) - m\sigma \int d\mathbf{v}\, \mathbf{v} v\, f(\mathbf{v} + \mathbf{V}, \mathbf{r}) \tag{8}$$

and

$$\frac{dE}{dt} = \frac{1}{2} m \int d\mathbf{v}\, v^2 \int d\mathbf{v}'\, \sigma(\mathbf{v}' \rightarrow \mathbf{v})\, f(\mathbf{v}' + \mathbf{V}, \mathbf{r}) - \frac{1}{2} m\sigma \int d\mathbf{v}\, v^3 f(\mathbf{v} + \mathbf{V}, \mathbf{r}) - \mathbf{U}.\mathbf{F} \tag{9}$$

2.2 The distribution function

Now according to Mott-Smith[2], the distribution function can be written as

$$f(\mathbf{v}, \mathbf{r}) = f_\alpha(v, \mathbf{r}) + f_\beta(v, \mathbf{r}) \tag{10}$$

where

$$f_\alpha(\mathbf{v}, x) = n_\alpha(x)\left(\frac{m}{2kT_\alpha}\right)^{3/2} \exp\left\{-\frac{m}{2kT_\alpha}(\mathbf{v} - \mathbf{i}\, U_\alpha)^2\right\} \tag{11}$$

and similarly for $f_\beta(v,x)$. **i** is a unit vector in the x-direction. We have specialized here to a one-dimensional system depending only on the x-variable in position. T_α and U_α are independent of x and are the upstream parameters; likewise T_β and U_β are the downstream parameters.

It was shown by Mott-Smith that $\nu_\alpha(x) = n_\alpha(x)/n_o$, where $n_o = n_\alpha(-\infty)$, obeys the following differential equation,

$$\frac{d\nu_\alpha}{dx} + \frac{B}{\lambda_g}\nu_\alpha(1 - \nu_\alpha) = 0 \tag{12}$$

where B is a function of the Mach number M and λ_g is the mean free path in the gas. More precisely,

$$M^2 = \frac{(a-2)}{a} \cdot \frac{u_\alpha^2}{c_\alpha^2} \tag{13}$$

where $a = 2\gamma/(\gamma-1)$, γ being the ratio of the specific heats, $c_\alpha^2 = kT_\alpha/m$ and

$$\lambda_g = 1/\sqrt{2}\, n_o \sigma_g \tag{14}$$

where σ_g is the atomic cross section of the gas atom. We should point out that there are some very troublesome misprints in Mott-Smith's paper and we give the correct expression for B in the Appendix.

The parameters of interest for the gas are as follows:

Gas temperature Tg:

$$\frac{T_g}{T_\alpha} = \frac{1}{\nu_\alpha + \nu_\beta}\left[\nu_\alpha + \nu_\beta \frac{c_\beta^2}{c_\alpha^2}\right] + \frac{1}{3}\frac{\nu_\alpha \nu_\beta}{(\nu_\alpha + \nu_\beta)^2}\frac{(u_\alpha - u_\beta)^2}{c_\alpha^2} \tag{15}$$

Gas velocity U:

$$\frac{U}{u_\alpha} = \frac{1}{\nu_\alpha + \nu_\beta}\left(\nu_\alpha + \nu_\beta \frac{u_\beta}{u_\alpha}\right) \tag{16}$$

The relationship between u_β/u_α, c_β/c_α, etc in terms of the Mach number are given in the next section.

Using the Legendre polynomial expansion for s(v'Æv), we can write

$$\sigma(\mathbf{v}' \to \mathbf{v}) = \sum_{l=0}^{\infty} \frac{2l+1}{4\pi} \sigma_l(v' \to v) P_l(\Omega'.\Omega) \tag{17}$$

where $\Omega.\Omega'$ is the cosine of the angle between **v** and **v'**. Then

$$F_x = m\int_0^\infty dvv^3 \int_0^\infty dv'v'^2 \sigma_1(v' \to v). \int d\Omega' \mathbf{i}.\Omega' f(v', \mathbf{i}.\Omega', x) - m\sigma\int_0^\infty dvv^4 \int d\Omega' \mathbf{i}.\Omega' f(v, \mathbf{i}.\Omega', x) \tag{18}$$

and

$$\frac{dE}{dt} = \frac{1}{2} m\int_0^\infty dvv^4 \int_0^\infty dv'v'^2 \sigma_0(v' \to v) \int d\Omega' f(v', \mathbf{i}.\Omega', x) - \frac{1}{2} m\sigma\int_0^\infty dvv^5 \int d\Omega\, f(v, \mathbf{i}.\Omega, x) \tag{19}$$

where we have noted that in the exponential of the Maxwellian we can write

$$(\mathbf{v} + \mathbf{i}V - \mathbf{i}u_\alpha)^2 = v^2 + \mathbf{i}.\Omega v(V - u_\alpha) + (V - u_\alpha)^2 \tag{20}$$

and

$$(\mathbf{v} + \mathbf{i}V - \mathbf{i}u_\beta)^2 = v^2 + \mathbf{i}.\Omega v(V - u_\beta) + (V - u_\beta)^2 \tag{21}$$

i.e. the distribution function depends on **v** and **i.Ω**.

2.3 gas-particle scattering model and final expression for force and heat transfer

In order to proceed, it is necessary to adopt a suitable gas-particle scattering model. For simplicity we assume that complete accommodation takes place on the particle surface and that the gas atoms are re-emitted with the particle surface temperature Tw. In that case[4]

$$\sigma_0(v' \to v) = 2\sigma_P \left(\frac{m}{2kT_\omega}\right)^2 vv' \exp\left(-\frac{mv^2}{2kT\omega}\right) \tag{22}$$

and

$$\sigma_1(v' \to v) = -\frac{8\sigma_P}{9} \left(\frac{m}{2kT_\omega}\right)^2 vv' \exp\left(-\frac{mv^2}{2kT\omega}\right) \tag{23}$$

where σ_P is the geometric cross section of the particle.

Using these expressions in eqns (18) and (19) we obtain integrals that have already been obtained[5] and therefore can write

$$F_x = 2\sigma_p k T_\alpha n_o \left[\frac{\sqrt{\pi}}{3} \left\{ f(x) s_\alpha + g(x) s_\beta \left(\frac{T_\beta}{T_\alpha} \right)^{\frac{1}{2}} \right\} \left(\frac{T_\omega}{T_\alpha} \right)^{\frac{1}{2}} + \frac{1}{\sqrt{\pi}} \left\{ f(x) g_o(s_\alpha) + g(x) \frac{T_\beta}{T_\alpha} g_o(s_\beta) \right\} \right] \tag{24}$$

where

$$f(x) = \frac{1}{\left(1 + \exp\left(\frac{Bx}{\lambda_g}\right)\right)} \tag{25}$$

$$g(x) = \frac{\left(u_\alpha / u_\beta\right)}{1 + \exp\left(-Bx / \lambda_g\right)} \tag{26}$$

$$s_\alpha = \frac{1}{\sqrt{2}\, c_\alpha}\left(u_\alpha - V\right) \tag{27}$$

$$s_\beta = \frac{1}{\sqrt{2}\, c_\beta}\left(u_\beta - V\right) \tag{28}$$

$$g_o(s) = \left(s + \frac{1}{2s}\right) e^{-s^2} + \left(s^2 + 1 - \frac{1}{4s^2}\right)\sqrt{\pi}\ \mathrm{erf}(s) \tag{29}$$

As shown by Mott-Smith using continuity of mass, momentum and energy across the shock, we can write

$$\frac{u_\beta}{u_\alpha} = \frac{1}{a-1}\left[1 + a\frac{c_\alpha^2}{u_\alpha^2}\right] \tag{30}$$

$$\frac{u_\beta^2}{c_\beta^2} = \frac{u_\alpha^2 / c_\alpha^2 + a}{(a-2)u_\alpha^2 / c_\alpha^2 - 1} \tag{31}$$

$$\frac{T_\beta}{T_\alpha} = \frac{c_\beta^2}{c_\alpha^2} = \frac{\left(u_\alpha / c_\alpha\right)^2 + a}{(a-1)^2}\left[a - 2 - \left(\frac{c_\alpha}{u_\alpha}\right)^2\right] \tag{32}$$

Thus we can express them in terms of the upstream Mach number, M, via equation (13).

Similarly, we can write

$$\frac{dE}{dt} = m n_o \sigma_p \left(\frac{2kT_\alpha}{m}\right)^{3/2} \cdot \left[\frac{f}{\sqrt{\pi}}\left(\frac{T_\omega}{T_\alpha} g_1(s_\alpha) - g_2(s_\alpha)\right) + \frac{g}{\sqrt{\pi}}\left(\frac{T_\beta}{T_\alpha}\right)^{3/2}\left(\frac{T_\alpha}{T_\beta}\frac{T_\omega}{T_\alpha} g_1(s_\beta) - g_2(s_\beta)\right)\right]$$

$$- m n_0 \sigma_p \left(\frac{2kT_\alpha}{m}\right)^{3/2} \frac{(f + g\, u_\beta / u_\alpha)}{(f+g)\sqrt{\pi}} \frac{u_\alpha}{c_\alpha} \left\{ \frac{\sqrt{\pi}}{3} f \left(\frac{T_\omega}{T_\alpha}\right)^{1/2} s_\alpha + \frac{\sqrt{\pi}}{3} g \left(\frac{T_\beta}{T_\alpha}\right)^{1/2} \left(\frac{T_\omega}{T_\alpha}\right)^{1/2} s_\beta + \frac{1}{\sqrt{\pi}} f g_o(s_\alpha) + \frac{1}{\sqrt{\pi}} g \frac{T_\beta}{T_\alpha} g_o(s_\beta) \right\} \tag{33}$$

Equations (24) and (33) constitute complete expressions for the force and energy exchange between gas and particle as a function of Mach number. $g_1(s)$ and $g_2(s)$ are defined as

$$g_1(s) = e^{-s^2} + \left(s + \frac{1}{2s}\right)\sqrt{\pi}\,\mathrm{erf}(s) \tag{34}$$

$$g_2(s) = \frac{1}{2}\left(s^2 + \frac{5}{2}\right)e^{-s^2} + \frac{1}{2}\left(s^3 + 3s + \frac{3}{4s}\right)\sqrt{\pi}\,\mathrm{erf}(s) \tag{35}$$

III. The Equation of Motion

In order to obtain the velocity of the particle as it moves through the shockwave under the influence of the force mentioned earlier, it is necessary to solve the equation of motion

$$M_p \frac{dV}{dt} = F_x(V, T_\omega) \tag{36}$$

where we have noted that Fx depends on V and the particle surface temperature Tw. Mp is the particle mass. Since V = dx/dt, we can also write eqn (36) as

$$M_p V \frac{dV}{dx} = F_x(V, T_\omega) \tag{37}$$

In fact this is more convenient because we are interested in the way in which V varies across the shock.

To obtain T_ω it is necessary to observe that

$$\frac{dE}{dt} = -K_p \int_A \underline{n} \cdot \nabla T \, dA \tag{38}$$

where Kp is the thermal conductivity of the particle and A its surface area.

Now the conduction equation for the particle is

$$K_p \nabla^2 T = \rho_p C_p \frac{\partial T}{\partial t} \tag{39}$$

where K_p, ρ_p and C_p are, respectively, the thermal conductivity, density and specific heat of the particle.

Integrating this equation over the volume of the particle v_p and using Green's theorem, we find

$$K_p \int_A \underline{n} \cdot \nabla T \, dA = \rho_p C_p \int_{v_p} d\underline{r} \frac{\partial T}{\partial t} \tag{40}$$

whence

$$\frac{dE}{dt} = -\rho_p C_p v_p \frac{d}{dt} \tilde{T}_\omega \tag{41}$$

where

$$\tilde{T}_\omega \equiv \frac{1}{v_p} \int_{v_p} d\underline{r} \, T \tag{42}$$

is the volume averaged temperature of the particle.

In principle, it is now necessary to solve the conduction equation for the particle to obtain $\tilde{T}_\omega$ in terms of T_ω, but for a good thermal conductor and such a small particle it will be a very good approximation to set $\tilde{T}_\omega = T_\omega$. We therefore write for $\tilde{T}_\omega$, the following equation,

$$m_p C_p \frac{d\tilde{T}_\omega}{dt} = -\frac{dE}{dt} \tag{43}$$

or in terms of x, we can write

$$m_p C_p V \frac{dT_\omega}{dx} = -\frac{dE}{dt} \tag{44}$$

Since dE/dt is a function of V and T_ω, eqns (37) and (44) constitute a pair of coupled first order differential equations for the velocity and temperature of the particle.

If we write,

$$F_x = 2\sigma_p k T_\alpha n_o \Lambda_o \left(M_p, T_\omega / T_\alpha\right) \tag{45}$$

and

$$\frac{dE}{dt} = m n_o \sigma_p \left(\frac{2kT_\alpha}{m}\right)^{3/2} \Lambda_1 \left(M_p, T_\omega / T_\alpha\right) \tag{46}$$

then eqs (43) and (44) reduce to

$$\frac{l_v}{\lambda_g} \frac{dM_p}{dx_B} = \frac{\Lambda_1 \left(M_p, T_\omega / T_\alpha\right)}{M_p} \tag{47}$$

$$\frac{l_v}{\lambda_g} \frac{d}{dx_B}\left(\frac{T_\omega}{T_\alpha}\right) = -\frac{\sqrt{2}}{\zeta} \frac{\Lambda_1 \left(M_p, T_\omega / T_\alpha\right)}{M_p} \tag{48}$$

where $\zeta = C_p m/k$,

$$l_v = \frac{m_p}{2 m n_o \sigma_p} = \frac{m_p}{2\rho_g \sigma_p} = \frac{2}{3} R \frac{\rho_p}{\rho_g} \tag{49}$$

$x_B = x/\lambda_g$, and $M_p = V/C\alpha$. (R=particle radius)

The solution of eqs (47) and (48) subject to given initial conditions will determine the velocity and temperature of the particle through the shock region. It will be noted that there are two length scales; the mean free path λ_g and the particle relaxation length l_v for velocity relaxation and $l_T = l_v \zeta / \sqrt{2}$ for temperature relaxation. There lengths are analogous to Marble's l_v and l_T.[1]

Finally, we comment on the effect of diffusion on the particles. A measure of its importance in the Knudsen regime would be given by the ratio of the root mean square thermal velocity $(8kT_g/\pi m_p)^{1/2}$ to the convective velocity

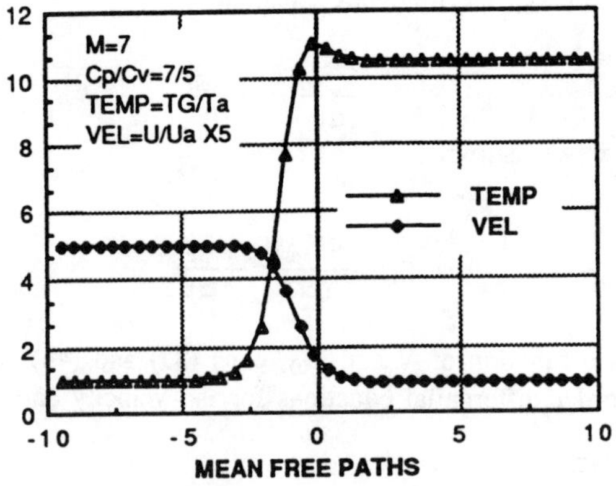

Fig.1 Gas temperature and velocity.

V. This ratio is approximately

$$\approx \left(\frac{m_g}{m_p}\right)^{1/2} \left(\frac{T_g}{T_\alpha}\right)^{1/2} \frac{1}{M_p}$$

But

$$\left(\frac{T_g}{T_\alpha}\right)^{1/2}_{max} = \frac{\sqrt{a}}{a-1} M$$

and so

$$\approx \left(\frac{m_g}{m_p}\right)^{1/2} \frac{\sqrt{a}}{a-1} \frac{M}{M_p}$$

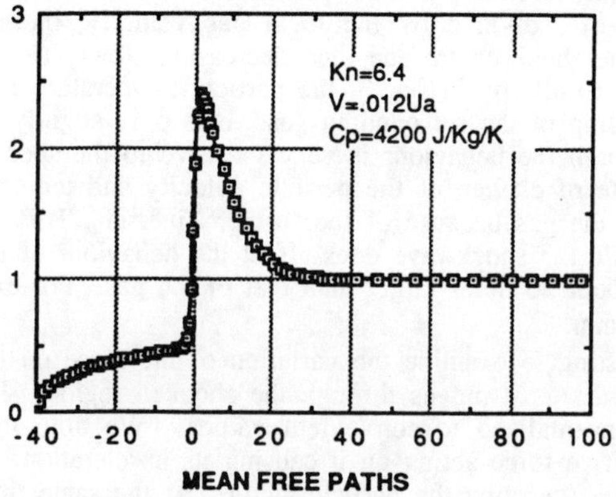

Fig.2 Particle/local gas velocity.

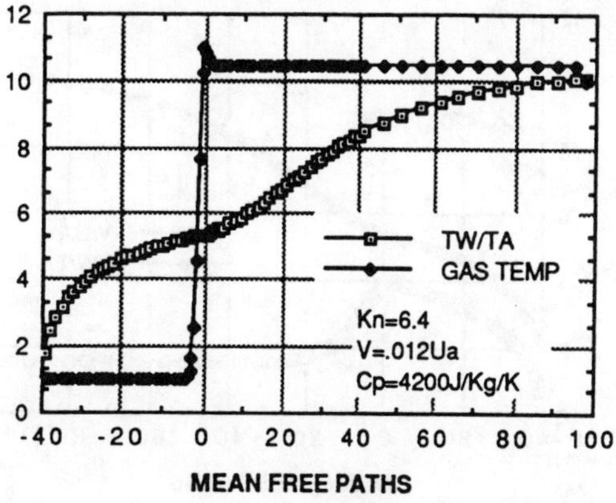

Fig.3 Gas and particle temperatures.

For most substances this is generally very much less than unity and therefore we many confidently neglect the effect of diffusion.

IV. Numerical Study and Conclusions

Since the gas is assumed to be unaffected by the presence of the particles, we can calculate its behaviour, i.e., temperature, density, velocity, etc., directly from the bimodal Mott-Smith distribution. The average velocity and gas temperature are shown in fig. 1 for M=7 and g=7/5 (air). We note the characteristic behaviour of increasing temperature and decreasing velocity through the shock front.

To show how the particle behaves as it moves through the shock, we consider a water droplet of radius .01 mm and C_p=4200 J $Kg^{-1}K^{-1}$ with Knudsen number equal to 6.4. It is assumed that the particle starts with a temperature equal to the upstream gas temperature and a speed of .012 U_α at 40 mean free paths from the centre of the shock. By solving the equations of motion, we can obtain the subsequent velocity and temperature of the particle. Figure 2 shows the particle velocity, divided by the local gas velocity, increasing up to a maximum near the shock centre and then decreasing slowly to the down-stream gas velocity ub. Similarly, in fig. 3, the particle temperature increases due to the frictional heating of the surrounding gas. It is only slightly affected by the shock front although the behaviour is very sensitive to the specific heat of the particle. The rate of change of the particle velocity and temperature is much less than that of the gas because of the former's inertia. This is illustrated in fig. 4. Thus while the shockwave does affect the behaviour of the particle, the spatial scale is about 20 times larger than that of the gas. For example, l_v=5.3 mm and l_T=27.6 mm.

It is interesting to examine the variation of the force on the particle and its heat transfer rate as it moves through the shocked region. Figure 5 shows these quantities normalized to convenient values. We note that the particle starts with a positive force acting on it causing its acceleration. This is due to the fast moving gas carrying the particle along. At the same time, the particle is heating up and receiving energy ($\dot{E}$<0). As the particle approaches the shock

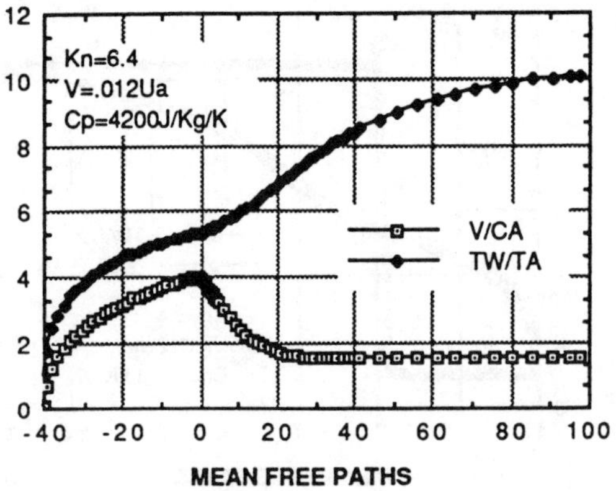

Fig.4 Particle velocity and temperature, C_p=4200 J $Kg^{-1}K^{-1}$.

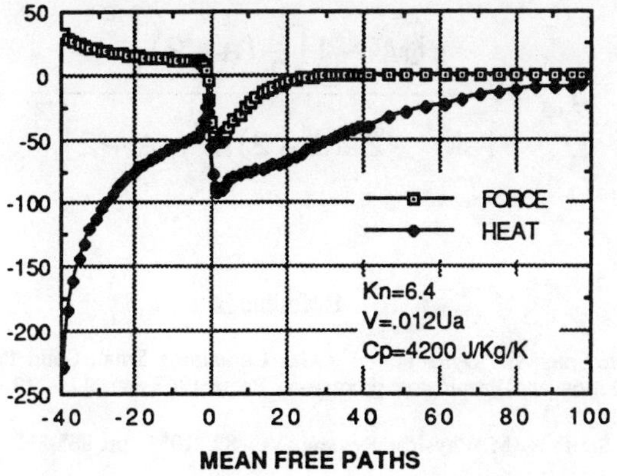

Fig.5 Force and heat transfer rate, C_p=4200 J Kg^{-1} K^{-1}.

centre, the force changes sign due to the decelerating gas and due to a form of non-linear thermophoresis arising from the strong temperature gradient in the gas. The rate of heat transfer also changes sign and the particle starts to cool. The main value of this study has been to highlight the approach needed to study the motion of small paticles through a shock front and to introduce two relaxation lengths l_V and l_T which govern the rapidity of velocity and temperature changes across the shock front. The work is therefore complementary to that of Marble. Nevertheless, we still require a formalism for dealing with particles that are intermediate in size between the Knudsen and Clausius regimes.

Appendix

The expression for B which appears in the Mott-Smith Solution is:

$$B = \frac{2}{3(a-3)}\left(\frac{a-2}{\pi a}\right)^{1/2} \cdot \frac{\left[aM^4 + 2a(a-2)M^2 - a + 2\right]^{3/2}}{\left[M^2-1\right]\left[M^2+a-2\right]\left[aM^2+a-2\right]} A(u_{\alpha\beta})$$

$$A(u) = \left(\frac{1}{u}+\frac{1}{u^3}-\frac{3}{4u^5}\right)u^3 e^{-u^2} + u^2\left(2u+\frac{3}{u}-\frac{3}{2u^3}+\frac{3}{4u^5}\right)\frac{\sqrt{\pi}}{2}\,\mathrm{erf}(u)$$

$$u_{\alpha\beta} = \frac{\left[M^2 - 1\right]\left[\frac{a}{2}(a-2)\right]^{1/2}}{\left[aM^4 + 2a(a-2)M^2 - a + 2\right]^{1/2}}$$

References

[1]Marble,Frank E. "Dynamics of a Gas Containing Small Solid Particles," Combustion and Propulsion, Pergamon Press 1963, pp. 175-213.

[2]Mott-Smith,H. M. Physical Review, Vol 82, 1951, pp. 885-892.

[3]Rosen,P. Journal of Chemical Physics, Vol 22, 1954, pp. 1045-1049.

[4]Williams,M. M. R. Zeitschrift fur Naturforschung,Vol. 27a, 1972, pp. 1798-1803.

[5]Williams,M. M. R. Journal of Physics D:Applied Physics, Vol. 6, 1973, pp. 744-758.

Method of Characteristics Description of Brownian Motion Far from Equilibrium

P. Riesco-Chueca,* R. Fernández-Feria,† and J. Fernández de la Mora‡
Yale University, New Haven, Connecticut

Abstract

The far-from-equilibrium Brownian motion of small particles or heavy molecules inmersed in a carrier gas may be described by closed hydrodynamic equations in the limit where their speed of thermal agitation is much smaller than their convective velocity (a hypersonic closure). The corresponding hyperbolic governing equations are studied in this paper through the method of characteristics. In two-dimensional and axisymmetric problems, the original system of seven partial differential equations is reduced to a set of seven ordinary differential equations along five different characteristic directions that we determine algebraically. Three of the characteristic paths coincide with the particle trajectories, and the two other pairs originate from two distinct signal propagation modes. The method is implemented numerically in two problems (Prandtl-Meyer expansion and Gaussian wake), leading to an excellent agreement between the numerical results from the characteristic equations and analytical solutions derived by asymptotic integration of the kinetic Fokker-Planck equation. The Prandtl-Meyer problem illustrates the failure of the hypersonic closure, when used blindly in the vicinity of singular points, though the associated errors in the density are tolerably small.

I. Introduction

The motion of a far-from-equilibrium phase of small particles or heavy molecules inmersed in a light carrier gas cannot be described by the standard near-equilibrium Chapman-Enskog closure of the

---

Presented as an Invited Paper.

* Graduate student, Department of Mechanical Engineering.
† Post-doctoral fellow, Department of Mechanical Engineering.
‡ Associate Professor, Department of Mechanical Engineering.

hydrodynamic equations for binary gas mixtures. Instead, more complex kinetic formulations such as those provided by the Fokker-Planck equation have to be used.[1] Fortunately, in many problems of interest, the smallness of the particles' thermal speed compared to typical values of the mean velocity, opens the door to a hypersonic closure of the hydrodynamic equations, even in situations where the carrier gas is subsonic.[2,3] The thesis of Fernández-Feria[2] (see also Ref. 3) reviews previous instances where the hypersonic condition of gases has been used to close the hydrodynamic equations. In particular, the moment equations can be closed by dropping the heat flux term in the equation for the pressure tensor while retaining (in contrast to previous deterministic formulations[4]) the pressure term in the momentum equation, thus accounting for the Brownian motion or diffusion of the particles[2,3]:

$$D\lambda_p + \nabla\cdot\mathbf{U}_p = 0 \tag{1}$$

$$D\mathbf{U}_p + \frac{k}{m_p}[(\boldsymbol{T}_p\cdot\nabla)\lambda_p + \nabla\cdot\boldsymbol{T}_p] = (\mathbf{U} - \mathbf{U}_p)/\tau \tag{2}$$

$$D\boldsymbol{T}_p + (\boldsymbol{T}_p\cdot\nabla)\,\mathbf{U}_p + ((\boldsymbol{T}_p\cdot\nabla)\,\mathbf{U}_p)^T = 2(T\boldsymbol{I} - \boldsymbol{T}_p)/\tau \tag{3}$$

where $\lambda_p = \log(\rho_p/\rho_{p0})$, ρ_{p0} is a reference density for the particles, $\mathbf{U}_p$ and $\mathbf{U}$ are the mean particle and fluid velocity, $D \equiv \partial/\partial t + \mathbf{U}_p\cdot\nabla$ and τ is the relaxation time, related to the first approximation of the diffusion coef-ficient D given by the Chapman-Enskog theory for binary mixtures[5] by

$$\tau = m_p D(n + n_p)/(nkT)$$

The quantities n and n_p are the number densities of carrier gas and particles, respectively, k is Boltzmann's constant, m_p is the mass of the particles, and the superscript T denotes a transposed tensor. It can be shown[2] that, away from singular points, Eqs. (1-3) yield errors of the order of M_p^{-3} for the density and the mean velocity and $O(M_p^{-1})$ for the temperature, where

$$M_p = U_p/(kT_p/m_p)^{1/2} >> 1 \tag{4}$$

is a sort of a Mach number or speed ratio of the particles and $T_p \equiv \mathrm{Trace}(\boldsymbol{T}_p)/3$. Moreover, these errors are considerably reduced [to $O(M_p^{-4})$ and $O(M_p^{-2})$, respectively] when the heat flux term vanishes initially.

In this paper, the hyperbolic nature of Eqs. (1-3) when M_p is sufficiently large, is exploited to reduce them from a system of partial differential equations into a set of ordinary differential equations along the characteristics of the problem. In the case of two-dimensional (2D) and axisymmetric steady problems, the characteristic directions at a given point are obtained explicitly as a function of the local particle density, velocity and temperature tensor. We shall show in Sec. II, that three of the seven characteristic degenerate into the particle trajectories, and the remaining two pairs intersect the streamlines at angles whose interpretation is similar to that of the Mach angle in supersonic gas dynamics: the sine of these angles is the ratio between the transversal propagation of perturbations at a speed $c_i \approx (\gamma_i k T_{nn}/m_p)^{1/2}$ and the particles' mean velocity U_p. T_{nn} is the component of the temperature tensor perpendicular to the streamline and γ_i is 1 for one pair of characteristics and 3 for the other. The expression for c_i is an approximation based on the smallness of M_p^{-1}.

The fact that the characteristic directions are known algebraically allows for a relatively simple numerical implementation of the method following the algorithm described in the Appendix. In Sec. III, the method is applied to two problems: The dispersion of an initially concentrated distribution of particles (we shall refer to it as the wake problem) and the two-dimensional version of this same problem when the initial distribution of particles is a step-function (which we shall denote as the a Prandtl-Meyer problem by analogy to the Prandtl-Meyer expansion of a pure gas into a vacuum). Explicit (asymptotic) kinetic solutions of the Fokker-Planck equation for these two problems are given here for the first time. The excellent agreement obtained when comparing the two pairs of kinetic and hydrodynamic solutions for both problems confirms the reliability (whenever no strong singularities are present) of the hypersonic approximation (1-3) as well as the usefulnessof the method of characteris-tics as a tool to analyze Brownian motion in more complex gas flow configurations. On the other hand, the limitations of Eqs. (1-3) used blindly in the vicinity of singular points are illustrated in the Prandtl-Meyer problem.

II. Method of Characteristics

A. Description of the method

The hypersonic truncation of the moment equations carried out in Eqs. (1-3) relies on the smallness of the particles' thermal speed compared to their mean velocity. Notice that, following the development in Ref. 2, the heat flux term, which is $O(M_p^{-1})$ compared to the convective term, has been dropped in Eq. (3), but the pressure term (proportional to k/m_p) in Eq. (2) has been retained so as to account for the Brownian motion of the particles, even though it is also small compared to the leading $D\mathbf{U}_p$ term. The close connection between this pressure term and Brownian motion is physically evident from the fact that both are direct consequences of thermal agitation. Bringing Eqs. (1-3) to normalized form displays the small parameter underlying the truncation. The variables are replaced by their dimensionless counterparts

$$\mathbf{U}_p \leftarrow \mathbf{U}_p/U_R, \boldsymbol{T}_p \leftarrow \boldsymbol{T}_p/T_R,\ \mathbf{U} \leftarrow \mathbf{U}/U_R,\ T \leftarrow T/T_R,\ \mathbf{x} \leftarrow \mathbf{x}/L_R \quad (5)$$

where U_R, L_R, T_R are characteristic values of the velocity, geometric scale, and temperature in the flow, respectively. Only steady problems are considered and the following parameters are introduced:

$$S = U_R\tau/L_R, \qquad \varepsilon^2 = kT_R/(m_pU_R^2) \quad (6)$$

where S is the so-called Stokes number and ε, proportional to the inverse of M_p, is very small. Equations (1-3) then become

$$D\lambda_p = -\nabla\cdot\mathbf{U}_p \quad (7)$$

$$D\mathbf{U}_p + \varepsilon^2 [(\boldsymbol{T}_p\cdot\nabla)\lambda_p + \nabla\cdot\boldsymbol{T}_p] = (\mathbf{U} - \mathbf{U}_p)/S \quad (8)$$

$$D\boldsymbol{T}_p+(\boldsymbol{T}_p\cdot\nabla)\mathbf{U}_p+[(\boldsymbol{T}_p\cdot\nabla)\mathbf{U}_p]^T=2(T\boldsymbol{I} -\boldsymbol{T}_p)/S \quad (9)$$

Under the assumption that the geometry of the flow is 2D or axisymmetric, Eqs. (7-9) can be rewritten in a canonical matrix form. Let (x,y) be the axial and radial coordinates in axisymmetric problems (or any plane coordinates in 2D), and let z stand for the azymuthal coordinate or the direction perpendicular to the plane, respectively. From symmetry considerations, $T_{xz} = T_{yz} = 0$, and the velocity component along z is zero. If (u,v) are the components of the particle velocity at (x,y) and (u',v') their carrier gas counterparts, Eqs. (7-9) can be written as

$$\boldsymbol{A}\cdot\partial\omega/\partial x + \boldsymbol{B}\cdot\partial\omega/\partial y = \mathbf{b} \quad (10)$$

where

$$\omega=\{\lambda_p,u,v,T_{xx},T_{xy},T_{yy},T_{zz}\}^T$$

$$\mathbf{b}=\{-\beta v/y,(u'-u)/S-\varepsilon^2\beta T_{xy}/y,(v'-v)/S-\varepsilon^2\beta(T_{yy}-T_{zz})/y,$$
$$2(T-T_{xx})/S,-2T_{xy}/S,2(T-T_{yy})/S,2(T-T_{zz})/S-2v\beta T_{zz}/y\}^T$$

β=0,1 for 2D and axisymmetric geometries, respectively, and

$$\boldsymbol{A} = \begin{bmatrix} u & 1 & 0 & 0 & 0 & 0 & 0 \\ \varepsilon^2 T_{xx} & u & 0 & \varepsilon^2 & 0 & 0 & 0 \\ \varepsilon^2 T_{xy} & 0 & u & 0 & \varepsilon^2 & 0 & 0 \\ 0 & 2T_{xx} & 0 & u & 0 & 0 & 0 \\ 0 & T_{xy} & T_{xx} & 0 & u & 0 & 0 \\ 0 & 0 & 2T_{xy} & 0 & 0 & u & 0 \\ 0 & 0 & 0 & 0 & 0 & 0 & u \end{bmatrix}$$

$$B = \begin{bmatrix} v & 0 & 1 & 0 & 0 & 0 & 0 \\ \varepsilon^2 T_{xy} & v & 0 & 0 & \varepsilon^2 & 0 & 0 \\ \varepsilon^2 T_{yy} & 0 & v & 0 & 0 & \varepsilon^2 & 0 \\ 0 & 2T_{xy} & 0 & v & 0 & 0 & 0 \\ 0 & T_{yy} & T_{xy} & 0 & v & 0 & 0 \\ 0 & 0 & 2T_{yy} & 0 & 0 & v & 0 \\ 0 & 0 & 0 & 0 & 0 & 0 & v \end{bmatrix}$$

Equation (10) is a first-order system of nonlinear partial differential equations where all derivatives enter linearly.The slope y' of its characteristic direction results from solving the associated eigenvalue problem

$$\lambda \cdot B = y' \lambda \cdot A \tag{11}$$

If Eq. (10) is multiplied on the left by one eigenvector λ, and y' is rewritten as

$$y' = \tan\theta \tag{12}$$

the following results:

$$\lambda \cdot A \cdot (\cos\theta\ \omega_x + \sin\theta\ \omega_y) = \lambda \cdot \mathbf{b}\ \cos\theta \tag{13}$$

Then, along each characteristic direction θ (determined by each of theeigenvalues y'), an ordinary differential equation can be written as

$$\lambda \cdot A \cdot d\omega/ds = \lambda \cdot \mathbf{b}\ \cos\theta \tag{14}$$

where s is the length measured along the characteristic defined by y'.

Although the characteristic equation associated with the eigenvalue problem (11) is of order seven, the determinant of B - y'A may be factorized exactly to yield the following equation for y':

$$(v-y'u)^3\{(v-y'u)^2-\varepsilon^2(T_{yy}-2y'T_{xy}+y'^2T_{xx})\}$$
$$\{(v-y'u)^2-3\varepsilon^2(T_{yy}-2y'T_{xy}+y'^2T_{xx})\}=0 \tag{15a}$$

Explicit expressions for each one of the seven eigenvalues and corresponding eigenvectors can be derived from Eqs. (11) and (15a):

$$y' = \frac{uv-\gamma\varepsilon^2T_{xy} \pm \varepsilon\sqrt{[\gamma^2\varepsilon^2(T_{xy}{}^2-T_{xx}T_{yy})+\gamma(v^2T_{xx}+u^2T_{yy}-2uvT_{xy})]}}{u^2-\gamma\varepsilon^2T_{xx}} \tag{15b}$$

where γ takes the values 0, 1, or 3.

In particular, the eigenvalue $y'_0=\tan\theta_0 = v/u$ is triple, indicating that the particle trajectory is a triple characteristic. An alternative implicit expression for the remaining four eigenvalues y' is provided by the equation

$$\mu = \arcsin(1/M') \tag{16a}$$

where, for a given characteristic, $y' = \tan(\theta_0 \pm \mu)$ and M' is a Mach number based on the temperature normal to the characteristic, defined as

$$M' = U_p/\sqrt{(\gamma k T_{pmm}/m_p)} \tag{16b}$$

with $U_p=\sqrt{(u_p{}^2+v_p{}^2)}$, and the transversal speed of disturbances is, as one would expect, proportional to the square root of the component of the temperature tensor normal to the characteristic direction

$$T_{pmm}=\mathbf{m}\cdot\boldsymbol{T}\cdot\mathbf{m}=\sin^2\theta T_{xx}-2\sin\theta\cos\theta T_{xy}+\cos^2\theta T_{yy}$$

where **m** is the unit vector normal to the characteristic direction given by y'.Equation (16a) is implicit because T_{pmm} is dependent on y'. It is informative to write the equation for y' in this form because it is strictly parallel to the definition of Mach angle in supersonic equilibrium flow of simple gases. An approximate form of Eq. (16a) can be helpful in the hypersonic limit where ε is very small. Then, with errors of order ε^2, the four characteristics cross the particle streamline at symmetric angles $\pm\mu_1$ and $\pm\mu_2$: $\theta_{1,2}$ = arctan $y'_{1,2} \approx \theta_0 \pm \mu_1$, $\theta_{3,4}$= arctan $y'_{3,4} \approx \theta_0 \pm \mu_2$. The twin characteristics are deflected with relation to θ_0 by a small angle μ_i (of order ε) related to the Mach number of the particles by

$$\mu_i = \arcsin(1/M_i)\ ,\ i=1,2 \tag{17a}$$

where two modes of propagation of signals give rise to corresponding Mach numbers. In physical magnitudes these Mach numbers may be expressed as

$$M_i = U_p/(\gamma_i k T_{pnn}/m_p)^{1/2})\ ,\ \gamma_1=1, \gamma_2=3 \tag{17b}$$

where the transversal speed of disturbances is now based on the normal component of the temperature tensor,

$$T_{pnn}=\mathbf{n}\cdot\boldsymbol{T}\cdot\mathbf{n} =\sin^2\theta_0 T_{xx}-2\sin\theta_0\cos\theta_0 T_{xy}+\cos^2\theta_0 T_{yy}$$

where **n** is the unit vector normal to the particle streamline. It may be observed that Eqs. (16b-17b) are exactly coincident with the usual definition of the Mach angle if T_{pnn} is replaced by the scalar temperature and γ_i takes the value of the ratio of specific heat coefficients (γ = 5/3 for monatomic gases). It is clear that the two modes arise because of the breakdown of the isotropy in the temperature tensor, and that further modes would be present in a 3D problem lacking symmetry. Values of 1 and 3 for the coefficient γ correspond, respectively, to isothermal propagation and to a problem with only 1 degree of freedom (one-dimensional). This last situation arises when the mechanical work put by the pressure term in a particular direction is not redistributed by colisions, heating up only the corresponding component of the temperature tensor.

As sketched in the Appendix, the seven differential equations (along different paths) obtained from Eq. (14) can be solved numerically by means of a characteristic mesh. The only additional difficulty with respect to traditional characteristic solutions is the increassed complexity of the grid. In the next section the characteristic method is applied to two problems for which asymptotic analytical solutions are available.

III. Examples: Gaussian Wake and Prandtl-Meyer Expansion

In this section, two problems are considered for which asymptotic analytical solutions are available. The building unit is an expression describing the spread of particles emanating from a steady point source in a uniform background gas at temperature T_b which travels at velocity U_b in the x direction.[3] This exact solution results from a numerical convolution integral of a known solution of the Fokker-Planck equation for the evolution in time of a pulse of particles.[6,7] In the limit $\varepsilon<<1$, the convolution integrals reduce asymptotically to simple algebraic expressions for n_p and U_p. These results may also be obtained by a near-axis boundary layer analysis of the hypersonic Eqs. (1-3), which provides also explicit expressions for T_{xx} and T_{yy}.[3] The asymptotic results for the steady source problem are summarized here: If particles at the source are produced at a rate n' and with a Maxwellian distribution with mean velocity U_0 (also in the x direction) and temperature T_0, the particle phase magnitudes at a point (x, y) [where $y \leq O(\varepsilon x)$] can be written in terms of parametric variables as[3]

$$x=x_0(s); \qquad u=u_0(s); \qquad T_{xx}=\xi(s); \qquad T_{yy}=d(s)/a(s)$$

$$n=\frac{N'}{u_0(s)}[2\pi\varepsilon^2 a(s)]^{(1-n_d)/2}\exp(-\frac{y^2}{2a(s)\varepsilon^2}); \qquad v=\frac{c(s)}{a(s)}y \qquad (18)$$

where $n_d = 2$ or 3 in 2D and axisymmetrical problems, respectively. In the preceding equations, the same normalization defined in Eq. (5) is used, with $U_R = U_b$, $L_R = U_b\tau$, $T_R = T_b$, so that $S = 1$, and the following functions are introduced:

$$x_0(s) = s + \delta(1 - e^{-s}); \qquad u_0(s) = x_0'(s) = 1+\delta e^{-s} \qquad (19a,b)$$

$$a(s) = \alpha\,(1 - e^{-s})^2 + 2s - 3 - e^{-2s} + 4e^{-s} \qquad (19c)$$

$$c(s) = a'(s)/2 = (1 - e^{-s})\,(1 - e^{-s} + \alpha e^{-s}) \qquad (19d)$$

$$d(s) = \alpha[1-4e^{-s}+(2s+3)e^{-2s}]+2s-4+8e^{-s}-(2s+4)e^{-2s} \qquad (19e)$$

$$\xi(s) = e^{-2s}[e^{2s}-1+\alpha(1+\delta)^2+4\delta(e^{s}-1)+2\delta^2 s]/u_0(s) \qquad (19f)$$

with $\alpha = T_0/T_b$, $\delta = (U_0 - U_b)/U_b$, $N' = n'\tau$. No closed-form expression for T_{xy} has been yet obtained, but in the vicinity of the source ($s<<1$), $T_{xy} = \alpha y/[(1 + \delta)s]$ for $\alpha \neq 0$ and $T_{xy} = 5y/[3(1 + \delta)]$ otherwise. Equations (18) can be considered a fundamental solution, from which more complicated problems can be solved by convolution. The fundamental solution for $N' = 1$ is denoted with a subscript F: $n_F = n$, $\mathbf{U}_F = u\mathbf{e}_x + v\mathbf{e}_y$, $\boldsymbol{T}_F = T_{xx}\mathbf{e}_x\mathbf{e}_x + T_{yy}\mathbf{e}_y\mathbf{e}_y + T_{xy}(\mathbf{e}_x\mathbf{e}_y + \mathbf{e}_y\mathbf{e}_x)$.

A. Dispersion of an Initially Gaussian distribution of particles (Gaussian Wake)

Particles are seeded in the plane $x = 0$ at a dimensionless rate N' with a Gaussian concentration profile, so that at $s = 0$,

$$n = \frac{N'}{u_0(s)}(2\pi\varepsilon^2 a_0)^{(1-n_d)/2}\exp(-\frac{y^2}{2a_0\varepsilon^2}), \quad u_0 = 1+\delta, \quad v=0$$

$$T_{xx} = T_{yy} = T_{zz} = \alpha, \quad T_{xy} = 0 \qquad (20)$$

where a_0 is a constant. As particles evolve in the uniform (T_b,U_b) back-ground, their Brownian motion broadens the concentration profile. From a convolution argument, the field quantities are

$$n(\mathbf{x}) = \int ds^i(\mathbf{x}')\, n^i(\mathbf{x}')u_0^i(\mathbf{x}')\, n_F(\mathbf{x}-\mathbf{x}') \qquad (21a)$$

$$n(\mathbf{x})\,\mathbf{U}(\mathbf{x}) = \int ds^i(\mathbf{x}')\, n^i(\mathbf{x}')u_0^i(\mathbf{x}')\, n_F(\mathbf{x}-\mathbf{x}')\, \mathbf{U}_F(\mathbf{x}-\mathbf{x}') \qquad (21b)$$

$$n(\mathbf{x})\boldsymbol{T}(\mathbf{x}) = \varepsilon^{-2}[\int ds^i(\mathbf{x}')n^i(\mathbf{x}')u_0^i(\mathbf{x}')n_F(\mathbf{x}-\mathbf{x}')\mathbf{U}_F(\mathbf{x}-\mathbf{x}')\mathbf{U}_F(\mathbf{x}-\mathbf{x}') - \mathbf{U}(\mathbf{x})\mathbf{U}(\mathbf{x})] + \int ds^i(\mathbf{x}')\, n^i(\mathbf{x}')u_0^i(\mathbf{x}')\, \boldsymbol{T}_F(\mathbf{x}-\mathbf{x}') \qquad (21c)$$

where the superscript i refer to the initial condition given in Eqs. (20), and the integration is carried over the seeding plane [$ds^i(\mathbf{x}')$ is the differential element of surface centered around $\mathbf{x}'$]. Performing the integrals (21), the following is obtained for the particle magnitudes

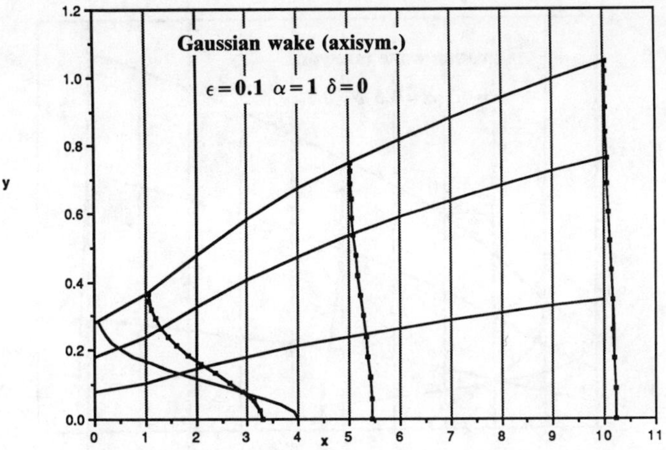

Fig. 1. Trajectories and density profiles for axisymmetric Gaussian wakes. Marked curves correspond to Eq. (22). —— numerical integration of the hypersonic equations by the method of characteristics. $\varepsilon=0.1$; $\alpha=1$; $\delta=0$.

downstream from the seeding plane:

$$n=\frac{1}{u_0}[2\pi\varepsilon^2(a+a_0)]^{(1-n_d)/2}\exp\left[-\frac{y^2}{2(a+a_0)\varepsilon^2}\right] \tag{22a}$$

$$\mathbf{U}=u_0\mathbf{e}_x+\mathbf{e}_y\frac{yc}{(a+a_0)},\quad T_{yy}=\frac{d}{a}+\frac{a_0c^2}{a(a+a_0)},\quad T_{xx}=\xi \tag{22b}$$

No expression for T_{xy} is given because it is not available in the fundamental solution. As before, a, c, d, ξ, and u_0 are functions of x, implicitly defined as $x_0(s)=x$ [Eqs. (19)].

Figures 1 and 2 show density profiles and trajectories as calculated from Eqs. (22) compared with the results from the numerical integration of the hypersonic equations using the method of characteristics (see the Appendix). Two different initial conditions (both with $\varepsilon=0.1$) are displayed for axisymmetric flows. A very close agreement is observed in the density profiles for both cases, as well as for the corresponding 2D situations (not shown). The agreement in the velocity and the temperature, not shown, is even better than for the densities. The mesh size used in the numerical method was $\Delta x=0.01$ for $x\le 5$ and $\Delta x=0.1$ for $x>5$. The convergence of the numerical method (see the Appendix) was very fast: typically, after two iterations the error was smaller than $0.1(\Delta x)^2$.

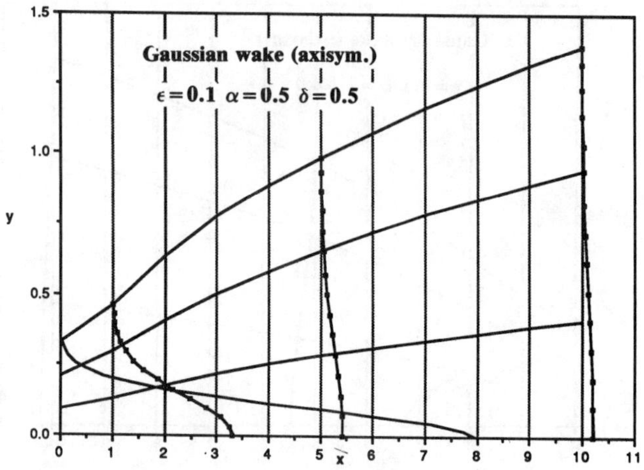

Fig. 2 Like Fig. 1, but with ε=0.1; α=0.5; δ=-0.5.

B. Dispersion of a Step-like Particle Distribution (Prandtl-Meyer Problem)

In this section we consider the analog of a 2D Prandtl-Meyer expansion.Particles are uniformly seeded only in the upper half of the plane x=0, so that there is a discontinuous step in their concentration at x=0. The properties of the carrier fluid are uniform everywhere. Although it is not included here, the axisymmetric counterpart of this problem corresponds to the expansion of particles uniformly seeded in a circle at plane x=0. With the help of Eqs. (24) and (25), and carrying out similar integrations as before, the following asymptotic results are obtained:

$$n = \mathrm{erfc}(-\eta)/2, \qquad T_{xx} = \xi \tag{23a}$$

$$U = u_0 e_x + e_y \frac{c}{a}\left(\frac{2\varepsilon^2 a}{\pi}\right)^{1/2} \exp(-\eta^2)/\mathrm{erfc}(-\eta) \tag{23b}$$

$$T_{yy} = \frac{d}{a} + \frac{c^2}{a}\left[1 - \frac{2\eta}{\sqrt{\pi}}\,\frac{\exp(-\eta^2)}{\mathrm{erfc}(-\eta)} - \frac{2}{\pi}\left(\frac{\exp(-\eta^2)}{\mathrm{erfc}(-\eta)}\right)^2\right] \tag{23c}$$

where we have defined the variable

$$\eta = y/\sqrt{(2a\varepsilon^2)} \tag{23d}$$

and erfc is the complementary error function.[8] To avoid the singularity at the origin, the numerical integration is started slightly downstream from x=0 (at x=0.25 in Fig. 3) using Eqs. (23) as initial conditions (from the starting behavior of the fundamental solution T_{xyF}, it

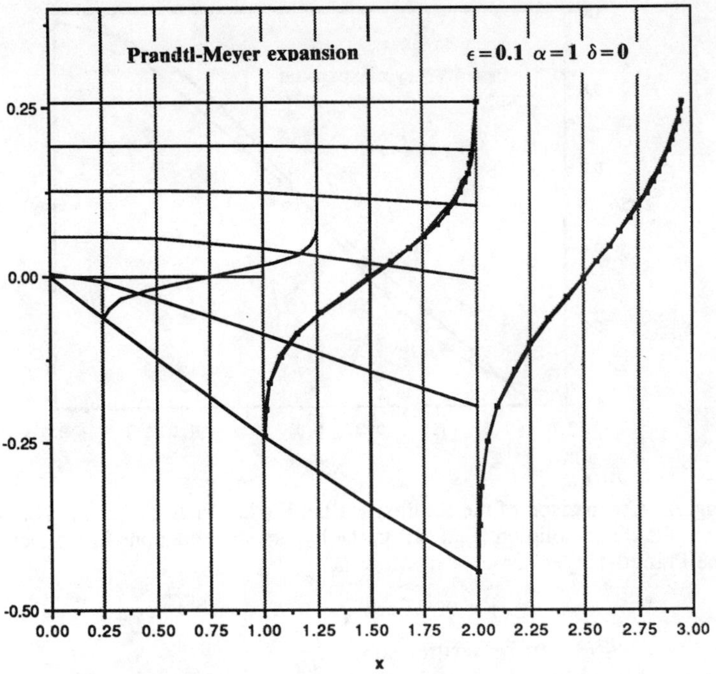

Fig. 3. Trajectories and density profiles for a Prandtl-Meyer expansion with $\varepsilon=0.1$, $\alpha=1$ and $\delta=0$. Marked curves correspond to the analytical solution (Eqs. 23) of the Fokker-Planck equation; —— numerical integration by the method of characteristics, started at x=0.25 using the analytical solution (23).

follows that $T_{xy} \approx 2\alpha v/(1+\delta)$ for small x) and the results for $\varepsilon=0.1$ are presented in Fig. 3, together with the solution (23). The mesh size used in the numerical integration was $\Delta x=0.01$. The agreement between both solutions for the velocities and temperatures (not shown) is also excellent.

While Eqs. (23) are only restricted to small values of ε, a note of caution is required when using the hypersonic truncated set of Eqs. (1-3) in problems such as the Prandtl-Meyer expansion where very large gradients occur near the origin[2,3] (this is why our numerical integration has to be started downstream from x=0). Indeed, from Eqs. (1-3) and through a boundary layer analysis near y=0, a solution can be obtained for $x \ll 1$ for $v=\mathbf{U_p}.\mathbf{e_y}$, $\boldsymbol{\lambda}$, and $\mathbf{T}_{yy}$ in terms of the similarity variable κ

$$\kappa=y/(\sqrt{3}\ \varepsilon x), \tag{24a}$$

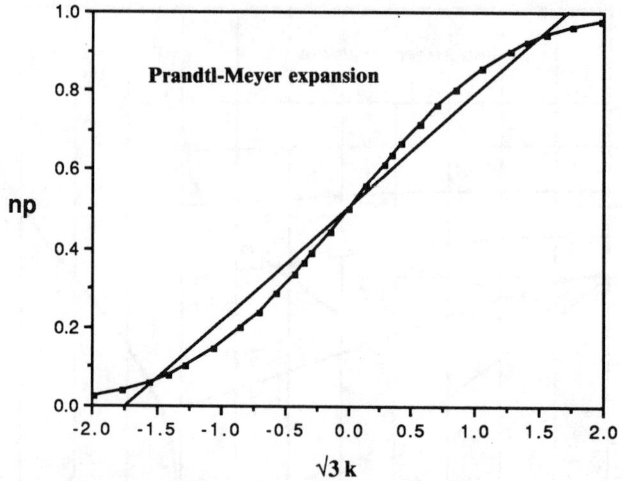

Fig. 4. Comparison of the solution via the Fokker-Planck equation [Eq. (23)] with the direct solution [Eq.(24)] to the hypersonic equations in the vicinity of the Prandtl-Meyer corner.

which for $\kappa \leq \sqrt{3}$ can be written as

$$w_0=\frac{v}{\varepsilon}=\frac{\sqrt{3}}{2}(\kappa-1);\quad \exp(\lambda_0)=\frac{1}{2}(1+\kappa);\quad T_{yy0}=\frac{1}{4}(\kappa+1)^2 \tag{24b}$$

This solution, though rigorously derived from Eqs. (1-3), is incorrect because the temperature and density jumps at the start of the expansion give rise to a singular transversal heat flux that cannot be neglected in the energy equation. Nonetheless, as shown in Fig. 4, even in this irregular case the hypersonic solution (24) is not intolerably different from the asymptotic solution (23).

Appendix: Sketch of the Numerical Method

After solving the eigenvalue problem (11) with y' given by Eq. (15), Eq. (13) yields the following three equations on the trajectories (defined by $y'=\tan\theta_0=v/u$):

$$U\, dT_{zz}/ds_0 = 2\,(T - T_{zz})/S \tag{A.1}$$

$$C_1\frac{d\omega}{ds_0}=\frac{-\beta v}{y}+\frac{\sin\theta_0\,(T-T_{xx})}{ST_a}-\frac{\cos\theta_0\,(T-T_{yy})}{ST_b} \tag{A.2}$$

$$C_2\frac{d\omega}{ds_0}=\frac{-\cos\theta_0}{S}\left[\frac{T_b}{T_a}(T-T_{xx})+2T_{xy}+\frac{T_a}{T_b}(T-T_{yy})\right] \tag{A.3}$$

where

$$\mathbf{C}_1=(U,\frac{1}{\cos\theta_0}+\frac{T_{xx}\tan\theta_0}{T_a},\frac{-T_{xy}}{T_b},\frac{u\tan\theta_0}{2T_a},0,-\frac{u}{2T_b},0) \tag{A.4}$$

$$\mathbf{C}_2=(0,T_{xy}-\frac{T_{xx}T_b}{T_a},T_{xx}-\frac{T_{xy}T_a}{T_b},\frac{-uT_b}{2T_a},u,-\frac{uT_a}{T_b},0) \tag{A.5}$$

$$U=\sqrt{(u^2+v^2)}=u/\cos\theta_0;\quad T_a=\cos\theta_0 T_{xy}-\sin\theta_0 T_{xx} \tag{A.6,7}$$

$T_b=\cos\theta_0 T_{yy}-\sin\theta_0 T_{xy}$, (A.8)
and s_0 is the coordinate along the trajectory. The other four equations along the characteristics defined by $y'_i=\tan\theta_i$, i=1,2,3,4, may be written as

$$\mathbf{D}_i\cdot\frac{d\omega}{ds_i}=\{-\frac{\beta v}{y}\lambda_{1i}+\frac{u'-u}{S}-\frac{\varepsilon^2\beta}{y}T_{xy}+\lambda_{3i}[\frac{v'-v}{S}-\frac{\varepsilon^2\beta}{y}(T_{yy}-T_{zz})]+$$
$$+2\lambda_{4i}\frac{T-T_{xx}}{S}-2\lambda_{5i}\frac{T_{xy}}{S}+2\lambda_{6i}\frac{T-T_{yy}}{S}+\lambda_{7i}[2\frac{T-T_{zz}}{S}-2v\beta\frac{T_{zz}}{y}]\cos\theta_i \tag{A.9}$$

with

$\mathbf{D}_i=\{u\lambda_{1i}+\varepsilon^2(T_{xx}+T_{xy}\lambda_{3i}),\ \lambda_{1i}+u+2T_{xx}\lambda_{4i}+T_{xy}\lambda_{5i},\ u\lambda_{3i}+v\lambda_{5i}$
$+2T_{xy}\lambda_{6i},\ \varepsilon^2+u\lambda_{4i},\ \varepsilon^2\lambda_{3i}+u\lambda_{5i},\ u\lambda_{6i},\ u\lambda_{7i}\}$

$\lambda_{1i}=[a_i^2+\varepsilon^2(b_i y_i'-c_i)]/(2a_i y'_i)$

$\lambda_{3i}=-[a_i^2+\varepsilon^2(3b_i y_i'-c_i)]/(2\varepsilon^2 c_i y'_i)$; $\lambda_{4i}=\varepsilon^2 y_i'/a_i$

$\lambda_{5i}=-[a_i^2+\varepsilon^2(3b_i y_i'+c_i)]/(2c_i a_i)$; $\lambda_{6i}=[a_i^2+\varepsilon^2(3b_i y_i'-c_i)]/(2c_i y_i' a_i)$

$a_i=v-y_i'u$; $b_i=T_{xy}-y_i'T_{xx}$; $c_i=T_{yy}-y_i'T_{xy}$

To solve these equations numerically, a grid is adapted to the particle flow as follows: in the first place a set of curves $x=\xi_k(y)$ (k = 0,1,2...) is defined in such a way that they are not tangent to the trajectories at any point. For instance, in the case of the examples in Sec. III, these curves are vertical straight lines $x=x_k=$ const. The first of these lines $x=\xi_0(y)$ has to be coincident with the boundary at which initial conditions are given, so that ω is known at k = 0. Along the first line, grid points (0, m) are defined (see Fig.5); the next row of grid points (1, m) is located by intersecting the particle trajectories through the corresponding points in the previous row (0, m) with the curve $\xi_1(x)$. The properties ω at the new points are still unknown, but an iterative technique allows to determine them: the iteration is started by assuming $\omega(1, m)=\omega(0, m)$, from which the slope of the characteristics through (1,m) can be obtained. The intersection of

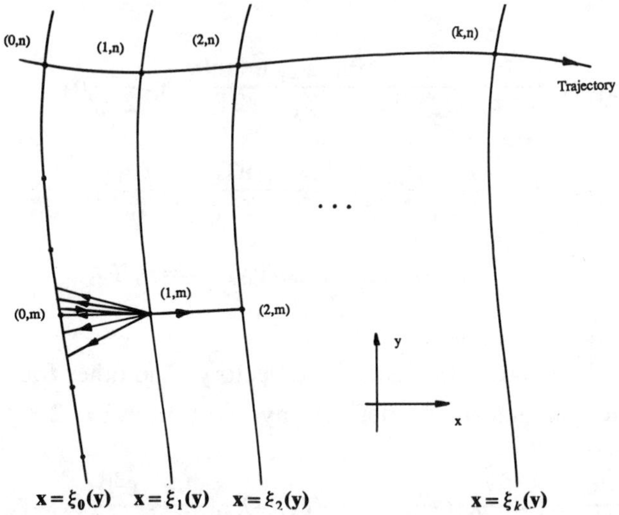

Fig. 5. Sketch of the grid and the procedure used in the numerical implementation of the method of characteristics.

the curve $\xi_0(x)$ with straight lines shot from (1,m) with the characteristic slopes identifies five *predecessor* points whose properties can be evaluated by means of a spline interpolation along the $\xi_0(x)$ curve. Now, a finite-difference formulation of Eq. (A9) can be written linking the properties at (1,m) with those at the predecessor points along $\xi_0(x)$. A first nontrivial approximation for $\omega(1,m)$ is thus obtained, and the process of backward shooting, interpolation, and finite-difference reevaluation of $\omega(1,m)$ can be repeated until a convergence criterion is met. In our case, the iterations have been stopped when the euclidean norm of the difference between two consecutive values of $\omega(1,m)$ is less than $(\Delta s_0)^2$, where Δs_0 is the increment in the streamline coordinate s_0 between the two points (0,m) and (1,m). This process is repeated each time when jumping from a curve ξ_{k-1} into a new ξ_k.

Acknowledgment

This work was supported by U.S. Department of Energy Grant DE-FG0287ER13750 and National Science Foundation Grant CBT-86-12143.

References

[1]Fernández de la Mora, J. and Fernández-Feria, R., "Kinetic Theory of Binary Gas Mixtures with Large Mass Disparity," Physics of Fluids, Vol. 30, pp.740-751, March 1987.

[2]Fernández-Feria, R., "On the Gas Dynamics of Binary Gas Mixtures with Large Mass Disparity," Ph.D. Dissertation, Yale University, New Haven, CT, 1987.

[3]Riesco-Chueca, P. and Fernández de la Mora, J., "Brownian Motion Far from Equilibrium: a Hypersonic Approach," (to be published in Journal of Fluid Mechanics, 1989).

[4]Friedlander, S. K., Smoke, Dust and Haze, Wiley, New York, 1977.

[5]Chapman, S. and Cowling, T. G., The Mathematical Theory of Non-Uniform Gases, Cambridge Univ. Press, Cambridge, UK, 1970.

[6]Rice, S. A. and Gray, P., The Statistical Mechanics of Simple Liquids, Interscience, New York, 1965.

[7]Nguyen, T. K.and Andres, R. P., "Fokker-Planck Description of the Freejet Deceleration Flow," Progress in Astronautics and Aeronautics: Rarefied Gas Dynamics, Vol.74, AIAA, New York, 1981.

[8]Abramowitz, M. and Stegun, I. A., Handbook of Mathematical Functions, Dover, New York, 1972.

Chapter 4. Clusters

Phase-Diagram Considerations of Cluster Formation When Using Nozzle-Beam Sources

E. L. Knuth* and W. Li†
University of California, Los Angeles, California
and
J. P. Toennies‡
Max-Planck-Institut für Strömungsforschung, Göttingen, Federal Republic of Germany

Abstract

In order to help understand the formation of clusters when using nozzle-beam sources, procedures for constructing and applying isentropic-expansion paths on the pressure-temperature phase diagram for the beam gas are summarized and exemplified for the case of He. The importance of using real-fluid properties in constructing expansion paths is emphasized. In the case of ^4He, it is found that, for a significant fraction of interesting source conditions, the real-fluid properties result in isentropic paths approaching the vapor-liquid line from the liquid side. Procedures for identifying appropriate source conditions for specific goals, such as expanding through the critical point, approaching the vapor-liquid line from the liquid side, and locating the beginning of phase transition relative to the orifice, are included. Recent observations of the properties of He clusters produced using a low-temperature nozzle-beam source are discussed with the aid of an appropriate phase diagram.

Introduction

With interest increasing in van der Waals molecules and clusters, the use of supersonic nozzle beams with high source pressures and low source temperatures is escalating. The interest in ^4He beams with these extreme source conditions is particularly keen[1–3]; the reasons include: 1) the simplicity of ^4He, 2) the search for evidence of the ^4He dimer, 3) the fact that ^4He clusters must be liquid, and 4) the fact that ^4He acts as a superfluid under appropriate tem-

* Professor, Department of Chemical Engineering.
† Ph.D. Student, Department of Chemical Engineering.
‡ Director, Molecular Interactions Department.

perature and flow conditions. At these extreme source conditions, fluids do not behave as thermally perfect gases so that real-fluid properties (density, enthalpy, entropy, sound speed, etc.) must be used in modeling the expansion. Although the procedures for using real-fluid properties are exemplified here by analyses of expansions of ^4He from source pressures above the ^4He critical pressure, they are applicable directly to any beam gas for which real-fluid properties are available for the pressures and temperatures of interest.

Procedures and Results

For most of the pressure-temperature (P-T) range of interest here, the relevant ^4He properties have been tabulated by McCarty.[4] Near the λ line, the tabulated data were supplemented using thermodynamic-property relations and the thermal equation of state given also by McCarty. Pressure-temperature variations during isentropic expansions were established by determining first the entropy value for an arbitrarily chosen source condition and then determining the local temperatures yielding this entropy value for a series of decreasing values of local pressure. The results of these calculations are shown on the accompanying P-T phase diagram, Fig. 1. Included here are 11 real-fluid isentropes, all starting from a source pressure P_o = 20 bar, but each starting from a different source temperature T_o (namely, 15, 12, 10, 9, 8, 7, 6, 5, 4, 3, and 2.61 K). For comparison, the straight line (slope = 5/2), which would be realized if one attempted to model the expansion with a perfect gas, has been drawn for the source condition 20 bar, 6 K.

Isentropes in the single-phase regions of Fig. 1 are found to be quantitatively useful in planning and interpreting experiments with high source pressures. However, quantitative modeling of an expansion that has reached the liquid-vapor line requires significant additional (kinetic) considerations. In typical molecular-beam systems, the characteristic time for the expansion is relatively short (in comparison with the characteristic time for phase changes) so that an expansion isentrope intersecting the liquid-vapor line will cross this line

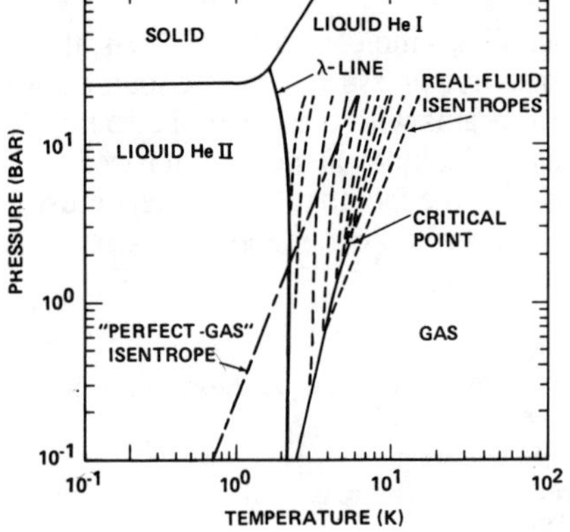

Fig. 1 Real-fluid and "perfect-gas" isentropes on the P-T phase diagram for ^4He.

resulting in either supersaturated vapor or overexpanded liquid, as the case may be, and the initiation of phase transition is delayed. Nevertheless, the earliest possible location of the initiation of phase transition is of considerable interest. A suitable measure of this location is provided by the local Mach number. With the aid of Fig. 1 and the energy-conservation equation

$$h_o = h + u^2/2 \tag{1}$$

this Mach number was determined by 1) identifying the state at the intersection of the isentrope and the liquid-vapor line, 2) determining the enthalpy h for this state, 3) calculating the local speed u from Eq. (1), 4) determining the sound speed for the identified state using Ref. 4, and 5) calculating the Mach number from its definition. For $P_o = 20$ bar, it is found that the phase transition is initiated downstream from the throat if $T_o > 6$ K and upstream from the throat if $T_o > 6$ K. The Mach number calculated for ^4He at the point at which the isentrope crosses the vapor-liquid phase line is shown as a function of T_o in Fig. 2.

Also of interest is the mass flux at the orifice. For an isentropic expansion from a given combination of P_o and T_o, the sonic point (located at the orifice) was found by trial and error--it corresponds to that state for which the local sound speed equals the local fluid speed calculated from Eq. (1). The orifice mass flux is then simply the product of density and sound speed at this sonic point. Values of the mass flux for ^4He at $P_o = 20$ bar calculated assuming an ideal gas are compared with the mass flux for the real fluid in Fig. 3. It is seen that real-gas effects are significant and reduce the mass flux through the orifice, in comparison with that for the ideal gas, by a factor that varies from

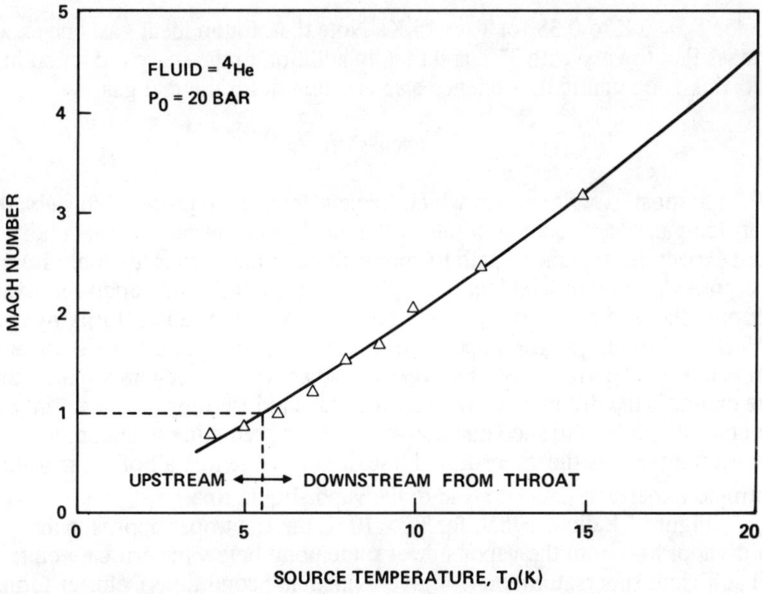

Fig. 2 Mach number at the point at which the isentrope crosses the liquid-vapor line as a function of source temperature.

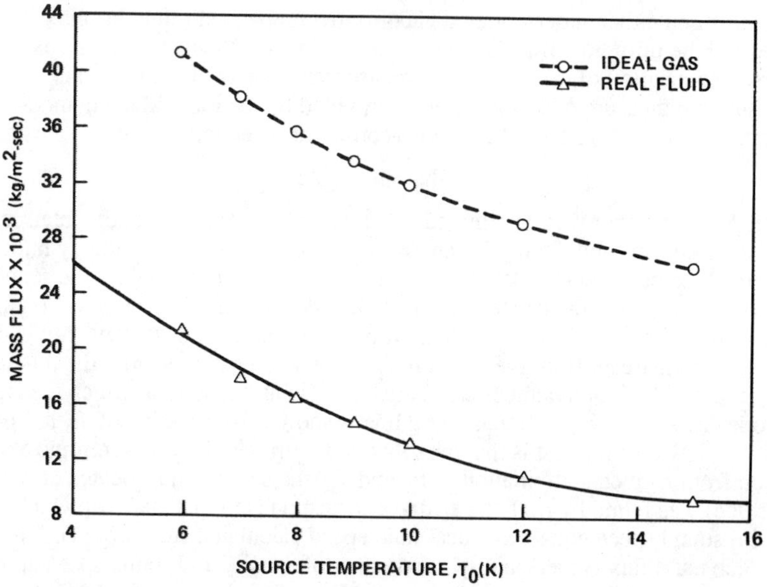

Fig. 3 Real-fluid [4]He mass flux at the nozzle throat as a function of source temperature compared with that calculated for an ideal gas. The flux is reduced by more than a factor of 0.5 and the simple $T_o^{-1/2}$ dependence no longer holds.

0.52 for $T_o = 6$ K to 0.35 for $T_o = 15$ K. Note that, for an ideal gas, one expects the mass flux to vary with $T^{-1/2}$ and that, in addition to the sharp reduction in value, this temperature dependence also is different for the real gas.

Discussion

In most experiments in which freejets are used to produce clusters, the expansion path was assumed to follow the ideal-gas isentrope, so that one would expect the expansion path to approach the liquid-vapor line only from the vapor side. Figure 1 indicates that, for the range of initial conditions examined, only the isentropes for $T_o \geq 15$ K are approximated satisfactorily by the "perfect-gas" isentrope; the slopes of the other 10 isentropes differ significantly from that of the "perfect-gas" isentrope--the lower the source temperature, the more dramatic the difference. Without the real-fluid isentropes, one might even encounter difficulties in such qualitative efforts as predicting whether the expansion intercepts the vapor-liquid line or not! Note that all of the real-fluid isentropic-expansion paths do lead to the vapor-liquid line.

Figure 1 indicates that, for $T_o > 10$ K, the isentropes approach the liquid-vapor line from the vapor side at some point below the critical point; after sufficient supersaturation (for given kinematic conditions), cluster formation is initiated. For $T_o < 10$ K, the isentropes approach the liquid-vapor line from the liquid side at some point below the critical point; after sufficient overexpansion, bubble formation (cavitation) is initiated. In the latter case, continued expansion results eventually in clusters contained in a vapor phase; the extent of departure from the liquid-vapor line is limited initially by the cooling that accompanies the phase change. This cooling is expected to be slowed (if not arrested) at temperatures in the vicinity of the λ line (not shown in Fig. 1). However, the vapor generated by this phase change would continue to expand to values of P and T within the region where, under equilibrium conditions, one finds He II. As a very specific example, note that, for $P_o = 20$ bar, one can approach the same point on the liquid-vapor line either from the liquid side starting from $T_o = 5$ K or from the vapor side starting from $T_o = 15$ K. In the former case, the fluid is liquid He I prior to initiation of a vapor phase; in the latter case, the fluid is vapor prior to initiation of a liquid phase.

Figure 1 indicates also that expanding into the liquid He II region is more difficult to achieve than expected on the basis of "perfect-gas" isentropes and as suggested in Ref. 2. This difficulty can be thought of as a consequence of the heat-capacity behavior near the λ line. It is found that, starting from 20 bar and 2.61 K, the isentrope intersects the λ line at 3.27 bar and 2.2 K.

If ^4He is to act as a superfluid, it not only must exist with a state to the left of the λ line in Fig. 1, but also must have a flow velocity less than some critical value that is fixed primarily by the characteristic flow dimensions. For an orifice of 5 μm, typically used in our experiments, the critical flow velocity is predicted to be 0.1 m/s, which is three orders of magnitude less than the velocity of the gas, which, for $T_o = 4$ K and $P_o = 20$ bar, is calculated to be 208 m/s. Since our orifices are short (≈ 20 μm) there is some question, however, as to whether the same critical velocities apply since the time spent in the orifice, ≈ 0.2 μs, may be too short to produce a significant concentration of vortices, which are thought to destroy superfluidity. Experiments are planned to study the nature of the flow in this region.

Presently we are investigating the implications of expanding along different isentropes on the properties of atoms and clusters in nozzle beams with high-pressure sources. Early nozzle-beam experiments by Becker et al.[6] indicated two velocity peaks corresponding to atoms (faster) and clusters (slower). Buchenau et al.[1,2] measured velocity distributions in nozzle beams with $P_o = 20$ bar and values of T_o such that the expansion approaches the liquid-vapor line from the liquid side and report three velocity peaks. Comparisons of the speeds corresponding to these peaks with speeds deduced from (real-fluid) enthalpy changes suggest that the additional (slowest) peak is due to clusters resulting from the disintegration of that liquid He produced in the expansion to the liquid side of the liquid-vapor line. These clusters appear to be a unique feature of beams formed by expansions to the liquid side of the liquid-vapor line.

One might explore systematically the consequences of using source conditions such that the fluid expands through the critical point. Starting at various combinations of P_o, T_o along the isentrope passing through the critical point results in expansion through the critical point at various locations relative to the throat; starting at 9 bar and 7 K locates the critical point at the throat. In recent experiments ^4He was expanded not only from the liquid side of the liquid-vapor line but also through the critical point.[7] It was observed that the speed ratio of the slowest peak is several times greater when expanding through the critical point than when expanding from the liquid side of the liquid-vapor line.

Figure 1 already has been found to be useful in interpreting some recent experiments on the deflection of He clusters by a monoenergetic electron beam.[3] These experiments reveal what appear to be two types of clusters, which are designated α and β. The signal for the α type of cluster is formed only for values of T_o significantly less than that for which fluid expands through the critical point. The signal from the β type of cluster appears at higher temperatures and peaks at a value of T_o = 10 K for which the fluid expands through the critical point. Thus, Fig. 1 and the observations concerning speed ratio suggest that the β clusters might be the result of disintegration of the liquid He produced in the expansion of the real fluid to the liquid side of the liquid-vapor line. It is interesting to note that the intensity enhancement by expanding through the critical point occurs simultaneously with a dramatic increase in speed ratio discussed earlier. This suggests that Mach number focusing[8] of the high-speed-ratio component may be the reason for the intensity enhancement.

These special features of cluster beams which are expanded through the critical point, are expected to be general phenomena and are anticipated also for other beam gases.

Acknowledgments

The authors thank H. Buchenau, K. Martini, A. Scheidemann, and C. Winkler for discussions of the experimental results.

References

[1]Buchenau, H., Götting, R., Minuth, R., Scheidemann, A., and Toennies, J. P., "Charakterisierung von Heliumclustern in einem Molekularstrahl," *Flow of Real Fluids*, edited by G. E. A. Meier and F. Obermeier, Springer-Verlag, Berlin 1985, pp. 157-169.

[2]Buchenau, H., Götting, R., Scheidemann, A., and Toennies, J. P., "Experimental Studies of Condensation in Helium Nozzle Beams," *Rarefied Gas Dynamics*, Vol. II, edited by V. Boffi and C. Cercignani, B. G. Teubner, Stuttgart, 1986, pp. 197-207.

[3]Martini, K., Toennies, J. P., and Winkler, C., "Deflection and Excitation of Helium Clusters by a Monoenergetic Electron Beam: Evidence for Two Different Types of He Clusters," to be published.

[4] McCarty, R. D., "Thermodynamic Properties of Helium 4 from 2 to 1500 K at Pressures to 10^8 Pa," *Journal of Physical Chemical Reference Data*, Vol. 2, 1973, pp. 923-1042.

[5]Wilks, J., *The Properties of Liquid and Solid Helium*, Clarendon Press, Oxford, 1967.

[6]Becker, E. W., Bier, K., and Henkes, W., "Strahlen aus kondensierten Atomen und Molekeln im Hochvakuum," *Zeitschrift für Physik*, Vol. 146, 1956, pp. 333-338.

[7]Buchenau, H., Knuth, E. L., Toennies, J. P., and Winkler, C., to be published.

[8]Sharma, P. K., Knuth, E. L., and Young, W. S., "Species Enrichment Due to Mach-Number Focusing in a Molecular-Beam Mass-Spectrometer Sampling System," *Journal of Chemical Physics*, Vol. 64, June 1976, pp. 4347-4351.

Fragmentation of Charged Clusters During Collisions of Water Clusters with Electrons and Surfaces

A. A. Vostrikov,* D. Yu. Dubov,* and V. P. Gilyova*
Siberian Branch of the USSR Academy of Sciences, Novosibirsk, USSR

Abstract

We present a number of new results on water clusters properties obtained by the nozzle molecular beam method. Ion ejection from cluster was observed due to cluster ionization by electron impact: $(H_2O)_N + e \rightarrow OH^+, H_2O^+, H^+(H_2O)_x$, x = 0 - 4. Formation and separation of charged clusters $H^+(H_2O)_i$ and $OH^-(H_2O)_j$ has been found to take place as a result of the scattering of neutral clusters $(H_2O)_N$ by solid metal, semiconductor or dielectric surfaces. Clusters were produced in freejet of superheated water vapor expanding out of a sonic nozzle into vacuum. Scaling laws for the condition of expansion with condensation and dependence of the mean size of water clusters on the nozzle source conditions are presented.

Introduction

The kinetics of cluster formation and their properties is of considerable current interest. Water clusters are the subject of the present investigations. Processes related to water condensation play an important role in atmospheric phenomena as well as in a number of technological applications.

This paper presents results of the investigation of cluster ion formation from neutral water clusters by electron impact in collision

*Institute of Thermophysics.

between intersecting molecular and electron beams and by surface collisions.

Experiments

Presently, the gasdynamic molecular beam method is the principle experimental method for obtaining molecular clusters and investigating their properties. Experimental techniques used in this work are the same as for molecular gas cluster investigations.[1,2] A schematic representation of the apparatus for the formation of a molecular beam of water clusters is shown in Fig. 1. A water vapor supersonic jet was obtained by heating water in an evaporator followed by superheating the vapor in a nozzle source and a subsequent free expansion of the vapor into a vacuum. Vapor pressure in the source P_0 was measured with a membrane pressure transducer and vapor temperature T_0 by thermocouple. The accuracy of the measurements of P_0 and T_0 was not lower than 2% and 0.3% respectively. The source construction could provide superheated vapor with $P_0 = 2 \cdot 10^3$ to $5 \cdot 10^5$ Pa and $T_0 = 290$ to 480 K. A sonic nozzle with a diameter $d_* = 1$ mm was used.

As follows, from the results of calculation[3] and experiments,[4] the particle velocity in the beam after expansion with condensation ($P_0 \cdot d_* \gtrsim 3 \cdot 10^4$ Pa·mm) is close to the maximum velocity v of the gasdynamic expansion, which is equal to 1.3 km/s under our conditions. Thus the kinetic energy of water clusters was estimated as

$$\epsilon(N) = N \cdot mv^2/2 = 0.17 \cdot N \text{ eV} \tag{1}$$

where N is the size of cluster (number of molecules in cluster).

To reduce background gas effects the molecular beam was interrupted by a mechanical chopper with a frequency of 20 Hz and registration was carried out by a lock-in amplifier.[1] Cryogenic and turbomolecular pumping enable the pressure in the registration section to be kept at $2 \cdot 10^{-4}$ Pa.

Water Cluster Generation in the Freejet Downstream of a Sonic Nozzle

From the viewpoint of the analysis of the developed condensation onset condition, i.e., determining $P_{0,c}$ at $T_0 =$ const. or $T_{0,c}$

at $P_0 = \text{const.}$, when mean cluster size begins to increase sharply, flows from sonic nozzles are characterized by the maximum values of $P_{0,c}$ and minimal $T_{0,c}$. In addition, if condensation proceeds at any nozzle configuration with the same d_*. Transition to developed condensation in the water vapor freejet was identified (as before for gases[5]) as the beginning of a sharp increase of molecular beam intensity $J(P_0, T_0)$ caused by increase of $\bar{N}$ and beam enrichment with clusters. Conditions of transition to developed condensation are shown in Fig. 2, curve 1 in the coordinates log P_0 - log T_0. The figure shows also the line of phase transition for water, curve 2, and conditions in the source at which saturation is already obtained at the nozzle throat, curve 3. Figure 2, curve 1 is described by the relationship $P_{0,c} = A \cdot T_0^{\gamma/(\gamma-1)}$, where A is a constant; γ, an effective specific heat ratio, is equal to 1.32. Taking into account the relationship $P_{0,c} \sim d_*^{-0.6}$, which holds true for all investigated

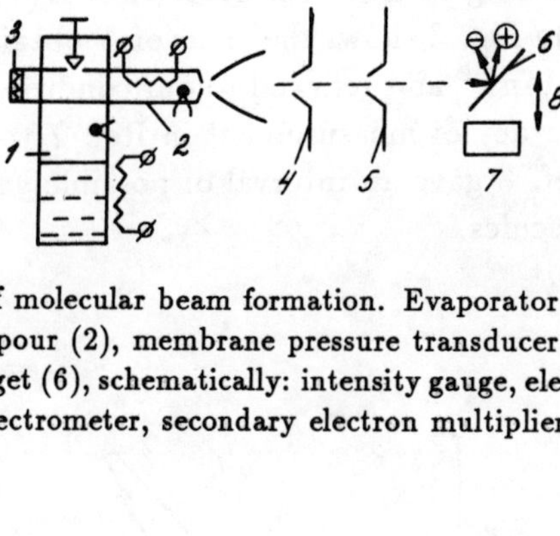

Fig. 1 Scheme of molecular beam formation. Evaporator (1), nozzle source of superheated vapour (2), membrane pressure transducer (3), skimmer (4), collimator (5), target (6), schematically: intensity gauge, electron gun with ion collector, mass spectrometer, secondary electron multiplier (7), and traverse gear (8).

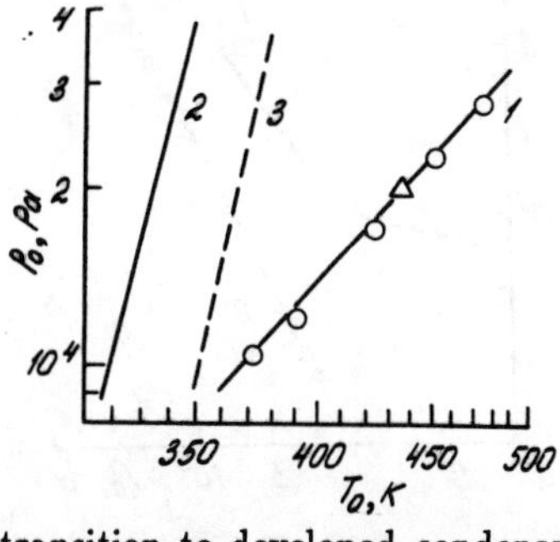

Fig. 2 Condition of transition to developed condensation (1), the curve of phase transition of H_2O (2), conditions of attaining saturation on sonic nozzle throat (3).

molecular gases,[2,5] the following relationship may be recommended to determine the condition of condensation onset in a water vapor freejet

$$P_{0,c} = A \cdot T_0^{4.1} \cdot d_*^{-0.6} \tag{2}$$

with $A = 2.6 \cdot 10^{-7}$ Pa.K$^{-4.1}$.mm$^{0.6}$.

Figure 3, curve 1 shows a typical dependence of beam intensity (mass flow in a beam, since the cluster has completely disintegrated inside intensity gauge and all molecules are detected independently) on source pressure. Mean cluster size N, which is also shown here in points 2 and 3, was determined according to Eq. (1) from measurements of negative and positive cluster ion energy via a retarding potential technique.[6] Water cluster ions were obtained by intersecting a molecular beam with an electron beam of energy $E_e \leq 0.2$ eV (points 2) and 30 eV (points 3)[6]. The changing T_0, $J(P_0)$, and $\bar{N}(P_{0,c})$ shift according to Eq. (2). Note that $\bar{N}(P_0)$ is about ten molecules. Dot 4 in Fig. 3 shows the value of N obtained by electron diffraction method in[7,8] and reduced to our conditions according to Eq. (2). The accuracy of measurement in Ref. 7 was estimated as 70% , whereas Ref. 8 gave an interval of possible values from $N =$ 1500 to 5000 molecules.

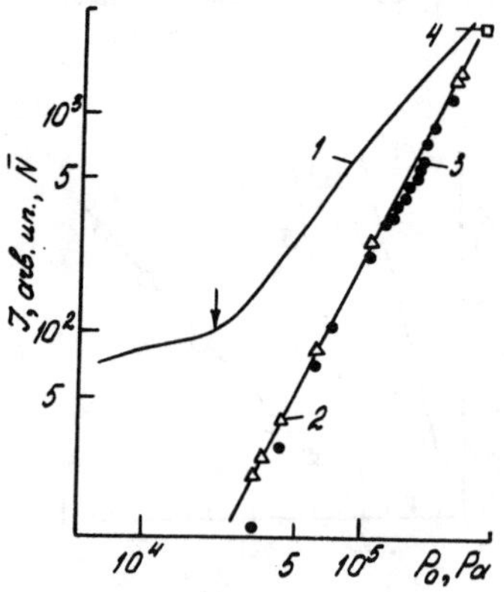

Fig. 3 Intensity of molecular beam J (1) and mean size of cluster ions $\bar{N}$ (2-4) vs pressure P_0.

Charge Formation and Separation at the Scattering of Water Clusters by Solid Surfaces

The surface electrification was initially detected at normal incidence of the clustered beam on a plane target made of stainless steel. The current on target observed in this case depended on the mean cluster size N and on the potential applied between the target and the grid in front of it.[9] To investigate the effect in detail the experimental apparatus was altered in order to measure the scattered ion current (Fig. 4). A collimated beam (1) was scattered by a plane target (2), which could rotate so that the angle of beam incidence γ measured from the normal could change from 0° to 80°. The following materials were used as targets: polished metals (gold, steel, duralumin); single crystals of germanium (face 100 and 110); glass-fibre-base laminate. All the targets were heated to 420-450 K. Charged fragments reflected from the target impinged on a collector (3) with two coaxial screens (4) (a grid with high transparency was used as an inner screen). The collector and both screens were made of stainless steel. Voltage $|U| \leq 300$ V could be applied between the collector and target. The collector signal formed the input to an electrometer (5), followed by lock-in amplifier. A secondary electron multiplier (SEM) was also used as an ion detector. In this case the neutral water cluster beam was directed onto the first dynode made of aluminum ($\gamma = 45°$), the multiplier itself working in the usual way.

First the absence of ions in the incident beam was checked. For this purpose a diaphragm was placed in front of the detector, and a potential relative to the vapor source ± 1 keV was applied to the diaphragm. No alteration in the signal magnitude was observed.

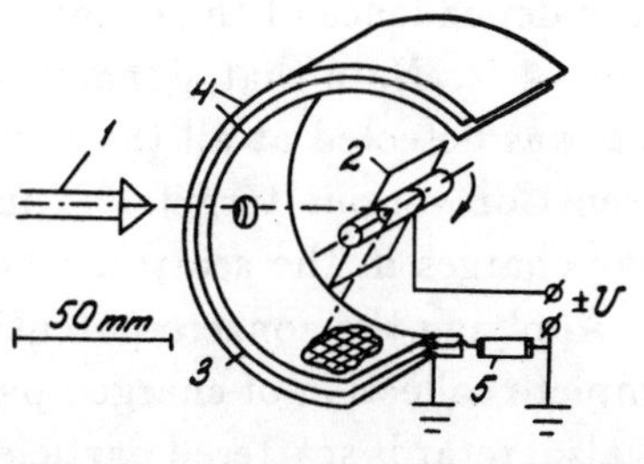

Fig. 4 Sheme of ion current measurement. Clustered molecular beam (1), scattering target (2), ion collector (3), screems (4), electrometer (5).

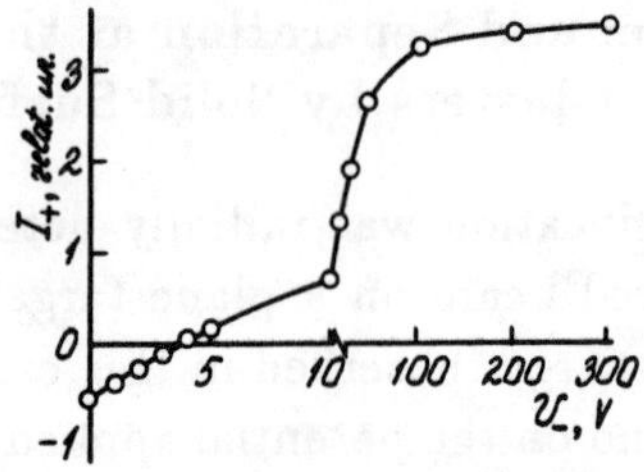

Fig. 5 Current on collector I versus target-collector potential. $\bar{N} = 900$, steel target, $\alpha = 70°$.

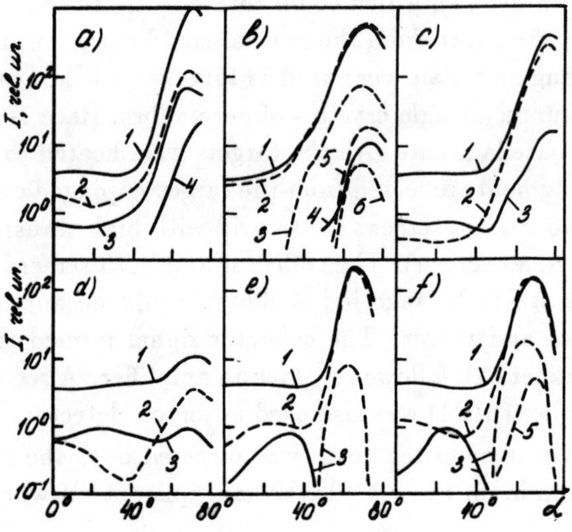

Fig. 6 Current on collector vs angle α. Targets made of dural/umin (a), steel (b), gold (c), glass-fibre-base laminate (d), germanium 110 (e), germanium 100 (f). Curves 1,4 - $U = \pm 300$ V; 2,5 - $U = -300$ V; 3,6 - $U = 0$ V. Curves 1-3 - $\bar{N} = 1200$, 4a -$\bar{N} = 270$, 4b, 5b, 6b, 5c - $\bar{N} = 350$. Continuous lines denote positive currents, dashed lines - negative currents.

Figure 5 shows the dependence of the collector current I on the collector-target potential U. Note that signal $I \neq 0$ is observed at $U = 0$. Such a signal was detected at all the surfaces investigated, its sign being different for various targets. Thus the currents of negative and positive charges in the scattered beam appear to be unequal at $U = 0$ V. Applying the nonzero potential U to the target provides a more complete collection of charged particles of one sign to the collector and, also, retards scattered particles of opposite sign. Thus, the current obtained at saturation corresponds to complete current of charges of one sign. Note that the required electric field ($\sim 10 - 100$ V/cm) is insufficient to desorb an ion from surface, so the discussion centers on scattered charged particles. Figure 5 shows that saturation was obtained at $|U| \sim 100$ V, i.e., kinetic energy of the particles does not exceed 100 eV. Direct measurements of scattered particle energy are discussed later.

Figure 6 shows dependences of I on incidence angle α, recorded for various targets and various cluster size N. Sharp maxima at tangential ($\alpha_{max} \simeq 70°$) incidence of clusters are characteristic for all curves. These seem to be connected with some features in the scattering of neutral clusters, i.e., large probabilities for reflection from the surface for clusters or fragments with a large tangential component of velocity.[6,10] Maxima smoothing on the target made of glass-fibre-base laminate seems to be due to a significant roughness (porosity) of this target surface. It is interesting that the sign of the current I_0 (at $U = 0$ V) on germanium surfaces depends on the angle of incidence of the cluster beam on the target (curves 6e and 6f). Also note that the type of surface material affects both the absolute current magnitude and the relationship between the magnitudes of currents of different sign.

Measurements of currents at various cluster sizes have shown that signal magnitude increases sharply with increase of $\bar{N}$ in the investigated range of average sizes (100 - 2000 molecules/cluster). Figure 7 shows the collector and SEM currents versus N. All the curves in the figure are normalized to the cluster flow in the beam $J_c(J_c \simeq J(P_0)/\bar{N})$, i.e., they represent the probability P to detect a unit charge with the collector on impact of a cluster of size $N \simeq \bar{N}$ with the target. By comparing absolute values of the recorded currents with ion currents formed as a result of cluster ionization by electron impact,[9] the absolute magnitude of P was evaluated. Note that the maximum value P in this experiment reached 10^{-4}.

Measurements of scattered ion energy have been carried out by the retarding potential method. For this purpose one more grid (not shown in Fig. 4) was placed between the inner collector screen and the target, and a potential relative to the target, cutting off the charges of one sign (± 300 V), was applied to this grid. The energy of passing ions was measured by varying the potential between the collector and the target U. The current of passing ions appeared to decrease twice at retarding voltages from 14 V (at $\bar{N} \simeq 1000$)

to 25 V (at $\bar{N} \simeq 2700$). The measurements were carried out at tangential beam incidence. The magnitude of the voltage did not depend much on ion sign or type of target. Measurements of neutral cluster velocity in the scattered flow v_s have proven[6,10] to be 2 - 6 times lower than that of the incident beam v and rises with the increase of $\bar{N}$ and α. For example, $v_s \simeq 0.5 \cdot v$ for water clusters at $\bar{N} \sim 1000$ and $\alpha \simeq 70°$.[10]

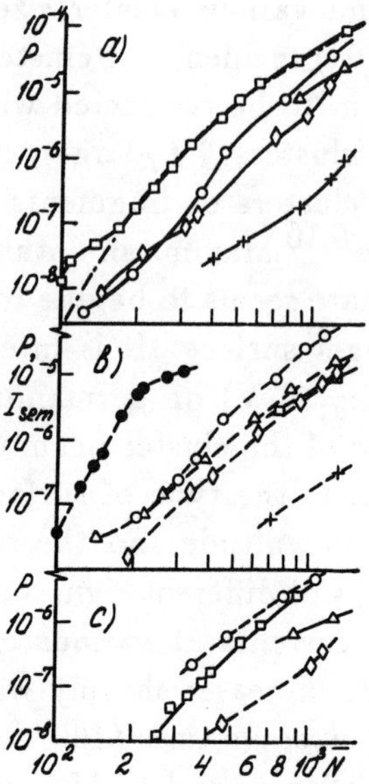

Fig. 7 Collector and SEM currents normalized on cluster flow vs $\bar{N}$. $\alpha = 70°$; $U = \pm 300$ V (a), - 300 V (b), 0 V (c). Target made of duralumin (□), steel (○), gold (△), glass-fibre-bass laminate (+), germanium 100 (◇); SEM signal (•). Continuous lines - positive currents, dashed lines - negative currents, dot-dash line - calculation (5).

Table 1 Characteristic properties for scattering by various materials

	Gold	Steel	Duralumin	Germanium 110	Germanium 100	Glass-Fiber-Base Laminate
α_{max}	77° (75°)	71° (70°)	71° (69°)	66° (65°)	69° (67°)	72° (71°)
$\Delta\alpha$	16°	18° (20°)	15° (17°)	13° (12°)	13° (14°)	28°
$I_+(I_-)$ at $\alpha=\alpha_{max}$	410 (280)	550 (600)	960 (120)	230 (210)	170 (170)	11.6 (3.9)
$I_+(I_-)$ at $\alpha=0°$	3.6 (0.9)	3.2 (2.6)	3.9 (2.2)	4.2 (0.4)	4.4 (0.6)	3.8 (0.4)
$I_0(\alpha_{max})$	+16	-80	+70	-10	-20	+1.2

Thus, according to our data on ion energy, the size of charged particles lies in the range 300-600 molecules, i.e., they are rather large cluster ions. Characteristic properties for scattering by different targets are collected in Table I. Here $N = 1200$, $\Delta\alpha$ is the peak width (fwhm), I_+ and I_- are relative values of currents on total collection of ions ($|U| = 300$ V). It is seen from Table I, that the difference of the currents $I_+ - I_-$ and the current I_0 without the electric field extracting ions (the lowest line) are different. This means that cluster ions are partially lost after reflection, positive ones being lost to a larger extent. Obviously, the rate of slow cluster ion neutralization may depend significantly on the presence of an absorbed layer. Slow cluster ions may serve as an effective tool for diagnosing conditions on the surface.

Thus it has been experimentally found that following the collision of neutral water clusters with various surfaces, scattering of cluster ions of both signs takes place and a current to the target appears. In similar experiments with clusters $(CO_2)_n$ and $(N_2O)_n$ the cluster ionization was not observed. Appearance of cluster ions in the scattered beam due to water cluster scattering seems to be due to formation of ion pairs in the cluster and their separation during a collision with a surface. We believe that formation of ion pairs proceeds due to collision kinetic energy ϵ in the reaction of ion dissociation (autoprotolysis) of excited H_2O molecules (as a reaction in condensed water[11]).

$$\begin{aligned}(H_2O)_N + \text{ surface} \rightarrow\ & (H_2O)^*_k(H_2O)_{N-k-s} \\ \rightarrow\ & H_3O^+(H_2O)_{i-1} \cdot OH^-(H_2O)_{N-i-s-1}\end{aligned} \tag{3}$$

Here s is the number of molecules evaporated in cluster collision with the surface. Charge separation can occur in two processes. The first one (and it seems to be the principle one) is asymmetric neutralization of ions in the cluster by the surface during the time of interaction with the surface:

$$H_3O^+(H_2O)_i \cdot OH^-(H_2O)_j + \text{ surface} \begin{array}{l} \nearrow H_3O^+(H_2O)_{i+j}OH \\ \searrow OH^-(H_2O)_{i+j+1}H \end{array} \tag{4}$$

This is supported by the large number of scattered cluster ions at tangential incidence. The second pathway may be connected with destruction of a cluster, fragments of which contain ions with different signs (at normal incidence of cluster), and partial neutralization of these ions:

$$H_3O^+(H_2O)_i \cdot OH^-(H_2O)_j \rightarrow H_3O^+(H_2O)_i + OH^-(H_2O)_j \quad (5)$$

In the first case (asymmetric neutralization without strong destruction) for the total scattered cluster ion current, I, we have $I = I_+ - I_-$, where

$$I_+ = J_c \, P_i \, \gamma_- \, (1 - \gamma_+) \quad (6)$$

$$I_- = J_c \, P_i \, \gamma_+ \, (1 - \gamma_-) \quad (7)$$

Here P_i is the probability of formation of an ion pair; γ_+ and γ_- are probabilities of neutralization of positive and negative ions during the time of the cluster - surface interaction. Experimentally measurable magnitudes $I_{+,-}(\alpha)$ (Fig. 6) and $P(\bar{N})$ (Fig. 7) are the result of the two processes - formation of ion pairs in cluster and separation of the charges of pairs. In the general case ion pair formation is possible both during collision of a cluster with a surface (nonequilibrium mechanism) and during the moving of a cluster in jet and beam ($\sim 10^{-3}$s) (establishing equilibrium concentration). The equilibrium concentration of ion pairs in condensed water is

$$q(T) = A \cdot \exp(-E_a/2kT) \quad (8)$$

where A is the constant, E_a energy of autoprotolysis reaction activation, and k the Boltzmann constant. For instance, in liquid water $A = 1.4 \cdot 10^{-4}$, $E_a = 0.58$ eV.[11] According to[8] clusters of the size $N = 300$ - 1000 in a beam, have the temperature $T \simeq 180$ K, which results in $q = 10^{-12}$ for liquid water and in $q = 10^{-16}$ for ice. Apparently, E_a is larger in clusters than in the condensed phase, so that $q \lesssim 10^{-12}$, whereas in the given experiments the limiting value is $q = P(\bar{N})/\bar{N} = 10^{-7}$ (Fig. 7). Consequently ion pair formation before collision of the cluster with the surface may be neglected. To evaluate the threshold size N_* from which the ionization proceeds, it should be noted that the cluster size distribution function in an incident beam $f(N)$ has a significant width.[3,12] Then the observed magnitude $P(\bar{N}) = \int_2^\infty P_*(N) \cdot f(N) \cdot dN$, where $P_*(N)$ is the true probability of ion detection on the collector at scattering of a cluster with the size N. Taking for $f(N)$ the dependence $f(N) = 4N/\bar{N}^{-2} \exp(-2N/\bar{N})$,[7] and for $P_*(N)$ - the form $P_*(N) = P^\circ(N/N_* - 1)^\beta$, we obtain

$$P(\bar{N}) = P^\circ \Gamma(\beta+1) \left(\frac{\bar{N}}{2 \cdot N_*}\right)^{\beta-1} \cdot \left[(\beta+1)\frac{\bar{N}}{2 \cdot N_*} + 1\right] \exp\left(-\frac{2N_*}{\bar{N}}\right) \tag{9}$$

Comparing the dependence of $P(\bar{N})$ calculated by Eq. (9) with experiment gave $\beta = 2 - 3$, $N_* = 300 \pm 70$. To illustrate this the dot-dash line in Fig. 7 shows the calculated form of $P(\bar{N})$ for $N_* = 350$, $\beta = 2$, approximating the experimental curve $P(\bar{N})$ for the duralumin surface.

Now consider the ion separation resulting from asymmetric neutralization of ions in a cluster. For this purpose let us introduce characteristic times τ_+ and τ_- of ion neutralization on the surface and write the equation for the probability $P^c_{+,-}$ of charge conservation in the cluster

$$dP^c_{+,-}/dt = -P^c_{+,-}/\tau_{+,-}\ ; \quad P^c_{+,-}(t=0) = 1 \tag{10}$$

if ions are formed in the first moment of collision, and

$$dP^c_{+,-}/dt = t_c^{-1} - P^c_{+,-}/\tau_{+,-}\ ; \quad P^c_{+,-}(t=0) = 0 \tag{11}$$

if ions are formed with constant probability during the whole time of collision, t_c, (the time of interaction of the cluster with the surface). The collision time may vary from "infinity" (cluster capture) to 10^{-13}s (period of intermolecular vibrations) and evidently $t_c/\tau_{+,-} \gg 1$. Thus in the first case total neutralization could take place. In the second case $P^c_{+,-}(t_c) = 1 - \exp(-t_c/\tau_{+,-})\tau_{+,-}/t_c$; $\gamma_{+,-} = 1 - P^c_{+,-}(t_c)$; and at $t_c/\tau_{+,-} \gg 1$ $\gamma_{+,-}(1-\gamma_{-,+}) \simeq \tau_{-,+}/t_c$. The second pathway is more probable, with regard to Eqs. (6) and (7) and the currents $I_{+,-}$ are equal to

$$I_{+,-} = J_c \cdot P_i \cdot \tau_{+,-}/t_c \tag{12}$$

It follows from Eq. (12) that the difference between the currents I_+ and I_- in Figs. 6 and 7 is caused by different time of neutralization of ions with different signs, i.e., $\tau_+ \neq \tau_-$. The magnitude of the ratio τ_+/τ_- depends on the target material and the angle of cluster incidence on the target. Furthermore, Fig. 7 shows that $I_+(\bar{N})/I_-(\bar{N}) \simeq$ const. Consequently the times τ_+ and τ_- either depend equally on N or do not depend on it at all. On the whole it follows from Eq. (12) that the behavior of $I_{+,-}(\bar{N}, \alpha)$ is determined by $P_i(N, \alpha)$, $t_c(N, \alpha)$ and by the ratio $t_c/\tau_{+,-}$.

The increase of $P(\bar{N})$ with the increase of $\bar{N}$ observed in Fig. 7 may be caused both by the increase of P_i due to the energy increase at collision of a cluster with a surface and by the increase of the probability of noncompensated cluster ion emission due to the decrease of the time of cluster-surface interaction. The conclusion that t_c may decrease with an increase of N can be drawn from observing the increase of the scattered cluster velocity v_s with the increase of $\bar{N}$.[6]

Ion Ejection from Clusters at Ionization by Electron Impact

Interest in the problem of cluster ionization by electron impact is to a large extent caused by a wide use of the mass-spectrometric diagnostic method for clusters. The ionization of clusters has appeared to have a much more complicated character than that of separate molecules. In the given paper, small cluster ions were measured by a monopolar mass-spectrometer placed along the molecular beam. Ion currents of separate components were measured by SEM. Corrections for the dependence of the electron emission coefficient on various types and sizes of the cluster ions[13] were not made.

Figure 8 shows the results of measuring the total ion current I_Σ^+ and the currents of separate ion components originating after ionization of H_2O molecules and clusters in the ion source of the mass-spectrometer. All the curves are brought into coincidence in the region of P_0 before the transition to developed condensation in the jet. The conditions of the experiment are the following: $T_0 = 454$ K, $E_e = 90$ eV, for the curve $2' - 20$ eV. Figure 8 shows that in the region of expansion with condensation the density of ion components exceeds the density of the neutral molecular component of

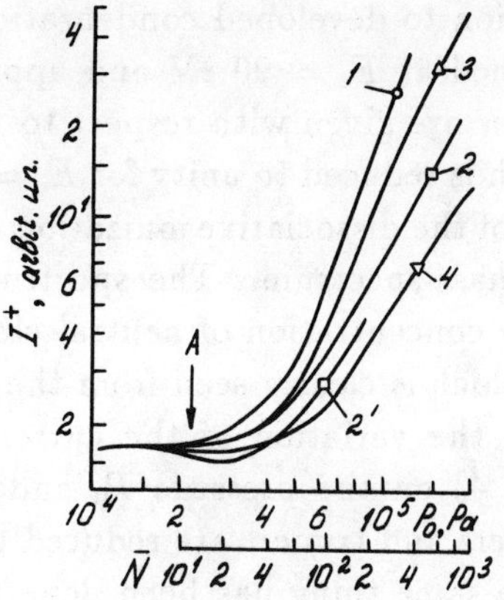

Fig. 8 Total current of ions I_Σ^+ (1), and currents of H_2O^+(2,2′), H^+ (3), OH^+ (4) vs pressure P_0 and cluster size $\bar{N}$.

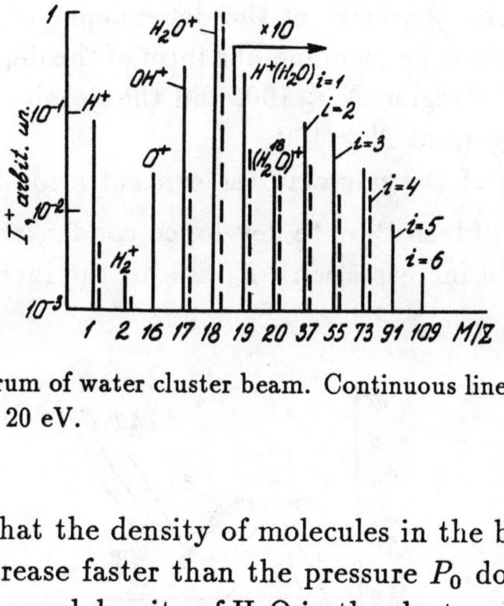

Fig. 9 Mass-spectrum of water cluster beam. Continuous lines - $E_e = 90$ eV, dashed lines - $E_e = 20$ eV.

the beam. Note that the density of molecules in the beam at $T_0 =$ const. cannot increase faster than the pressure P_0 does.[3] A faster increase of the observed density of H_2O in the clustered beam seems to be due to molecular ion ejection from clusters. The presence of ion fragments in this region implies a high effectiveness of dissociation of molecules ionized in a bound state in the cluster. The high probability of H^+ ion ejection was unexpected, since the molecule of water possesses of high energy of proton affinity (~ 7 eV). Thus, for example, mostly protonated clusters $H^+(H_2O)_i$ are detected after ionization of water clusters by electron impact.[14] Figure 9 shows the mass-spectrum of components of a clustered beam observed in the region of transition to developed condensation ($\bar{N} = 20$). The spectrum was obtained at $E_e = 90$ eV and approximately 20 eV; intensities of the lines are given with respect to the lines of H_2O^+, the intensity of which is reduced to unity for $E_e = 90$ eV and 20 eV. All the components of the dissociative ionization of water molecules are present in the mass-spectrum. The spectrum of microcluster ions reflects the true concentration of neutral clusters in the beam only qualitatively, which is clearly seen from the next figure.

Figure 10 shows the variation of the currents of cluster ions $H^+(H_2O)_i$ ($i = 1 - 4$) on the pressure P_0 and the size $\bar{N}$. The curves of I^+ for dimers and trimers are reduced to one value in the maximum of I^+, the same thing has been done for $H^+(H_2O)_3$ and $H^+(H_2O)_4$. The dependence of the currents of dimers (curves 1, 1′) and trimers (curves 2, 2′) on P_0 were measured for two electron energies: 90 eV (curves 1, 2) and 20 eV (curves 1′, 2′). The increase of energy E_e is seen to result in the shift of the curves of I^+ toward large $\bar{N}(P_0)$. The character of the dependence of $I^+(\bar{N})$ has not changed in this case, i.e., a dome-like form of the dependence $I^+(\bar{N})$ is observed in the region $\bar{N} \leq 100$, and the density of microcluster ions increases again at $\bar{N} \geq 100$.

Calculations of the microcluster concentration for variation of P_0 in the region of transition to developed condensation[3] show that the concentration increases with P_0 due to the increase of the su-

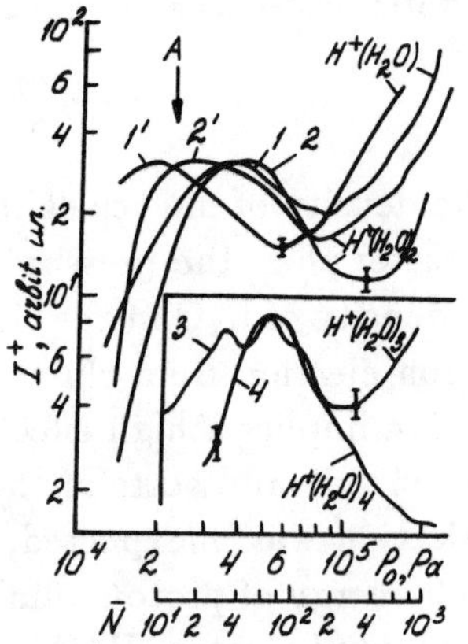

Fig. 10 Current of microcluster ions $H^+(H_2O)_i$ vs P_0 and cluster size $\bar{N}$. Curves (1-4) - $E_e = 90$ eV; (1′, 2′) - 20 eV.

persaturation in the jet; it decreases with the growth of clusters and the associated sharp decrease of supersaturation.

The shift of the dome-like part of the curves $I^+(\bar{N})$ in Fig. 10 toward larger values of $\bar{N}$ for variation of the energy E_e from 20 to 90 eV seems to be related with the process of molecule evaporation from ionized clusters. In this case the measured density of microcluster ions is determined by the expression:

$$\begin{aligned} I_i^+ &\sim \sigma_i(i, E_e) \cdot I_i \cdot (1 - \beta_i(i, E_e, \tau)) \\ &+ \sum_{N>i} \alpha_i(N, E_e, \tau) \cdot \sigma_N(N, E_e) \cdot I_N \end{aligned} \tag{13}$$

Here I_i is the true concentration of clusters of the ith size in the beam; I_N the concentration of clusters of the size N; β_i the probability of evaporation of a molecule from a cluster of the size i; α_i the probability of molecule evaporation from cluster of the size $N > i$ transforming them into the size i; τ the time to observation, i.e., the time from the moment of electron impact resulting in the formation of an ion, to the moment the ion is observed; and δ_N the ionization cross section of the cluster of size N. The influence of molecule evaporation from clusters after ionization at the expense of electron energy transferred to molecules in the cluster at collision has been estimated in Ref. 12. It has been found that the evaporation process can essentially distort measured concentrations of small clusters.

Since the evaporation probabilities α and β seem to be increasing functions of the energy E_e, then electron energy variation can reveal the presence of a contribution to the ion current of the ith component from larger clusters. (In this case, the correction for the value of the ionization cross section $\sigma^+(N, E_e)$ should, of course, be taken into account). As for clusters of small sizes[12] $\sigma^+(N, E_e) = N \cdot \sigma^+(1, E_e)$, it is sufficient to normalize ion currents to the current of a molecular component, as it is done in Fig. 9.

The decrease of the normalized signal of the ith component due to the decrease of E_e (as for dimers, trimers or 4-mers ($i = 3$) in Fig. 9) shows that the prevailing contribution to the total signal I_i^+ belongs to the last sum in the right side of Eq. (13), i.e., to large size clusters partialy evaporated at ionization. Consequently, the lines observed in the cluster ion spectrum poorly reflect the true concentration of the ith component in the beam at the given degree of condensation. The suggested controlling method may be of use in investigating the processes with participation of microclusters, when the necessary information about neutral cluster concentration is obtained exclusively from mass-spectrometric measurements.[15]

The increase of ion currents of microclusters at $\bar{N} \gtrsim 100$ (see Fig. 10) is not related to the process of cluster generation in the jet. The measured currents seem to have been completely initiated by microcluster ejection. Note that evaporation of microclusters $H^+(H_2O)_i$, $i = 3-8$, was also detected at irradiation of the surface of ice cooled to 193 K in vacuum by electrons of energy 80 eV.[16]

A problem arises concerning the mechanism of ion ejection from clusters. Kinetic energy of dissociative ionization products is known to increase sharply with the increase of the ionizing electron energy and, for example, already at E_e -- 30 eV this energy is equal to several eV for a significant part of the charged fragments of H_2O molecules.[17] However, the measurements of the present paper have shown that the probability of ejection does not decrease with the decrease of E_e. Ejection of molecular ion H_2O^+ (see Fig. 8) cannot be explained by this either. An effect of negative hydration of ions in water is known.[18,19] However, the appearance of molecular ions H_2O^+ and their fragments in our case seems to be the consequence of ejection of these molecules after their excitation by electron impact over the ionization threshold. A molecule in the highly excited state does not already "fall" within the cluster structure due to the "instantaneous" enlargement of its sizes. In this case, the "electron-exchange repulsion"[20] appearing between the excited molecule and molecules in the ground state may be the force causing ejection of a molecule from the cluster. This mechanism is corroborated by the results of the paper,[21] where emission of molecules H_2O and NH_3 from pure ice and its mixture with ammonia ($T = 90 - 130$ K) was detected for electron excitement of these molecules by laser emission at energies lower than the ionization threshold, and also by the fact that a significant part of molecular and fragment ions are formed due to ionization of the overexcited state.[22,23] In our case, molecular ions and their fragments may appear during the excitation reaction:

$$(AB)_N + e \rightarrow AB^{**}(AB)_{N-1} + 2e \begin{array}{l} \nearrow (AB)_{N-1} + AB^+ \\ \searrow (AB)_{N-1} + AB^{**} \rightarrow A^+ + B \end{array} \tag{14}$$

Ejection of microclusters seems to be related to the peculiarities of structural relaxation of cluster after formation of ions in it. Most likely the first solvate shell is formed quickly ($10^{-10} - 10^{-12}$s) due to strong interaction of an ion with its closest neighbors (3-5 molecules[24]), whereas the time of total relaxation related, for example, to reorientation of water molecules exceeds 10^{-1}s,[11] at a water temperature of $T \lesssim 200$ K ,[8] which is characteristic for water clusters. This may cause weakening of the bonds on the boundary of the first solvate shell and ion ejection together with this shell.

References

[1]Vostrikov, A. A., Kusner, Yu. S., Rebrov, A. K., and Semyachkin, B. E., "Molecular Beam Generator," *Experimental Methods in Rarefied Gas Dynamics*, Novosibirsk, USSR, 1974, pp. 29-38 (in Russian).

[2]Vostrikov, A. A., "Investigation of Formation and Properties of Clusters N_2O," *Soviet Physics - Technical Physics*, Vol. 29, No. 2, 1984.

[3]Vostrikov, A. A. and Dubov, D. Yu., "Real Cluster Properties and Condensation Model," ITF Siberian Branch Academy of Science, USSR, Novobirsk, Preprint no. 112-84, 1984, (in Russian).

[4]Dreyfuss, D. and Wachman, H. Y., "Measurement of Relative Concentration and Velocities of Small Clusters ($n \leq 40$) in Expanding Water Flows," *Journal of Chemical Physics*, Vol. 76, No. 4, 1982, pp. 2031-2042.

[5]Vostrikov, A. A., Revrov, A. K., and Semyachkin, B. E., "Condensation of SF_6, CF_2Cl_2, CO_2 in Expanding Jets," *Soviet Physics - Technical Physics*, Vol. 25, No. 11, 1980.

[6]Vostrikov, A. A., Mironov, S. G., and Semyachkin, B. E., "Scattering of Clusters by a Surface," *Soviet Physics - Technical Physics*, Vol. 27, No. 6, 1982, pp. 705-708.

[7]Stein, G. D. and Armstrong, J. A., "Structure of Water and Carbon Dioxide Clusters Formed Via Homogeneous Nucleation on Nozzle Beams," *Journal of Chemical Physics*, Vol. 58, No. 5, 1973, pp. 1999-2003.

[8]Torchet, G., Schwartz, P., and Farges, J., "Structure of Solid Water Clusters Formed in a Free Jet Expansion," *Journal of Chemical Physics*, Vol. 79, 1983, pp. 6196-6202.

[9]Vostrikov, A. A., Dubov, D. Yu., and Predtechensky, M. R., "Water Clusters: Electron Attachment, Ionization, Electrification at Destruction," *Soviet Physics - Technical Physics*, Vol. 32, No. 4, 1987.

[10]Dreyfuss, D. and Wachman, H. Y., "Scattering of Water Cluster Beams from Surfaces," *Progress in Astronautics and Aeronautics*, Vol. 74, Pt. 1, AIAA, New York, 1981, pp. 183-197.

[11]Eisenberg, D. and Kauzmann, W., *The Structure and Properties of Water*, Oxford Univ. Press, New York, 1969.

[12]Vostrikov, A. A., Dubov, D. Yu., and Predtechensky M. R., "Electron-Cluster Interaction," ITF Siberian Branch Academy Science, Novosibirsk, USSR, Preprint No. 150-86, 1986 (in Russian).

[13]Vostrikov, A. A. and Predtechensky, M. R., "Recording of Cluster Ions by a Secondary Electron Multiplier," *Soviet Physics - Technical Physics* USA, Vol. 31, No. 4, 1986, pp. 450-452.

[14]Hummel, A. C., Haring, R. A., Haring, A., and De Vries, A. E., "Mass Spectra of Nozzle-produced Small Molecular Clusters of H_2O, NH_3, CO and CH_4," *International Journal of Mass Spectrometry Ion Processes*, Vol. 61, No. 1, 1984, pp. 97-112.

[15]Märk, T. D., "Cluster Ions: Production, Detection and Stability," *International Journal of Mass Spectrometry Ion Processes*, Vol. 79, No. 1, 1987, pp. 1-59.

[16]Prince, R. H. and Floyd, G. R., "Production of Ionized Clusters by Electron Bombardment of Condensed Polar Solvents," *Chemical Physics Letters*, Vol. 43, No. 2, 1976, pp. 326-331.

[17]Ehrhardt, H. and Kresling, A., "Die dissoziative Ionization von N_2O_2, H_2O, CO_2 and Äthan," *Z aturforsch*, Vol. 22A, No. 12, 1967, pp. 2036-2043.

[18]Karyakin, A. B., Petrov, A. B., Germit, Yu. B., and Zubilin, M. E., "Study of Ion Hydration in Water Solutions with Respect to Absorption Spectra in Infrared Range," *Teor. i exper. Khim.*, Vol. 2, No. 6, 1966, pp. 918-927 (in Russian).

[19]Goncharov, V. V., Romanova, I. I., Samoylov, O. Ya., and Yashkichev, B. I., "Quantitative Characteristics of the Closest Hydration of Some Ions in Diluted Water Solutions Self-diffusion Activation Energy," *Zh. Str. Khim.*, Vol. 8, No. 5, 1967, pp. 613-617 (in Russian).

[20]Watson, W. D. and Salpeter, E. E., "Molecule Formation on Interstellar Grains," *The Astrophysical Journal*, Vol. 174, No. 2, Pt. 1, 1972, pp. 321-340.

[21]Nishi, N., Shinohara, H., and Okuyama, T., "Mass-spectrometric Investigation of Photosplitting-out, Photodissociation, Photochemistry of Surface Molecules of Ices Containing Ammonia and Pure Water Ices," *II Symposium "Kinetic mass-spectrometry, book of Abstracts*, Moscow, 1984, pp. 10-11.

[22]Makarov, V. I. and Polak, L. S., "Excited and Overexcited Electron States of Atoms and Molecules and Their Role in Radiation Chemistry," *Khim. Vys. Energiy*, Vol. 4, No. 1, 1970, pp. 3-23 (in Russian).

[23]Shutten, J., de Heer, F. J., Moustafa, H. K., Boerboom, A. J. H., and Kistemaker, J., "Gross- and Partial- Ionization Cross Section for Electron in Water Vapor in the Energy Range 0.1-20 KeV," *Journal of Chemical Physics*, Vol. 44, No. 10, 1966, pp. 324-326.

[24]Good, A., Durden, D. A., and Kebarle, P., "Ion-molecular Reaction in Pure Nitrogen Containing Traces of Water at Total Pressure 0.5-4 torr. Kinetics of Clustering Reactions Forming $H^+(H_2O)_n$," *Journal of Chemical Physics*, Vol. 52, No. 1, 1970, pp. 212-221.

Homogeneous Condensation in H_2O - Vapor Freejets

C. Dankert* and H. Legge*
DFVLR Institute for Experimental Fluid Mechanics, Göttingen, Federal Republic of Germany

Abstract

Experimental studies of large free jets with stagnation conditions near the vapour pressure line of pure water vapour were conducted in a low density wind tunnel. The condensation onset points on the centerline of the supersonic free jets are detected by laser light scattering, the flow velocities are measured by the electron beam technique, and pitot pressure surveys are taken on axis.

Nomenclature

A	= stream tube area
c_p	= specific heat
d^*	= sonic nozzle diameter
H	= enthalpy
H_{cond}	= condensation enthalpy
m	= mass
m_{cond}	= condensed mass
M	= Mach number
p	= pressure

*Member of Professional Staff.

p_0 = stagnation pressure

p_K = chamber pressure

p_{stat} = static pressure

p_{t2} = pitot pressure

R = gas constant

T = temperature

T_0 = stagnation temperature

T_{stat} = static temperature

U = velocity

U_{lim} = maximum isentropic velocity, see Eq. (8)

x, y = coordinates

x/d^* = downstream position in free jets

$(x/d^*)_W$ = Wilson point, condensation onset point

κ = ratio of specific heats

λ = wavelength

ρ = density

Introduction

Water vapour expansions in supersonic jets were of special interest to us when we studied the expansion process of satellite thrusters blowing into a vacuum. These engines normally use hydrazine or a two-component fuel to produce thrust. One of the exhaust components is water vapour. This report deals with light scattering of laser light [1 – 3] in condensing flow and velocity measurements by means of an electron beam [4] to obtain information concerning a possible velocity gain due to the condensation heat. Such a condensation process might act like an after-burner on the satellite thruster; more velocity gives more thrust, and this results in higher efficiency of the engine.

We used free jet expansions of pure water vapour from sonic orifices to obtain results free from disturbing influences of the boundary layer and shocks in nozzles and to compare the present to previous free jet results of other test gases. [2, 3, 7, 9]

Three experimental tools were used to study the water vapour expansion during condensation:
1) a pitot tube,
2) an electron beam, and
3) laser light scattering.

Theory

The flow-field of an isentropic free jet can be calculated on the axis by the stream tube theory [5] and the free jet theory of Ashkenas and Sherman. [6] In free jet expansions the cooling rates near the orifices are very high, reaching up to 200 K per x/d^*. Figure 1 shows the temperature gradients in the sub- and supersonic parts of free jets, [7] normalized with the orifice diameter and the stagnation temperature, parameter is the ratio of specific heats. The ratio of specific heats of water vapour is dependent on temperature and pressure as shown in Fig. 2.

Flows with heat addition and mass reduction due to condensation were studied by Zierep. [5] With the assumption that the size of the free jet is not influenced by condensation the three basic equations can be set up for condensing flows:

Continuity:

$$\frac{1}{\rho}\frac{\mathrm{d}\rho}{\mathrm{d}x} + \frac{1}{U}\frac{\mathrm{d}U}{\mathrm{d}x} = -\frac{1}{\dot{m}}\frac{\mathrm{d}\dot{m}_{\mathrm{cond}}}{\mathrm{d}x} \tag{1}$$

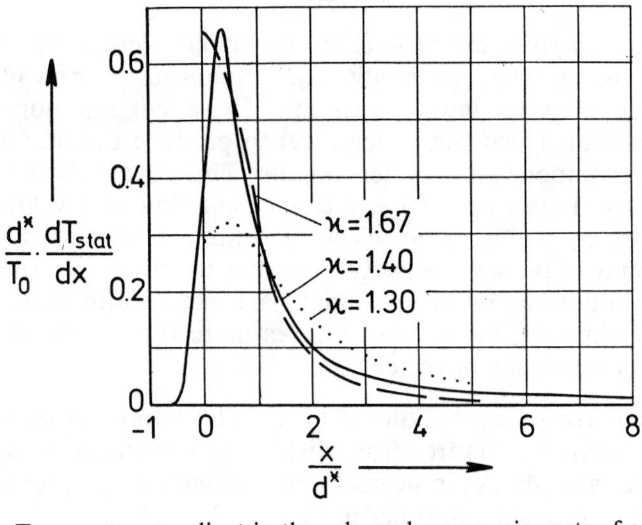

Fig. 1 Temperature gradient in the sub- and supersonic parts of a free jet.

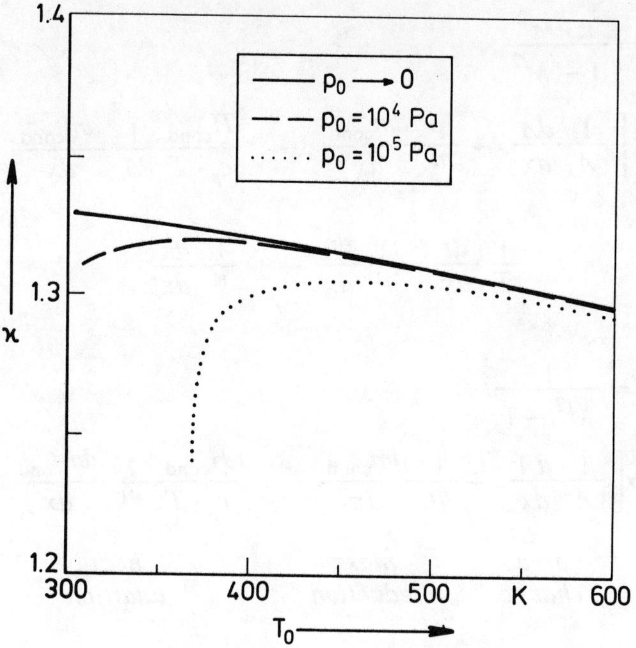

Fig. 2 Ratio of specific heats as function of temperature and pressure of water vapour.

momentum:

$$\frac{1}{\kappa \cdot M^2} \cdot \frac{1}{p} \frac{dp}{dx} + \frac{1}{U} \frac{dU}{dx} = 0 \tag{2}$$

energy:

$$c_p \frac{dT}{dx} + U \frac{dU}{dx} = \frac{H_{cond}}{m} \frac{dm_{cond}}{dx} \tag{3}$$

This system [Eqs.(1-3)] of coupled differential equations can be solved [5] if the stream tube area $A(x)$ is given.

The solutions for density, pressure, temperature, and velocity are

$$\frac{1}{\rho} \frac{d\rho}{dx} = \frac{M^2}{1 - M^2} \times \left[\frac{1}{A} \frac{dA}{dx} + \frac{1}{m} \frac{dm_{cond}}{dx} - \frac{1}{M_2} \frac{H_{cond}}{c_p \cdot T} \frac{1}{m} \frac{dm_{cond}}{dx} \right] \tag{4}$$

$$\frac{1}{p}\frac{dp}{dx}=\frac{\kappa\cdot M^2}{1-M^2}\times\left[\frac{1}{A}\frac{dA}{dx}+\frac{1}{m}\frac{dm_{\text{cond}}}{dx}-\frac{H_{\text{cond}}}{c_p\cdot T}\frac{1}{m}\frac{dm_{\text{cond}}}{dx}\right] \qquad (5)$$

$$\frac{1}{T}\frac{dt}{dx}=\frac{1}{p}\frac{dp}{dx}-\frac{1}{\kappa}\frac{d\kappa}{dx} \qquad (6)$$

$$\frac{1}{U}\frac{dU}{dx}=\frac{1}{M^2-1}\times\left[\underbrace{\frac{1}{A}\frac{dA}{dx}}_{\text{area change}}+\underbrace{\frac{1}{m}\frac{dm_{\text{cond}}}{dx}}_{\text{mass reduction}}-\underbrace{\frac{H_{\text{cond}}}{c_p\cdot T}\frac{1}{m}\frac{dm_{\text{cond}}}{dx}}_{\text{heat addition}}\right] \qquad (7)$$

In hypersonic flows ($M^2 \gg 1$) with small condensed mass fraction, the heat release, the condensed mass, static pressure jump, and the static temperature can be calculated. [8]

Experiments

The measurements are conducted in the low-density wind tunnel of the Deutsche Forschungs- und Versuchsanstalt für Luft- und Raumfahrt, Göttingen. A schematic view of the experimental setup is given in Fig. 3. The nozzle throat diameters are 2, 4.6, and 5 mm. Applying an electric stream generator varies the stagnation temperature between 300 and 800 K, and the stagnation pressure between 0.3 and 1.1 x 10^5 Pa. The pressure ratio of stagnation to background pressure ist about $p_0/p_K = 3000$ using roots pumps, and about 50000 using cryopumps at 80° K. Three different measurement techniques are used to study the H_2O - vapour free jet expansion (see Fig. 3).

Pitot Tube

In the supersonic flowfield a Pitot pressure probe can be moved along the jet axis and perpendicular to it.

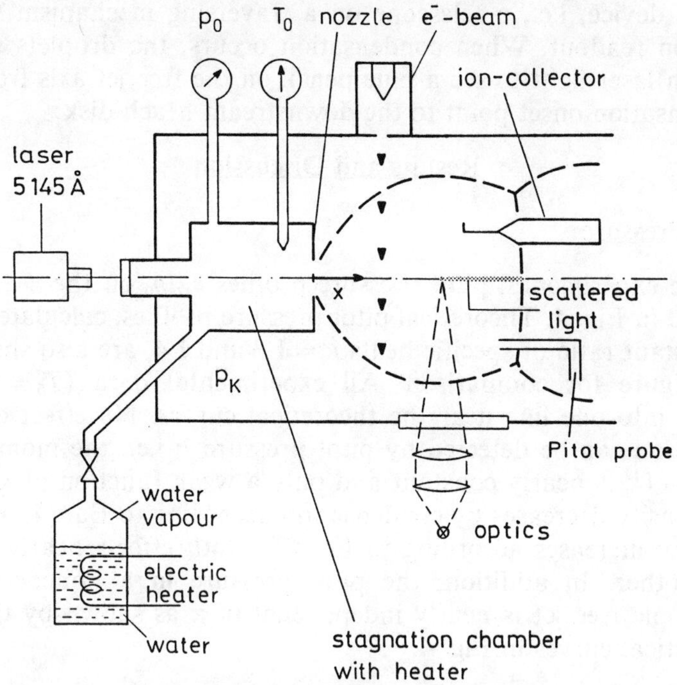

Fig. 3 Experimental setup for measuring pitot pressure, flow velocity with electron beam and ioncollector, and condensation onset by scattered laser light.

Electron Beam

Flow velocities are measured by means of an ion-time-of-flight technique [4, 8]; the pulsed electron beam ($20\,kV, 0.2\,mA$) passes through the free jet as shown in Fig. 3, and the ion collector is installed further downstream on the jet axis and can be positioned at different locations on the flow axis. In each of these positions the flight time is measured, from which the velocity near the collector is calculated.

Laser Light Scattering

The light source is an argon ion laser (4 W type) with a wavelength of $\lambda = 5145\mathring{A}$ and an adjusted output of 1 W. The laser is located outside the wind tunnel and aimed through a heated window and through the stagnation chamber, and the nozzle throat, at the axis of the free jet. The scattered light is observed under 90° through a second window at room temperature by a simple observation device, i.e., a telescope on a traversing mechanism with a position readout. When condensation occurs, the droplets scatter enough laser light to see a blue pencil on the free jet axis from the condensation onset point to the downstream Mach disk.

Results and Discussion

Pitot Pressure

The experimental pitot pressure profiles p_{t2}/p_0 on the x axis are plotted in Fig. 4. Theoretical pitot pressure profiles, calculated with a constant ratio of specific heats $\kappa = 1.3$ and 1.4, are also shown in this figure for comparison. All experimental data ($T_0 = 693\ K$) merge into one line near the theoretical curves. No effect of condensation can be detected by pitot pressure [9]; i.e., the momentum flux $\rho \cdot U^2$ is nearly constant and only a weak function of κ, since the density decreases by condensation according to Eq. (4) and the velocity increases according to Eq. (7). Both effects nearly cancel each other. In addition, the pitot pressure in a noncondensing isentropic free jet is nearly independent of κ as shown by the two theoretical curves in Fig. 4.

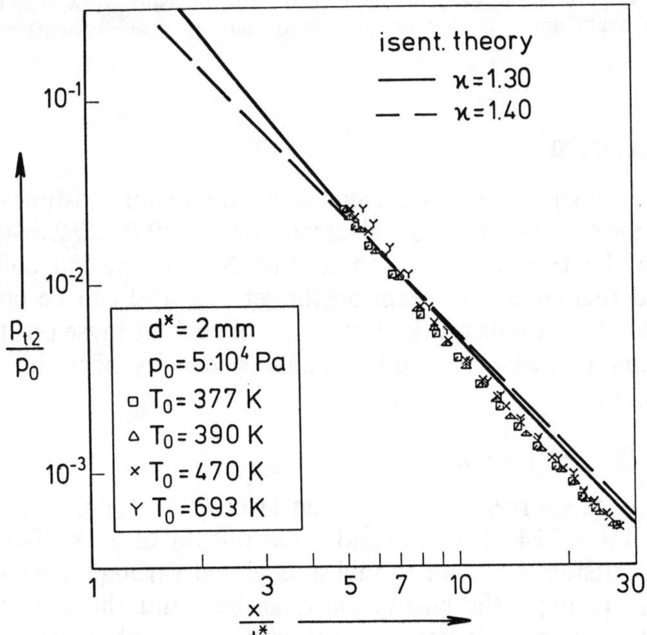

Fig. 4 Pitot pressures p_{t2}/p_0 along the axis of water vapour free jets.

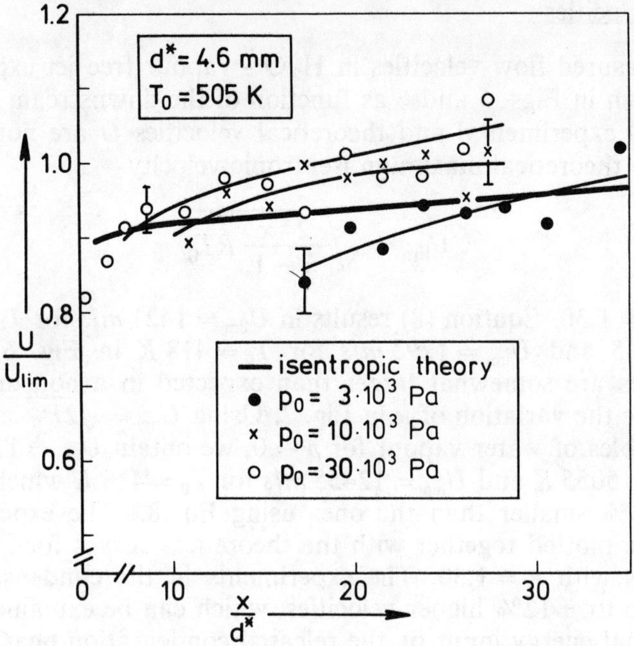

Fig. 5 Flow velocities along the axis of condensing water vapour free jets, $T_0 = 505° K$.

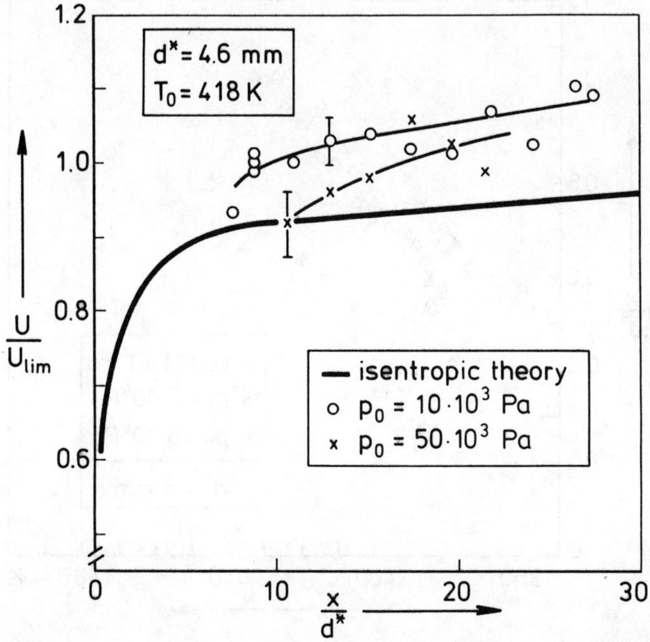

Fig. 6 Flow velocities along the axis of condensing water vapour free jets, $T_0 = 418° K$.

Flow Velocities

The measured flow velocities in H_2O - vapour free jet expansions are shown in Figs. 5 and 6 as function of the downstream position x/d^*. All experimental and theoretical velocities U are normalized with the theoretical maximum isentropic velocity

$$U_{\lim} = \sqrt{\frac{2 \cdot \kappa}{\kappa - 1} RT_0} \tag{8}$$

with $\kappa = 1.30$. Eqation (8) results in $U_{\lim} = 1421\ m/s$ for $T_0 = 505K$ in Fig. 5 and $U_{\lim} = 1293\ m/s$ for $T_0 = 418\ K$ in Fig. 6. (These velocities are somewhat larger than expected in a noncondensing flow; see the variation of κ in Fig. 2. Using $U_{\lim} = \sqrt{2H_0}$ and standard tables of water vapour for $p \to 0$, we obtain $U_{\lim} = 13745\ m/s$ for $T_0 = 5055\ K$ and $U_{\lim} = 12455\ m/s$ for $T_0 = 418\ K$ which are 3.4 and 3.8% smaller than the ones using Eq. 8.) The experimental data are plotted together with the theoretical curves for isentropic free jets with $\kappa = 1.30$. The experiments in the condensed flows show up to +12% higher velocities, which can be explained by an additional energy input by the released condensation heat (see Eq.

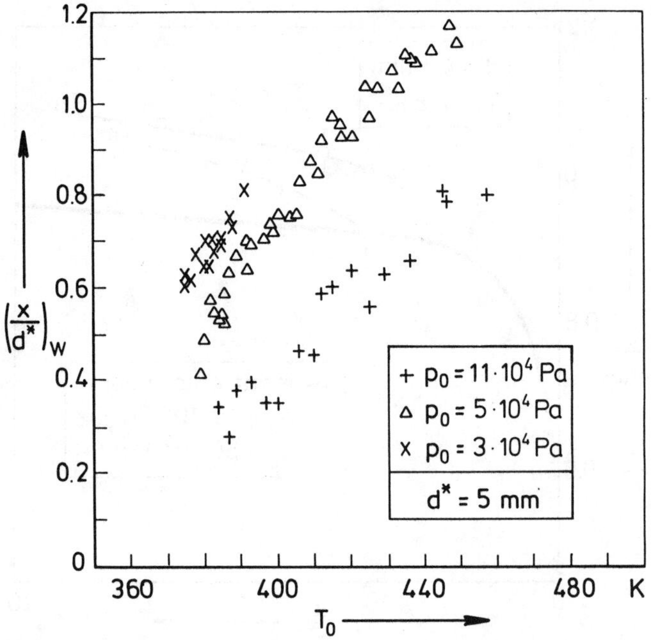

Fig. 7 Condensation onset points for different stagnation pressures.

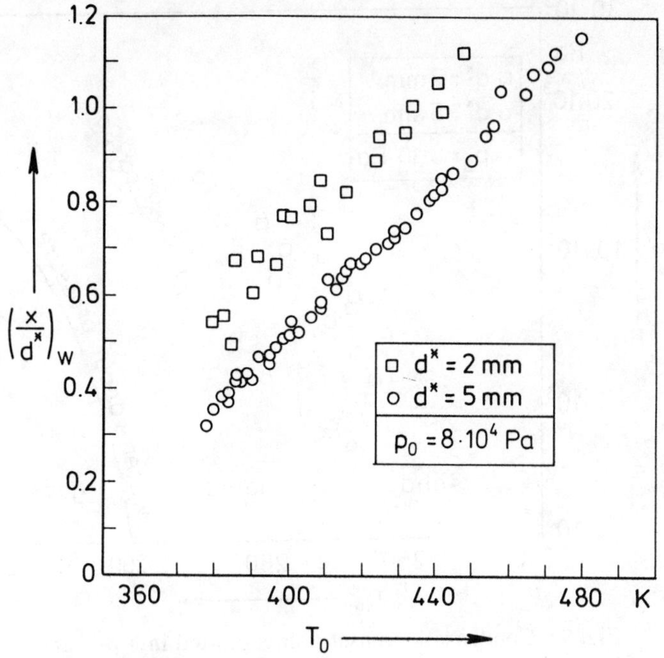

Fig. 8 Condensation onset points for different nozzle diameters.

(7), last term). From the velocity gain the condensed mass fraction can be calculated; the obtained mass fractions of 4 - 9% are in agreement with electron beam experiments using x-ray Bremsstrahlung [9]. All present experimental velocities in the condensed free jets are obtained at $10 < x/d^* < 25$. It is obvious that the beginning of condensation must be further upstream in case of the higher stagnation pressures.

Condensation Onset

Figures 7 and 8 show the experimental condensation onset (Wilson) points in H_2O - vapour free jets measured by laser light scattering for two nozzle sizes, four stagnation pressures, and several stagnation temperatures. Figure 7 demonstrates that the higher the pressure, the earlier the condensation onset. Figure 8 shows that the smaller the nozzle, the larger the relaxation and therefore the later the condensation onset. In Figure 9 condensation onset points for two nozzles are plotted in a p-T phase diagram, calculated with a constant ratio of specific heats of $\kappa = 1.30$. All

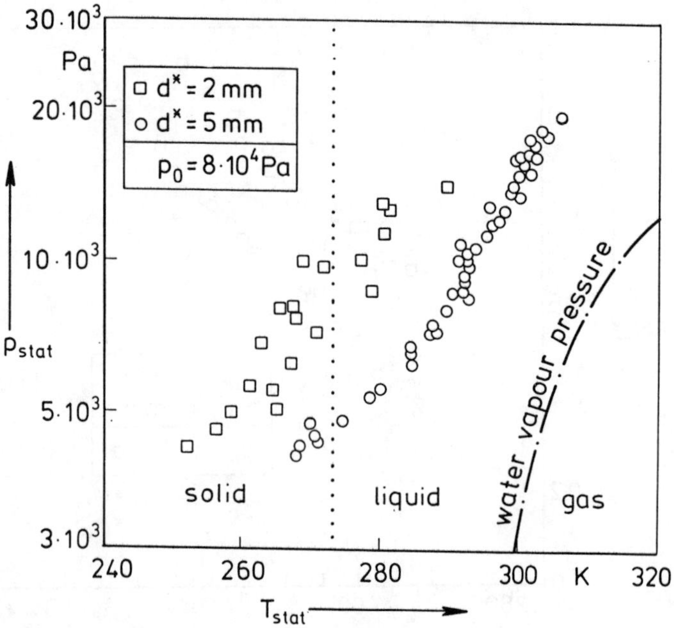

Fig. 9 Condensation onset points plotted in a p-T graph (assuming $\kappa = 1.3$ for the flow expansion).

data merge into two curves in the liquid and solid phase for equilibrium, indicating that homogeneous condensation onset is strongly dependent on the expansion isentrope. These results are consistent with previous experiments in argon[3, 7, 10] and Nitrogen.[1, 2, 8]

References

[1] Williams, W. D. and Lewis, J. W. L., "Experimental Study of Condensation Scaling Laws for Reservoir and Nozzle Parameters and Gas Species," AIAA-Paper 76-53, AIAA 14th Aerospace Meeting, Washington DC, 1976

[2] Dankert, C., "Condensation Onset in Free Jets Measured by Laser Light Scattering," Rarefied Gas Dynamics, Vol. II, 1984, pp. 983 -990.

[3] Teshima, K., Abe, K., and Nishino,T., "Structures of Condensing Freejets of Argon," Rarefied Gas Dynamics, Vol. II, 1986, pp. 544-553.

[4] Dankert, C., Bütefisch, K.-A., "Geschwindigkeitsmessungen in Überschallströmungen," Vol. 52, VDI-Verlag 5/86, Forschung im Ingenieurwesen, 1986, pp. 133 -

[5] Zierep, J., "Theory of Flows in Compressible Media with Heat Addition," AGARD, Vol. 191, 1974.

[6] Ashkenas, H. and Sherman, F.S., "The Structure and Utilization of Supersonic Free jets in Low Density Wind Tunnels," Rarefied Gas Dynamics , Vol. II, 1966, pp. 84 - 105.

[7] Koppenwallner and G., Dankert, C., "Homogenous Condensation in N_2, and H_2O free Jets," Journal of Physical Chemistry , Vol. 91, 1987, pp. 2482-2486.

[8] Dankert, C. and Koppenwallner, G., "An Experimental Study of Nitrogen Condensation in a Free Jet Expansion," Rarefied Gas Dynamics , Vol II., 1979, pp. 1107 - 1117.

[9] Yarykin, Y.N., Skovordko, P.A., Gorchakova, N.G., Kharmor,G.A. and Nerashev, O.A., "The Effect of Homogeneous Condensation on Gasdyamics and IR Radiation of Carbon Dioxide and Water Vapour Free Jets," Rarefied Gasdynamics, Vol II, 1984, pp. 951 - 958.

[10] Stein, G.D., "Argon Nucleation in a Supersonic Nozzle," Office of Naval Research Rept. AD-A 007 357/7 Gi ,1974.

Formation of Ion Clusters in High-Speed Supersaturated CO_2 Gas Flows

P. J. Wantuck*
Los Alamos National Laboratory, Los Alamos, New Mexico
and
R. H. Krauss† and J. E. Scott Jr.‡
University of Virginia, Charlottesville, Virginia

Abstract

The rate of ion-neutral clustering has been studied in a freely expanding CO_2 jet using a molecular beam sampling method. Trimer ion formation proceeds at a rate determined by a three-body reaction mechanism, and the rate constant for the overall reaction has been found to be 4.4 (±2.9) x 10^{-25} cm^6/s at a temperature of approximately 40 ° K. Dimer ion formation proceeds at a rate proportional to the cube of the neutral gas density, suggesting several plausible reaction mechanisms that cannot be distinguished in these experiments. Time-of-flight (TOF) measurements, made for the various ion species produced in the expansion, show that both dimer and trimer ion TOF traces are single-peaked and indicate the ion speeds are close to the gas flow speed (650 m/s). The fragment ions (C^+, O^+, CO^+), which are products of dissociative ionization, are observed to move at speeds much faster than that of the jet. Monomer ion TOF traces are bimodal, i.e., the early arriving packet is composed of ions moving at approximately four to five times jet speed, whereas the late-arriving monomer ions move at approximately jet speed. The presence of fast monomer ions precludes a definitive investigation of the rate of dimer ion formation.

Introduction

This investigation was motivated primarily by experimental results obtained previously at the University of Virginia[1] in which

*Research Scientist.
†Associate Research Professor of Engineering Physics.
‡Professor of Mechanical and Aerospace Engineering.

the velocity distribution in freely expanding jets of UF_6 and CO_2 was measured using an ion time-of-flight (TOF) method. The observed ion TOF distributions were much narrower than expected, and the TOF peak heights were observed to increase in a nonlinear manner with increasing flow density. The suggested interpretation of these observations was that ion-induced clustering was occurring within or downstream of the electron beam. This hypothesis was consistent with stopping potential measurements, which showed that the ions had kinetic energies well in excess of that expected for a monomer ion moving at the freejet speed.

Since the nature of the intermolecular attractive forces for homogeneous nucleation is weaker than those occurring in the heteromolecular process, clusters formed through ion-neutral interactions, under the same conditions, will occur more readily than those formed through purely homogeneous nucleation. Once a cluster ion is established, by attaching more neutral species, it can form successively larger clusters. For these experiments, a dimer ion is defined as a cluster comprised of one neutral CO_2 molecule (a monomer) attached to a singly ionized CO_2 molecule $[CO_2^+{\cdot}(CO_2)]$. A trimer ion is a monomeric ion bound to two neutral CO_2 molecules $[CO_2^+{\cdot}(CO_2)_2]$, and a quatromer ion is a cluster comprised of three neutral molecules attached to a CO_2 monomer ion $[CO_2^+{\cdot}(CO_2)_3]$.

The experiments carried out in this investigation involve bombarding fast, low-density, supersaturated CO_2 gas flows with a beam of high-energy electrons. The formation of CO_2 clusters is studied by extracting a molecular beam from the flow in which the clusters are being formed, using a mass-sensitive detector to characterize and measure the molecular beam flux. The data required for rate constant determination are obtained by measuring cluster fluxes in the sampled beam as a function of the cluster size and the number of ion-neutral encounters downstream of the ion-formation region. The number of encounters is varied, for a fixed gas density level, by varying the ion-neutral interaction flight path length, i.e., the distance between the electron beam and the sampling point. A pulsed electron beam TOF method has also been used to determine the speeds of the various ion species formed in the freely expanding jet.

Description of Apparatus

The apparatus employed in these experiments, shown schematically in Fig. 1, has been described in detail by Wantuck.[2] The vacuum system is comprised of two interconnected vacuum chambers, viz., the freejet chamber, which contains the nozzle, the electron gun and electron beam-detecting Faraday cup, and the cryopump, and the mass spectrometer chamber, which houses the ion extraction and analysis apparatus used to characterize the molecular/ion beam. The freejet chamber is evacuated by means

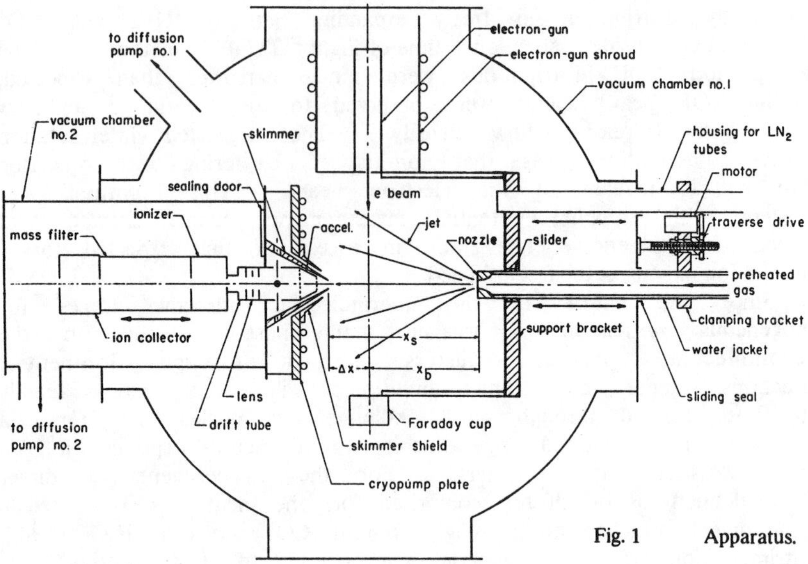

Fig. 1 Apparatus.

of a liquid nitrogen-trapped oil diffusion pump. The mass spectrometer chamber is inserted into the freejet chamber, and the molecular beam skimmer and the cryopump are attached to an aluminum plate, which covers the interior end part of the mass spectrometer chamber. The cryopump is a copper disk that is cooled by liquid nitrogen circulating through a coil of copper tubing soldered to its front surface. The mass spectrometer chamber is evacuated by a liquid nitrogen-trapped oil diffusion pump.

Gas flow regulation is provided by two valves in series, one of which is used to regulate the nozzle stagnation pressure, and the other to serve as a shutoff valve. The stagnation temperature of the gas is maintained at 353 °K (to avoid homogeneous condensation in the jet) by means of a constant-temperature bath; the water from this bath is used to heat the water-jacketed nozzle supply line. The converging nozzle is the same one used by Fisher et al.,[3] having a throat diameter d_t of 1.98 mm. Nozzle stagnation pressure is measured by two independent, absolute diaphragm-type transducers.

The electron gun has been described by Wantuck.[2] It consists of a tungsten filament, a grid electrode (used to turn the beam on and off), two beam accelerating electrodes, and two pairs of electrostatic plates (for beam deflection). Typical filament, grid, and anode potentials are -5.0, -5.0, and 3.8 kV, respectively. The beam diameter is estimated from CO_2 fluorescence to be 2 mm. The electron gun, its liquid nitrogen-cooled cylindrical shroud, and the Faraday cup are all mounted on a special support bracket that can be traversed along the jet axis enabling variation of the distance Δx between the ion source and the skimmer tip. The nozzle supply line can be traversed, independently, along the jet axis in order to vary the distance x_b between the ion source and the nozzle discharge.

The molecular/ion beam is formed by a classical skimmer, made from aluminum, with internal and external angles of 30 and 35 deg, respectively, and with an aperture diameter of 1 mm. A liquid nitrogen-cooled conical shield is placed over the skimmer in order to reduce the effect of skimmer interference. To enable the ions to be effectively separated by the quadrupole mass filter, an accelerator is located at one end of a drift tube positioned approximately 9 mm downstream of the skimmer. An einzel lens is used in the drift tube to focus the accelerated ions at the entrance of the mass filter. A commercial mass spectrometer (Extranuclear Labs), consisting of an electron impact ionizer, a quadrupole-type mass filter, and an electron-multiplier ion collector is used to mass filter the various molecular beam species.

Experimental Procedure

The experimental procedure has been described in detail by Wantuck.[2] Since quadrupole mass filters are mass discriminative (i.e., they do not detect all masses with equal sensitivity), and since the quadrupole sensitivity is inversely proportional to the resolution,[4] the resolution of the mass filter was selected to ensure that the transmission function of the filter was approximately independent of mass. To verify this independence, the filter resolution dependence of the ratio of the trimer-to-dimer peak heights (used to determine the rate constant) was measured. The experimental results indicated that this ratio was indeed independent of resolution at sufficiently small values of the resolution. Since the dimer and trimer ions, with masses of 88 and 132 amu, respectively, are small compared to the design range of the filter used in these experiments (up to 1010 amu), mass discrimination effects are not expected to be appreciable at the low filter resolutions used in these experiments.

Discrimination of homogeneously nucleated clusters, which are subsequently ionized by the electron beam, from those clusters formed through ion-neutral interactions is not possible with the apparatus shown in Fig. 1. Hence, it is essential to perform the experiments in such a manner that homogeneous nucleation processes are negligible. Using condensation scaling laws for CO_2 given by Golomb et al.,[5] it is possible to estimate the stagnation pressure at which, for a given stagnation temperature, homogeneously formed dimers will appear in expansions through the 1.98-mm nozzle used in these experiments. The result is that CO_2, heated to a temperature of 353 ° K, exhausting from a 1.98-mm nozzle, will begin to form dimers at a source pressure of 120 torr. Since the maximum source pressure used in these experiments is 30

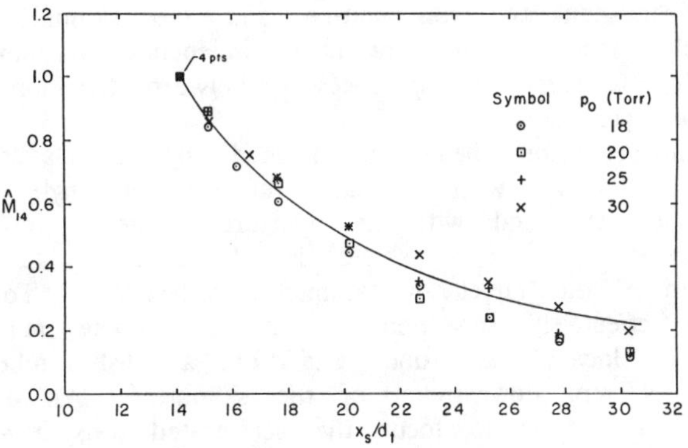

Fig. 2 Normalized monomer current $[i_0\ (x_s)/i_0\ (x_s = 28\ \text{mm})]$, $\hat{M}_{14}$, vs reduced nozzle skimmer separation distance x_s/d_t; various p_0, x_b = 26mm, d_t = 1.98 mm, i_e = 70 μa.

torr, the formation of clusters as a result of homogeneous nucleation processes can be neglected. This conclusion is confirmed by the results of many other investigators (e.g., Hagena and Obert[6]).

A singularly important condition to be satisfied in the performance of these experiments is the determination of a suitable combination of nozzle stagnation pressure p_o, electron beam-nozzle separation distance x_b, and a range of nozzle-skimmer separation distances x_s, such that both skimmer interference effects and background interference effects were reduced to acceptable levels while cluster-ion currents with an acceptable signal-to-noise ratio were simultaneously obtained. Typical results obtained from experiments, designed to meet these constraints, are shown in Fig. 2. It is well known that, in the region downstream of the nozzle exit, the flow streamlines are straight and appear to diverge from a virtual source located near the nozzle exit. The gas density on the axis of symmetry in this far-field region decreases as the inverse square of the distance from the location of the effective gas source point (see e.g., Ashkenas and Sherman[7]). This behavior is demonstrated by the experimental results shown in Fig. 2 by means of a curve, corresponding to a simple inverse-square dependence on nozzle-skimmer separation distance, which has been drawn through the monomer data sets. These results indicate that at least necessary conditions are satisfied in these experiments for the absence of appreciable skimmer and/or background interference effects; i.e., the sampling integrity is satisfactory.

Axial velocity components of the monomer, dimer, and trimer ions, measured using the pulsed electron beam TOF method, are presented in Fig. 3, which shows the ion flight-path length (Δ x) dependence of the time interval (Δ t_m) between the electron beam

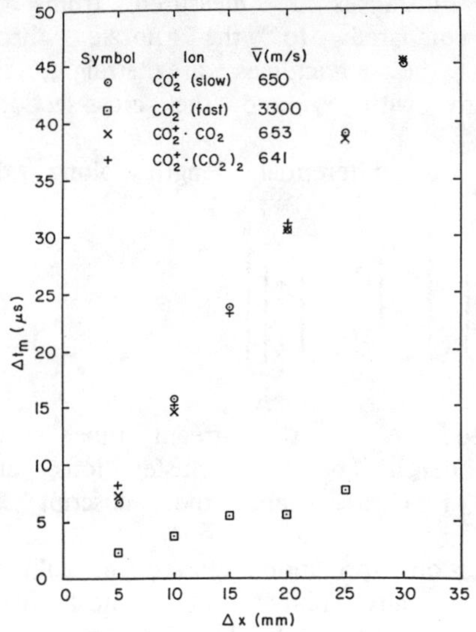

Fig. 3 TOF signal peak time of arrival Δt_m vs flight-path length Δx; p_0 = 18 torr, x_b = 26 mm, d_t = 1.98 mm, i_e = 70 μa, Δt_{eb} = 8 μs.

pulse and the arrival of the maximum ion current at the skimmer entrance. The axial speed of the ions, indicated by $(\Delta x/\Delta t_m)$, can be obtained from least-squares fitting of the data. It is evident from the data shown in Fig. 3 that the CO_2 monomer ions move with two distinct axial speeds. The early-arriving TOF peak, hereafter referred to as "fast," corresponds to an axial speed of approximately 3300 m/s, whereas the late-arriving, or "slow," ion packet has an axial speed of 650 m/s. The speeds of the dimer and trimer ions are 653 and 641 m/s, respectively, in good agreement with values reported previously by Fisher et al.[3] The slow cluster ion speeds are also in qualitative agreement with the limiting speed achievable through steady, adiabatic expansion of a freejet (683 m/s for a stagnation temperature 353 °K). Since the existence of the fast CO_2 ion precludes an unambiguous determination of a rate constant for dimer formation, and because both the dimer and trimer ions do not exhibit this behavior, the experimental efforts were limited to the determination of the rate of trimer formation from dimer ion nuclei.

Kinetic Formulation of Ion-Neutral Clustering

Consider the clustering process as it proceeds along an arbitrary stream tube centered axially in the jet. The number of neutral molecules comprising the cluster will be denoted by j, i.e., for a dimer ion j=1 and for a trimer ion j=2 (for the monomer, j=0). Once a stable j-cluster ion is formed, the only significant loss process for the cluster ion is assumed to be that due to growth of the cluster to the j+1 size. Consideration of the clustering process is restricted to distances x, measured from the nozzle exit, which are large compared to the nozzle throat diameter. Under such conditions, the streamlines are straight, the stream tube radius increases linearly with x, and the cross-sectional area varies as x^2.

Conservation of species for a differential length along the stream tube requires that

$$\frac{d(n_j \bar{u} A_t)}{dx} = \left[\left| \frac{dn_j}{dt} \right|_g - \left| \frac{dn_j}{dt} \right|_\ell \right] A_t \tag{1}$$

where $\bar{u}$ is the average flow speed, A_t is the stream tube cross-sectional area, n_j is the number density of the j-cluster ions, and the subscript g denotes gain of j clusters and the subscript ℓ denotes loss of j clusters.

Early stages of stable cluster-ion formation require a collision of the excited cluster ion with a third body. The most likely third-body candidate is another neutral CO_2 molecule. A large CO_2 cluster ion, however, may possess a sufficient number of internal degrees of freedom to successfully distribute its excess energy of formation among the available modes. Under such circumstances, the clustering mechanism becomes effectively second order; i.e., only binary collisions are required for the production of higher-order cluster ions. There is also the possibility, however, that the successful formation of a stable cluster ion will require more than one interaction with another neutral. For these reasons, the cluster ion growth will be expressed in terms of a general kinetic equation (see for example, Hasted[8]) of the form,

$$\left. \frac{dn_j}{dt} \right|_g = k_j n_{j-1} n_p^{v_j} \tag{2}$$

where v_j denotes the order of reaction for the neutral reactant participating in the formation of the j-cluster ion.

Similarly, the loss process for a j-cluster ion may be expressed as

$$\left. \frac{dn_j}{dt} \right|_\ell = -k_{j+1} n_j n_p^{v_{j+1}} \tag{3}$$

where v_{j+1} denotes the order of reaction for the neutral reactant participating in the formation of the (j+1)-cluster ion. The value of v_j is given by the difference between the overall order of reaction describing the formation of the j-cluster ion and the overall order measured for the formation of the (j-1)-cluster ion.

Hence, for a three-body reaction, v_j has a value of two. Substituting Eq. (2) and (3) into Eq. (1), incorporating the previously noted assumption regarding the streamtube cross-sectional area A(x), and noting that, for the range of x of interest, $\bar{u}(x)$ is approximately constant,[3] one obtains

$$\frac{dn_j}{dx} + \frac{2n_j}{x} = \frac{1}{\bar{u}} [k_j n_{j-1} n_p^{v_j} - k_{j+1} n_j n_p^{v_{j+1}}] \quad (4)$$

The molecular beam ion current i_j, corresponding to the collection rate for the j-cluster ion, is assumed to be directly proportional to the number flux of j-cluster ions passing through the skimmer aperture; i.e.,

$$i_j = qC_j n_j \bar{u} A_s \quad (5)$$

where C_j is a constant of proportionality, q denotes the electronic charge per ion, and A_s is the area of the skimmer aperture. As noted previously, the dimer-to-trimer current ratio is essentially constant for a wide range of (low) settings of the mass filter resolution.

Substitution of Eq. (5) into Eq. (4) gives

$$\frac{di_j}{dx} + \frac{2i_j}{x} = \frac{1}{\bar{u}} [k_j \frac{C_j}{C_{j-1}} i_{j-1} n_p^{v_j} - k_{j+1} i_j n_p^{v_{j+1}}] \quad (6)$$

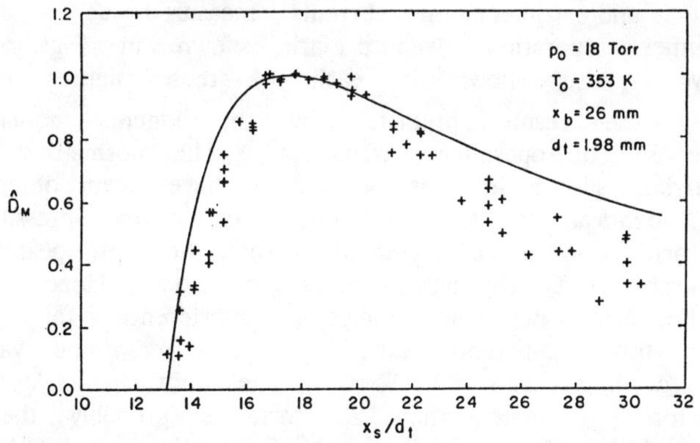

Fig. 4 Normalized dimer current $[i_1(x_s)/i_{1max}]$, $\hat{D}_M$, vs reduced nozzle skimmer separation distance x_s/d_t; p_0 = 18 torr; solid lines generated by kinetic model equations, T_0 = 353 °K, k_1 = 10^{-39} cm^9/s, k_2 = 10^{-25} cm^6/s, x_b = 26 mm, d_t = 1.98 mm.

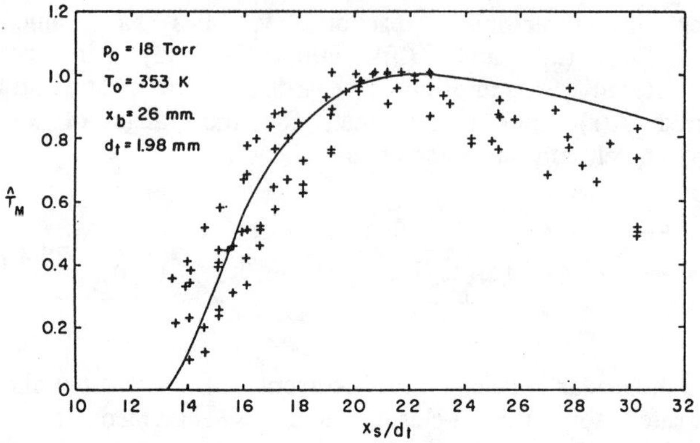

Fig. 5 Normalized trimer current $[i_2(x_s)/i_{2_{max}}]$, $\hat{T}_M$, vs reduced nozzle skimmer separation distance x_s/d_t. Details as in Fig. 4.

Since the radial velocity component for each cluster-ion species is inversely proportional to the square root of the cluster mass, Eq. (6) must be modified for each species. Hence, Eq. (6) can be written, e.g., for j=2,

$$\frac{di_2}{dx} + \left(\frac{2}{\sqrt{3}}\right)\frac{i_2}{x} = \frac{1}{u}\left[k_2 \frac{C_2}{C_1} i_1 n_p^{\,v_2} - k_3 i_2 n_p^{\,v_3}\right] \qquad (7)$$

Experimental Results

Dimer and trimer ion currents measured as a function of nozzle-skimmer separation distance are shown in Figs. 4 and 5, respectively. Also shown in each of these figures is a curve representing the results predicted by the kinetic model with an appropriate set of operating parameters. The normalized dimer and trimer current shown in Figs. 4 and 5 have been observed to be essentially identical to those obtained for source pressures of 18 and 20 torr. As p_o is increased further, however, peaking of both cluster species shifts to higher values of x_s/d_t. Hence, a condition corresponding to negligible sampling interference for dimer and trimer ion fluxes apparently exists, at x_b=26 mm, and values of p_o equal to or less than 20 Torr. The shift in cluster ion peak location, for p_o greater than 20 torr, is probably the result of increasing skimmer interference in conjunction with higher local background pressure.

Two conclusions can be drawn from the results presented in Figs. 4 and 5. First, measurement of dimer and trimer ion currents can be made under conditions of negligible skimmer and background pressure interference for nozzle stagnation pressures equal to or less than 20 torr for x_b equal to approximately 13 nozzle diameters. Although it is not possible to conclude that these measurements are completely disturbance-free, it is safe to assume that the interference effects are sufficiently small such that sampling of the cluster-ion formation rate process is valid under these conditions of operation. Second, the qualitative agreement between the model-predicted curves and experimental measurements, at least for that range of x_s/d_t, where cluster currents are rising to their peak values, is reasonable. This agreement lends support, both for use of the model equations to estimate a trimer rate constant, and for the premise that cluster sampling at these values of source pressure is reasonably free of skimmer interference.

Neglecting the loss of trimers through quatromer formation, and solving Eq. (7) for the trimer rate constant k_2, results in the following expression:

$$k_2 = \frac{\bar{u}}{n_p^{v_2}} \frac{C_2}{C_1} \left\{ \frac{1}{i_1} \frac{di_2}{dx} + \frac{2}{\sqrt{3}\,x} \frac{i_2}{i_1} \right\} \tag{8}$$

The rate constant is a function of the local static temperature on the jet centerline, which, in the jet far field, varies very slowly with x_s. Hence, the trimer rate constant should be approximately constant over this range. The proportionality constants C_1 and C_2 are, to good approximation, equal to the electron multiplier gain for the dimer and trimer ions, respectively. The jet flow speed $\bar{u}$ can be determined from the results of the TOF experiments and the number density n_p can be estimated using the relations developed by Ashkenas and Sherman.[7] The remaining unknown quantities in Eq. (8), viz., the order of the reaction for the neutral CO_2 species participating in the trimer formation process v_2 and the dimer and trimer currents i_1 and i_2, are determined in the following manner.

The nozzle source pressure dependence of the normalized trimer ion current is shown in Fig. 6. As indicated by the best-fit line through the data, trimer formation varies at a rate proportional to the sixth power of the local neutral density. It is evident from Fig. 6 that the resolution of the sixth power dependence of trimer ion formation on local flow density is quite good. The variation of dimer ion current with increasing nozzle source pressure is presented in Fig. 7. These data indicate that dimer ion formation proceeds at a rate proportional to the fourth power of the local neutral density. Since the overall order of the trimer ion formation reaction is six, and the overall order of the dimer ion formation reaction is four, the order of the trimer reaction with respect to the neutral monomer v_2 is two. This type of dependence is consistent with a three-body reaction scheme, viz., the collision of a dimer ion with a neutral molecule to

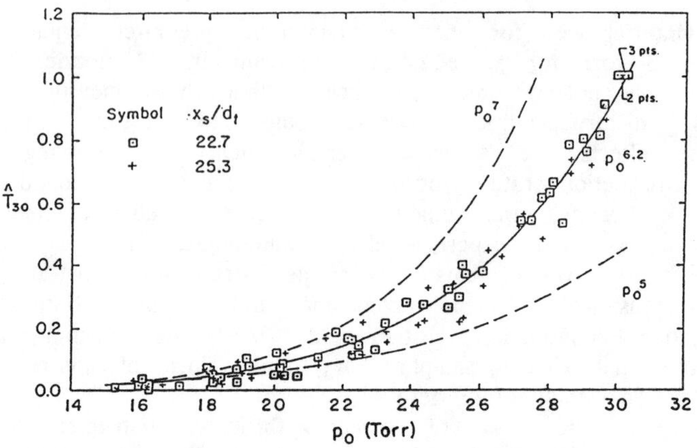

Fig. 6 Normalized trimer current $[i_2(p_0)/i_2(p_0 = 30\ \text{torr})]$, $\hat{T}_{30}$, vs nozzle stagnation pressure p_0; x_s = 45 and 50 mm, x_b = 26 mm, d_t = 1.98 mm, i_e = 70 μ a. Dashed lines represent projected trimer formation scaling as p_0^5 and p_0^7.

produce an excited trimer ion, which is subsequently stabilized in a collision with another neutral species.

For a particular x_s, the large number of individual cluster-ion current measurements made along the jet centerline are reduced to a single value, by averaging all the currents recorded at that position. The normalized average dimer and trimer ion currents are shown in Figs. 8 and 9, respectively. (The error bars attached to several points represent the standard deviation of the cluster-ion current data at that x_s.) Also shown in Figs. 8 and 9 are curves corresponding to the normalized dimer and trimer currents calculated using the kinetic model equations. (Selection of the parameters in the kinetic model has been discussed in detail by Wantuck.[2]) The calculated results are seen to be in reasonable agreement with results obtained experimentally in that range of x_s/d_t where the dimer and trimer current is increasing. Deviation of the average currents and the model-generated results presented in Figs. 8 and 9 in the jet far field for both cluster species (see also Figs. 4 and 5) is likely to be due to increased sampling interference which results from an increase in background pressure as x_s is increased. Reduction of sampled molecular beam flux due to increasing background pressure has been observed previously by several investigators.[9]

Trimer reaction rate constants, calculated using average cluster ion currents in Eq. (8), for those values where normalized experimental trimer and dimer ion currents compare well with normalized model-predicted results, are presented in Table 1.

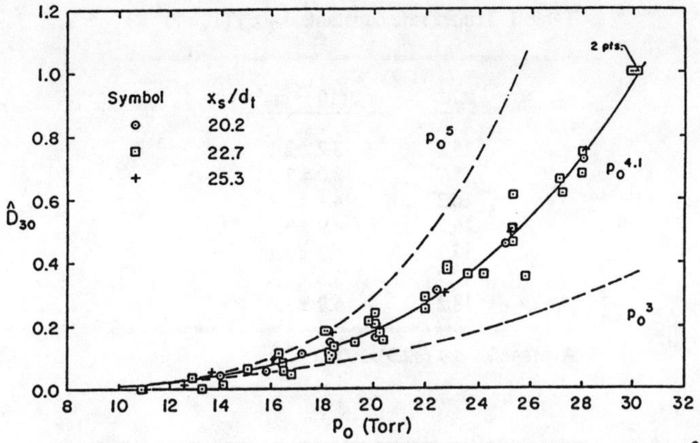

Fig. 7 Normalized dimer current $[i_1\ (p_0)/i_1\ (p_0 = 30\ \text{torr})]$, $\hat{D}_{30}$, vs nozzle stagnation pressure p_0; x_s = 40, 45 and 50 mm; x_b = 26 mm; d_t = 1.98 mm; i_e = 70 μ a. Dashed lines represent projected dimer formation scaling as p_0^3 and p_0^5.

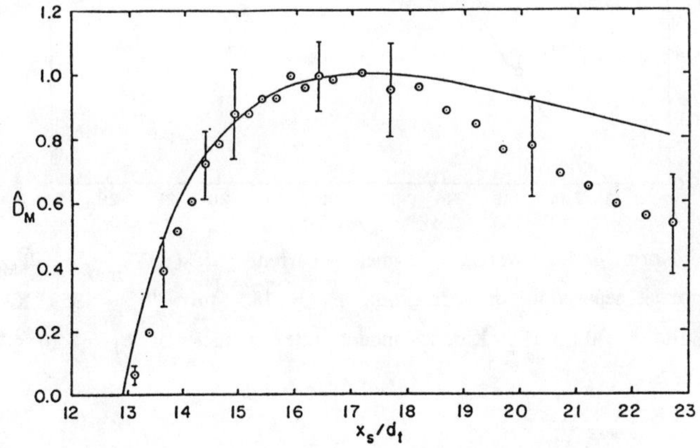

Fig. 8 Normalized average dimer current $[\bar{\imath}_1\ (x_s)/\bar{\imath}_{1max}]$, $\hat{D}_M$, vs reduced nozzle skimmer separation distance x_s/d_t; p_0 = 18.1 torr; T_0 = 353 ° K; x_b = 25.6 mm; d_t = 1.98 mm; i_e = 70 μ a. Kinetic model rate coefficients: $k_1 = 10^{-39}$ cm^9/s, $k_2 = 10^{-25}$ cm^6/s.

The average rate constant has a value of 4.4 (± 2.9) x 10^{-25} cm^6/s. This rate constant value may seem large, considering the low gas temperatures encountered in these expanding jet flows. However, as noted by Bates[10,11] and Herbst,[12,13] the rate constant for an association or cluster reaction increases dramatically with decreasing temperature, scaling as T^{-n}, where n is typically equal

Table 1 Trimer rate constants, $k_2(x_s)$ (T_s <40° K)

x_s/d_t	k_2, (10^{-25} cm^6/s)
15.2	3.2 ± 2.1
15.7	3.0 ± 1.9
16.2	4.1 ± 2.7
16.7	4.9 ± 3.1
17.2	4.2 ± 2.7
17.7	5.3 ± 3.3
18.2	5.9 ± 3.8

Average k_2=4.4 (±2.9)x10^{-25} cm^6/s

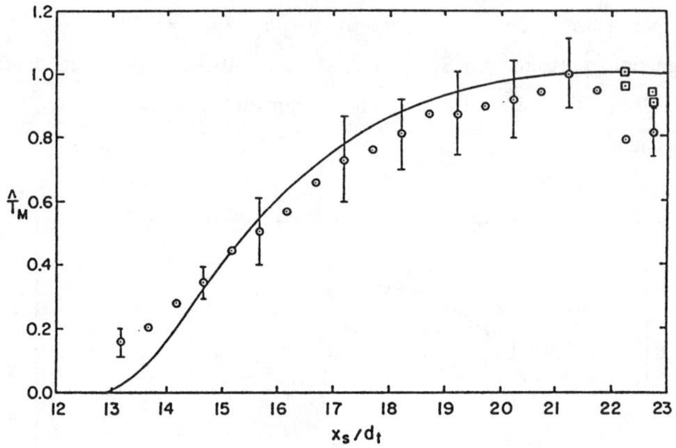

Fig. 9 Normalized average trimer current [$\bar{I}_2(x_s)/\bar{I}_{2max}$], $\hat{T}_M$, vs reduced nozzle separation distance x_s/d_t; p_0 = 18.1 torr; T_0 = 353 ° K, x_b = 25.6 mm; d_t = 70 μ a. Kinetic model rate coefficients: k_1 = 10^{-39} cm^9/s, $k_2 = 10^{-25}$ cm^6/s.

to 2. For example, Rowe et al.[14] measured a rate constant for the cluster reaction $N_2^+ + 2N_2 \rightarrow N_4^+ + N_2$ and found that it could be represented by the power law $6 \times 10^{-29}(300/T)^{1.85}$. Thus, for a far-field jet temperature of 20 ° K, the rate constant for the N_2 reaction increases to approximately 10^{-26} cm^6/s. Herbst[12] lists other three-body reactions for which the temperature dependence of the rate constant is even stronger (n=2.5-3.5).

The increase in rate constant with decreasing temperature is attributable to a decrease in the internal energy of the molecule as well as to an increase in the molecular polarizability.[15] The less excited an ion-neutral complex, the longer the complex exists. Hence, there is the probability that stabilization by collision with a third body increases.

Concluding Remarks

The freejet medium, combined with molecular beam sampling, is a powerful method for the study of various ion molecule reactions, including clustering. The major difficulty in the experiments reported here lies in achieving adequate pumping capability, such that cluster current sampling is accomplished under conditions of reduced background density and negligible skimmer interference. For the present experiments, the nozzle stagnation pressure is limited to 20 torr, and the minimum nozzle-skimmer separation distance is limited to approximately 13 nozzle diameters. Quatromer (and higher mass) ion formation processes are not resolvable under these operating conditions. A study of large polymer formation kinetics would determine the stage at which the cluster-ion formation becomes effectively second order. The performance of such a study would require an increase in the jet flowfield density, while at the same time rendering interference effects negligible. The use of high-energy electrons to ionize the gas, although necessary to allow penetration of the electron beam into the jet, causes a high degree of internal molecular ion excitation. The degree of excitation can influence the clustering process. Consequently, alternate means of ionizing the gas, e.g., photoionization, would be advantageous. Use of this alternate ionizing source may also eliminate the fast monomer peak by changing the properties of the plasma in the ion formation region, enabling a study of CO_2 dimer ion formation.

Acknowledgments

We are indebted to G. C. Clements and R. F. Jackson for their fine craftsmanship in the construction of the apparatus and to C. G. Pope for his help in collecting and reducing the copious quantities of data. Support of this research by the U.S. Department of Energy under Contract DE-AC05-820R20900 is gratefully acknowledged.

References

[1] Fisher, S. S., Hawsey, R. A., Krauss, R. H., and Scott, J. E., "Ion Time-of-Flight Velocity Measurements in UF_6 and CO_2 freejet Expansions," Rarefied Gas Dynamics, edited by R. Camparque, Commissariat a l'Energie Atomic, Paris, France 1979, Vol. 2, pp. 1163-1174.

[2] Wantuck, P. J., "Rate Constant Determination for the Formation of Ion-Nucleated Trimer Ions in High Speed, Super-saturated CO_2 Gas Flows," Ph.D. Dissertation, Engineering Physics Department, Univ. of Virginia, Charlottesville, VA, 1986.

[3] Fisher, S. S., Yurkanin, D. J., Graybeal, G. A., and Scott, J. E., "Measurement of the Impact Probe Rarefaction Effect in High Mach Number, $\gamma = 7/5$ CO_2 Flows," in Rarefied Gas Dynamics, edited by H. Oguchi, Univ. of Tokyo Press, 1984, Vol. 1, pp. 405-413.

[4] Peden, J. A., "The Quadrupole Approach," Industrial Research, Vol. 5, 1970, p. 50.

[5] Golomb, D., Good, R. E., Bailey, A. B., Bushy, M. R., and Dawborn, R., "Dimers, Clusters, and Condensation in Free Jets," Journal of Chemical Physics, Vol. 57, 1972, pp. 3844-3852.

[6] Hagena, O. F. and Obert, W., "Cluster Formation in Expanding Jets: Effect of Pressure, Temperature, Nozzle Size, and Test Gas," Journal of Chemical Physics, Vol. 56, 1972, pp. 1793-1802.

[7] Ashkenas, H. and Sherman, F.S., "The Structure and Utilization of Supersonic Freejets in Low Density Wind Tunnels," Rarefied Gas Dynamics, edited by J.H. deLeeuw, Vol. II, 84, Academic, New York, 1966, Vol. II, pp. 84-105.

[8] Hasted, J. B., Physics of Atomic Collisions, American Elsevier, New York, 1972, p. 16.

[9] Anderson, J. B., "Molecular Beams from Nozzle Sources," in Molecular Beams and Low Density Gas Dynamics edited by P. P. Wegener, Dekker, New York, 1974, pp. 1-91.

[10] Bates, D. K., "Temperature-Dependence of Ion-Molecule Association," Journal of Chemical Physics, Vol. 71, 1979, pp. 2318-2319.

[11] Bates, D. K., "Density of Quantum States of Ion-Molecule Association Complex and Temperature-Dependence of Radiative Association Complex," Journal of Chemical Physics, Vol. 73, 1980, pp. 1000-1001.

[12] Herbst, E., "A Statistical Theory of Three-Body Ion-Molecule Reactions," Journal of Chemical Physics, Vol. 70, 1979, pp. 2201-2209.

[13] Herbst, E., "Refined Calculated Ion-Molecule Association Rates," Journal of Chemical Physics, Vol. 72, 1980, pp. 5284-5285.

[14] Rowe, B. R., Dupeyrat, G., Marquett, J. B., and Gaucherel, P., "Study of the Reactions of $N_2^+ + 2N_2 \rightarrow N_4^+ + N_2$ and $O_2^+ + 2O_2 \rightarrow O_4^+ + O_2$ from 20 to 160K by the CRESU Technique," Journal of Chemical Physics, Vol. 80, 1984, pp. 4915-4921.

[15] Jackson, J.D., Classical Electrodynamics, Wiley, New York, 1975, pp. 155-158.

Molecular Dynamics Study of Dynamic-Statistic Properties of Small Clusters

S. F. Chekmarev* and F. S. Liu*
USSR Academy of Sciences, Novosibirsk, USSR

Abstract

To study dynamic-statistic properties of clusters, an approach based on a combination of the conventional and stochastic molecular dynamics methods is used. In the paper the technical aspects of the approach are discussed, and its applications to an investigation of frequency (phonon) spectra and probabilities of decay of small clusters are given. Specifically, the calculations are made for an argonlike substance with atoms interacting through a pairwise Lennard-Jones potential.

With the trimer ($n = 3$) being an example, it is shown that at a low temperature ($T = 3°K$) the frequency spectra of individual free clusters are usually nonequilibrium, but when averaged over the canonical ensemble of clusters, they take the equilibrium form. The frequencies of monomer emission from free small liquid clusters $\nu_e^1(n, E)(3 \leq n \leq 13)$ are calculated, and it is found that they obey the RRK theory of monomolecular decay suitably interpreted. It is shown that the canonical ensemble averages $\nu_e^1(n, T)$ differ from the microcanonical ones $\nu_e^1(n, E)$ with the energies E being equal to their averages for the temperature given.

I. Introduction

Under conditions typical for condensation phenomena, frequencies of molecular motions in clusters usually are much higher than of collisions between the clusters and particles of the surroundings

---

*Institute of Thermophysics.

(vapor and solvent molecules). Taking this into account we use the following approach to investigate cluster properties. Clusters are regarded as free systems (being in a vacuum). With cluster size (number of particles n) and cluster energy E given, the characteristics of the clusters are calculated as the averages over the corresponding microcanonical ensemble. The effect of the surroundings for every cluster size is reduced to the existence of a canonical ensemble of clusters for the temperature of interest T. Cluster characteristics corresponding to certain values of n and T can be calculated as the averages over the canonical ensemble. Such an approach makes it possible to take into consideration dynamic processes in clusters as well as cluster statistics.

To average characteristics of a free cluster over the canonical ensemble, we first have to write down the Boltzmann distribution as a function of cluster energy E:

$$f_0(E,T) = Z_n^{-1}(T)\Omega(E)\exp(-E/kT) \quad (1)$$

where $Z_n(T)$ is the cluster partition function, k is the Boltzmann constant, and $\Omega(E)$ is an area of hypersurface $H(\vec{R}_n, \vec{P}_n) = E$ in phase space, where $H(\vec{R}_n, \vec{P}_n)$ is the Hamiltonian function of the cluster (here and later in this paper we use abbreviations $\vec{R}_n = \{\vec{r}_1, \cdots, \vec{r}_n\}$ and $\vec{P}_n = \{\vec{p}_1, \cdots, \vec{p}_n\}$, where $\vec{r}_i$ and $\vec{p}_i$ represent coordinates and momenta of the ith particle, respectively). As is known, $\Omega(E)$ can be written in explicit form, when particle vibrations are being represented by a collection of harmonic oscillators (see, for e.g., Refs. 1 and 2). Unfortunately, this approximation generally is not valid for the Lennard-Jones cluster considered. Thus, to calculate Eq. (1) a numerical method is needed.

Taking the previously mentioned factors into account we use the following procedure to realize the approach given earlier. Characteristics of free clusters are calculated by the conventional molecular dynamics method (CMD), and the canonical ensembles of clusters are generated using the stochastic molecular dynamics method (SMD). If the characteristics of free clusters are not of special interest, they are averaged over canonical ensembles immediately; in another case before averaging over canonical ensembles their averaging over microcanonical ensembles have been made (these variants are represented in Secs. III and IV, respectively).

In the next section we give a brief description of the CMD and SMD methods used (Sec. II), discuss a formation of canonical ensembles of clusters by the SMD method (Sec. III), and, finally, illustrate a combined use of the CMD and SMD methods for an investigation of frequency spectra (Sec. IV) and probabilities of decay (Sec. V) for small clusters of an argonlike substance.

II. Molecular Dynamics Methods

Conventional Molecular Dynamics Method

An evolution of a free (being in a vacuum) cluster of n atoms is described by the system of Newtonian equations,

$$m\frac{d\vec{v}_i}{dt} = \vec{F}_i, \quad \frac{d\vec{r}_i}{dt} = \vec{v}_i \tag{2}$$

Where m is the atomic mass, v_i and r_i are velocities and coordinates of the ith atom, and $\vec{F}_i$ is the force on the ith atom due to the other $n-1$ atoms:

$$\vec{F}_i = -\frac{\partial}{\partial \vec{r}_i} U(\vec{R}_n) \tag{3}$$

The potential $U(\vec{R}_n)$ is a pairwise additive

$$U(\vec{R}_n) = \Sigma_{i<j=1} u(|\vec{r}_j - \vec{r}_i|) \tag{4}$$

where $u(r)$ is the Lennard-Jones potential, which is allowed to be truncated:

$$u(r) = \begin{cases} 4\epsilon[(\sigma/r)^{12} - (\sigma/r)^6] & \text{at} \quad r \leq r_0 \\ 0 & \text{at} \quad r > r_0 \end{cases} \tag{5}$$

All of the calculations were conducted for an argonlike substance with $m = 6.68 \times 10^{-23}$ g, $\epsilon = 119.8$ K and $\sigma = 3.405$ Å.

For numerical integration of Eq. (2) the finite-difference scheme of Schofield[3] was used. Through time interval $\sim 10^4 \Delta t$ (with the time step $\Delta t = 0.1$, see next section) it conserved momenta, angular momenta and energy of clusters within $\sim 10^{-4}\%$, $\sim 10^{-3}\%$, and $\sim 10^{-2}\%$, respectively.

Stochastic Molecular Dynamics Method

Here the cluster is assumed to be embedded in a viscous medium with constant temperature T. An evolution of a n atom cluster is described by the system of Langevin equations:

$$m\frac{d\vec{v}_i}{dt} = \vec{F}_i - \alpha\vec{V}_i + \vec{\phi}_i(t), \qquad \frac{d\vec{r}_i}{dt} = \vec{v}_i \tag{6}$$

Where $\vec{F}_i$ are the interatomic forces [Eq. (3)], α is a coefficient of viscous friction, and $\vec{\phi}_i(t)$ are random forces, which obey the conditions $< \vec{\phi}_i(t) >= 0$ and $< \phi_i^j(t) \cdot \phi_i^j(t+\tau) >= 2kT\delta_{ii'}\delta_{jj'}\delta(\tau)$ (corner brackets in equations denote the averages over an ensemble of phase trajectories of the cluster, or, briefly speaking, over an ensemble of clusters).

Equation (6) assumes that every atom, irrespective of its position in the cluster, experiences, on the average, the identical influence of the medium. Therefore, in contrast to the case of Brownian particles, the equations do not describe correctly an evolution of the cluster embedded in a real medium. Nevertheless, in the long time limit they yield correct results. A distribution function of representative points of the cluster in phase space based on an ensemble of phase trajectories corresponding to Eq. (6) obeys the Fokker-Planck equation. The last, in turn, at $t \to \infty$ has a solution corresponding to the canonical distribution[4]

$$f_0(\vec{R}_n, \vec{P}_n) = Z_n^{-1}(T)\ \exp\{-(\sum_1^n \vec{p}_i^2/2m + V_n)/kT\} \tag{7}$$

where $V_n = U - U_0$ is the potential energy of a cluster of size n measured from an energy of the ground state of the cluster ($U_0 < 0$).

Thus, the SMD method based on the Langevin equations can be used to generate canonical ensembles of clusters. For the numerical integration of Eq. (6) the scheme from Ref. 6 was used.

Scaling

The calculations were conducted in terms of nondimensional variables. For mass, distance, energy, velocity, and time scales the values of $m, \sigma, \epsilon, v_0 = \sqrt{48\epsilon/m}$ and $t_0 = \sigma/v_0 = \sqrt{m\sigma^2/48\epsilon}$, respectively, were chosen, which make it possible to reduce a number of operations when integrating Eq. (2) or (6). According to the values of the time step of the integration Δt, factors for viscous friction $\gamma = \alpha/m$, and angular frequency of vibration ω are expressed in units of t_0 : $\Delta\bar{t} = \Delta t/t_0, \bar{\gamma} = \gamma t_0$, and $\bar{\omega} = \omega t_0$ (the bar denoting nondimensional values will be omitted). For the argonlike substance considered, $v_0 \approx 1.09 \times 10^5$ cm/s and $t_0 \approx 3.1 \times 10^{-13}$ s.

Using the corresponding state principle,[6] the results given later can be applied to other noble gases. For this purpose one has to take corresponding meanings of m, σ, and ϵ.

III. SMD Generating of Canonical Ensembles of Clusters

Let us show first how partial energies of the cluster behave when it is embedded in a medium with a constant temperature. We shall be interested in translational (E_t), rotational (E_r), and vibrational (E_v) energies of the clusters, dividing the last into two parts, namely, kinetic energy of atoms (E_{vk}) and their potential energy (E_{vp}).

The cluster considered as a unit behaves like a Brownian particle of mass $M = mn$. Neglecting "switch-on" effect, the ensemble average value of the translational energy obeys the equation[3]

$$< E_t >= \frac{3}{2}kT\{1 - \exp(-2\gamma t)\} \tag{8}$$

where t is the time elapsed from the moment of cluster embedding. Equation (8) shows that $\tau_t = 1/2\gamma$ characterizes the time of translational relaxation of the cluster, and the factor in front of the curly bracket characterizes an equilibrium value of translational energy. Following the theory of molecular gas relaxation,[7] it is possible to assume that for the other partial energies similar equations will be valid with their particular values of a relaxation time and an equilibrium energy.

Figures 1a and 1b illustrate a process of cluster heating; here $n = 2$ (dimers), $r_0 = \infty$, and number of clusters in an ensemble $N = 200$. Initial values of all partial energies were zero. The solid lines correspond to the SMD calculations and the dashed ones to Eq. (8), with the factor before the curly brackets replaced by proper equilibrium values of the energies: $< E_r >= kT, < E_{vk} >= kT/2$, and $< E_{vp} >= kT/2$ (the last relation assumed vibrations to be harmonic).

Figure 1a shows that, with time going on, all of the energies achieve their equilibrium values, and for the temperature given, the assumption that the vibrations are harmonic seems to be true. Figure 1b shows the process in smaller scale [as it follows from Eq. (8), lowering of γ results in stretching of the process in time]. As one can see, the time of rotational relaxation is about τ_t, whereas that of vibrational relaxation is higher than τ_t. These conclusions are consistent with concepts of molecular gas relaxation theory.[7] The ensemble averages of partial energies at $t/\Delta t = 200$ depending on the number of clusters in the ensemble are shown in Fig. 2.

To form a canonical ensemble of N clusters under the temperature given, there is no need to repeat the process of cluster heating N times. For this purpose it is possible to use one SMD trajectory in phase space, considering representative points in the trajectory as being components of the ensemble.

The ensemble will be more representative the weaker the correlations between its components. A study of atom velocity autocorrelation functions of clusters $\sum_1^n < \vec{v}_i(t')\vec{v}_i(t) >$ has shown that the correlations fail approximately as $\exp(-a|t'-t|)$ where $a \sim \gamma$. Taking this into account, the following procedure was used to form

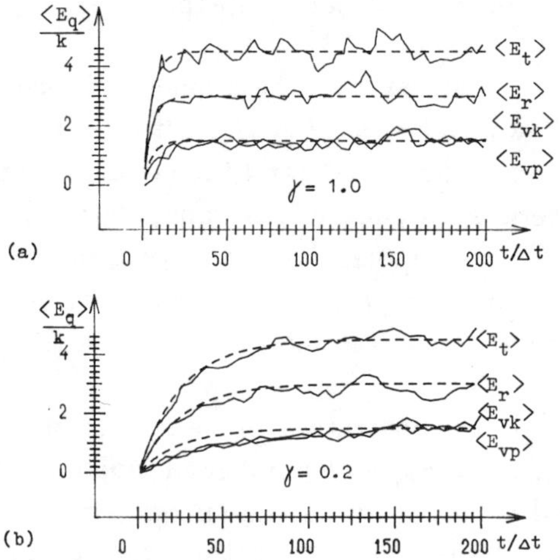

Figure 1: Cluster heating ($n = 2$, $T = 3°K$, $N = 200$, $t = 0.1$). Solid lines, SMD calculations; dashed lines, Eq. (8).

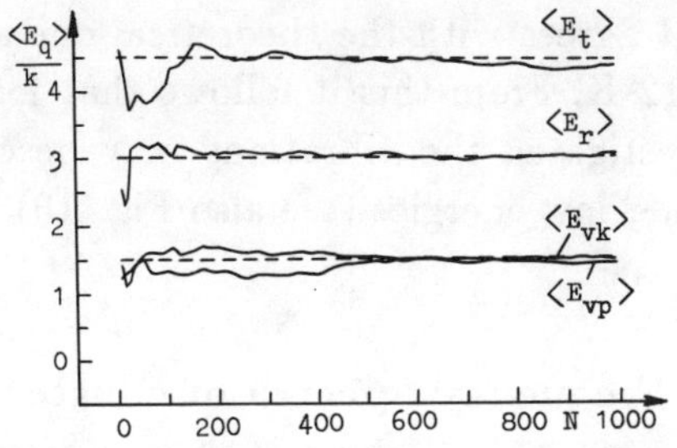

Figure 2: Canonical ensemble averages of partial energies vs number of clusters N ($n = 2$, $T = 3°K$, $t = 1.0$, $\Delta t = 0.1$, $t/\Delta t = 200$). Solid lines, SMD calculations; dashed lines, theoretical meanings.

the canonical ensemble. At $t = 0$, with velocities being zero, atoms were arranged so as to satisfy roughly the condition $E = 0$. Then using the SMD method the cluster was subjected to heating up to the temperature of interest. Starting from the moment $t = 2/\gamma$, at the end of every time interval equal to $1/\gamma$ the representative points in the SMD trajectory were stored and further used as components of the ensemble. This procedure went on until the number of desired components were obtained, or a decay of the cluster occurred. In the latter case a new SMD trajectory was initiated and so on.

Now, consider the equilibrium distribution functions of cluster partial energies corresponding to distinct degrees of freedom of the cluster. They can be obtained by an integration of Eq. (1) [or Eq. (7)] with respect to all the partial energies besides the qth:

$$f_0(E_q) = (kT)^{-(m_q+1)}\Gamma^{-1}(m_q+1)E_q^{m_q} \exp(-E_q/kT) \qquad (9)$$

where $\Gamma(\ldots)$ is the gamma function. Index q takes meanings t, r and vk, which earlier corresponded to translational energy, rotational energy and the kinetic part of vibrational energy, respectively. Accordingly, $m_t = 3/2, m_r = 3/2$ and $m_{vk} = (3n-6)/2$ (here clusters are assumed to be nonlinear). If vibrations are harmonic, Eq. (9) is valid for the potential part of the energy as well, then $m_{vp} = m_{vk}$.

Typical results of the SMD calculations are given in Figs. 3 and 4 by solid lines. The dashed lines correspond to Eq. (9). As one can see, the distributions of translational and rotational energies, as well as the distribution of the kinetic part of vibrational energy, show good agreement with the theoretical distributions. Regarding potential energy, it agrees with the theoretical one at $T = 3°$Kand does not at $T = 12°$K. From this it follows that for the Lennard-Jones clusters investigated the vibrations may be considered harmonic only at rather low energies (see also Fig. 10).

IV. Frequency Spectra of Clusters

One of the most important dynamic-statistic characteristics of the clusters is spectral density of vibrational energy. For an individual free cluster it can be easily found with the help of the CMD method.[8] But, unfortunately, in the general case there is no confidence that it correctly represents the equilibrium spectral density function corresponding to a certain temperature that is of interest.

Here we shall illustrate the possibilities of a combination of the CMD and SMD methods to solve this problem. As an example, the case of low temperature is considered. It allows us, on the one hand, to obtain distinct results (due to a possibility to separate vibrational and rotational motions) and, on the other hand, to exhibit an essence of the problem (due to poor energy redistribution between vibrational modes). To find the spectral density function, we use the direct method of spectral analysis, i.e., the Fourier transformation of atom velocities depending on time.

Let us take a cluster from the canonical ensemble of clusters of size n corresponding to the temperature given (T) and, assuming it to be free, calculate its phase trajectory within the time interval $(0, t_s)$ by the CMD method. Then, the spectral density function of the cluster may be determined by the expression[9]

$$\rho(\omega, E_k) = \frac{2}{t_s} \sum_{k=1}^{n} \sum_{j=1}^{3} \frac{m}{2} \left(\int_0^{t_s} u(t) v_k^j(t) e^{-i\omega t}\, dt \right)^2 \qquad (10)$$

where E_k is the kinetic energy of the cluster, and $u(t)$ is the so-called taper, which is used to reduce the side lobes of a window function due to the "square" shape of the data window accepted in Eq. (10) [as u(t) we used the Hanning taper].

Integration of $\rho(\omega, E_k)$ with respect to $\omega/2\pi$ over the interval $0 \leq \omega/2\pi \leq \infty$ gives the kinetic energy of cluster E_k. But, if we take a frequency Ω situated between rotational and vibrational frequencies

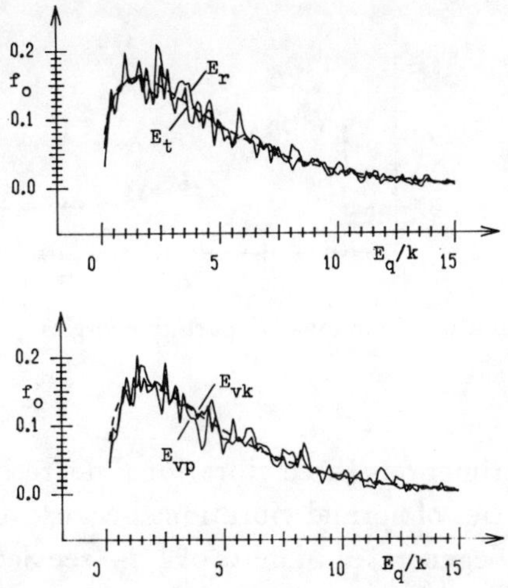

Figure 3: Distribution functions of partial energies (n = 3, T = 12°K, N = 2000). Solid lines, SMD calculations; dashed lines, Eq. (9).

as a lower limit, we shall obtain the kinetic part of vibrational energy of the cluster

$$E_{vk} = \int_{\Omega}^{\infty} \rho(\omega, E_k)\ \mathrm{d}\omega/2\pi$$

With cluster energy being small, the choice of Ω is not difficult because in this case rotational and vibrational spectra of clusters are widely separated.

Now, applying the relation (10) to all the clusters of the canonical ensemble, we can write

$$\int_{\Gamma} E_{vk} f_0\ \mathrm{d}\Gamma = \int_{\Omega}^{\infty} \rho(\omega, T)\ \mathrm{d}\omega/2\pi = kTs/2 \qquad (11)$$

where $f_0(\vec{R}_n, \vec{P}_n)$ is the canonical distribution [Eq. (7)], $\mathrm{d}\Gamma = \mathrm{d}\vec{R}_n\ \mathrm{d}\vec{P}_n$ is an element of phase volume, $s = 3n - 6$ is the number of vibrational degrees of freedom of a cluster of size n, and, finally,

$$\rho(\omega, T) = \int_{\Gamma} \rho(\omega, E_k) f_0\ \mathrm{d}\Gamma$$

is the equilibrium spectral density function of interest.

In Figs. 5 and 6 calculated results for the trimer ($n = 3$) with a nonlinear configuration of atoms ($U_0 = -3\epsilon, r_0 = \infty$) are shown.

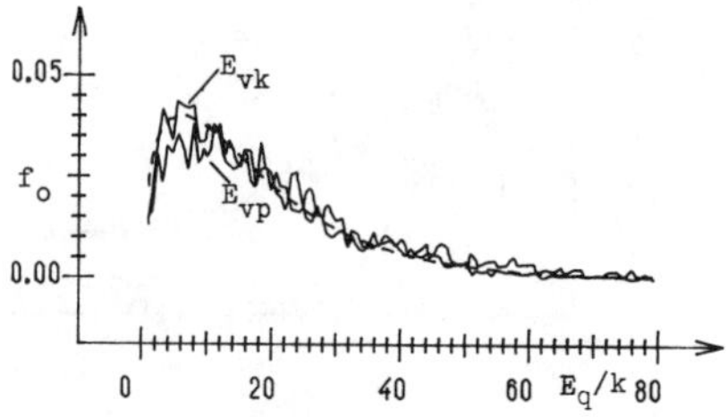

Figure 4: Distribution functions of partial energies ($n = 3$, $T = 12°$K, $N = 2000$).

In this case the trimer has three vibrational degrees of freedom, but only two frequencies of normal vibrations because one of the normal modes is doubly degenerate. In units of $1/t_0$ (see Sec. 2.3) $\omega_I \approx 1.34$ (the degenerate mode) and $\omega_{II} \approx 1.89$. As one can see from Eq. (11), under equilibrium conditions the first mode must have twice as much energy in comparison with the second $[< E_I >= 2 < E_{II} >]$.

At the temperature considered ($T = 3°$Kthe vibrations are harmonic with good accuracy (see Fig. 3). Therefore, the interaction between normal vibrational modes is small, and the frequency spectra of the free clusters do not display recognizable variations in time (we monitored the clusters up to ~ 250 periods of vibrations). As a rule, frequency spectra of individual clusters were nonequilibrium; i.e., they had the ratio of mode energies $< E_I > / < E_{II} >$ not equal to 2.

After averaging over the canonical ensemble (for this purpose we used the ensemble illustrated in Fig. 3), the frequency spectra takes the equilibrium form. Figure 5 shows how the energies of the first $< E_I >$ and the second $< E_{II} >$ modes change when the number of clusters in the ensemble used for the averaging (N) increases. As one can see, the larger N, the closer $< E_I >$ and $< E_{II} >$ approach to their equilibrium values kT and $kT/2$, respectively.

In Fig. 6 the frequency spectra are given: "pure" spectrum for an individual cluster ($N = 1$) and the spectra averaged over 50, 100, and 150 clusters. The areas under the first and the second peaks are equal to $< E_I >$ and $< E_{II} >$, respectively, and are shown in Fig. 5. Note that the widths of the peaks depend on the duration of time interval t_s; in Figs. 5 and 6 it was equal to $500\Delta t$ with $\Delta t = 0.1$. If t_s is higher, the peak widths will decrease as $1/t_s$; thus, in the limit $t_s \to \infty$ the peaks will take the form of δ functions (see, e.g., Ref. 9).

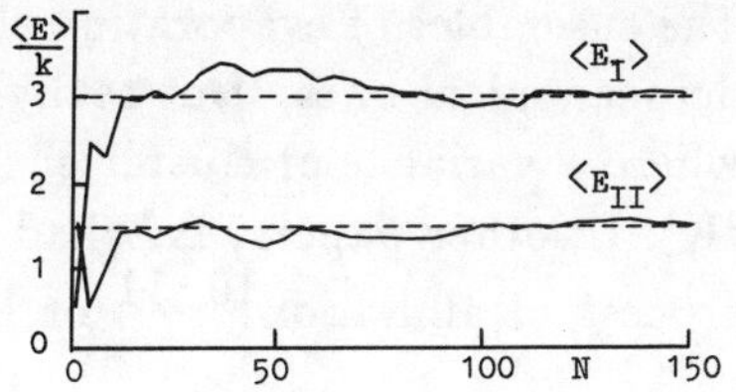

Figure 5: Canonical ensemble averages of $E\hat{I}$ and $E\hat{II}$ vs N ($n = 3$, $T = 3$°K).

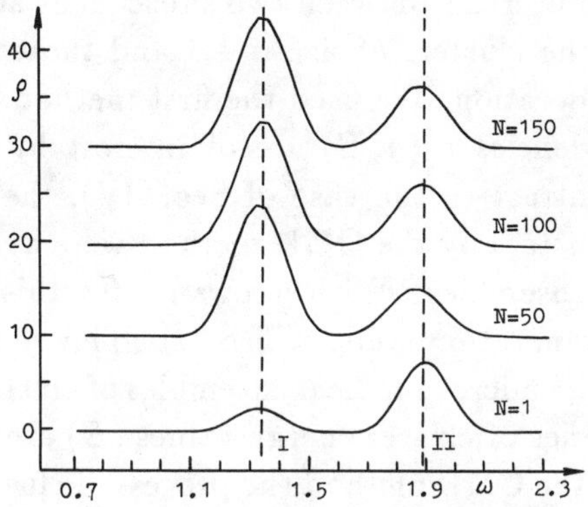

Figure 6: Canonical ensemble averages of frequency spectra vs N ($n = 3$, $T = 3$°K). The curves for $N = 50$, 100, and 150 are shifted up by 10, 20, and 30 units, respectively.

V. Evaporation of Clusters

In this section we shall consider another fundamental dynamic-statistic characteristic of clusters: a frequency (a probability per time unit) of an emission of subclusters from a cluster of the given size. For the sake of brevity, we shall confine ourselves to the consideration of monomers. The frequency of their emission from a cluster of size n will be denoted as $\nu_e^1(n, \alpha)$, where α takes on meanings of E or T depending on the ensemble considered.

An emission of monomers from free clusters (being in a vacuum) previously was studied with the CMD method by Brady et al.[10–12] and Bedanov et al.[13,14] In Refs. 10 and 11 the cluster of size n under examination was the product of the reaction $A_{n-1} + A \rightarrow A_n$, and in Refs. 14 and 15 that of $A_{n+1} \rightarrow A_n + A$; in Ref. 13 the Monte Carlo method was used to prepare the cluster.

In all of the papers the ensembles of evaporating clusters **nominally** were regarded as microcanonical ones. But really it was true only for the last paper, where a variance of cluster energies E was controlled ($|\Delta E| \leq 10°K$). In other papers, ΔE had uncertain values depending on the process of affiliation[10, 11] or emission[13, 14] of monomers.

To calculate ν_e^1, in Refs. 11-13 a rate of attenuation in time of clusters of size n in the ensemble was drawn, and in Refs. 14 and 15 the time intervals between two subsequent acts of monomer emission (from the clusters of size $n+1$ and those of size n) were taken into consideration. We used the first method.

Since dependencies $\nu_e^1(n, E)$ are of interest by themselves, in this case (in contrast to the case of Sec. IV), the data obtained for individual clusters by the CMD method were averaged over the microcanonical ensembles. With allowance for this the procedure of the cluster evaporation study will be as follows. First, for every cluster size (n) the microcanonical ensembles of clusters corresponding to certain series of cluster energy values (E) are formed. Then, with the help of the CMD method the process of cluster evaporation is investigated, and dependencies $\nu_e^1(n, E)$ are determined. Finally, using the SMD method the canonical ensemble of clusters related to the temperature of interest is formed, and $\nu_e^1(n, T)$ is calculated.

To form the microcanonical ensemble of clusters of size n, we could choose from the corresponding canonical ensemble the cluster whose energies are close to the desirable one. But, to collect in such a manner enough ensemble representative for our purposes (with a number of clusters $N \sim 200-300$), much computation time is needed. Therefore, we proceeded in the following way. Using the SMD method, the canonical ensemble of clusters for a suitable temperature was formed. Then, by successive transition into the co-

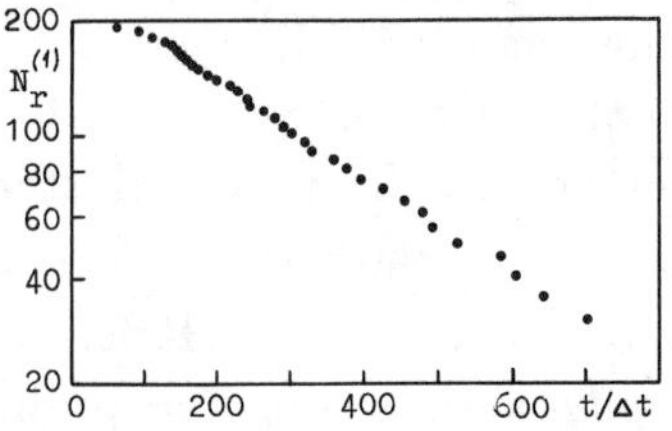

Figure 7: Decay curve for the microcanonical ensemble ($n = 13$, $E\hat{t} = E\hat{r} = 0$, $E = 4140°K$, $\Delta t = 0.05$.

ordinate system of every cluster their translational and rotational motions were eliminated (in this case the total energies of clusters E are reduced to their vibrational energies E_v). After that the clusters were adjusted for energy, multiplying all the atom velocities by the factor $\sqrt{(E - E'_{vp})/E'_{vk}}$, where the primes denote cluster characteristics before the adjustment. As a result, we had the ensemble of clusters of size n with $E_t = E_r = 0$ and $E = E_v = \text{const}$.

Cluster translation does not affect $\nu_e^1(n, E)$, but cluster rotation does, especially for small clusters. In outline, the rotational effects for clusters are the same as for large molecules (for the last see in Refs. 1 and 2). In this paper we shall consider only the previously mentioned ensembles with $E_r = 0$ and thus are not concerned with the rotational effects.

With the ensemble regarded as the microcanonical one being formed, an evolution of every cluster was calculated by the CMD method. At every time step the procedure of cluster analysis based on an estimation of the distances between all the pairs of atoms related to the cluster was performed. It was assumed that r_0 in Eq. (5) had a finite value ($r_0 = 2.6\sigma$). According to this a subcluster (specifically, a monomer) was considered to have been emitted if the distances between all the atoms of the subcluster and those of the maternal cluster exceeded r_0. The first moment when this condition had been satisfied was taken as the subcluster emission moment.

Figure 7 illustrates how a number of the clusters in the ensemble that have not experienced the emission of monomers N_r^1 decreases in time. As can be easily seen, after a certain transition period (with the duration $t_d \approx 10^2\Delta t$ in the case shown), the decay curve can be described by the relation

$$N_r^1 = N \exp\{-\beta(t - t_d)\}$$

From this it follows that at $t \geq t_d$ the process of monomer emission is a stationary random process of the Poisson type. The frequency of the emission ν_e^1 being equal to β can be easily calculated through the decay curve slope.

The character of the decay curve makes it possible to conclude that at $t \leq t_d$ the initial artificial distribution of cluster representative points in phase space (being a result of the adjusting procedure) goes to the equilibrium one. This conclusion is corroborated by the time behavior of distribution functions of bond lengths in clusters

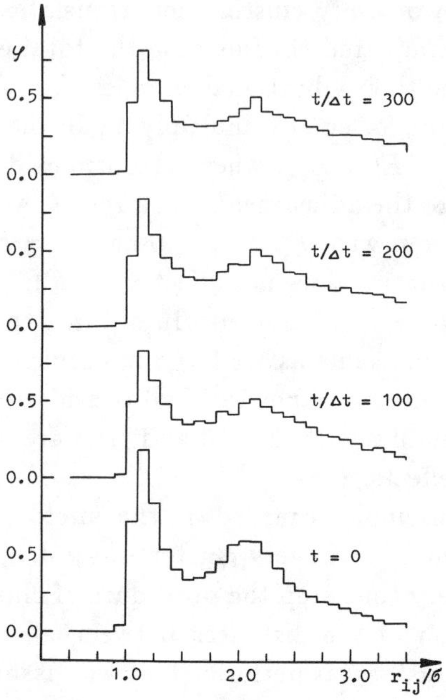

Figure 8: Distribution functions of bond lengths in clusters for the conditions of Fig. 7.

shown in Fig. 8 (the calculations were performed with respect to the clusters that had not experienced the subcluster emission by the given time). As one can see, the rearrangement of atoms in the clusters comes to an end, in the main, by the moment $t = t_d$.

It is valuable to note here that when the time-step of numerical integration of Eq. (2) is altered, the moments of monomer emission for individual clusters drastically change in a random manner, but the decay curves and, as a consequence, the values of ν_e^1 do not exhibit any considerable changes. This testifies to the stochastic nature of atom motions in clusters. The calculations have been performed for the cluster size n being equal to 3, 4, 9, and 13.

In Fig. 9 the dependencies of ν_e^1 on E are shown; here $E_t = E_r = 0$ and $E = E_v$, and U_0 are the minimum energies of the clusters for the pentagonal growth scheme[15] (they are given in Table 1). In addition, the results of Ref. 13 for $n = 4$ are presented (it was assumed here that E is equal to $E_{BDT} + 3\epsilon$, where E_{BDT} is the kinetic energy E used in Ref. 13); as one can see, they show good agreement with our results.

Initially, we attempt to describe the data of Fig. 9 using the classical theory of monomolecular reactions of Rice, Ramsperger, and Kassel (RRK)[1,2]. In accordance with that theory

$$\nu_e^1 = \nu_0 n^{2/3}(1 - E_c/E)^{s-1} \tag{12}$$

where E_c is the critical (minimum) energy required for a monomer to leave the cluster, $s = 3n - 6$, and ν_0 is a proportionality constant with the dimension of frequency (for the explicit purposes we withdraw the factor $n^{2/3}$ characterizing the cluster surface area). However, we did not succeed in these attempts either with the theoretical values of s and E_c or with any empirical ones.

In our opinion, the reason of the correlation between the formula (12) and the data of Fig. 9 being absent is that, in contrast to the assumptions of the RRK theory, the vibrations in clusters under the values of U_0 given are not harmonic in the whole range of cluster energies considered. The last circumstance is illustrated by Fig. 10, where the averages of E_{vk} and E_{vp} vs. E are given by ($E = E_v =< E_{vk} > + < E_{vp} >$). These values were obtained by twofold averaging: both over the time interval $t_d \leq t \leq t_e$ for every cluster (t_e is the evaporative lifetime of the cluster) and over the microcanonical ensembles of the clusters.

As one can see, the relations $< E_{vk} > \approx < E_{vp} > \approx E/2$ being an evidence of the harmonic character of the vibrations holds true only for very low energies. For all other energies, in particular corresponding to those in Fig. 9, there is a considerable discrepancy between $< E_{vk} >$ and $< E_{vp} >$. Note that for $n = 9$ and $n = 13$

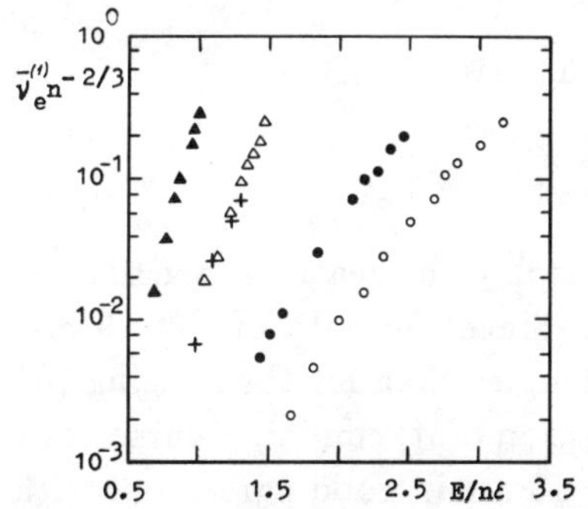

Figure 9: Frequencies of monomer emission $\bar{\nu}^{(1)} = \nu_e^{(1)}\sqrt{m\sigma^2/E}$ vs cluster energies ($E\hat{t} = E\hat{r} = 0$). △, $n = 3$; △, $n = 4$; •, $n = 9$; ∘, $n = 13$; +, Ref. 13.

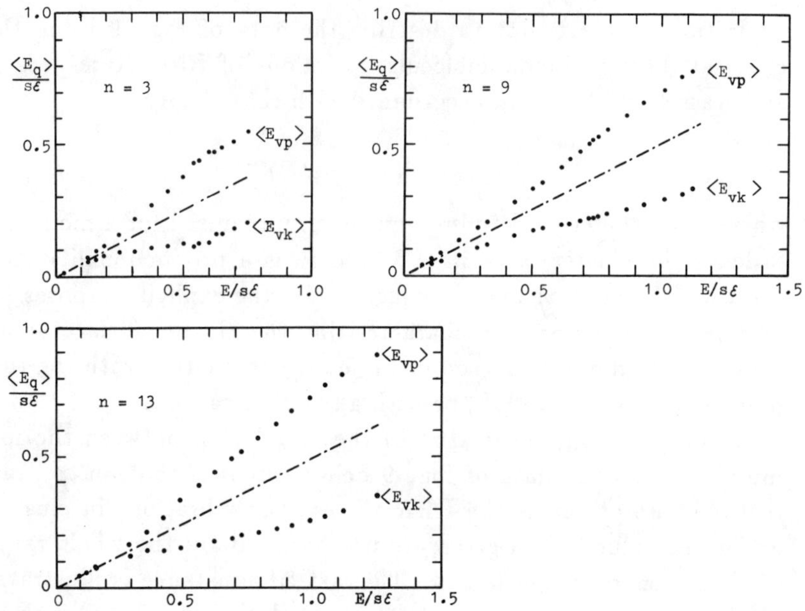

Figure 10: Microcanonical ensemble averages of $E\hat{vk}$ and $E\hat{vp}$ vs cluster energy E ($E\hat{t} = 0$, $E\hat{r} = 0$). Dashed-dotted line, harmonic vibrations.

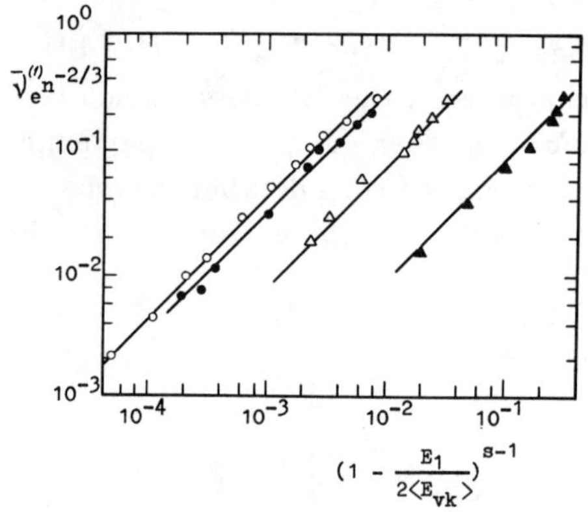

Figure 11: Data of Fig. 9 in terms of RRK theory.

the curves given display the features peculiar to the phenomenon of cluster melting. Since the data of Fig. 9 are related to higher values of cluster energies than for the melting phenomena, we shall not go into a discussion of it, confining ourselves to the remark that our results for $n = 13$ are in good agreement with those of Refs. 17 and 18 (results for free clusters of the size $n = 9$ with the energy being constant are not known to us).

Table 1

Minimum energies of clusters, and parameters of eq.(13)

n	$U_o/\mathcal{E}$	s	$E_1/\mathcal{E}$	$\bar{\nu}_o$ *
3	-3.0	3	0.76	0.78
4	-6.0	6	1.5	8.2
9	-24.113	21	2.83	30.0
13	-44.327	33	3.26	43.0

* Normalization is the same as in Fig. 9.

Presuming that the discrepancy between $< E_{vk} >$ and $< E_{vp} >$ may be attributed not so much to the anharmonic character of the vibrations but to the cluster structure isomerization phenomenon, instead of Eq. (12) we used the relation

$$\nu_e^1 = \nu_0 n^{2/3}(1 - E_1/2 < E_{vk} >)^{s-1} \tag{13}$$

which agrees with Eq. (12) in the harmonic vibrations limit ($E = 2 < E_{vk} >$). The representation of the data of Fig. 9 in the form of Eq. (13) is shown in Fig. 11; the values of E_1 and ν_0 providing the best agreement for the results are given in Table 1. The values of E_1 being smaller than the theoretical ones E_c (for the last see Ref. 15) testify that the clusters are "shaggy" rather than compact.

Though in outline this approach appears to be similar to that of Refs. 11-13, really it is closer to the approach of Refs. 14 and 15 where the Boltzmann entropy hypothesis was used to describe the results of the CMD calculations. Indeed, if we suppose that $E_1 \ll 2 < E_{vk} >$ and $s \gg 1$, then, using the relation $\ln(1-x) \approx -x$ (with $x = E_1/2 < E_{vk} >$), expression (13) can be written in the form

$$\nu_e^1 = \nu_0 n^{2/3} \exp(-E_1/k(T))$$

where $(T) = 2(E_{vk})/s$ is the mean value of cluster temperature.

Finally, let us dwell on a difference between $\nu_e^1(n, E)$ and $\nu_e^1(n, T)$. The latter can be found by averaging the dependencies $\nu_e^1(n, E)$ over the canonical distributions obtained with the help of the SMD method. In the cases investigated the values of $\nu_e^1(n,T)$ turned out to be slightly higher than the corresponding values $\nu_e^1(n,E)$ with the energy E related to $< T >= T$. For example, for $n = 13$ at $T = 60°$Kthe calculations give $\nu_e^1 = 1.14 \times 10^{11}$ s^{-1}. On the other hand, at $< T >= 60°$Kin accordance with Fig. 10(c) we have $E =$ 3810°Kand, then, using the data of Fig. 9 obtain $\nu_e^1 \approx 1.0 \times 10^{11}$ s^{-1}. This discrepancy between the values of ν_e^1 as a result of the difference between the canonical and microcanonical ensembles will change with cluster size as $1/\sqrt{n}$.

Acknowledgments

We wish to thank Drs. V. A. Gaponov and M. S. Ivanov for their useful discussions.

References

[1]Nikitin, E. E., *Theory of Elementary Atom-Molecular Processes in Gases*, Chemistry, Moscow, 1970 (in Russian).

[2]Robinson, P. J. and Holbrook, K. A., *Unimolecular Reactions*, Wiley, London, 1972.

[3]Schofield, P., "Computer Simulation Studies of the Liquid State," *Computer Physics Communications*, Vol. 5, May 1973, pp. 17-23.

[4]Haken, H., *Synergetics*, Springer Verlag, Berlin, 1978.

[5]Biswas, R. and Hamann, D. R., "Simulated Annealing of Silicon Atom Clusters in Langevin Molecular Dynamics," *Physical Review B: Solid State*, Vol. 34, July 1986, pp. 895-901.

[6]Landau, L. D. and Lifshitz, E. M., *Statistical Physics*, Nauka, Moscow, 1964 (in Russian)

[7]Gordietz, B. F., Osipov, A. I. and Shelepin, L. A., *Kinetic Processes in Gases and Molecular Lasers*, Nauka, Moscow, 1980 (in Russian)

[8]Dickey, J. M. and Paskin, A., "Size and Surface Effects on the Phonon Properties of Small Particles," *Physical Review B: Solid State*, Vol. 1, Jan. 1970, pp. 851-857.

[9]Bendat, J. S. and Piersol, A. G., *Engineering Applications of Correlation and Spectral Analysis*, Wiley, New York, 1980.

[10]Brady, J. W., Doll, J. D. and Thompson, D. L., "Cluster Dynamics: A Classical Trajectory Study of $A + A_n \rightleftharpoons A_{n+1}$," *Journal of Chemical Physics*, Vol. 71, Sept. 1979, pp. 2467-2472.

[11]Brady, J. W., Doll, J. D. and Thompson, D. L., "Cluster Dynamics: Further Classical Trajectory Studies of $A + A_n \rightleftharpoons A_{n+1}$," *Journal of Chemical Physics*, Vol. 73, Sept. 1980, pp. 2767-2772.

[12]Brady, J. W., Doll, J. D. and Thompson, D. L., "Cluster Dynamics: A Classical Trajectory Study of $A_n^* \rightarrow A_{n-1} + A$," *Journal of Chemical Physics*, Vol. 74, Jan. 1981, pp. 1026-1028.

[13]Bedanov, V. M., "Vapor Equilibrium Pressure over Embryonic Clusters," *Chemical Physics*, Vol. 6, July 1987, pp. 997-999 (in Russian).

[14]Bedanov, V. M., "Numerical Simulation of Evaporation of Lennard-Jones Clusters and Calculation of Nucleation Rate in Supersaturated Vapor," *Chemical Physics*, Vol. 7, March 1988, pp. 412-419 (in Russian).

[15]Hoare, M. R. and Pal, P., "Physical Cluster Mechanics: Statics and Energy Surfaces for Monatomic Systems," *Advances in Physics*, Vol. 20, March 1971, pp. 161-196.

[16]Jellinek, J., Beck, T. L. and Berry, R. S., "Solid-Liquid Phase Changes in Simulated Isoenergetic Ar_{13}," *Journal of Chemical Physics*, Vol. 84, March 1986, pp. 2783-2794.

[17]Blaisten-Barojas, E., Gazon, I. L. and Avalos-Boria, M., "Melting and Freezing of Lennard-Jones Clusters on a Surface," *Physical Review B: Solid State*, Vol. 36, Dec. 1987, pp. 8447-8455.

Chapter 5. Evaporation and Condensation

Angular Distributions of Molecular Flux Effusing from a Cylindrical Crucible Partially Filled with Liquid

Y. Watanabe,* K. Nanbu,† and S. Igarashi‡
Tohoku University, Sendai, Japan

Abstract

Evaporation into vacuum from a cylindrical crucible partially filled with liquid is studied by use of the direct simulation Monte Carlo method. The evaporation rate is obtained in the near-continuum to free molecular regime for $\ell/a = 1$, 3, and 5, where a is the inner radius of the crucible and ℓ is the distance from the crucible exit to the liquid level. When ℓ is small, the evaporation rate does not change so much with the Knudsen number. The flow structure near the crucible is similar to that of the usual freejet expansion. In the far field the density and radial velocity of gas and the probability of finding a molecule in unit solid angle are calculated as a function of the polar angle. These flow properties in the far field are also obtained in the case of sonic efflux from a circular orifice.

Introduction

The vapor deposition in vacuum chamber is a familiar technique for fabrication of thin films. In most applications nonuniformity of the thickness distribution of deposited films must be less than ±1%. The thickness distribution can be predicted if the angular distribution of molecular flux effusing from a source is known. In the gas source molecular-beam-epitaxy (MBE) technique, material gas is supplied from nozzle or orifice. Since the kind of the

*Research Associate, Institute of High Speed Mechanics.
†Professor, Institute of High Speed Mechanics.
‡Research Associate, Institute of High Speed Mechanics.

material that is in gas phase at an ordinary temperature is limited, however, the evaporation from a crucible filled with molten material is still widely used.

When the vapor pressure of material is low and hence the flow of vapor is free molecular, the angular distribution of molecular flux depends only on the geometry of the crucible.[1,2] When the vapor pressure is high, molecular collisions occur in the inside and outside of the crucible. Previously we treated such case for a two-dimensional crucible.[3,4] From a practical point of view, we here study the vapor flow from a cylindrical crucible partially filled with liquid. The direct simulation Monte Carlo method[5] is used to evaluate the evaporation rate in the near-continuum to free molecular regime and to examine the flow structure near to and far from the crucible. It is shown that the Knudsen number and the position of the liquid level have large effects on the angular distribution of molecular flux. The angular distribution for the case when gas comes into vacuum through a circular orifice with sonic velocity is also given.

Procedure for Numerical Calculation

Figure 1 shows a cylindrical crucible that is set up vertically in vacuum. Molten material is in the inside. The liquid surface is at $x = -\ell$. The inner radius of the crucible is a. Initially the shutter on the top is closed and the inner space is filled with saturated vapor with the number density n_0 and temperature T_0. At time $t = 0$ the shutter is impulsively opened. A steady flow of vapor is established after a short transient. It is assumed that the temperatures of the liquid and the wall of the crucible are always kept at T_0. The lowering of the liquid level due to evaporation is disregarded. The steady flow is governed by the depth-to-radius ratio ℓ/a and the Knudsen number $Kn = \lambda_0/2a$, where λ_0 is the mean free path for the saturated vapor with density n_0 and temperature T_0. Molecules are assumed to be Maxwellian.

Direct Simulation Monte Carlo Method

In order to employ the simplest scattering law for hard sphere molecule we apply the variable diameter hard sphere model[6] to Maxwellian molecules. The mean free path λ_0 for this model takes the form

$$\lambda_0 = 2/(\sqrt{3\pi}\, n_0\bar{\sigma}_0) \tag{1}$$

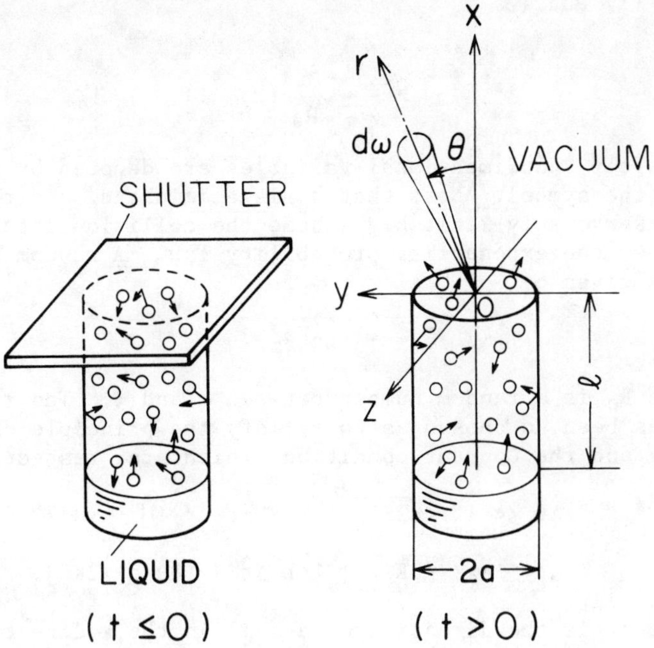

Fig. 1 Cylindrical crucible and coordinates system.

where $\overline{\sigma}_0$ is the average total cross section at temperature T_0. The total collision cross section for a collision pair with relative velocity g is

$$\sigma_T = \overline{\sigma}_0 \sqrt{6RT_0} \, /g \tag{2}$$

where R is the gas constant per unit mass. The molecular collision is simulated by use of the Belotserkovskiy-Yanitskiy scheme.[7,8] The collision rate ν in a cell is

$$\begin{aligned} \nu &= n(N - 1) g \sigma_T \, /2 \\ &= n(N - 1)\overline{\sigma}_0 \sqrt{6RT_0} \, /2 \end{aligned} \tag{3}$$

where n and N are the number density and the number of simulated molecules in the cell, respectively. Here Eq. (2) is used. The collision rate for Maxwellian molecules does not depend on g. The numerical calculation is carried out by use of nondimensional variables. Multiplying ν by $\lambda_0/(2RT_0)^{1/2}$, we have the nondimensional collision rate from

Eqs. (1) and (3)

$$\hat{\nu} = \frac{1}{\sqrt{\pi}} \frac{n}{n_0} (N - 1)$$

Hereafter, nondimensional variables are denoted by a caret over the symbol. Note that $\overline{\sigma}_0$ is eliminated. In the Belotserkovskiy-Yanitskiy scheme the collision interval $\hat{\tau}$ follows the exponential probability law. A random sample of it is given by

$$\hat{\tau} = - (\ell n\ R_n)/\hat{\nu}$$

where R_n is a random number between 0 and 1. The time step Δt has been so chosen as to satisfy the principle of uncoupling and the Courant condition, which are, respectively,

$$\Delta t << \lambda_0 / \sqrt{2RT_0} \quad [\text{or } \Delta\hat{t} << 1]$$

$$\Delta t \lesssim \Delta x / \sqrt{2RT_0} \quad [\text{or } \Delta\hat{t} \lesssim (\Delta x/a)/2Kn]$$

where Δx is the dimension of a cell in the x direction (Fig. 1). Our choice is $\Delta x/a = 0.2$ for any Kn, and $\Delta\hat{t} = 0.1$ for $Kn \leq 1$ and $\Delta\hat{t} = 0.01$ for $Kn = 10$.

Molecules incident on the wall of the crucible are diffusely reflected, and those incident on the liquid surface are always condensed. The velocities of evaporated molecules are sampled from the Maxwellian velocity at temperature T_0. The number of molecules evaporated from the liquid surface per unit time is

$$\dot{n}_0 = \pi a^2 n_0 \ (RT_0/2\pi)^{1/2}$$

The mass flux $\dot{m}_0$ is equal to $m\dot{n}_0$, m being the mass of a molecule. Since some of the evaporated molecules are recondensed because of the molecular collision or backward reflection on the wall, the net evaporation rate $\dot{m}$ is smaller than $\dot{m}_0$. Here evaporation is assumed to occur at the saturation density n_0 and temperature T_0.

The domain of calculation is in the inside and outside of the crucible. The outside domain is the cylinder whose radius and height are L. The size L should be so large that the free molecular flow prevails in the outside of this cylinder. Our choice is $L/a = 20$, as in the two-dimensional case.[4] Since the flow is axisymmetrical, we only have to make the network of cells on the x-y plane in Fig. 1. Two kinds of the network are used: the Cartesian network with the cell size $\Delta x \times \Delta y$ is for collision calculation, and the polar network of $\Delta r \times \Delta\theta$ is for data sampling. Our choice is

$$\Delta x = \Delta y = \Delta r = 0.2a$$

$$\Delta\theta = 3 \text{ deg}$$

The weighting factor is not used. The same network is also employed in case of the circular orifice. It is assumed that gas comes into vaccum through the orifice with sonic speed. Note that in this case a is the radius of the orifice and n_0 and T_0 are the density and temperature of the gas at the sonic condition. The sampled data are the density n, the x and r components v_x, v_r of the flow velocity, the parallel and perpendicular temperatures T_x, T_p, the net evaporation rate $\dot{m}$, and the angular distribution of molecular flux.

Angular Distribution of Molecular Flux

The molecule that has crossed over the boundary of the calculation domain runs on a straight line because of no more collision. We consider N such molecules. The number of the molecules scattered into the solid angle $d\omega$ in Fig. 1 is written as

$$dN = N\, f(\theta)\, d\omega \tag{4}$$

The function $f(\theta)$ is called angular distribution of molecular flux. The fraction $f(\theta)\, d\omega$ is the probability that a molecule chosen at random is scattered into $d\omega$. We consider the distribution $f(\theta)$ at a very far distance from the crucible. The distribution $f(\theta)$ is subject to the normalization condition

$$2\pi \int_0^{\pi} f(\theta) \sin\theta d\theta = 1$$

The upper limit of integration is not $\pi/2$ but π. This is because the molecule that has a negative x component velocity in its final collision is scattered into $d\omega$ of $\theta > \pi/2$. We have determined $f(\theta)$ from Eq. (4) by counting the number of molecules ΔN scattered into the range $(\theta - \Delta\theta/2, \theta + \Delta\theta/2)$ at the radial distance of $r = 20a$. Note that there are no molecular collisions for $r > 20a$. As for N and $\Delta\theta$, our choice is $N > 10^6$ and $\Delta\theta = 2.5$ (deg).

The polar angle θ sampled for each molecule is, in fact, the direction of the molecular velocity. It happens that a molecule whose velocity has the polar angle θ does not appear in the solid angle $d\omega$ in Fig. 1. This is because the position of the final collision of the molecule is not the origin 0. For this reason the distribution $f(\theta)$ obtained here holds good at a distance so far that the collision dominated flow region near the crucible can be regarded as a point source at the origin. At a large distance r, the probability $f(\theta)\ d\omega$ is proportional to the number flux $nv_r r^2\ d\omega$, where n is the number density and v_r is the radial component of the flow velocity. We also sample nv_r at $r = 20a$ and compare it with $f(\theta)$. If nv_r were sampled at $r = \infty$, it would exactly be proportional to $f(\theta)$. It is possibe, however, that the exact proportionality does not hold because $f(\theta)$ is the distribution at $r = \infty$, but nv_r is that at $r = 20a$.

Results and Discussion

Field Near the Crucible

Figure 2 shows the flow properties along the axis of the crucible for Kn = 0.1 and $\ell/a = 5$ and 1. Here v_x is the velocity component in the direction of the axis, n is the number density, and T_x, T_p, and T are the parallel, perpendicular, and overall temperatures, respectively. Note that the liquid level is at the position of the ordinate and the exit of the crucible is at x = 0. The vapor flow is strongly accelerated near the exit. The axial velocity and overall temperature are almost frozen at $x/a = 6$. The frozen limit of v_x for $\ell/a = 5$ is smaller than that for $\ell/a = 1$. This is because the flow undergoes a greater frictional resistance on the longer wall of the crucible. The behavior of T_x and T_p is similar to that found in axisymmetric expansion of freejets. Let $|\ T_x - T_p\ |$ be a measure of the degree of nonequilibrium. In case of $\ell/a = 5$ the flow is nearly in equilibrium up to $x/a = -2$, but in case of $\ell/a = 1$ nonequilibrium begins at the very liquid level. The freezing of the parallel temperature starts at $x/a = 1$.

Figure 3 shows the number density along the radial lines with the polar angles 0 and 20 deg. The straight lines show $n \propto r^{-2}$. It is seen that the density is inversely proportional to r^2 in the far field. Figure 4 shows the net evaporation rate $\dot{m}$ (evaporation rate minus recondensation rate) as a function of the Knudsen number. Here, $\dot{m}_0$ is the evaporation rate in the idealistic case when no recondensation occurs. The ratio $\dot{m}/\dot{m}_0$ is always less than unity. As is shown in the two-dimensional case,[3] the rate $\dot{m}$ decreases with increasing ℓ, i.e., lowering the liquid level. This is because of the chance that an evaporated

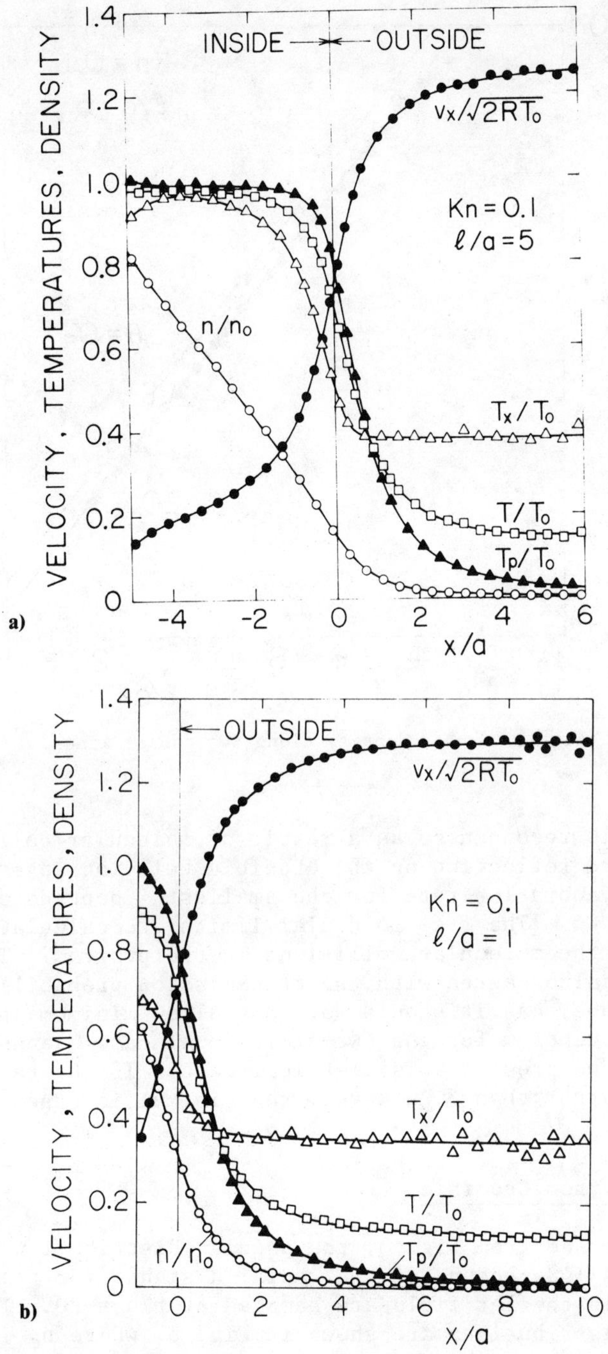

Fig. 2 Flow properties along the centerline for Kn = 0.1: a) $\ell/a = 5$; b) $\ell/a = 1$.

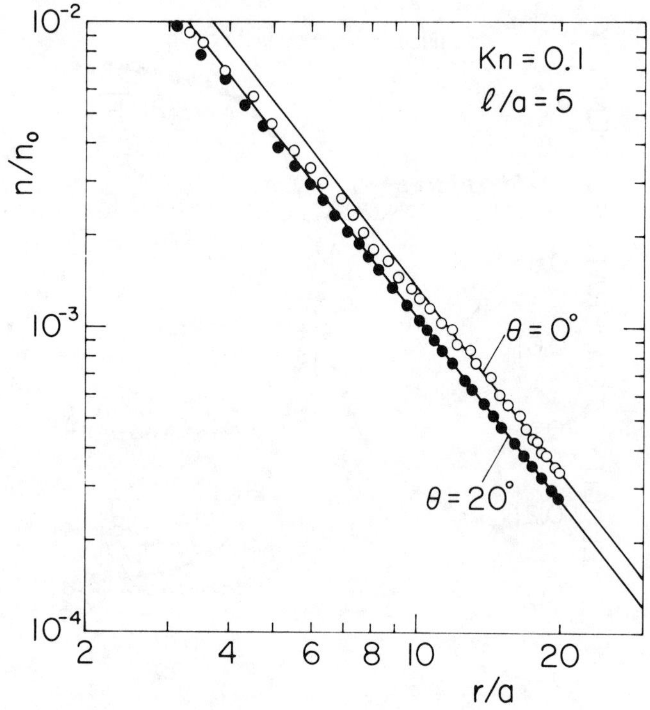

Fig. 3 Number density along two radial lines.

molecule is recondensed as a result of molecular collision or backward reflection on the crucible wall increases with ℓ. The evaporation rate for the smallest ℓ depends only weakly on Kn. The free molecular limits are calculated by bypassing the molecular collisions in the program. These limiting values agree with the transmission probabilities of a short tube,[9] as it should do. A shallow minimum appeared in the curve of $\dot{m}$ for the two-dimensional case[3] cannot be found in the present axisymmetrical case. If the ratio ℓ/a is much larger than 5, however, the minimum is expected to appear.

Field Far from Crucible

As we see from Fig. 3, the angular distribution of the density is independent of the radial distance for r/a > 10. All data in the far field are sampled at r/a = 20. The density distributions are shown in Fig. 5, where n_c is the number density on the centerline θ = 0. It is seen that the effect of ℓ/a is small for Kn = 0.1 but is large for Kn = 1.

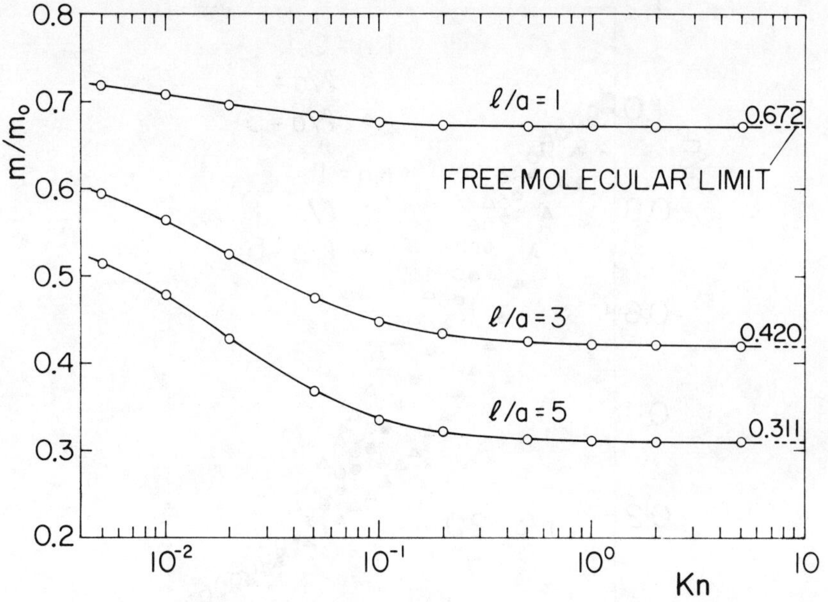

Fig. 4 Evaporation rate.

Figure 6 shows the radial component v_r of the flow velocity. Since the velocity is already frozen near x/a = 6 (Fig. 2), the velocity v_r in Fig. 6 represents the terminal velocity. The molecular collisions accelerate the flow. The collision dominated region, where the density is comparatively high, is wider for smaller Kn, so that v_r is larger for Kn = 0.1. Since the density decreases with increasing θ, so does v_r.

Figure 7 shows the angular distributions $f(\theta)$ for ℓ/a = 1 and 5. It is seen that in case of larger ℓ/a, the distribution becomes sharper with increasing Kn, but the opposite is true in case of smaller ℓ/a. This can be explained as follows. There are two factors that make the distribution sharp: one is the inertia effect of the flow, which increases with the decrease of Kn, and the other is the collimation effect of the crucible wall which is stronger for greater ℓ/a. When ℓ/a is large, the collimation effect at Kn = 10 is greater than the inertia effect at kn = 0.1. When ℓ/a is small, the inertia effect at Kn = 0.1 exceeds the collimation effect at Kn = 10. Figure 8 shows the effect of ℓ/a on $f(\theta)$ at fixed Kn. The curves represent $f(\theta)$, and the symbols represent the normalized nv_r. The two results are in good agreement. In case of Kn = 0.1, the effect of ℓ/a on $f(\theta)$ is small, since the collimation effect of the wall is almost smeared by molecular collisions. In

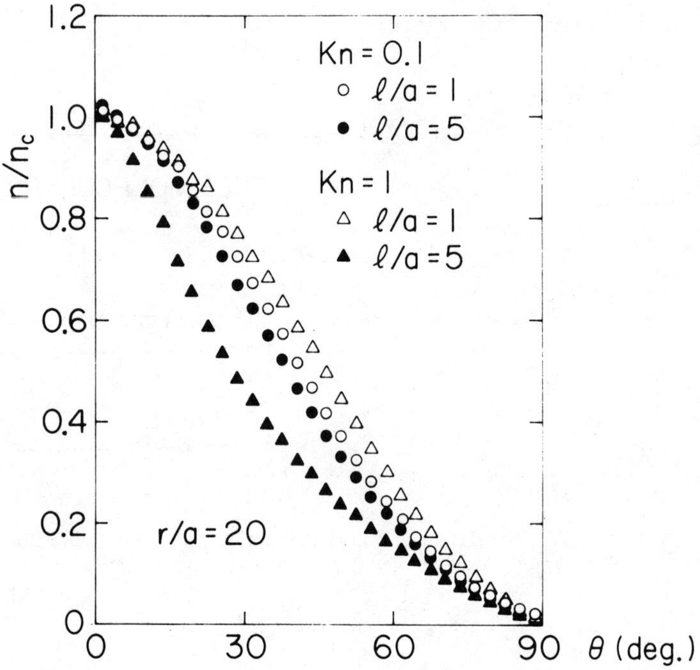

Fig. 5 Angular distributions of number density.

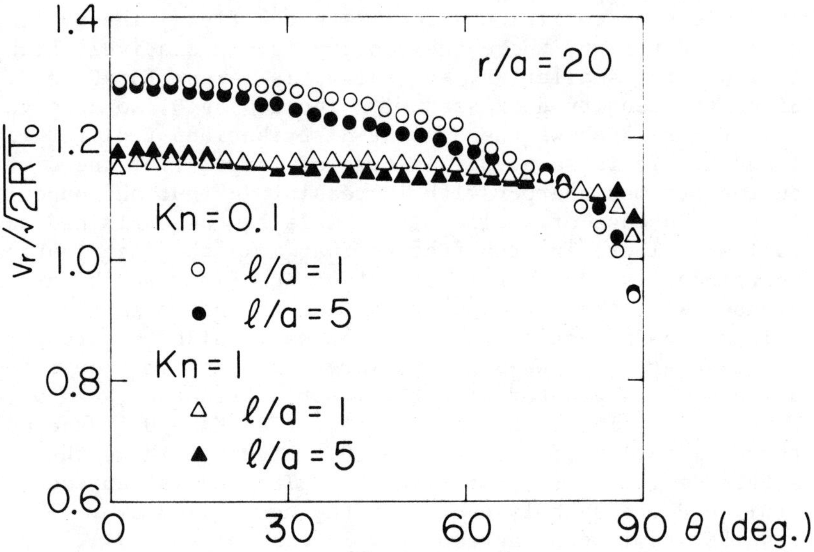

Fig. 6 Angular distributions of radial velocity.

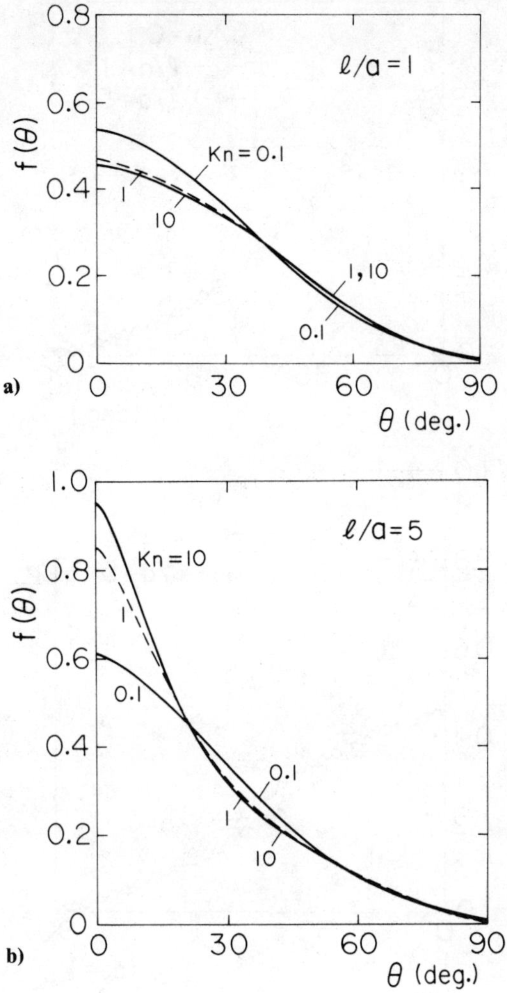

Fig. 7 Angular distributions of molecular flux: a) $\ell/a = 1$; b) $\ell/a = 5$.

case of Kn = 10, molecules are strongly collimated for larger ℓ by the molecule-surface collision.

Circular Orifice

In contrast to that for the crucible, gas comes into vaccum with sonic velocity. The density and radial flow velocity in the far field are shown in Fig. 9 for Kn = 0.01. (The Knudsen number is usually small in the gas source MBE.) The solid curve represents the best-fit equation for the

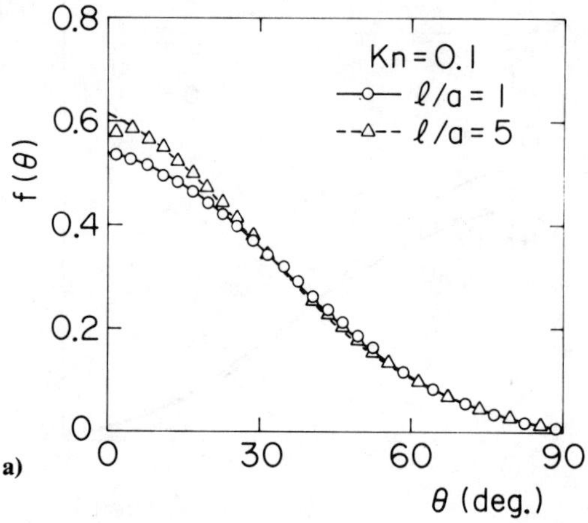

a)

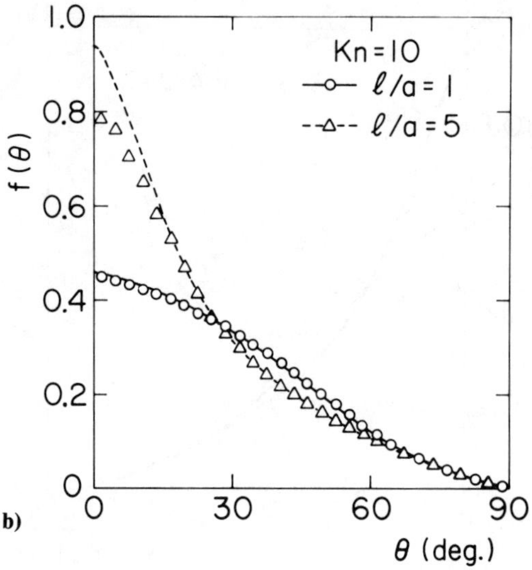

b)

Fig. 8 Angular distributions of molecular flux: a) Kn = 0.1; b) Kn = 10.

direct simulation Monte Carlo data

$$\frac{n}{n_c} = \begin{cases} \cos\theta \cos^2(1.15\theta) & (\ 0 < \theta < \theta_0\ ) \\ 0.370 \exp[\ -\ 0.370(\theta - \theta_0)\ ] & (\ \theta_0 < \theta < \pi/2\ ) \end{cases}$$

where $\theta_0 = 2\pi/9 (= 40$ deg). The dashed curve is the continuum solution of Ashkenas and Sherman[10] and includes the correction of displacing apparent point source. We see

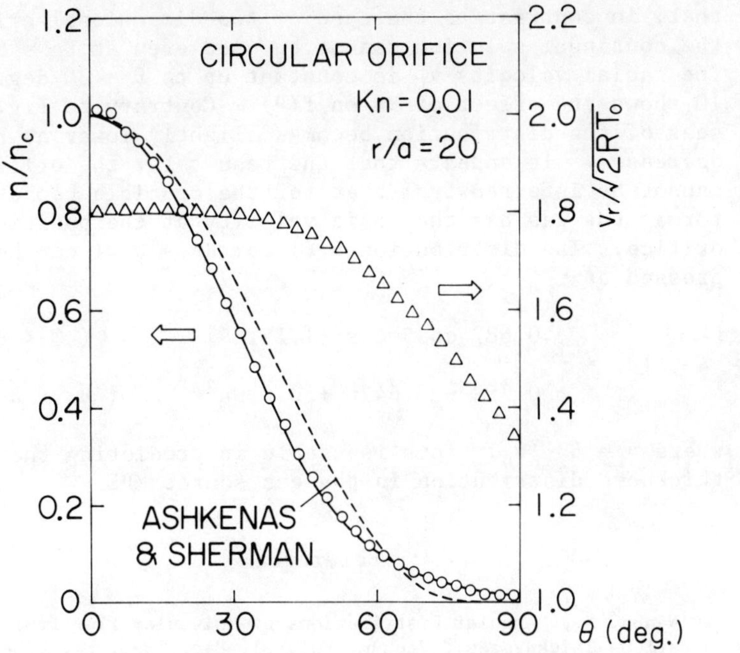

Fig. 9 Angular distributions of number density and radial velocity for the orifice.

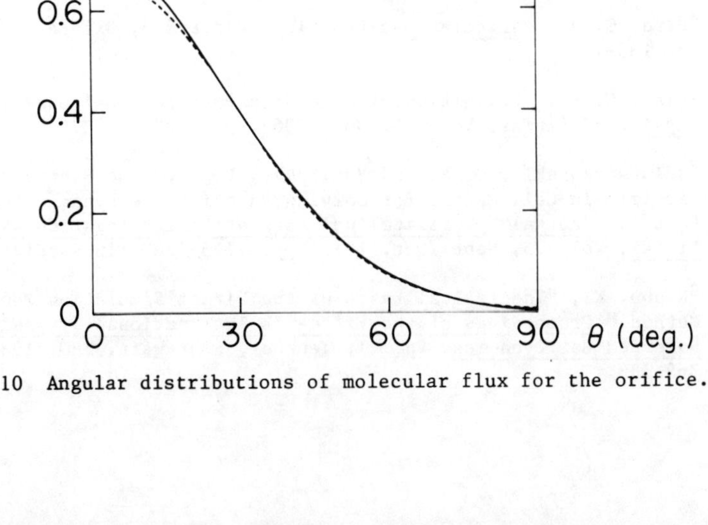

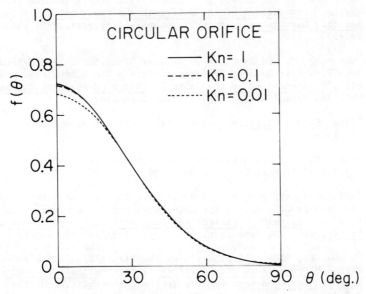

Fig. 10 Angular distributions of molecular flux for the orifice.

that, in contrast to the case of two-dimensional orifice,[11] the continuum solution cannot be used even at Kn = 0.01. The radial velocity v_r is constant up to θ = 30 deg. Figure 10 shows the effect of Kn on f(θ). Contrary to Fig. 7a, the peak of the distribution becomes slightly lower as Kn decreases. It appears that the result for the orifice cannot be inferred from that for the crucible because in the former the gas has the sonic velocity at the position of the orifice. The distribution f(θ) for Kn = 0.01 can be expressed as

$$f(\theta) = \begin{cases} 0.682 \cos\theta \cos^2(1.151\,\theta) & (\,0 < \theta < \theta_0\,) \\ 0.252 - 0.642\eta + 0.420\eta^2 & (\,\theta_0 < \theta < \pi/2\,) \end{cases}$$

where $\eta = \theta - \theta_0$. This is usable in predicting the film thickness distribution in the gas source MBE.

References

[1]Nanbu, K., "Angular Distributions of Molecular Flux from Orifices Various Thicknesses," Vacuum, Vol. 35, Dec. 1985, pp. 573-576.

[2]Nanbu, K. and Watanabe, Y., "Thickness Distribution of Films Fabricated by the Molecular Beam Epitaxy Technique," Vacuum, Vol. 36, June 1986, pp. 349-354.

[3]Watanabe, Y. and Nanbu, K., "Evaporation from a Furnace into a Vacuum," Proceedings of the Fifteenth International Symposium on Rarefied Gas Dynamics, Vol. 2, Teubner, Stuttgart, FRG, 1986, pp. 261-270.

[4]Watanabe, Y. and Nanbu, K., "Angular Distributions of Molecular Beam Flux (1st Report, Evaporation from a Two-Dimensional Crucible)," Transactions of the Japan Society of Mechanical Engineers, Vol. 54, Feb. 1988, pp. 459-465 (in Japanese).

[5]Bird, G. A., Molecular Gas Dynamics, Clarendon, Oxford, UK, 1976, pp. 133-153.

[6]Bird, G. A., "Definition of Mean Free Path for Real Gases," Physics of Fluids, Vol. 26, Nov. 1983, pp. 3222-3223.

[7]Belotserkovskiy, O. M. and Yanitskiy, V. Y., "The Statistical Particle-in-Cell Method for Solving Rarefied Gas Dynamic Problems, Part 1," Zhurnal Vychislitel'noi Matematiki i Matematicheskoi Fiziki, Vol. 15, Sept./Oct. 1975, pp. 1195-1208 (in Russian).

[8]Nanbu, K., "Theoretical Basis of the Direct Simulation Monte Carlo Method," Proceedings of the Fifteenth International Symposium on Rarefied Gas Dynamics, Vol. 1, Teubner, Stuttgart, FRG, 1986, pp. 369-383.

[9]Bird, G. A., Molecular Gas Dynamics, Clarendon, Oxford, UK, 1976, p. 104.

[10]Ashkenas, H. and Sherman, F. S., "The Structure and Utilization of Supersonic Free Jets in Low Density Wind Tunnels," Proceedings of the Fourth International Symposium on Rarefied Gas Dynamics, Vol. 2, Academic, New York, 1966, pp. 84-105.

[11]Anderson, J. B., Foch, J. D., Shaw, M. J., Stern, R. C., and Wu, B. J., "Monte Carlo Simulation of Free Jet Flow from a Slit," Proceedings of the Fifteenth International Symposium on Rarefied Gas Dynamics, Vol. 1, Teubner, Stuttgart, FRG, 1986, pp. 442-451.

Numerical Studies on Evaporation and Deposition of a Rarefied Gas in a Closed Chamber

T. Inamuro*
Mitsubishi Heavy Industries, Ltd., Yokohama, Japan

Abstract

The behavior of a rarefied gas evaporating from one portion of the walls of a two-dimensional chamber and depositing on another portion is investigated by numerical analyses. Two numerical methods suitable for free molecular flows and transition regime flows, respectively, are used. For the free molecular flows, an integral equation for the mass flux of incident molecules on walls is solved by using the boundary element method; for the transition regime flows, the Boltzmann-Krook-Welander equation is solved by using the finite-difference method. As examples of practical engineering applications, the flows in a vacuum vapor deposition apparatus and a cryopump are studied. The desired results can be obtained with limited computation time.

I. Introduction

Investigation of rarefied gas flows accompanied by evaporation and deposition of gas molecules is not only an interesting physical subject but also an important engineering problem. For instance, it is necessary to investigate these flows when designing devices for making thin films, such as a vacuum vapor deposition apparatus, and in designing vacuum chambers, such as a cryopump. In the vacuum vapor deposition apparatus, it is important to estimate the thickness and distribution of the films. In the cryopump, it is necessary to predict the pumping speeds and the heat loads on the pump.

Recently, Nanbu[1] analyzed rarefied gas flows in a model of an apparatus for making thin films by using the direct simulation Monte Carlo method. In addition, for free molecular flow, the test particle Monte Carlo method has been used for a long time.[2] However, calculations by the Monte Carlo method gen-

---

*Currently, Visiting Associate, California Institute of Technology, Pasadena, CA.

erally are very costly in computation time. It would be desirable to use a more suitable method for our problem in order to obtain accurate results with limited computation time.

In the present paper, we calculate practical engineering problems by suitable numerical methods for both free molecular and transition regime flows and show the usefulness of the methods.

II. Problem and Method of Solution

2.1. Problem and Assumptions

Consider a rarefied gas evaporating from one portion of the walls of a two-dimensional chamber and depositing on another portion. We analyze the steady problem under the following assumptions: 1) The behavior of the gas is described by the Boltzmann-Krook-Welander (B-K-W) equation,[3] sometimes called the BGK model equation, in the case of transition regime flows. 2) The gas molecules, on leaving the evaporation surface, have the stationary Maxwellian distribution corresponding to the saturated gas at the surface temperature. 3) On the deposition surface, the fraction α of the incident gas molecules is deposited on the surface, and the remaining fraction $(1 - \alpha)$ is reflected diffusely. α is called the condensation coefficient and varies with surface location.

2.2. Free Molecular Flows

As in Ref. 4, first we show that the problem can be reduced to an integral equation for the mass flux of incident molecules on the walls, except the surface where the molecules evaporate. Let X_i be a boundary point, ξ_i the molecular velocity, n_i the unit normal to the boundary at X_i pointing toward the gas, and $f(X_i, \xi_i)$ the velocity distribution function of the gas molecules. The mass flux $\sigma(X_i)$ of incident molecules on the boundary point is given by

$$\sigma(X_i) = -\frac{2}{\pi} \int_{\xi_i n_i < 0} \xi_j n_j \, f(X_i, \xi_i) \, d\xi_1 \, d\xi_2 \, d\xi_3 \tag{1}$$

where the constant $2/\pi$ is multiplied for later convenience. In free molecular flows, $f(X_i, \xi_i)$ in Eq. (1) can be replaced by $f(\tilde{X}_i, \xi_i)$, where $\tilde{X}_i$ is the boundary point from which the molecules with velocity ξ_i incident on X_i originate (see Fig. 1). With $\sigma(\tilde{X}_i)$ we can express $f(\tilde{X}_i, \xi_i)$ on the surface as follows:

On the deposition surface,

$$f(\tilde{X}_i, \xi_i) = (1 - \alpha)\frac{\sigma(\tilde{X}_i)}{4R^2T_B^2} \exp\left[-\frac{\xi_i^2}{2RT_B}\right] \tag{2}$$

where T_B is the temperature of the surface and R is the gas constant.

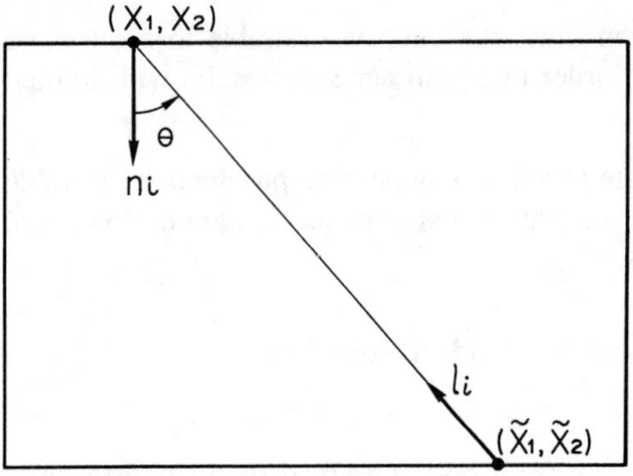

Fig. 1 Notations: $X_i, \tilde{X}_i, n_i, l_i, \theta$.

On the evaporation surface,

$$f(\tilde{X}_i, \xi_i) = \frac{\sigma_W}{4R^2T_W^2} \exp\left[-\frac{\xi_i^2}{2RT_W}\right] \tag{3}$$

where

$$\sigma_W = \frac{(2RT_W)^{1/2}}{\pi^{3/2}} \rho_W$$

T_W is the temperature of the surface and ρ_W the saturated gas density at the temperature T_W.

Hence, substituting Eqs. (2) and (3) into Eq. (1), we obtain the integral equation for the mass flux $\sigma(X_i)$:

$$\sigma(X_i) = \frac{1-\alpha}{2} \int_{\text{depo.}} l_j n_j \sigma(\tilde{X}_i)\, d\theta + \frac{1}{2} \int_{\text{evap.}} l_j n_j \sigma_W\, d\theta \tag{4}$$

where $l_i = \xi_i/\xi$ with $\xi = |\xi_i|$ and θ is the angle between n_i and $-l_i$.

Note that the integral equation (4) contains only the temperature T_W of the evaporation surface. Therefore, the mass flux of incident molecules on a surface depends only on T_W and does not depend on the temperature of the deposition surface.

We solve the integral equation (4) by using the boundary element method.[5] The walls are divided into small straight-line elements, and the mass flux σ is assumed to be linear within an element. That is,

$$\sigma(s) = \sigma_k + \frac{s}{s_k}(\sigma_{k+1} - \sigma_k) \tag{5}$$

where s is a linear coordinate in the element ($s = 0$ at point k); σ_k and σ_{k+1} the mass fluxes at points k and $k+1$, respectively; and s_k the length of the element (Fig. 2).

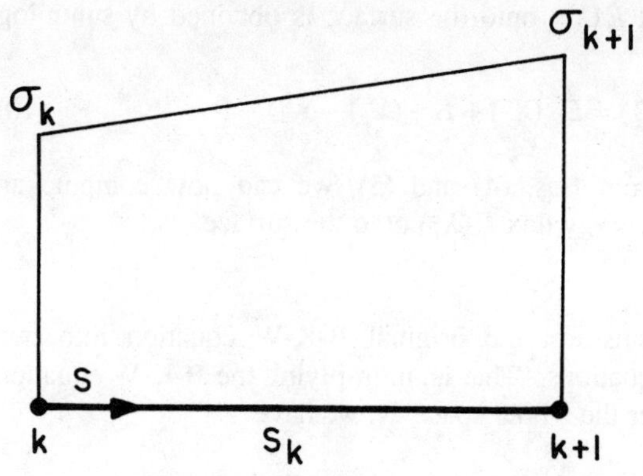

Fig. 2 Boundary elements.

Substituting Eq. (5) into Eq. (4), we obtain linear simultaneous equations for the unknowns σ_k. Solving these equations numerically, we can obtain an approximation for the mass flux $\sigma(X_i)$ of incident molecules on the boundary surface.

Using the resulting approximation for $\sigma(X_i)$, we then can calculate the energy flux. The inward energy flux $E_{\rm in}(X_i)$ to the surface is expressed as a function of $f(X_i, \xi_i)$ as follows:

$$E_{\rm in}(X_i) = -\int_{\xi_i n_i < 0} \frac{1}{2}\xi_i^2 \xi_j n_j\, f(X_i, \xi_i)\, d\xi_1\, d\xi_2\, d\xi_3 \tag{6}$$

Here consider only monatomic molecules that possess no internal energy. Now, replacing $f(X_i, \xi_i)$ in Eq. (6) by $f(\tilde{X}_i, \xi_i)$ and substituting Eqs. (2) and (3) into Eq. (6), we obtain

$$E_{\rm in}(X_i) = \frac{3}{8}\sqrt{2\pi R}\ (1-\alpha) \int_{\rm depo.} l_j n_j \sqrt{T_B(\tilde{X}_i)}\, \sigma(\tilde{X}_i)\, d\theta$$

$$+ \frac{3}{8}\sqrt{2\pi R} \int_{\rm evap.} l_j n_j \sqrt{T_W}\, \sigma_W\, d\theta \tag{7}$$

On the other hand, the outward energy flux from the surface is

$$E_{\rm out}(X_i) = -\int_{\xi_i n_i > 0} \frac{1}{2}\xi_i^2 \xi_j n_j\, f(X_i, \xi_i)\, d\xi_1\, d\xi_2\, d\xi_3 \tag{8}$$

Substituting Eq. (2) into Eq. (8), we have

$$E_{\rm out}(X_i) = -\frac{3(1-\alpha)}{4}\sqrt{2\pi R T_B(X_i)}\, \sigma(X_i) \tag{9}$$

Therefore, the net energy flux $E(X_i)$ onto the surface is obtained by summing Eqs. (7) and (9). That is,

$$E(X_i) = E_{\text{in}}(X_i) + E_{\text{out}}(X_i) \tag{10}$$

Then, using σ_k calculated from Eqs. (4) and (5), we can now compute an approximation E_k for the net energy flux $E(X_i)$ onto the surface.

2.3. Transition Regime Flows

Following Ref. 6, we transform the original B-K-W equation into two coupled integro-differential equations. That is, multiplying the B-K-W equation by 1 or ξ_3^2 and integrating over the whole space ξ_3, we have

$$\frac{\partial}{\partial \bar{t}}\begin{bmatrix} g \\ h \end{bmatrix} + \zeta_1 \frac{\partial}{\partial \bar{x}_1}\begin{bmatrix} g \\ h \end{bmatrix} + \zeta_2 \frac{\partial}{\partial \bar{x}_2}\begin{bmatrix} g \\ h \end{bmatrix} = \frac{2}{\sqrt{\pi}}\left[\frac{\rho}{\rho_W}\right]\begin{bmatrix} g_e - g \\ h_e - h \end{bmatrix} \tag{11}$$

$$\begin{bmatrix} g_e \\ h_e \end{bmatrix} = \frac{1}{\pi}\frac{\rho}{\rho_W}\begin{bmatrix} \left(T/T_W\right)^{-1} \\ \dfrac{1}{2} \end{bmatrix} \exp\left[-\left(\frac{T}{T_W}\right)^{-1}\left\{\left(\zeta_1 - \frac{u_1}{\sqrt{2RT_W}}\right)^2 + \left(\zeta_2 - \frac{u_2}{\sqrt{2RT_W}}\right)^2\right\}\right] \tag{12}$$

$$\frac{\rho}{\rho_W} = \iint_{-\infty}^{\infty} g \, d\zeta_1 \, d\zeta_2 \tag{13a}$$

$$\frac{u_1}{\sqrt{2RT_W}} = \left[\frac{\rho}{\rho_W}\right]^{-1} \iint_{-\infty}^{\infty} \zeta_1 g \, d\zeta_1 \, d\zeta_2 \tag{13b}$$

$$\frac{u_2}{\sqrt{2RT_W}} = \left[\frac{\rho}{\rho_W}\right]^{-1} \iint_{-\infty}^{\infty} \zeta_2 g \, d\zeta_1 \, d\zeta_2 \tag{13c}$$

$$\frac{T}{T_W} = \frac{2}{3}\left[\frac{\rho}{\rho_W}\right]^{-1}\left[\iint_{-\infty}^{\infty}\left\{\left(\zeta_1 - \frac{u_1}{\sqrt{2RT_W}}\right)^2 + \left(\zeta_2 - \frac{u_2}{\sqrt{2RT_W}}\right)^2\right\} g \, d\zeta_1 \, d\zeta_2 + \iint_{-\infty}^{\infty} h \, d\zeta_1 \, d\zeta_2\right] \tag{13d}$$

where

$$g = 2RT_W \rho_W^{-1} \int_{-\infty}^{\infty} f \, d\xi_3 \tag{14a}$$

$$h = \rho_W^{-1} \int_{-\infty}^{\infty} \xi_3^2 f \, d\xi_3 \tag{14b}$$

$\bar{t} = (2RT_W)^{1/2} \lambda_W^{-1} t$, $\bar{x}_1 = \lambda_W^{-1} x_1$, $\bar{x}_2 = \lambda_W^{-1} x_2$ (x_1 and x_2 are the rectangular spatial coordinates), $\zeta_1 = (2RT_W)^{-1/2} \xi_1$, $\zeta_2 = (2RT_W)^{-1/2} \xi_2$, and $\lambda_W = (8RT_W/\pi)^{1/2} (A_{\text{col}} \rho_W)^{-1}$ is the mean free path of the saturated gas at temperature T_W ($A_{\text{col}} \rho_W$ is the collision frequency of the saturated gas molecules).

The boundary conditions are as follows:

On the evaporation surface,

$$\begin{bmatrix} g \\ h \end{bmatrix} = \frac{1}{\pi} \begin{bmatrix} 1 \\ \frac{1}{2} \end{bmatrix} \exp(-\zeta_1^2 - \zeta_2^2) \quad \text{for} \quad \zeta_i n_i > 0 \tag{15}$$

On the deposition surface,

$$\begin{bmatrix} g \\ h \end{bmatrix} = \frac{1-\alpha}{\pi} \left[\frac{\rho_B}{\rho_W} \right] \begin{bmatrix} \left[T_B/T_W \right]^{-1} \\ \frac{1}{2} \end{bmatrix} \exp\left[-\left[\frac{T_B}{T_W} \right]^{-1} (\zeta_1^2 + \zeta_2^2) \right]$$

$$\text{for } \zeta_i n_i > 0 \tag{16}$$

$$\frac{\rho_B}{\rho_W} = -2\sqrt{\pi} \left[\frac{T_B}{T_W} \right]^{-1/2} \int_{\zeta_i n_i < 0} \zeta_i n_i g \, d\zeta_1 d\zeta_2 \tag{17}$$

The boundary-value problem [Eqs. (11-17)] is solved numerically by using the finite-difference method.[7] We use an implicit method, except that the collision terms are evaluated explicitly. The finite-difference scheme for Eq. (11) is as follows:

$$\frac{g^{n+1} - g^n}{\Delta t} + \zeta_1 D_1 g^{n+1} + \zeta_2 D_2 g^{n+1} = \frac{2}{\sqrt{\pi}} \left[\frac{\rho^n}{\rho_W} \right] (g_e^n - g^{n+1}) \tag{18}$$

where Δt is the increment of time t; g^n, g^{n+1}, ρ^n, and g_e^n are approximations to $g(n\Delta t)$, $g[(n+1)\Delta t]$, $\rho(n\Delta t)$, and $g_e(n\Delta t)$, respectively; and D_1 and D_2 are spatial differencing operators. The same scheme is used for the unknown function h. The spatial differencing operators D_1 and D_2 are approximated by the following first- and third-order difference schemes to ensure numerical stability.

First-order upwind scheme for D_1:

$$D_1 g = \begin{cases} \dfrac{g_i - g_{i-1}}{\Delta \bar{x}_1} & \text{for} \quad \zeta_1 > 0 \\ \dfrac{g_{i+1} - g_i}{\Delta \bar{x}_1} & \text{for} \quad \zeta_1 < 0 \end{cases} \tag{19}$$

Third-order upwind scheme for D_1:

$$D_1 g = \begin{cases} \dfrac{11 g_i - 18 g_{i-1} + 9 g_{i-2} - 2 g_{i-3}}{6 \Delta \bar{x}_1} & \text{for} \quad \zeta_1 > 0 \\ \dfrac{-11 g_i + 18 g_{i+1} - 9 g_{i+2} + 2 g_{i+3}}{6 \Delta \bar{x}_1} & \text{for} \quad \zeta_1 < 0 \end{cases} \tag{20}$$

In Eqs. (19) and (20), $\Delta \bar{x}_1$ is the increment of the variable $\bar{x}_1$ and $g_i, g_{i-1}, \cdots$ are approximations to $g(i\Delta \bar{x}_1)$, $g[(i-1)\Delta \bar{x}_1], \cdots$.

The same schemes are used for D_2. The integrations over the velocity space in Eqs. (13) and (17) are approximated by using the trapezoidal rule.

III. Results and Discussion

3.1. Vacuum Vapor Deposition Apparatus

Let us consider a two-dimensional model of a vacuum vapor deposition chamber (see Fig. 3). $\overline{BC}$ is the evaporation surface (temperature T_W), $\overline{EF}$ is the deposition surface (temperature T_B, condensation coefficient α), and the remaining walls are surfaces at temperature T_B and have a condensation coefficient of $\alpha = 0$. In this apparatus, it is important to determine the mass

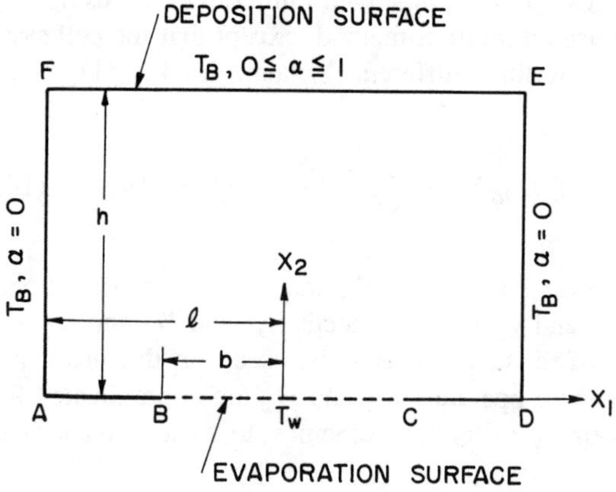

Fig. 3 Model chamber of vacuum vapor deposition.

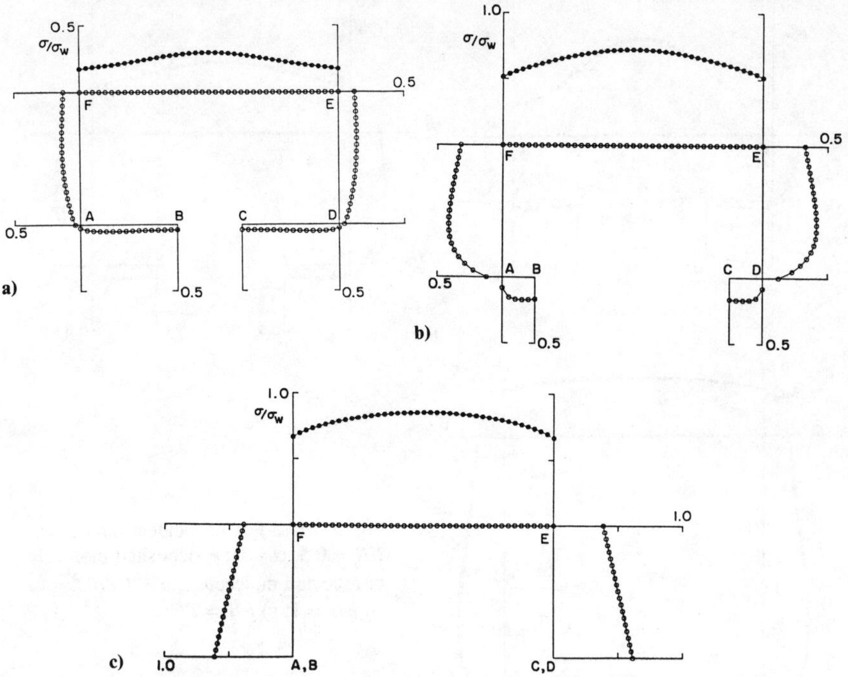

Fig. 4 Mass flux of incident molecules ($h/l = 1$, $\alpha = 1$): •, deposited molecules; o, reflected molecules. a) $b/l = 0.25$; b) $b/l = 0.75$; c) $b/l = 1$.

flux of the deposited molecules in order to estimate the thickness and distribution of thin films.

First, the calculated results in the case of free molecular flow are shown. Figures 4 and 5 show the mass flux of incident molecules on the surfaces for various chamber dimensions. Figure 6 shows the mass flux for various condensation coefficients when the chamber geometry is held fixed. The total mass flux of deposited molecules on the deposition surface is shown in Fig. 7 in the case of $b/l = 0.5$. It is seen from Figs. 4-6 that the thickness distribution on the deposition surface becomes more uniform as b/l and h/l increase or as α decreases. The relation between the thickness distribution and b/l is obvious. The other two relations are explained as follows. The molecules deposited on surface $\overline{EF}$ come from either the evaporation surface $\overline{BC}$ or the remaining surfaces $\overline{FA}$, $\overline{AB}$, $\overline{CD}$, and $\overline{DE}$. The molecules coming directly from surface $\overline{BC}$ are deposited more often in the center than on the sides of surface $\overline{EF}$. Conversely, the molecules coming from the other surfaces are deposited more often on the sides than in the center of surface $\overline{EF}$. Now, as h/l increases or α decreases, more molecules come from the remaining surfaces relative to those coming directly from surface $\overline{BC}$. Thus, the thickness distribution becomes more uniform as h/l increases or α decreases.

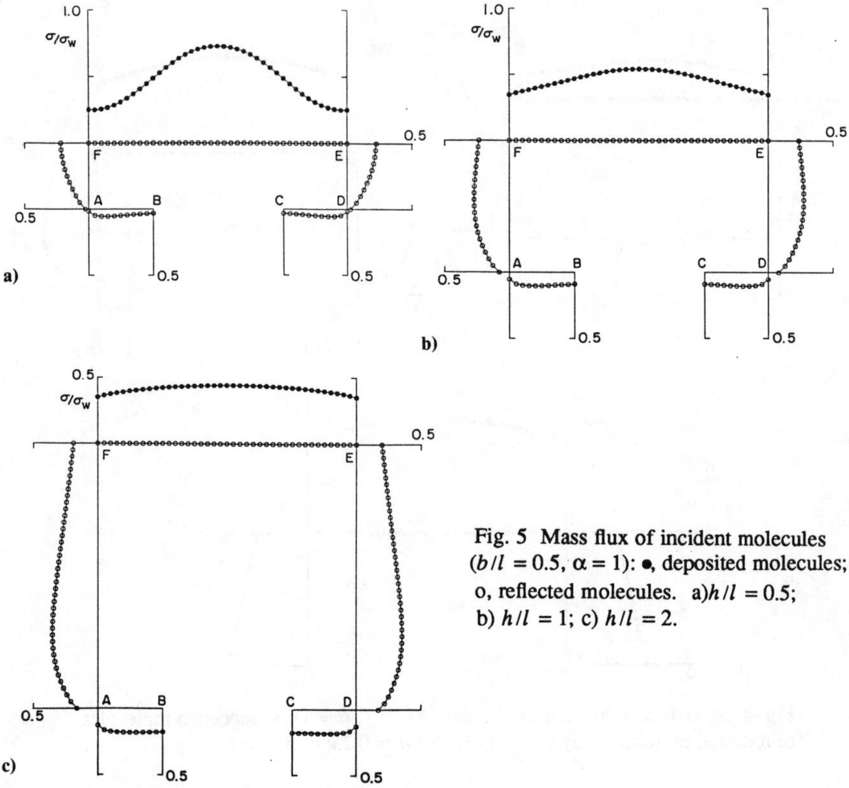

Fig. 5 Mass flux of incident molecules ($b/l = 0.5$, $\alpha = 1$): •, deposited molecules; o, reflected molecules. a) $h/l = 0.5$; b) $h/l = 1$; c) $h/l = 2$.

On the other hand, we can see from Fig. 7 that the total mass flux of deposited molecules decreases as h/l increases. The reason is that, as h/l increases, the molecules coming from surfaces $\overline{FA}$ and $\overline{DE}$ increase and some fraction of these molecules is lost through surface $\overline{BC}$.

Next, the calculated results for one of the preceding cases in transitional flow are shown. The conditions of the problems are $h/l = 0.5$, $b/l = 0.25$, and α of the deposition surface = 1. In order to obtain a reliable solution of the B-K-W equation, we need to decrease the mesh size in physical space as the Knudsen number decreases; i.e., several points are required in each mean free path. In the present calculations, we use a 20 x 10 grid in physical space and a 38 x 38 grid in velocity space ($-3 \le \zeta_1 \le 3$, $-3 \le \zeta_2 \le 3$). Figure 8 shows the velocity vectors and the density contours calculated by the first-order upwind scheme. The Knudsen number is defined by $Kn = \lambda_W / 2b$. Figures 9 and 10 show results calculated by the third-order upwind scheme. The mass flux distribution of molecules incident on half of the deposition surface is shown in Fig. 11. The mass flux distribution of the first-order scheme becomes discontinuous at $x_1/l = 0.25$ because the first-order scheme has numerical

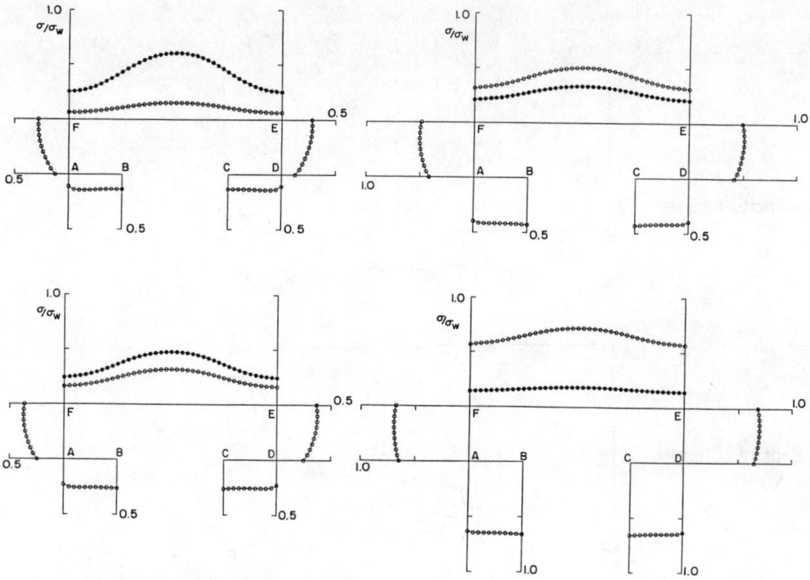

Fig. 6 Mass flux of incident molecules ($h/l = 0.5$, $b/l = 0.5$): •, deposited molecules; o, reflected molecules. a) $\alpha = 0.8$; b) $\alpha = 0.6$; c) $\alpha = 0.4$; d) $\alpha = 0.2$.

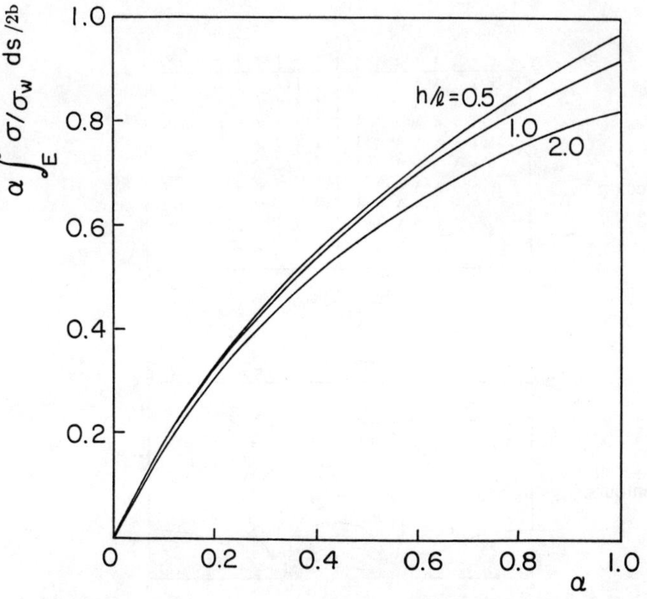

Fig. 7 Total mass flux of incident molecules on surface $\overline{EF}$ ($b/l = 0.5$).

a) velocity vectors.

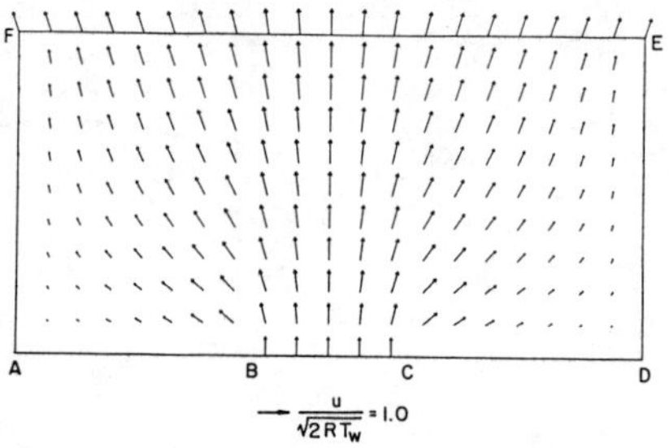

b) density contours.

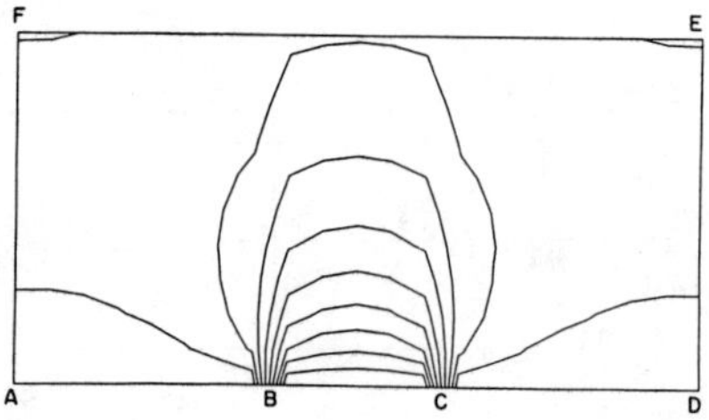

Fig. 8 Calculated results of first-order upwind scheme ($Kn = 4$).

a) velocity vectors.

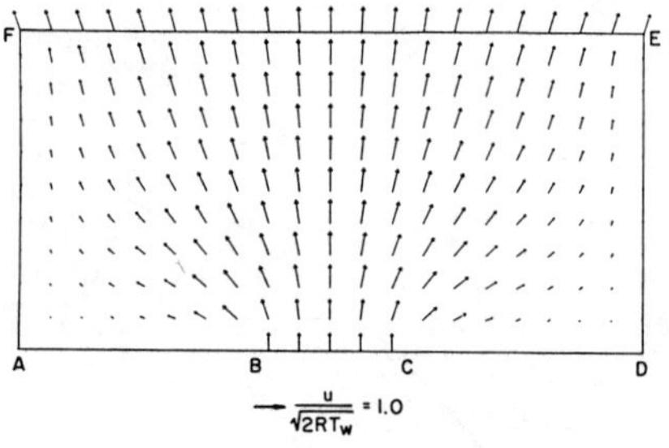

b) density contours.

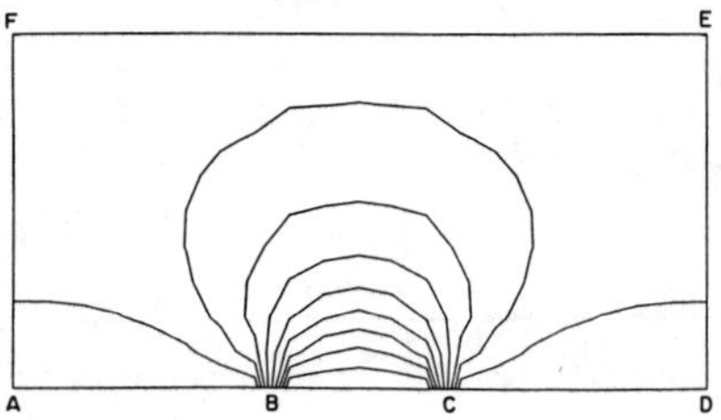

Fig. 9 Calculated results of third-order upwind scheme ($Kn = 4$).

a) velocity vectors.

b) density contours.

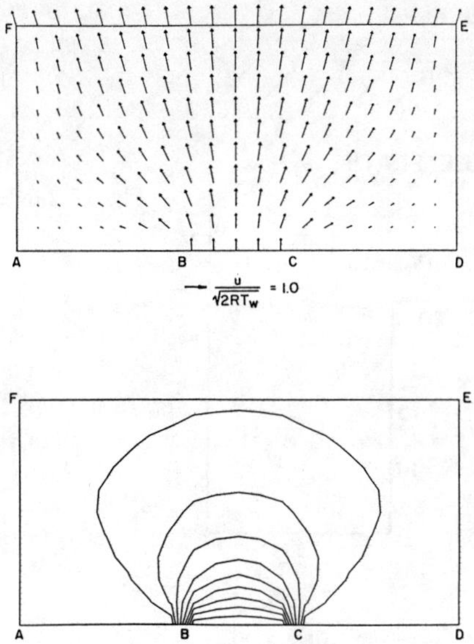

Fig. 10 Calculated results of third-order upwind scheme (Kn = 0.5).

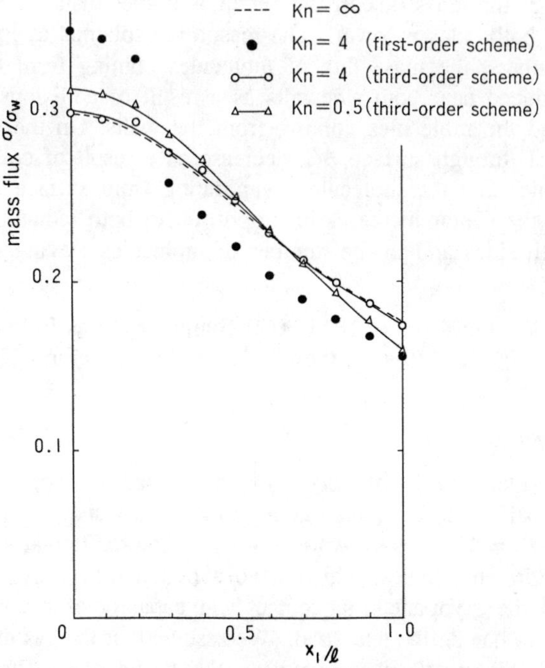

Fig. 11 Mass flux distribution on half of surface EF.

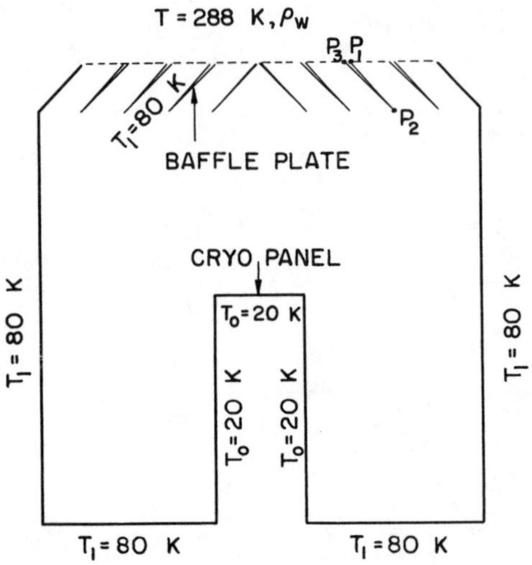

Fig. 12 Model chamber of cryopump.

dissipation as a result of truncation error terms and the dissipation varies in every spatial direction. Thus, in this case, the first-order scheme is inadequate. From the results of the third-order scheme, we can see that at a Knudsen number of 0.5, the mass flux in the center is higher than for $Kn = 4$, and the mass flux on both sides is lower. The reason is explained as follows. At small Knudsen numbers, the mass flux of molecules coming from the evaporation surface is reduced near both sidewalls as a result of collisions between those molecules and the molecules coming from the walls. On the other hand, the molecules lost through surface $\overline{BC}$ decrease as a result of collisions between those molecules and the molecules evaporating from surface $\overline{BC}$. Thus, the mass flux in the center increases in proportion to both reductions in the mass flux near both sides and in the number of molecules leaving through surface $\overline{BC}$.

Computation times on an IBM 3090 computer are as follows: 11 s for the case given in Fig. 4a; 419 s for the case given in Fig. 9; and 625 s for the case of Fig. 10.

3.2. *Cryopump*

Next, consider a model of a cryopump composed of cryopanels (temperature $T_0 = 20$ K), baffle plates (temperature $T_1 = 80$ K), and the remaining walls (temperature $T_1 = 80$ K) as shown in Fig. 12. The baffle plates are arranged so that the entering molecules, which have a high temperature, cannot impinge directly onto the cryopanels, since the heat capacity of the cryopanel is not large enough to handle the heat load. We assume that the gas molecules, which enter through the open spaces between the baffle plates, have a stationary Maxwellian distribution (temperature $T_W = 288$ K and density ρ_W).

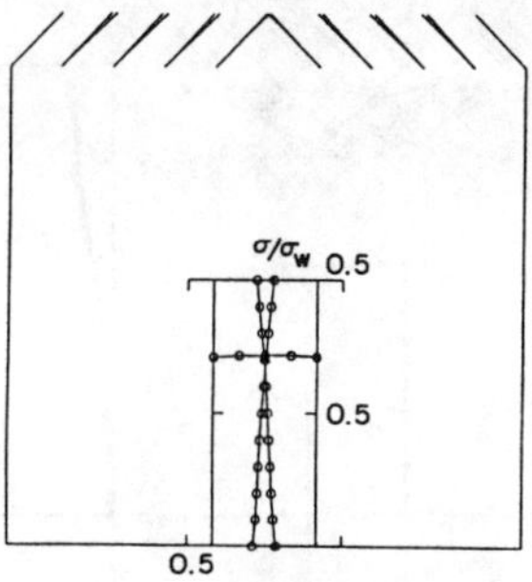

a) mass flux.

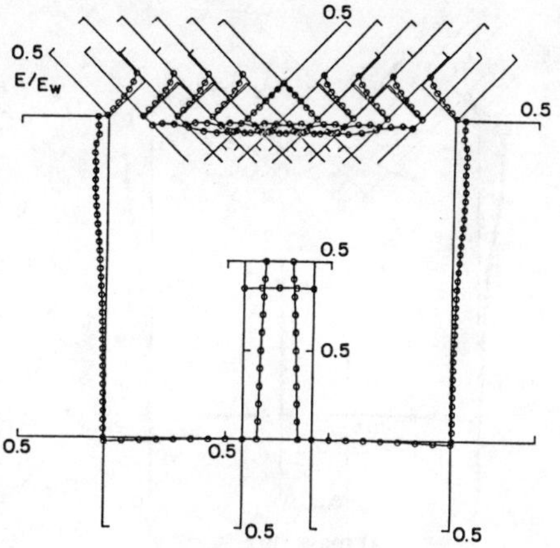

b) net energy flux.

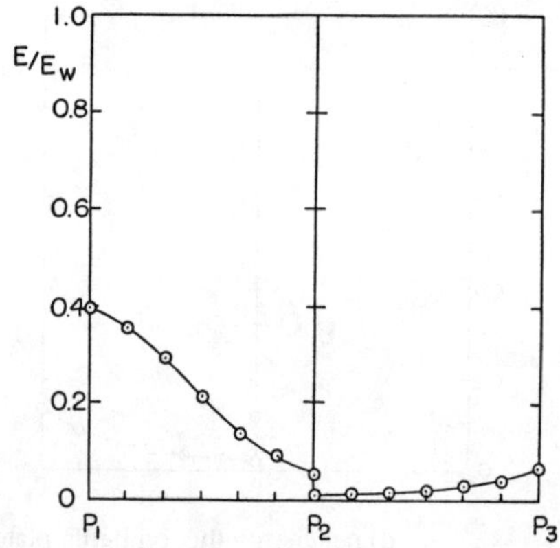

c) net energy flux on baffle plates ($\overline{P_1P_2}$ and $\overline{P_2P_3}$).

Fig. 13 Type A molecules.

Cryopumps usually operate at extremely low pressures in the range $p = 10^{-5} \sim 10^{-10}$ Torr. Thus, we can consider that the flow of gas in the cryopump is a free molecular flow.

We calculate the mass flux of deposited molecules and the energy flux onto the surfaces for two types of molecules: 1) molecules A, which are condensed on surfaces of temperature $T \leq 20$ K, and 2) molecules B, which are condensed on surfaces of temperature $T \leq 80$ K. That is, when molecules collide with a wall, type A molecules are deposited only on the cryopanel while type B molecules are deposited on all surfaces.

Figure 13 shows the calculated results for type A molecules. Although these molecules are deposited only on the cryopanels, there is a net energy flux onto all of the surfaces because the gas temperature at the entrance is higher than the wall temperature. The total mass flux of molecules deposited on the

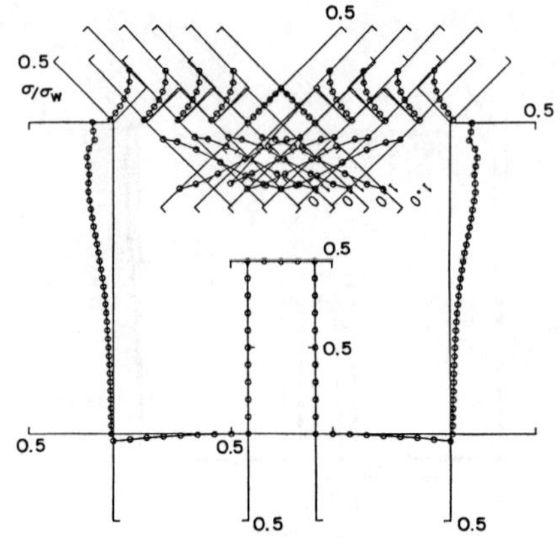

a) mass flux.

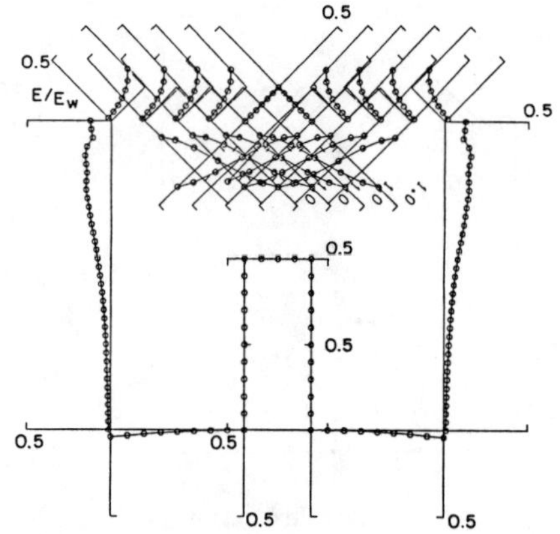

b) net energy flux.

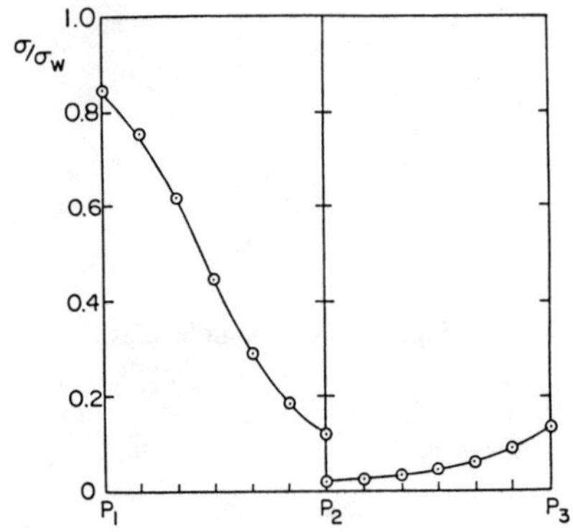

c) mass flux on baffle plates ($\overline{P_1P_2}$ and $\overline{P_2P_3}$).

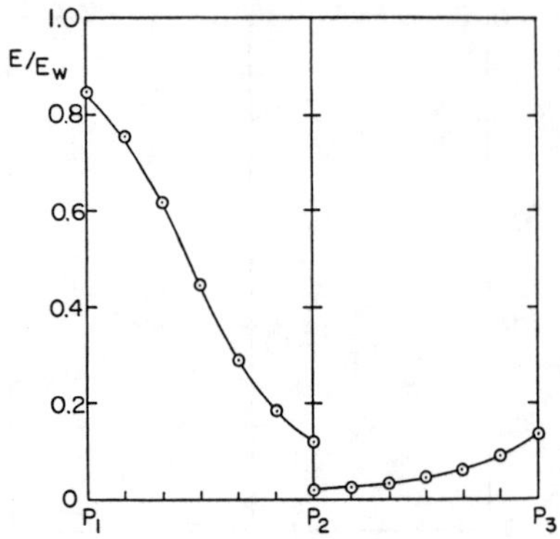

d) net energy flux on baffle plates ($\overline{P_1P_2}$ and $\overline{P_2P_3}$).

Fig. 14 Type B molecules.

cryopanels is 34.2% of the total mass flux of molecules entering the chamber. Figure 14 shows the calculated results for type B molecules. These molecules are deposited on all of the surfaces except the cryopanels, since they are assumed to be deposited on all of the walls and cannot pass directly to the cryopanels because of the obstruction of the baffle plates. It is clear that many molecules are deposited on the outside of the baffle plates and on the upper part of both sidewalls. From these results, the pumping speeds required and the heat load on the pump can be predicted. The code also can be extended easily to calculate more complex geometries. The computation times required 108 s for each type of molecule on the IBM 3090 computer.

IV. Acknowledgments

The author expresses cordial thanks to Professor Y. Sone of Kyoto University for suggesting this problem, helpful advice and encouragement. He also wishes to express gratitude to Professor B. Sturtevant of the California Institute of Technology for kind inspection of the manuscript.

VI. References

[1]Nanbu, K., "Rarefied Gas Dynamics Problems on Fabrication Processes of Semiconductor Films," in *Rarefied Gas Dynamics*, edited by V. Boffi and C. Cercignani, Teubner, Stuttgart, F.R.G., 1986, pp. 410-419.

[2]Davis, D. H., "Monte Carlo Calculation of Molecular Flow Rates through a Cylindrical Elbow and Pipes of Other Shapes," *Journal of Applied Physics*, Vol. 31., 1960, pp. 1169-1176.

[3]Kogan, M. N., "On the Equations of Motion of a Rarefied Gas," *Journal of Applied Mathematics and Mechanics*, Vol. 22, 1958, pp. 597-607.

[4]Maruyama, T., "Deposition of Free Molecular Flow Gas in a Rectangular Domain," M.Eng. Thesis, Kyoto Univ., Japan, 1984.

[5]Brebbia, C. A., *The Boundary Element Method for Engineers*, Pentech, London, U.K., 1978.

[6]Chu, C. K., "Kinetic-Theoretic Description of the Formation of a Shock Wave," *Physics of Fluids*, Vol. 8, 1965, pp. 12-22.

[7]Sone, Y., Aoki, K., and Yamashita, I., "A Study of Unsteady Strong Condensation on a Plane Condensed Phase with Special Interest in Formation of Steady Profile," in *Rarefied Gas Dynamics*, edited by V. Boffi and C. Cercignani, Teubner, Stuttgart, F.R.G., 1986, pp. 323-333.

Transition Regime Droplet Growth and Evaporation: An Integrodifferential Variational Approach

J. W. Cipolla Jr.*
Northeastern University, Boston, Massachusetts
and
S. K. Loyalka†
University of Missouri-Columbia, Columbia, Missouri

Abstract

A variational principle based on the integrodifferential form of the Boltzmann equation, which allows very general forms of the collision integral and of the boundary operator, has been used to compute the mass flux to a spherical particle. For simplicity the problem has been restricted to the one-speed, constant cross-section approximation, the black sphere problem of neutron transport theory. Using the Hilbert expansion of the solution far from the sphere as a trial function leads to a rational expression in the inverse Knudsen number for the extrapolation distance, which is simply related to the number flux to the sphere. A five-term expansion gives results that are no more than 6% in error when compared to an accurate numerical solution. The technique can be applied to other forms of the collision model and other boundary conditions.

Introduction

We consider the problem of computing the rate of evaporation from or condensation onto a spherical droplet using an integrodifferential variational principle.[1] Although the strength of using this approach is in its generality, we study the problem initially using a constant collision cross-section model, so that the problem is formally identically to the black sphere problem of one-speed neutron transport theory, for which results are readily available.[2,3]

Previous work using this technique[1,4–6] has been limited to planar, bounded geometries. In all of these cases quantities of physical interest have been related to the extremum of the variational functional so that the formalism gives very accurate results for the moments of the distribution function. In all cases treated so far in bounded planar geometries, the desired moments have been accurately predicted by means of closed form expressions rational in the inverse Knudsen number. In addition, we have recently computed the Milne extrapolation distance for both

* Department of Mechanical Engineering.

† Nuclear Engineering Program and Particulate Systems Research Center.

the constant cross-section and constant collision frequency (BGK) models using this technique in an *unbounded* but planar geometry.[7] We have found that it is necessary, in unbounded problems, to extract the asymptotic Chapman-Enskog solution from the distribution so that the reduced distribution decays strongly far from the surface. In planar geometries, the Hilbert expansion terminates, and the resulting problem is amenable to extremely accurate estimation using very simple trial functions. In particular, we have been able to generate four-figure accuracy in the extrapolation distance.[7]

Naturally, the situation is more complicated in the unbounded spherical case. In particular the asymptotic solution (Hilbert expansion) does not terminate, and extracting only a finite number of terms in addition to the collision invariants makes the kinetic equation inhomogeneous, introducing difficulties into the calculation procedure. We present here a theory that gives the spherical extrapolation distance as a rational function of the inverse Knudsen number based on the sphere radius, which is then easily related to the mass flux to the surface.[2]

Formulation

The problem is determined by the integrodifferential boundary-value problem

$$\mu\frac{\partial f}{\partial r}+\frac{(1-\mu^2)}{r}\frac{\partial f}{\partial \mu}=\frac{1}{2}\int_{-1}^{1}d\mu f(r,\mu)-f(r,\mu) \tag{1}$$

$$f(r,\mu)\sim 1-Q\sum_{n=0}^{\infty}\frac{n!P_n(\mu)}{r^{n+1}} \tag{2}$$

$$f^+(\delta,\mu)=0, \qquad (\mu>0) \tag{3}$$

where $f(r,\mu)$ is the perturbed distribution at the location r for particles with trajectories that make an angle θ with the radius and $\mu=\cos\theta$. The asymptotic distribution $f_{asy}(r,\mu)$ is the formal Hilbert expansion valid for large r, and $P_n(\mu)$ is the Legendre polynomial. The distribution f has been normalized so that the constant density far from the sphere is unity, and r is nondimensionalized with respect to the particle mean-free-path. The quantity δ is the inverse Knudsen number based on the sphere radius, and the quantity Q is related to the particle flow at the surface of the sphere and to the extrapolation distance λ. The quantity Q may be expressed in terms of the distribution f by using particle conservation [the first moment of Eq.(1)] as

$$Q=\frac{3}{2}\delta^2\int_0^1 f^-(\delta,-\mu)\mu\,d\mu=\frac{\delta^2}{\lambda+\delta} \tag{4}$$

In addition, in the free-molecular limit, $f^-=1$, and we have

$$Q_{\mathrm{fm}}=\frac{3}{4}\delta^2, \qquad \lambda_{\mathrm{fm}}=\frac{4}{3} \tag{5}$$

To use the variational principle, we set $f=h+1$ to obtain

$$Dh=Lh \tag{6}$$

$$h(r,\mu) \sim -Q\sum_{n=0}^{\infty}\frac{n!P_n(\mu)}{r^{n+1}}, \qquad r\to\infty \tag{7}$$

$$h^+(\delta,\mu) = -1, \qquad \mu > 0 \tag{8}$$

The operators L and D are given by the right- and left-hand sides of Eq. (1) respectively.

We now define the functional

$$J(\tilde{h}) = ((\tilde{h}, R(D-L)\tilde{h})) + ((R\tilde{h}^-, \tilde{h}^+ + 2))_B \tag{9}$$

where $\tilde{h}$ is a trial function, $Rf(r,\mu) = f(r,-\mu)$ and the inner products (with double parentheses) have been defined as

$$((f,g)) = 4\pi\int_\delta^\infty r^2 \mathrm{d}r \int_{-1}^{+1} \mathrm{d}\mu f(r,\mu)g(r,\mu) \tag{10}$$

$$((f,g))_B = 4\pi\delta^2\int_0^1 \mu\,\mathrm{d}\mu f(r,\mu)g(r,\mu) \tag{11}$$

It has been shown in Ref.1 that the functional is stationary if and only if $\tilde{h} = h$, where h is the solution of Eqs. (6–8). In addition, we can see that

$$J_{\mathrm{st}} = 4\pi\delta^2\int_0^1 h^-(\delta,-\mu)\mu\,\mathrm{d}\mu \tag{12}$$

and, using Eqs. (4) and (5), we find that

$$Q/Q_{\mathrm{fm}} = 1 + J_{\mathrm{st}}/(2\pi\delta^2) \tag{13}$$

so that Q and λ may be accurately approximated.

Results and Discussion

To retain the simplicity of the final result that is the hallmark of this technique, it has been necessary to use only the asymptotic form of the solution in the trial function. For this purpose we have used a series of trial functions in order to demonstrate the approach of the results to the accurate numerical solution of Ref.3 as more terms of the asymptotic expansion are included. The orthogonality properties of the terms in the trial function simplify some of the calculations. The general procedure is to assume a trial function consisting of a finite number of terms of the asymptotic solution, each multiplied by an unknown constant. This trial function is inserted into the definition of J and the variational parameters varied to render J stationary. This leads to linear algebraic equations for the variational parameters which are then used to form J_{st} and the physical quantity of interest. We present here the results obtained using three-, four-, and five-term trial functions.

For the three-term trial function we take

$$\tilde{h}_3 = \frac{\beta}{r} + \gamma\frac{\mu}{r^2} + \sigma\frac{3\mu^2 - 1}{r^3} \tag{14}$$

which leads to the linear equations

$$\begin{aligned} 6(\beta+1) + 4\gamma + 3\sigma &= 0 \\ 20(\beta+1) - (15+40\delta)\gamma - 16\sigma &= 40 \\ 15(\beta+1) - 16\gamma + (30+32\delta)\sigma &= 0 \end{aligned}$$

Then using $\tilde{h}_3$ in Eq. (12) and using Eqs. (13), (4), and (5), we find that $\lambda = -1/\gamma - \delta$. The solution of the linear equations gives

$$\gamma = -\frac{7680\delta + 5400}{7680\delta^2 + 10840\delta + 7881}$$

Thus we find

$$\lambda_3 = \frac{5440\delta + 7881}{7680\delta + 5400}$$

In addition, we present results for more complicated trial functions omitting the details

$$\tilde{h}_4 = \tilde{h}_3 + \Delta\frac{15\mu^3 - 9\mu}{r^4}$$

$$\lambda_4 = \frac{2437120\delta^2 + 7372288\delta + 9080685}{3440640\delta^2 + 8064000\delta + 7286760}$$

$$\tilde{h}_5 = \tilde{h}_4 + \Gamma\frac{105\mu^4 - 90\mu^2 + 9}{r^5}$$

Table 1 Computed values of the extrapolation distance*

δ	λ	λ_3	% error	λ_4	% error	λ_5	% error
0.0	1.3333	1.4594	9.4611	1.2462	-6.5334	1.2542	-5.9344
0.1	1.2838	1.3659	6.3967	1.2110	-5.6726	1.2182	-5.1074
0.2	1.2423	1.2931	4.0899	1.1788	-5.1151	1.1847	-4.6399
0.5	1.1464	1.1473	0.0780	1.0983	-4.1957	1.1004	-4.0094
0.7	1.0984	1.0847	-1.2450	1.0560	-3.8638	1.0566	-3.8045
1.0	1.0420	1.0184	-2.2625	1.0053	-3.5267	1.0048	-3.5660
1.5	0.9749	0.9480	-2.7542	0.9446	-3.1038	0.9439	-3.1752
2.0	0.9286	0.9037	-2.6805	0.9031	-2.7492	0.9026	-2.7950
2.5	0.8950	0.8732	-2.4345	0.8732	-2.4367	0.8731	-2.4473
5.0	0.8117	0.8009	-1.3261	0.7998	-1.4673	0.8005	-1.3858
∞	0.7104	0.7083	-0.2973	0.7083	-0.2973	0.7083	-0.2973

* Results are shown for three- four- and five-term trial functions. The inverse Knudsen number based on the sphere radius is given by δ, with $\delta = \infty$ corresponding to the planar case. The results for λ are from Ref. 3.

$$\lambda_5 = \frac{74868326640\delta^3 + 391214530566\delta^2 + 673622787720\delta + 34345808175}{105696460800\delta^3 + 478937088800\delta^2 + 617531059920\delta + 27385155000}$$

Each of these results has been obtained by solving the relevant linear system using MACSYMA, a symbolic manipulation program developed at the Massachusetts Institute of Technology.

These results have been evaluated for $0 \leq \delta < \infty$ and compared with the numerical results of Ref. (3). The comparison is tabulated in Table 1. It is interesting to note that the maximum error occurs for the free molecular limit, $\delta = 0$, since the trial function contains no half-range terms. Although the convergence toward the correct $\delta = 0$ value of 4/3 seems rather slow, it is clear that including more terms should lead to much better agreement. In the continuum limit $\delta \to \infty$, each of the expressions for λ approaches the value $17/24 = 0.7083$, a result that agrees with the integral variational principle. However, the current results are much easier to use for numerical computations, since they contain none of the complicated integrals normally found in these solutions. Finally, this technique can be extended to other collision models and boundary conditions. Fortunately, the algebraic complexity may be lessened by the use of symbolic manipulation programs such as MACSYMA.

Acknowledgments

It is a pleasure to acknowledge the helpful discussions with Professor Carlo Cercignani and the hospitality of the Politecnico di Milano, where portions of this work were carried out. In addition, this work was supported by the National Science Foundation. Some of the results contained in this work were obtained using MACSYMA, a large symbolic manipulation program developed at the Massachusetts Institute of Technology Laboratory for Computer Science and supported from 1975 to 1983 by the National Aeronautics and Space Agency Grant NSG 1323, by the Office of Naval Research Grant N00014-77-C-0641, by the U. S. Department of Energy Grant ET-78-C-02-4687, and by the US Air Force Grant F49620-79-ZC-020, and since 1982 by Symbolics, Inc. MACSYMA is a trademark of Symbolics, Inc.

References

[1] Cercignani, C "A Variational Principle for Boundary Value Problems in Kinetic Theory", Journal of Statistical Physics, Vol. 1, 1969, p. 297.

[2] Williams, M. M. R. *Mathematical Methods in Particle Transport Theory*, Butterworths, London, 1971, pp.340–342.

[3] For accurate numerical results see S. K. Loyalka "A Numerical Method for Solving Integral Equations of Neutron Transport", Nuclear Science and Engineering, Vol. 56, 1975, pp.317–319.

[4] Cipolla, J. W. Jr. and C. Cercignani, "Effect of Molecular Model and Boundary Conditions on Linearized Heat Transfer", in *Rarefied Gas Dynamics*, Edited by C. Cercignani, D. Dini, and S. Nocilla, Editrice Tecnico Scientifica, Pisa, Italy, Vol. II, 1971, p.767.

[5] Cipolla, J. W. Jr. "Heat Transfer and Temperature Jump in a Polyatomic Gas", International Journal of Heat and Mass Transfer, Vol. 14, 1971, pp.1599–1610.

[6] Cercignani, C. and M. Lampis, "Variational Approach to Rarefied Gas Flows in an External Force Field with Application to the Gas Centrifuge", in *Rarefied Gas Dynamics*, edited by H. Oguchi, University of Tokyo Press, 1984, pp. 89–98.

[7] Loyalka, S. K. and J. W. Cipolla Jr. "On Choice of Trial Functions in Integro-Differential Variational Principles of Transport Theory", Nuclear Science and Engineering, Vol. 99, 1988, pp.118–122.

Molecular Dynamics Studies on Condensation Process of Argon

T. Sano*
Tokai University, Kitakaname, Hiratsuka, Kanagawa, Japan
and
S. Kotake*
University of Tokyo, Hongo, Bunkyo-ku, Tokyo, Japan

Abstract

The condensation process of gaseous molecules of argon on their condensate is studied numerically by using the molecular dynamics method. No evidence of significant interfacial surface structure is shown, while inside the condensate ordered structures are observed. The warmer molecules coming to the condensate concentrate more densely near the interface than the cooler molecules accumulating arround the interface. The energy flow of condensation occurs more locally than that of evaporation does. Both flows have a peak just inside the interface of phase change. The total energy flow of condensation and evaporation changes locally and time dependently.

Introduction

In recent years, the dynamic processes of condensation, vaporization, solidification, and melting have received much attention in various fields of science and engineering technology as a means of understanding the fundamental mechanism of phase change as well as for actively controlling the material processing. The atomic or molecular aspects of these transition processes, however, are hardly observed experimentally even with advanced techniques of measurement. Numerical computations by using high-speed, large-memory computers with molecular dynamics or Monte Carlo methods can provide essential in-

*Professor, Faculty of Engineering.

formation not otherwise obtainable by experimental techniques to understand their dynamic processes and structures.[1,2]

In the present study, the condensation process of gaseous molecules on their condensate is studied numerically by using the molecular dynamics method. The argon system is chosen to simulate the condensation process because of its simple structure of molecule for solving the dynamic equations of motion.

Molecular Dynamics of Argon Molecules

Condensation or clustering takes place for molecules in which translational energy is reduced sufficiently through their collisions. Such a system of phase changing can be constructed by a molecular dynamics model with a heat sink. The dynamic behavior of molecular collisions can be described by Hamilton's equations of molecular motion with an appropriate interaction potential. The potential energy function may be assumed to be a pairwise additive of effective pair potentials. For spherical nonpolar molecules such as argon, the Lennard-Jones (6-12) potential gives a fairly good realistic representation of molecular interaction forces. For argon, the parameters used are σ=3.405 Å and ϵ/κ=119°K.

The present sysytem is placed in a box with the two square faces of dimensions L=45.8 Å in the x and y directions as shown in Fig. 1. These faces are constrained to

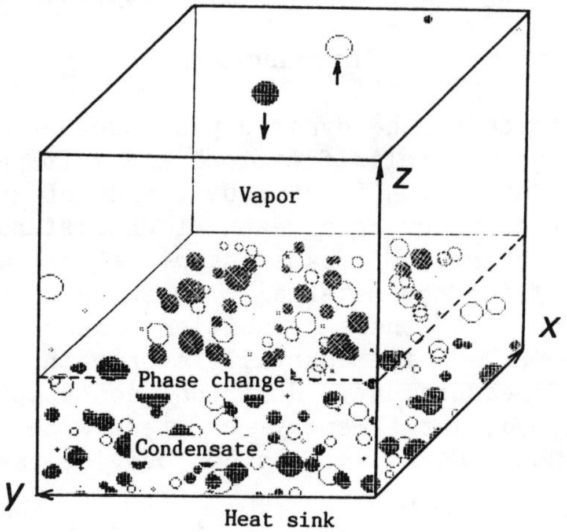

Fig. 1 Molecular system for phases change of condensation; heat sink is imposed at the cell bottom to cause the heat flux of condensation.

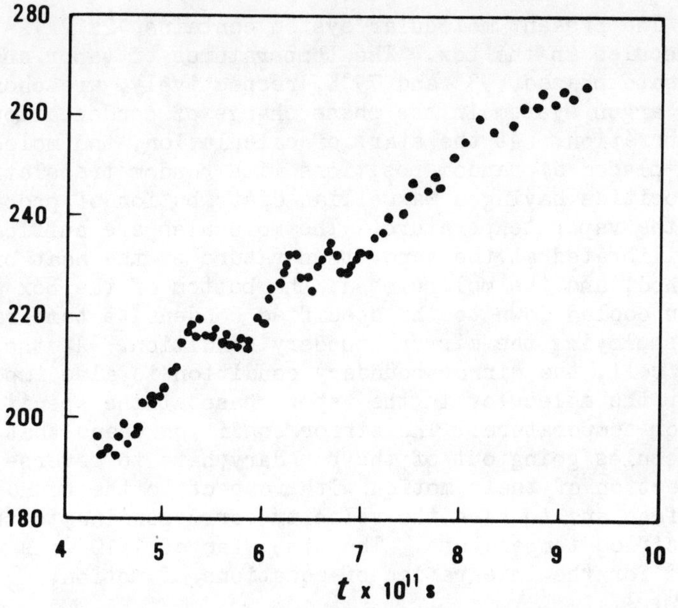

Fig. 2 Instantaneous molecular motions: molecules coming to the condensate are represented by solid circles, and those going out of the condensate are represented by open circles: $\Delta t = 2 \times 10^{-12}$ s.

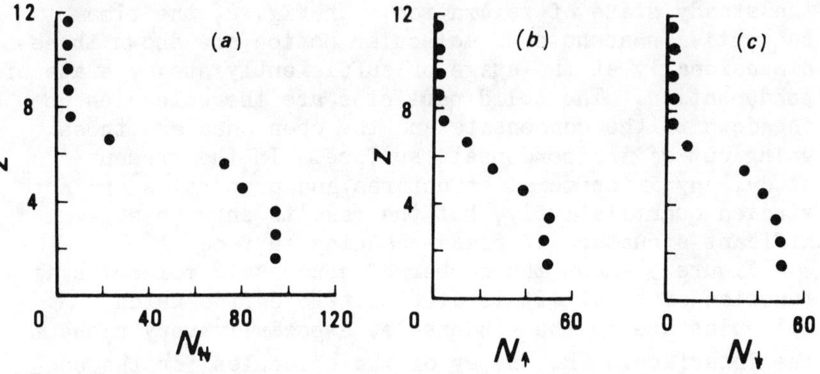

Fig. 3 The number of molecules accumulated in the condensate.

have the periodic conditions in order to include a realistic vapor and condensate system of infinite dimensions, eliminating the effect of the finite number of molecules considered. It is in the third direction (z) of dimension L=68.8 Å that the heat fluxes caused by the heat sink result in the phase change of condensation of molecules.

The present molecular system contains 12x12x12=1728 molecules in the box. The temperatures of vapor and condensate phases, 93 and 79°K, respectively, are chosen for the argon system in the phase change of condensation and evaporation. At the start of calculation, the molecules are placed at random positions with random translational velocities having a Maxwellian distribution of probability at the vapor temperature. The molecules are sufficiently equilibrated at the vapor temperature by the heat bath method, and the molecules at the bottom of the box are then cooled down to the specified condensate temperature by employing the mirror boundary condition. At the top of the cell, the mirror boundary condition is also imposed to keep the molecules in the vapor phase at the specified vapor temperature. The mirror condition means that the molecules going out of the boundary have to reverse the direction of their motion with respect to the boundary surface and to have the velocity corresponding to the specified temperature. The time step of $1x10^{-15}$ s was used for the integration of equations of motion.

Results and Discussion

Since the process is a suddenly started nonstationary relaxation from the high-energy to the low energy states, a very long computation has to be carried out to obtain the steady state of relaxation. In Fig. 2, the time sequential snapshots of molecular motion are shown three-dimensionally at the stage of sufficiently steady state of condensation. The solid particles are the molecules coming down to the condensate and the open ones are those going out of the condensate surface. In the present study, any interfacial structures and properties are not studied quantitatively, but the results show no significant structure of phase-changing surface.[1,2]

Figure 3 shows the number of condensate molecules as a function of time. It is difficult to define which molecules are in the condensate, especially very close to the interface. The number of the molecules for the condensate are counted in the cells of the first to the fifth layer meshes, which contains almost all molecules of the condensate as shown in Fig. 4. It is seen that the condensate molecules increase in the number, being linearly proportional to the time elapsed.

The distribution of the number density is shown in Fig. 4. The number density here is defined as the number of molecules in the layer of thickness 4.3 Å in the z direction. Figures 4b and 4c show the number density of

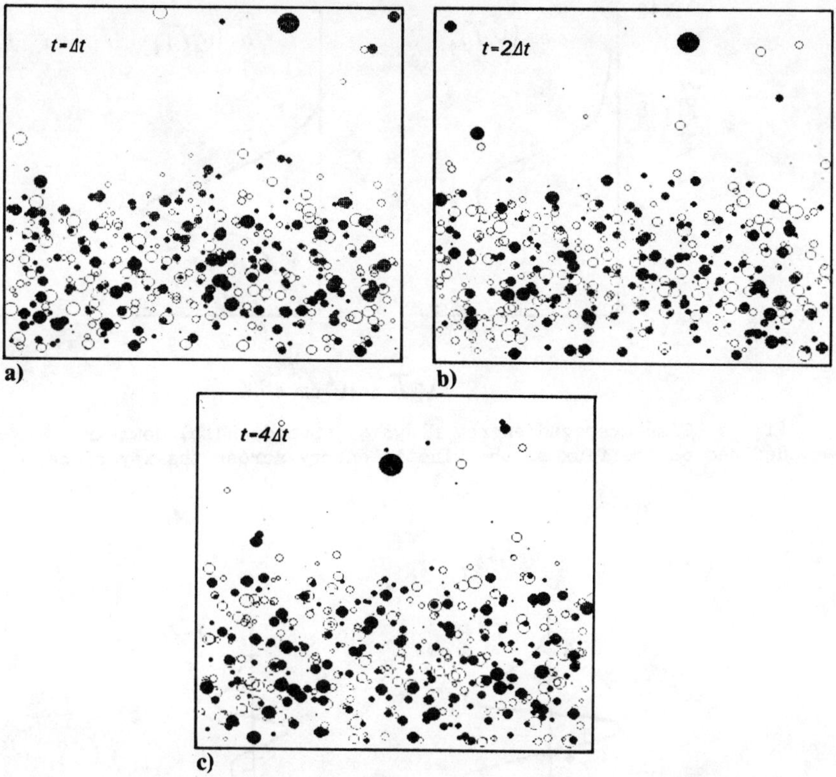

Fig. 4 Distribution of the number density in the z direction: a) the total number density; b) number density for molecules going out of the condensate; and c) number density for molecules coming to the interface.

molecules coming to and going out the condensate interface, respectively. The interface between the vapor and condensate phases lies in about z=5 at the time shown in the figure. The warmer molecules coming to the condensate concentrate more densely near the interface than the cooler molecules accumulating arround the interface. This is due to the difference in the translational energy of the molecules. The going-out molecules have less translational enrgy than the coming-in molecules because they lose their tanslational energy in the course of collision with the condensate molecules. The excess energy released by the coming molecules is to be transferred to the heat sink at the bottom cell.

The energy flows upward and downward are shown in Fig. 5. The flux is defined as the amount of the kinetic energy (temperature) transported by the z-component velocity across the x-y plane. At a x-y plane slightly

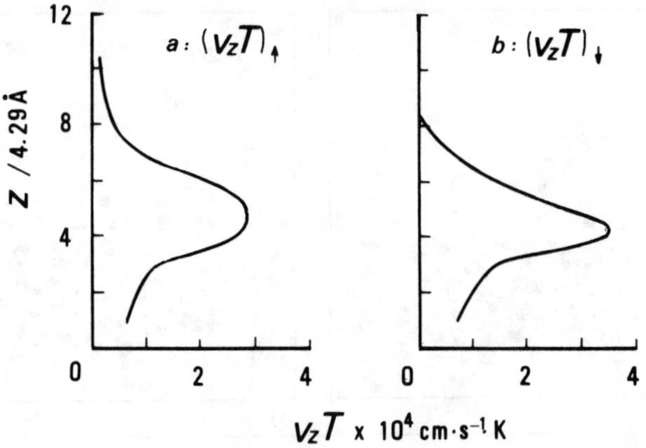

Fig. 5 Time-averaged energy flows a) upward and b) downward defined as the flux of the kinetic energy across the x-y plane.

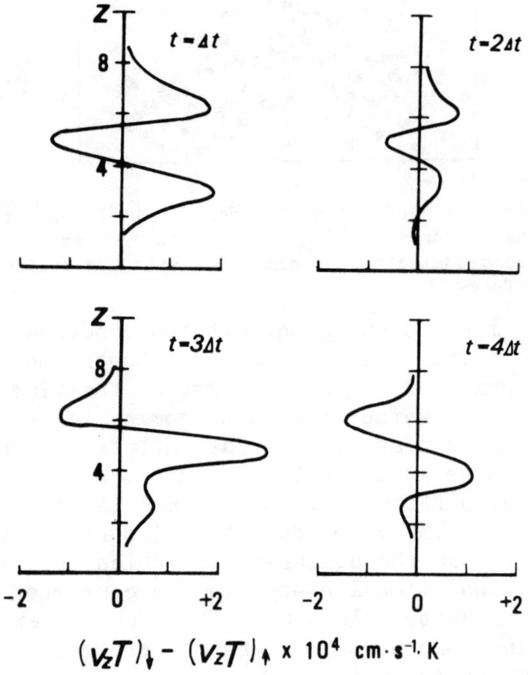

Fig. 6 Instantaneous energy flows in the z direction: $\Delta t = 2 \times 10^{-12}$ s.

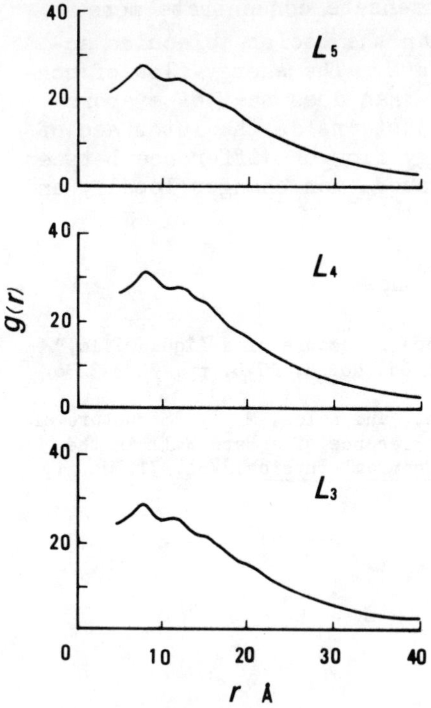

Fig. 7 The radial two-point distribution function.

inside the phase-changing interface, both fluxes take a peak value. The evaporation fluxes going out the interface are more widely distributed in the z direction, whereas the condesation fluxes occurs more locally just inside the interface. The total energy flow, the difference between condensation and evaporation fluxes, changes in its amount locally as well as time dependently as shown in Fig. 6.

In Fig. 7, the radial two-point distribution function of molecules is shown at three points in the z direction. It is seen that the condensate molecules tend to have more ordered structures as they move into the condensate farther from the interface.

Conclusion

The condensation process of gaseous molecules of argon on their condensate is studied numerically by using the molecular dynamics method with the Lennard-Jones (6-12) potential. Any evidence of significant interfacial surface structure is not shown in the results, but inside the condensate ordered structures are observed. The warmer molecules coming into the condensate concentrate more densely near the interface than the cooler molecules accumulating arround the interface. The energy flow of condensation occurs more locally than does that of evaporation. Both flows have a peak just inside the interface of phase change. The total energy flow of difference between fluxes of condensation and evaporation changes locally and time dependently.

References

[1]Rao, M and Levesque, D., "Surface Structure of a Liquid Film," Journal of Chemical Physics, Vol.65, No. 8, 1976, pp. 3233-3236.

[2]Rao, M., Berne, B., Percus, J.K., and Kalos, M.H., "Structure of a Liquid-vapor Interface in the Presence of a Hard Wall in the Transition Region," Journal of Chemical Physics, Vol. 71, No. 1, 1979, pp. 3802-3806.

Condensation and Evaporation of a Spherical Droplet in the Near Free Molecule Regime

J. C. Barrett* and B. Shizgal†
University of British Columbia, Vancouver, British Columbia, Canada

Abstract

The growth and evaporation of a liquid droplet in contact with its vapor dilutely dispersed in a second inert gas is investigated. In the near free molecule regime, where the droplet radius R is much smaller than the vapor-gas mean free path λ, the vapor flux Φ can be written to first order in $\alpha = R/\lambda$ as $\Phi = \Phi_{\rm fm}(1+\alpha\Gamma^{(1)})$, where $\Phi_{\rm fm}$ is the free molecule flux. The first correction term $\Gamma^{(1)}$ was found by using the free molecule solution in the collision operator of the linear Boltzmann equation to obtain the vapor molecule distribution function to first order in α. The correction term depends on the vapor-gas molecular interaction and the ratio Z of vapor to gas molecular mass. For the hard sphere interaction, the correction $\Gamma^{(1)}_{\rm HS}$ was reduced to a four-dimensional integral and for the r^{-4} interaction (Maxwell molecules), the correction $\Gamma^{(1)}_{\rm MM}$ was reduced to a five-dimensional integral. A table of values of $\Gamma^{(1)}_{\rm HS}$ and $\Gamma^{(1)}_{\rm MM}$ for Z up to 10 is presented. Although both decrease with increasing Z, the ratio $\Gamma^{(1)}_{\rm MM}/\Gamma^{(1)}_{\rm HS}$ is virtually independent of Z with the value 0.80. The approximate results of other authors are found to be inaccurate, often by a factor of two or more, in the near free molecule regime when Z is not small.

I. Introduction

The growth or evaporation of small liquid droplets in a vapor gas mixture is a fundamental problem in kinetic theory, of interest to researchers in atmospheric physics, colloid science, reactor safety, and

---

* Department of Chemistry.

† Department of Geophysics and Astronomy, and Department of Chemistry.

rarefied gasdynamics. The physical situation considered in this paper is that of a single droplet of radius R in contact with its vapor dilutely dispersed in a second gas taken to be at equilibrium. The droplet evaporation rate depends on the Knudsen number in the vapor gas mixture, here defined by

$$Kn = (3D_{12})/(\bar{c}R) \tag{1}$$

where D_{12} is the vapor-gas diffusivity and $\bar{c} = (8kT/\pi m_v)^{1/2}$ is the mean velocity for the vapor molecules of mass m_v at temperature T. Formulas for the growth rate in the limiting cases $Kn << 1$ and $Kn >> 1$ are well known,[1] and the main interest is in solving the Boltzmann equation to obtain results valid for all Kn. Approximate analytical results for the growth rate as a function of Kn have been obtained using Grad's moment method,[2] Lees' 2 stream Maxwellian approximation[3] and a variational technique[4] applied to the BGK model.[5] Numerical results in the Lorentz limit, where m_v is very much less than the gas molecule mass m_g, formed the basis of the Fuchs-Sutugin formula,[1] widely used in aerosol physics. This and other theoretical work, together with the experimental situation, has been reviewed by Wagner[6] and by Davis.[7] Recently, Monchick and Blackmore[8] presented numerical results for all mass ratios $Z = m_v/m_g$ and noted that, although many of the formulas agree in the slip flow regime, where Kn is small, they disagree in the near free molecule regime (large Kn), and this is also where experimental measurements are difficult. The disagreement in theoretical results is probably due to the difficulty of representing the vapor molecule velocity distribution function, which is very anisotropic in velocity space.

In the Lorentz limit, the first correction term to the free molecule flux, proportional to Kn^{-1}, is known exactly from Davison's work[9] on neutron transport. The purpose of this paper is to find this first correction term for all mass ratios Z. The treatment is similar to that used by Kelly and Sengers[10,11] for a droplet growing in its pure vapor and gives the term proportional to Kn^{-1} exactly in terms of a multiple integral. In the next section, we derive this correction term for hard sphere and Maxwell molecule interactions. Although the method used can be applied to any interaction, the results simplify in these two cases. Numerical results are presented in Sec. 3 and are compared with values obtained by other authors in this limit. In the concluding section we summarize our findings and give some suggestions for further work.

II. First Correction Term in the Large *Kn* Limit

For a spherical droplet of radius R, the evaporation rate (no. of molecules/s) is

$$\dot{N} = 4\pi R^2 \Phi = 4\pi R^2 \int \hat{\mathbf{R}} \cdot \mathbf{v} f(\mathbf{R}, \mathbf{v})\, d\mathbf{v} \tag{2}$$

where $\mathbf{R} = \hat{\mathbf{R}}R$ is a vector from the droplet center to a point on its surface, and $f(\mathbf{r}, \mathbf{v})$ is the distribution function for vapor molecule velocity $\mathbf{v}$ at position $\mathbf{r}$. We assume that the growth or evaporation is "quasi-stationary"; that is, it can be described by steady-state equations. We also assume that the vapor density is much smaller than that of the background gas, which is at equilibrium. Hence, vapor-vapor collisions can be neglected, and $f(\mathbf{r}, \mathbf{v})$ satisfies the linear steady-state Boltzmann equation[12]

$$\mathbf{v} \cdot \nabla f(\mathbf{r}, \mathbf{v}) = \int K(\mathbf{v}, \mathbf{v}') f(\mathbf{v}')\, d\mathbf{v}' - \nu(\mathbf{v}) f(\mathbf{v}) \tag{3}$$

Expressions for the scattering kernel $K(\mathbf{v}, \mathbf{v}')$ and the collision frequency $\nu(\mathbf{v})$ for vapor-gas collisions are given by Cercignani.[12] As is well known, these become infinite if the range of the intermolecular force is unresticted.[12] At present, we assume the range of the interaction is finite. At the droplet surface, we assume a mass accomodation coefficient (sticking probability) of unity. Then vapor molecules are emitted with an equilibrium Maxwellian distribution; thus, $f(\mathbf{R}, \mathbf{v}) = f_0$ for $\mathbf{R} \cdot \mathbf{v} > 0$, where

$$f_0 = n_v \left(\frac{m_v}{2\pi kT}\right)^{3/2} \exp\left(-\frac{m_v v^2}{2kT}\right) \tag{4}$$

In general, n_v will depend on R because of the Kelvin effect.[6] Since we ignore vapor-vapor collisions, we need only consider the evaporation of a droplet in pure gas. The free molecule solution for an evaporating droplet (i.e., the solution to $\mathbf{v} \cdot \nabla f = 0$ that satisfies the boundary condition on the surface of the droplet and goes to zero as $r \to \infty$) is

$$f^{(0)} = f_0 H(\hat{\mathbf{r}} \cdot \hat{\mathbf{v}} - [1 - R^2/r^2]^{1/2}) \tag{5}$$

where $H(x)$ is the Heaviside unit step function; $H(x) = 0$ for $x < 0$, and $H(x) = 1$ for $x > 0$. Equation (5) expresses the fact that the distribution has its equilibrium value in the "cone of influence" of the droplet (region I in Fig. 1) and is zero elsewhere. Following Kelly and Sengers,[10] we now substitute the free molecule solution in the right-hand side of the Boltzmann equation and put $f = f^{(0)} + f^{(1)}$ on the left-hand side. The equation for $f^{(1)}$ is then

$$\begin{aligned} \mathbf{v} \cdot \nabla f^{(1)}(\mathbf{r}, \mathbf{v}) &= \int K(\mathbf{v}, \mathbf{v}') f^{(0)}(\mathbf{r}, \mathbf{v}')\, d\mathbf{v}' - \nu(\mathbf{v}) f^{(0)}(\mathbf{r}, \mathbf{v}) \\ &\equiv g(\mathbf{r}, \mathbf{v}) \end{aligned} \tag{6}$$

and the boundary conditions are that $f^{(1)}$ vanishes on the droplet surface and at infinity. The evaporative flux is written

$$\Phi = \Phi_{\text{fm}}(1 - \alpha\Gamma^{(1)} + \cdots) \tag{7}$$

where $\Phi_{\text{fm}} = n_v \bar{c}/4$ is the free molecule flux, and $\alpha = Kn^{-1}$. The first correction to the free molecule flux represents the vapor molecules that are reflected back onto the droplet by a single collision with a gas molecule.[1,10,13] Thus it always leads to a reduction in the evaporative flux (so $\Gamma^{(1)}$ is always positive). The dimensionless first correction term is given by[10]

$$\alpha\Gamma^{(1)} = \frac{4}{n_v\bar{c}} \int_{\mathbf{R}\cdot\mathbf{v}<0} \mathbf{R}\cdot\mathbf{v} f^{(1)}(\mathbf{R},\mathbf{v})\, d\mathbf{v} \tag{8}$$

where the integral extends over values of $\mathbf{v}$ such that $\mathbf{R}\cdot\mathbf{v} < 0$. Unlike the Knudsen iteration procedure (where the distribution function is written as a power series in α), no assumptions are made concerning the functional dependence of the correction terms on $\alpha = Kn^{-1}$. Although the first correction term is indeed proportional to α, Kelly and Sengers[10] showed that the second correction term by this method varies as $\alpha^2 \ln\alpha$ in the case of a droplet in its pure vapor.

Equation (6) can be solved by integrating along the characteristics (which correspond to the trajectories of the vapor molecules after a single collision). The solution on the droplet surface can be written[10]

$$f^{(1)}(\mathbf{R},\mathbf{v}) = \int_0^\infty g(\mathbf{R}-\mathbf{v}\tau,\mathbf{v})\, d\tau \qquad \text{for} \qquad \mathbf{R}\cdot\mathbf{v} < 0 \tag{9}$$

and $f^{(1)} = 0$ for $\mathbf{R}\cdot\mathbf{v} > 0$.

The inhomogeneous term in Eq. (6), $g(\mathbf{r},\mathbf{v})$, is found by substituting Eq. (5) into the right-hand side of Eq. (3). The effect of the step function is to limit the integration over $\mathbf{v}'$ to directions originating from the droplet surface. Also, for $\mathbf{r}\cdot\mathbf{v} < 0$, the term $\nu f^{(0)}(\mathbf{r},\mathbf{v})$ vanishes; thus, we have

$$f^{(1)}(\mathbf{R},\mathbf{v}) = \int_0^\infty d\tau \int_{\mathbf{v}' \text{ from droplet}} d\mathbf{v}' K(\mathbf{v},\mathbf{v}') f_0(v') \tag{10}$$

Substituting Eq. (10) into Eq. (8) gives the first correction term. Referring to Fig. 1, we take the polar axis for the integrations over $\mathbf{v}$ as the line OA. Then $d\mathbf{v} = 2\pi v^2\, dv\, d(\cos\theta_a)$. For the integrations over $\mathbf{v}'$, we take AP as the polar axis; thus $d\mathbf{v}' = v'^2\, dv'\, d\phi'\, d\mu$, where $\mu = \hat{\mathbf{v}}\cdot\hat{\mathbf{v}}' = \cos\theta_{vv'}$ is the cosine of the scattering angle. For the vapor molecule to originate from the droplet, we must have $-1 \le \mu \le \mu_2 = \cos\theta_2$. Also, for $\mu_2 \ge \mu \ge \mu_1 = \cos\theta_1$, ϕ' must lie between $-\phi_m$ and ϕ_m (for $\mu \le \mu_1, \phi'$ is unresticted, ie., $\phi_m = \pi$). The expression for the first correction term can now be written

$$\alpha\Gamma^{(1)} = \frac{8\pi}{n_v\bar{c}} \int_0^{\pi/2} d\theta_a \sin\theta_a \cos\theta_a \int_0^\infty d\tau \int_{-1}^{\mu_2} d\mu \int_{-\phi_m}^{\phi_m} d\phi' \int_0^\infty dv v^3 \int_0^\infty dv' v'^2 K(\mathbf{v},\mathbf{v}') f_0(v') \tag{11}$$

We assume that the intermolecular potential is a function of the molecular separation alone; hence, the kernel $K(\mathbf{v},\mathbf{v}')$ depends only on v, v' and μ, and the integration over ϕ' can be performed. Then, changing variables from τ to r, where $r^2 = R^2 + v^2\tau^2 + 2Rv\tau\cos\theta_a$ (see Fig. 1), and introducing the dimensionless variables $x = \sin\theta_a$ and $y = R/r$, we find that

$$\Gamma^{(1)} = \int_0^1 \mathrm{d}x x \int_0^1 \frac{\mathrm{d}y}{y^2(1-x^2y^2)^{1/2}} \int_{-1}^{\mu_2} \mathrm{d}\mu \frac{\phi_m}{\pi} S(\mu) \tag{12}$$

with

$$S(\mu) = \frac{2\pi R}{\alpha}\left(\frac{m_v}{kT}\right)^2 \int_0^\infty \mathrm{d}v v^2 \int_0^\infty \mathrm{d}v' v'^2 K(v,v',\mu) \exp\left(-\frac{m_v v'^2}{2kT}\right) \tag{13}$$

Since $K(v,v',\mu)$ involves a double integral, in general, four integrations are needed to evaluate $S(\mu)$ for each value of μ. However, as shown in the Appendix, for hard spheres and Maxwell molecules it is possible to perform some of these integrations. For hard spheres, we obtain

$$\begin{aligned} S_{\text{HS}}(\mu) &= \frac{27\pi}{8}(1+Z)^{5/2} \\ &\quad Z^2 \int_0^1 \mathrm{d}t \frac{\{t(t-2)[(1+\mu)t(t-2)+2]\}^2}{([P(t)+2t(t-2)]P(t))^{5/2}} \end{aligned} \tag{14}$$

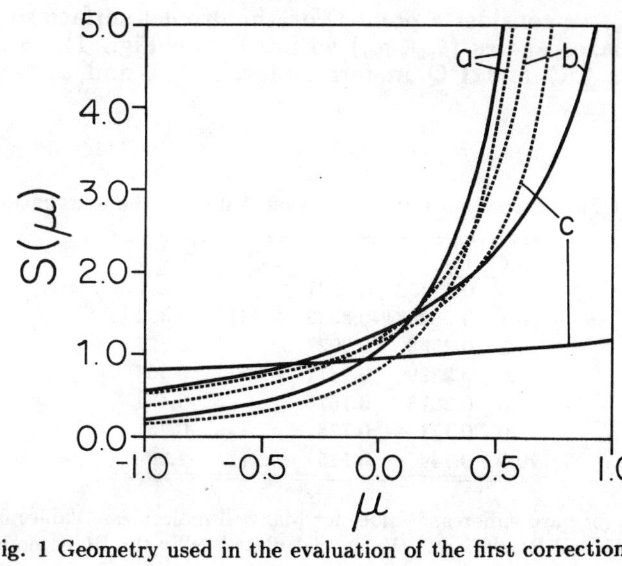

Fig. 1 Geometry used in the evaluation of the first correction term.

where

$$P(t) = (1+\mu)Zt(t-2) + 2(1+Z)$$

For Maxwell molecules, we have

$$S_{MM}(\mu) = \frac{3}{8\sqrt{2}A_1(5)} Z^{3/2}(1+Z)\int_0^{1/2} \frac{du}{[u(1-u)]^{3/2}\cos^3\theta)} \int_0^1 dt \frac{t^2(2-t)^2(2p^2+q^2)}{(p^2-q^2)^{5/2}} \quad (15)$$

where the constant $A_1(5) \simeq 0.42194$ is defined by Chapman and Cowling[14] and the functions $\theta(u)$, $p(\theta, t)$, and $q(\theta, t)$ are given in the Appendix [Eqs. (A.1-A.3)]. In using $u = 0$ for the lower limit of the integration over u, we have allowed the range of the Maxwellian interaction to extend to infinity, and we find that $S_{MM} \to \infty$ as $\mu_{vv'} \to 1$. However, we expect the contribution to $\Gamma^{(1)}$ to become zero as $\mu_{vv'} \to 1$, since this represents molecules emitted from the droplet and not scattered through an appreciable angle. It is possible to prove formally from Eqs. (11) and (14) that $\Gamma^{(1)}$ remains finite as the range of the Maxwellian interaction becomes infinite, but we do not present the details here. Finally, we need expressions for μ_1, μ_2, and ϕ_m. From Fig. 1, we have

$$\mu_1 = -y^2x - [(1-y^2x^2)(1-y^2)]^{1/2}$$
$$\mu_2 = y^2x - [(1-y^2x^2)(1-y^2)]^{1/2}$$

To find ϕ_m, we consider a point Q on the droplet surface so that the line PQ makes angles $(\theta_{vv'}, \phi_m)$ with AP (see Fig. 1). Expressing the vectors **OQ** and **PQ** in terms of $r, \theta_a, \theta_{vv'}$, and ϕ_m and using

Table 1 First correction terms for various values of the mass ratio Z

Z	$\Gamma^{(1)}_{HS}$ [a]	$\Gamma^{(1)}_{MM}$ [b]	$\Gamma^{(1)}_{M-B}$ [c]	$\Gamma^{(1)}_{BGK}$ [d]
0	0.29452	0.2337		0.297
0.1	0.28253	0.2238	0.334	0.302
0.5	0.2538	0.2027		0.325
1.0	0.2329	0.1872	0.297	0.363
2.0	0.2078	0.1676		0.454
5.0	0.171	0.138		0.751
10.0	0.144	0.115	0.211	1.26

[a]Values for hard spheres. [b]Values for Maxwell molecules. [c]Values from Monchick and Blackmore.[8] [d]Values calculated using the BGK model.[13]

Fig. 2 $S(\mu)$ v μ for a) $Z = 10$, b) $Z = 1$ and c) $z = 0.1$. The solid lines show values for hard spheres [Eq. (14)], and the dashed lines are for Maxwell molecules [Eq. (15)].

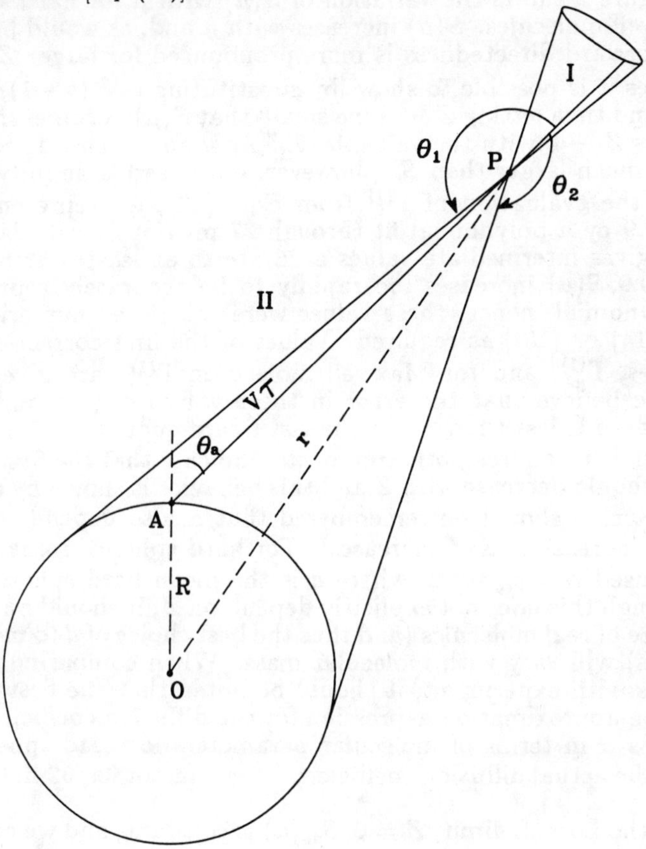

$\mathbf{OQ} \cdot \mathbf{PQ} = 0$, we find that

$$
\begin{aligned}
\phi_m &= \arccos\left(\frac{[1 - x^2y^2]^{1/2}\mu_{vv'} + [1 - y^2]^{1/2}}{xy[1 - \mu_{vv'}^2]^{1/2}}\right) \text{ for } \mu_2 > \mu_{vv'} > \mu_1 \\
&= \pi \text{ for } \mu_1 \geq \mu_{vv'} \geq -1 \\
&= 0 \text{ otherwise}
\end{aligned}
$$

III. Results and Discussion

The integrations in Eqs. (12), (14), and (15) were performed numerically using Gauss-Legendre quadrature. Typically, between 30 and 60 quadrature points were used in each dimension. Comparisons between numerical values using different numbers of quadrature points were used to estimate the accuracy of our results.

Figure 2 shows the variation of $S(\mu)$ with μ for hard spheres and Maxwell molecules. $S(\mu)$ increases with μ and, as would be expected, this forward directedness is more pronounced for larger Z. For hard spheres it is possible to show [by substituting $t' = (t-1)/\sqrt{Z}$ in Eq. (14) and then letting Z become small] that $S(\mu)$ becomes independent of μ as $Z \to 0$ with the value $9\pi/32$. As μ approaches 1, S_{MM} appears to be much larger than S_{HS}; however, both tend to infinity as $\mu \to 1$.

In the evaluation of $\Gamma^{(1)}$ from Eq. (12), we represent $S(\mu)$ for $\mu < 0.9$ by a polynomial fit through 27 previously calculated values. This gives intermediate values accurate to at least four figures. For $\mu > 0.9, S(\mu)$ increases too rapidly to be accurately represented by a polynomial; hence, these values were calculated numerically [from Eq. (14) or (15)] as required. Values of the first correction for hard spheres, $\Gamma^{(1)}_{HS}$, and for Maxwell molecules, $\Gamma^{(1)}_{MM}$, are given in Table 1. We believe that the error in these values due to our numerical procedure is less than ± 3 in the last figure quoted.

Simple mean free path arguments[1] indicate that the first correction term should decrease with Z and this behavior is shown by our results. However, it should be remembered that α also depends on Z (generally increasing as Z increases). For hard spheres, some authors[8,11] have used $\alpha = n_g \pi d^2 R$, where d is the mean hard sphere diameter. Although this does not explicitly depend on Z, it should be noted that the size of real molecules (and thus the best choice of d to model experiments) will vary with molecular mass. When comparing theoretical results with experiment, it should be noted that the first Chapman-Enskog approximation expression for the diffusion coefficient, used to express α in terms of molecular parameters for hard spheres, is less than the actual diffusion coefficient[14] (by a factor $9\pi/32$ in the Lorentz limit).

In the Lorentz limit, $Z = 0, S_{HS}(\mu)$ is constant, and we can perform the integrations in Eq. (12) analytically to obtain $\Gamma^{(1)}_{HS}(0) = 3\pi/32$, in agreement with the exact result of Davison[9] in this limit. This value can be obtained more simply by using the Lorentz form of the collision operator and also by an extension of the simple model discussed by Fuchs and Sutugin.[1] The value of $\Gamma^{(1)}_{MM}(0)$ was obtained using the Lorentz limit of the Boltzmann equation for Maxwell molecules. From the values in Table 1, we find that the ratio $\Gamma^{(1)}_{MM}/\Gamma^{(1)}_{HS}$ remains remarkably constant as Z varies, with the value 0.800 ± 0.007.

The third column in Table 1 gives values for $\Gamma^{(1)}$ from the results of Monchick and Blackmore[8] for hard sphere interactions. We derived these values from their table of values by a quadratic fit through the two smallest radii they consider (corresponding to $\alpha = 0.096(1 + Z)^{-1/2}$ and $0.19(1+Z)^{-1/2}$). Although not really justifiable (as noted above, the evaporative flux cannot be represented as a power series in α), we believe this procedure gives the limiting form of their results to within about 5%. It appears that Monchick and Blackmore predict too large a correction to the free molecular flux for small α, the error being greatest for large Z.

Brock[13] applied Knudsen iteration to the BGK model form of the Boltzmann equation to obtain an expression for $\Gamma^{(1)}$. Using this result together with Brock's[13] Eq. (16) relating the collision frequency to the hard sphere radii of the molecules, we obtain the values shown in the last column of Table 1. Although apparently quite accurate at $Z = 0$, the model is poor for larger values of Z, predicting that the magnitude of the correction increases with increasing Z. Furthermore, had we used the expression for the collision frequency in terms of the diffusion coefficient for Maxwell molecules [Brock's Eq. (14)], the values of $\Gamma^{(1)}_{\text{BGK}}$ would be multiplied by approximately 1.32, and the agreement for $Z = 0$ would no longer be so good.

There are some other comparisons we can make. Kelly and Sengers[11] considered droplet growth in a pure vapor; thus, their final values cannot be compared with ours (in fact, they find a positive first correction term). However, the value of their recollision integral, representing vapor molecules reflected back onto the droplet, is (with our definition of α) 0.234 ± 0.0025, which agrees well with our value of $\Gamma^{(1)}_{\text{HS}}(Z = 1) = 0.2329$. The Fuchs-Sutugin formula[1] uses the definition Eq. (1) of Kn and has the limiting behavior, for large Kn,

$$\Phi = \Phi_{\text{fm}}(1 - 0.283Kn^{-1}) \tag{16}$$

[this limit appears to be given incorrectly by some authors[4,8]]. Thus, this formula predicts that $\Gamma^{(1)}$ is independent of Z with the value 0.283. Sampson and Springer[3] use Lees' moment method to derive the expression $\Phi = \Phi_{\text{fm}}/(1 + 0.75Kn^{-1})$ for the vapor flux from an evaporating droplet. The predicted value of the first correction term, $\Gamma^{(1)} = 0.75$, is again independent of Z and is clearly too large.

IV. Summary and Conclusions

We have presented values for the first correction term $-\Gamma^{(1)}Kn^{-1}$ to the free molecular flux for hard spheres and Maxwell molecules. We found that $\Gamma^{(1)}$ decreases by a factor of about two as the vapor-to-gas molecular mass ratio Z increases from 0.1 to 10. Although the correction term for Maxwell molecules $\Gamma^{(1)}_{\text{MM}}$ is about 20% smaller than that for hard spheres, $\Gamma^{(1)}_{\text{HS}}$, the ratio $\Gamma^{(1)}_{\text{MM}}/\Gamma^{(1)}_{\text{HS}}$ is virtually independent of Z. We would expect $\Gamma^{(1)}$ for real molecular interactions to lie between the values of $\Gamma^{(1)}_{\text{HS}}$ and $\Gamma^{(1)}_{\text{MM}}$.

Our results provide a stringent test of other solutions in the near free molecule regime where the anisotropy of the distribution function is very pronounced. The Fuchs-Sutugin formula and Brock's results using the BGK approximation give reasonable values for small Z, but both overestimate the correction when Z is large. In addition, different expressions for the collision frequency in the BGK approximation lead to different values for $\Gamma^{(1)}$. The numerical results of Monchick and Blackmore predict values of $\Gamma^{(1)}$ between 20 and 50% too large. They solved the Boltzmann equation [Eq. (3)] by a variational method using a four term trial function, and their method appears to offer a promising way to obtain the evaporative flux for all Knudsen numbers and mass ratios. However, our results indicate more terms are needed to obtain accurate values for large Kn, even though their trial function included the free molecule solution.

Since Kn^{-1} is necessarily small in our treatment, significant differences in $\Gamma^{(1)}$ may still lead to only small changes in the total flux. This makes experimental testing of our results difficult. It is unlikely that the method used here can be extended to give higher-order terms explicitly[10] since the computations involved appear to be prohibitive.

Comparison with experiment is also complicated by uncertainties concerning the sticking probability,[6] s_A for vapor molecules on the liquid surface. For $s_A < 1$ and a concentration n_∞ of vapour molecules far from the droplet, the net evaporative flux to first order in Kn^{-1} is

$$\Phi = \frac{\bar{c}}{4} s_A \left(n_v - n_\infty - [s_A n_v + (1 - s_A) n_\infty] \Gamma^{(1)} Kn^{-1} \right) \qquad (17)$$

For $s_A \neq 1$ the flux is no longer proportional to $(n_v - n_\infty)$.

The method used here can be applied to a number of other transport problems in aerosol physics.[1] These include thermophoresis and diffusiophoresis, which have been investigated in the near free molecule regime by Brock,[15,16] using the BGK approximation. It would be interesting to compare exact results with Brock's approximate values.

Appendix

In general, we have[12]

$$K(v, v', \mu) = \left(\frac{1+Z}{2}\right)^3 \int\int B\left(\theta, \left[\frac{1+Z}{2\cos\theta}\right] V\right) F_0\left(\mathbf{v} + Z\mathbf{V} - \left[\frac{1+Z}{2\cos\theta}\right] V \hat{\mathbf{n}}\right) \frac{d\theta\, d\epsilon}{\cos^3\theta}$$

where $\mathbf{V} = \mathbf{v} - \mathbf{v}'$, and $\hat{\mathbf{n}}$ is a unit vector making angles θ, ϵ with $\mathbf{V}$. The expression $B(\theta, g) = g\sigma(\theta, g)$ (σ is the differential scattering cross section), and $F_0(\mathbf{v})$ is the equilibrium Maxwellian distribution for the gas molecules.

For hard spheres, $B(\theta, g) = gd^2 \sin\theta \cos\theta$, and we can perform the integrations to obtain

$$K(v, v', \mu) = \left(\frac{m_g}{2\pi kT}\right)^{1/2} d^2 n_g \left(\frac{1+Z}{2}\right)^2 \cdot \frac{1}{V} \exp\left(\frac{-m_g}{2kT}\right.$$
$$\left.\left[\frac{(v^2 - vv'\mu)^2}{V^2} + \left(\frac{Z-1}{2}\right)^2 V^2 + (Z-1)(v^2 - vv'\mu)\right]\right)$$

with $V^2 = v^2 + v'^2 - 2vv'\mu$.
The first Chapman Enskog approximation for the hard sphere diffusion coefficient[14] gives $\alpha = 32 n_g d^2 R/(9[1+Z]^{1/2})$. Using these results in Eq. (13) and setting $v' = \beta v$ gives

$$S(\mu) = \frac{9\sqrt{\pi}}{16}\left(\frac{m_g}{2kT}\right)^{5/2} Z^2(1+Z)^{5/2} \int_0^\infty \frac{d\beta \cdot \beta^2}{\sqrt{1+\beta^2-2\beta\mu}} \int_0^\infty dv \cdot v^4$$
$$\exp\left(-\frac{m_g v^2}{2kT}\left[Z + \frac{(1-\beta\mu)^2}{1+\beta^2-2\beta\mu} + \right.\right.$$
$$\left.\left.\left(\frac{Z-1}{2}\right)^2 (1+\beta^2-2\beta\mu) + (Z-1)(1-\beta\mu)\right]\right)$$

Performing the integration with respect to v, substituting $t = 2/(1+\beta)$, and noting the symmetry of the integrand about $t = 1$, gives the result in Eq. (14).

For Maxwell molecules, the vapor-gas intermolecular force is $m_v m_g k_f / r^5$, and$B(\theta, g)\, d\theta = \sqrt{m_g k_f (1+Z)/2}\, b\, db$. The expression for K becomes

$$K(v, v', \mu) = \frac{(1+Z)^{7/2} m_g^2 k_f^{1/2} n_g}{32(\pi kT)^{3/2}} \int\int db\, d\epsilon \frac{b}{\cos^3\theta} \exp\left(-\frac{m_g}{2kT}\right.$$
$$\left[Zv'^2 + 0.25(1+Z)^2 V^2 \sec^2\theta\right.$$
$$\left.\left.(1+Z)vv'\{\mu - \sqrt{1-\mu^2}\tan\theta\cos\epsilon\}\right]\right)$$

We use the expression for the diffusion coefficient for Maxwell molecules[14] to obtain

$$\alpha = \frac{8\pi}{3} A_1(5) n_g R (m_v m_g k_f)^{1/2} / ([1+Z][2\pi kT])^{1/2}$$

Using these results in Eq. (13), putting $v' = \beta v$, and performing the integration with respect to v gives

$$S(\mu) = \frac{3}{8\pi\sqrt{2A_1}} Z^{3/2}(1+Z)\int_0^\infty \frac{b\,db}{\cos^3\theta}\int_0^\infty \beta^2\,d\beta \int_0^{2\pi}\frac{d\epsilon}{(a_1 - a_2\cos\epsilon)^3}$$

where

$$a_1 = \beta\mu + 0.25[\beta^2 - 2\beta\mu + 1](1+Z)\sec^2\theta$$
$$a_2 = \beta\sqrt{1-\mu^2}\tan\theta$$

The integration with respect to ϵ can now be performed. Then, setting $t = 2/(1+\beta)$ as before gives

$$S(\mu) = \frac{3}{2\sqrt{2A_1}(5)} Z^{3/2}(1+Z)\int_0^\infty \frac{b\,db}{\cos^3\theta}\int_0^1 dt\cdot\frac{t^2(2-t)^2(2p^2+q^2)}{(p^2-q^2)^{5/2}}$$

where

$$p(\theta,t) = \frac{1+Z}{\cos^2\theta} + t(2-t)\left[\mu - 0.5(1+Z)\frac{(1+\mu)}{\cos^2\theta}\right] \quad (A.1)$$

$$q(\theta,t) = t(2-t)\sqrt{1-\mu^2}\tan\theta \quad (A.2)$$

Finally, changing variables from b to u, where $b^2 = 2\cot 2\psi$ and $u = \sin^2\psi$ gives the result in Eq. (15). The θ is given in terms of u by

$$\theta(u) = \sqrt{1-2u}\,\mathrm{K}(u^2) \quad (A.3)$$

where $\mathrm{K}(u^2)$ is the complete elliptic integral of the first kind.

Acknowledgment

This research is supported by a grant from the Natural Science and Engineering Research Council of Canada.

References

[1] Fuchs, N. A. and Sutugin, A. G., "High-Dispersed Aerosols," Topics in Current Aerosol Research, Vol.2, edited by G. M. Hidy and J. R. Brock, Pergamon, Oxford, U.K., 1971, pp.1-60.

[2] Sitarski, M. and Nowakowski, B., "Condensation rate of trace vapor on Knudsen aerosols from the solution of the Boltzmann equation," Journal of Colloid and Interface Science, Vol.72, 1979, pp.113-122.

[3] Sampson, R. E. and Springer, G. S.,"Condensation on and evaporation from droplets by a moment method," Journal of Fluid Mechanics, Vol.36, 1969, pp.577-584.

[4] Tompson, R. V. and Loyalka, S. K., "Condensational growth of a spherical droplet: free molecular limit," Journal of Aerosol Science, Vol.17, 1986, pp.723-728.

[5] Bahatnager, P. L., Gross, E. P., and Krook, M., "A model for collision processes in gases. I. Small amplitude processes in charged and neutral one component systems," Physical Review, Vol.94, 1954, pp.511-525.

[6] Wagner, P. E., "Aerosol growth by Condensation," Aerosol Microphysics II: Chemical Physics of Microparticles, edited by W. H. Marlow, Springer-Verlag, Berlin, FRG, 1982.

[7] Davis, E. J., "Transport phenomena with single aerosol particles," Aerosol Science and Technology, Vol.2, 1983, pp.121-144.

[8] Monchick, L. and Blackmore, R., "A variation calculation of the rate of evaporation of small droplets," Journal of Aerosol Science, Vol.19, 1988, pp.273-286.

[9] Davison, B., "Influence of a black sphere and of a black cylinder upon the neutron density in an infinite non-capturing medium," Proceedings of the Physical Society London, Vol.A64, 1951, pp.881-902.

[10] Kelly, G. E. and Sengers, J. V., "Kinetic theory of droplet growth in nucleation," Journal of Chemical Physics, Vol.57, 1972, pp.1441-1458.

[11] Kelly, G. E. and Sengers, J. V., "Droplet growth in a dilute vapour," Journal of Chemical Physics, Vol.61, 1974, pp.2800-2807.

[12] Cercignani, C., Theory and Application of the Boltzmann Equation, Scottish Academic Press, Edinburgh, Scotland, U.K., 1975.

[13] Brock, J. R., "Highly nonequilibrium evaporation of moving particles in the transition region of Knudsen number," Journal of Colloid and Interface Science, Vol.24, 1967, pp.344-351.

[14] Chapman, S. and Cowling, T. G., The Mathematical Theory of Non-Uniform Gases, Cambridge University Press, Cambridge, U.K., 1970.

[15] Brock, J. R., "The thermal force in the transition region of Knudsen number," Journal of Colloid and Interface Science, Vol.23, 1967, pp.448-452.

[16] Brock, J. R., "The diffusion force in the transition region of Knudsen number," Journal of Colloid and Interface Science, Vol.27, 1968, pp.95-100.

Theoretical and Experimental Investigation of the Strong Evaporation of Solids

R. Mager,* G. Adomeit,† and G. Wortberg†
Rheinisch-Westfälische Technische Hochschule Aachen, Aachen, Federal Republic of Germany

Abstract

The kinetic boundary layers (Knudsen layers) adjacent to two surfaces of evaporating and condensing Iodine have been investigated experimentally and theoretically. By means of fluorescence spectroscopy with a tunable dye laser, the translational temperature parallel and perpendicular to the flow, the density, and the macroscopic flow velocity have been measured. The experimental results are compared with theoretical calculations. An approximate solution of the Boltzmann equation has been obtained by means of a five-moment method, applying a bimodal distribution function and the model of Maxwellian molecules. Special attention has been paied to the formulation of the boundary conditions at the evaporating and condensing surfaces. The influence of the evaporation coefficient, which characterizes the properties of the evaporating surface, has been studied.

Introduction

In this study we are interested in the nonequilibrium flow regime between an evaporating and a condensing surface. Both surfaces are parallel to each other, and we are faced with a one-dimensional problem. If the distance between the surfaces is considerable large in terms of mean free path, an equilibrium layer separates the two nonequilibrium layers (Knudsen layers) adjacent to the surfaces. If the distance is rather small, the two Knudsen layers merge. In

---

* Assistant, Institut fuer Allgemeine Mechanik.
† Professor, Institut fuer Allgemeine Mechanik.

both cases we are interested in strong evaporation with a large temperature difference between the surfaces and therefore large deviation from equilibrium.

The previously mentioned problem suggests the application of the moment method in order to find an approximate solution of the Boltzmann equation. From the literature various approaches to the problem are known, which basically differ in the choice of an appropriate molecular velocity distribution function. The Ytrehus[1] approach resembles the Mott-Smith method, known from the shock-wave profile problem in gasdynamics. He used a weighted sum of Liu-Lees-type distribution functions. His work was mainly concerned with the equilibrium flow properties far away from the evaporating surface. In two recent papers[2,3] we used an ellipsoidal distribution function with the two different temperatures T_T and T_L perpendicular and parallel to the flow direction. The quantities mentioned may be obtained directly from measurements by means of laser fluorescence spectroscopy. Therefore, we were able to compare flow properties measured within the Knudsen layer with our theoretical calculations. Unfortunately, the continuous ellipsoidal distribution function may be matched to the boundary conditions at the evaporating surface in a rather artificial way only.

The following bimodal distribution function, first suggested by Koffman et al.[4], is most convenient for the formulation of the boundary conditions:

$$f_{\pm} = n_{\pm} \left(\frac{m}{2\pi k T_{\pm}}\right)^{3/2} \exp\{-\frac{m}{2kT_{\pm}}[(\xi_x - w_{\pm})^2 + \xi_y{}^2 + \xi_z{}^2]\} \tag{1}$$

$$f = f_+ \; ; \quad \text{for} \quad \xi_x > 0$$
$$f = f_- \; ; \quad \text{for} \quad \xi_x < 0$$

The six unknown parameters contained in the formula above may be obtained by applying a six-moment method. However, Koffman et al.[4] report in their paper that the temperature profile in the vapor for the continuum problem is inverted from what would seem physically reasonable.

In our present work we used the distribution function Eq. (1), but we put

$$w_+ = w_- = w \tag{2}$$

Therefore, a five-moment method is sufficient. We obtained physically reasonable results that are in good agreement with our measurements.

Experimental Technique

The evaporation of iodine was performed in a stainless steel chamber that was part of a high vacuum system. A flat thin layer of iodine deposited on a metal plate was held at a constant preset temperature. The evaporation was initiated by using liquid nitrogen to cool down a second metal plate facing the iodine. The distance between the evaporating and the condensing surface could be varied from 99 to 205 mm. The temperature of the evaporating surface was adjusted in different experiments in the range of 225 - 263°K. In equilibrium these temperatures correspond to a mean free path of 99 and 2.6 mm, respectively.

By means of fluorescence spectroscopy with a tunable continuous-wave dye laser, the profile of an absorption line could be obtained. The situation is complicated by the fact that each absorption line has a fine structure consisting of many lines. Because of the Doppler effect the absorption line profile is modified by the molecular motion. The probing laser beam could be directed parallel or perpendicular to the evaporating surface by a movable mirror. Therefore, at each measuring point two absorption profiles were obtained. Assuming an ellipsoidal distribution function and applying a trial- and -error method, the number density n, the macroscopic velocity u, and T_L and T_T could be obtained. A more detailed description of our experimental technique may be found in our previous papers[2,3] or in a forthcoming thesis.[5]

Theory

For a theoretical description of the processes occurring within the Knudsen layer, an approximate solution of the Boltzmann equation has been obtained by means of a five-moment method. The evaporation model is based on the bimodal distribution function described in Eq. (1) and putting $w_+ = w_- = w$. Because the problem is one-dimensional the five parameters n_+, n_-, T_+, T_-, and w are functions of only the distance x from the evaporating surface. Besides the conservation equations for mass, momentum, and energy, the non-collision-invariant equations for $Q=m\xi_x^2$ and $Q=m\xi_x\xi^2$ have been evaluated using the model of Maxwellian molecules.

In the case of an evaporating liquid, the basic assumption for the boundary conditions at the evaporating surface (x = 0) is that the distribution function of the molecules leaving the surface ($\xi_x > 0$) is not altered by the evaporation process. In terms of our distribution function Eq. (1), we obtain

$$n_+ = n_o, \quad T_+ = T_o, \quad w = 0 \quad \text{for } x = 0$$

where n_o is the equilibrium number density at $T = T_o$.

Because of the very low temperature of the condensing surface, we make the restricting assumption that no molecules are leaving the surface ($x = L$). Furthermore, we assume that T_- goes to a minimum at $x = L$. Therefore, we obtain

$$n_- = 0, \quad T_- \rightarrow \min \quad \text{for } x = L$$

If the distance between the two surfaces is large in terms of mean free path, the two Knudsen layers are separated by a quasiequilibrium layer. In this case we may formulate

$$T_+ = T_- = T_\infty; \quad n_+ = n_- = n_\infty; \quad w = u = u_\infty \quad \text{for } x \rightarrow \infty$$

If we prescribe the Mach number M_∞ at infinity, a relation between u_∞ and T_∞ exists. Ytrehus[1] has shown that for strictly one-dimensional flow the maximum possible Mach number is $M_\infty = 1$.

In our present work, however, we are interested in the evaporation from a solid surface. The flux of the molecules leaving the surface depends on the binding energy of the molecules to the surface and therefore on the surface conditions. This may be taken into account by an evaporation coefficient α $(0 < \alpha < 1)$, introduced by Knudsen. For $\alpha < 1$ and strong evaporation, the molecular flux evaporated from the surface is less than predicted by the half-space Maxwellian distribution, and for the evaporated molecular flux we may put

$$J_{e+} = \alpha \, J_{o+}$$

For the total flux in the general case we have to add the contribution of the reflected molecules. By equilibrium considerations we obtain the molecular fluxes depicted in Fig. 1. Assuming full thermal accommodation ($T_+ = T_o$), we obtain at the evaporating surface

$$n_+ = n_o \left(\alpha + (1-\alpha) \frac{n_-}{n_o} \sqrt{\frac{T_-}{T_o}} \right) \quad \text{for } x = 0$$

For convenience in computation nondimensional quantities are introduced. Usually quantities at

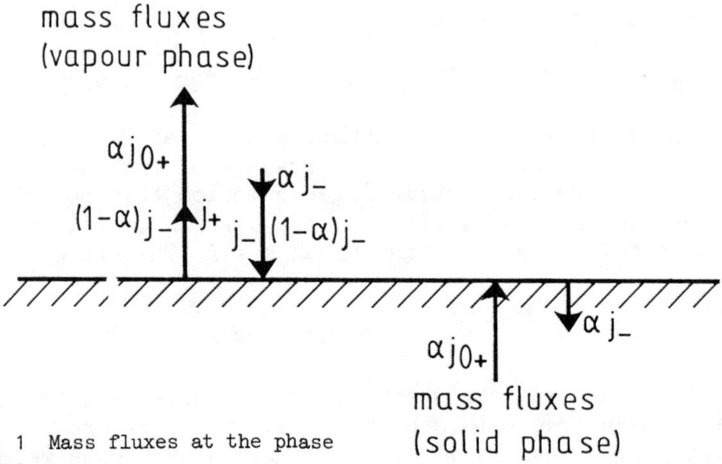

Fig. 1 Mass fluxes at the phase interface.

equilibrium (n_o, T_o) are used for normalization. Solutions that are independent of the evaporation coefficient α may be obtained if the quantity n_+ at $x = 0$ is used instead of n_o.

The application of the moment method yields five coupled nonlinear first-order ordinary differential equations. Since some of the boundary conditions are formulated at the evaporating surface and some at the condensing or at infinity, one has to apply a shooting method similar to the work of Koffman et al.[4]. Finally, it should be mentioned that, after the parametric functions n_+, n_-, T_+, T_-, and w are known, macroscopic quantities such as T_T, T_L, $T = 1/3\,(T_L + 2T_T)$, n, and u may be obtained by integration of the distribution function Eq. (1).

Results

At first we studied the case of quasiequilibrium flow at infinity. In Fig. 2 the dependence of the normalized quantities T'_∞, n'_∞ and of M on the normalized mass flow j' is shown. The mass flow has been normalized by the maximum possible flow at the evaporating surface when no molecules are returning to the surface. Therefore, the curves depicted in Fig. 2 are independent of the evaporation coefficient α. As far as these results are concerned, the results of our previous work[2,3] are identical with the work of Crout.[6] The differences between various theories are remarkably small.

Included in Fig. 2 are experimental results of our present work. For measured values to be converted into values normalized in the manner mentioned earlier, α has to

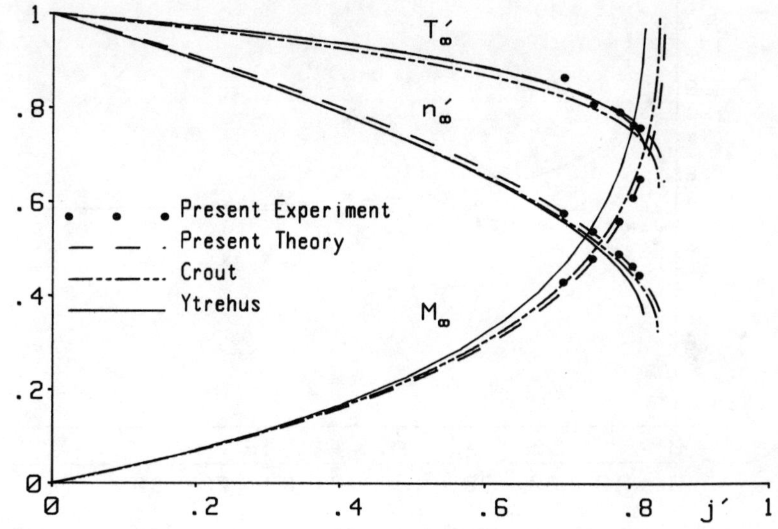

Fig. 2 Number density, temperature, and Mach number at infinity vs mass flux.

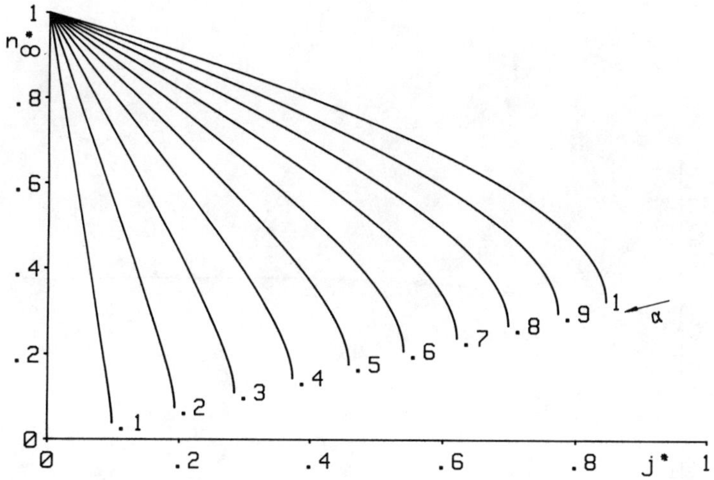

Fig. 3 Number density at infinity as function of mass flux and evaporation coefficient α.

be known. In Fig. 3 the dependence is shown of n^* normalized by the equilibrium number density on j^* normalized by the mass flux of the molecules with $\zeta_x > 0$ in equilibrium. The evaporation coefficient α enters here as a parameter and may therefore be determined from experiment by means of

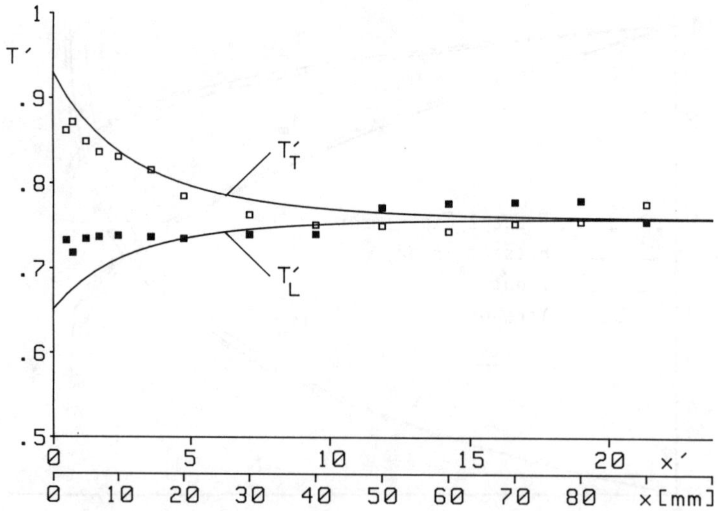

Fig. 4 Temperatures parallel to the evaporating surface T_T' and perpendicular to the surface T_L' vs distance. Condensing surface at infinity.

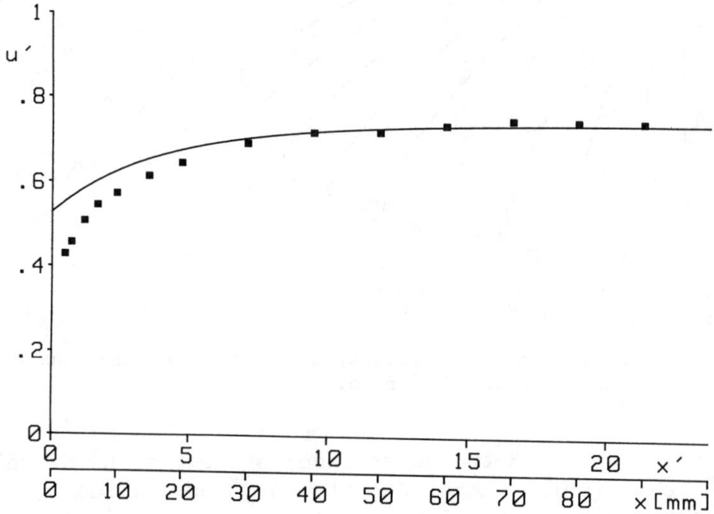

Fig. 5 Macroscopic flow velocity u' vs distance. Condensing surface at infinity.

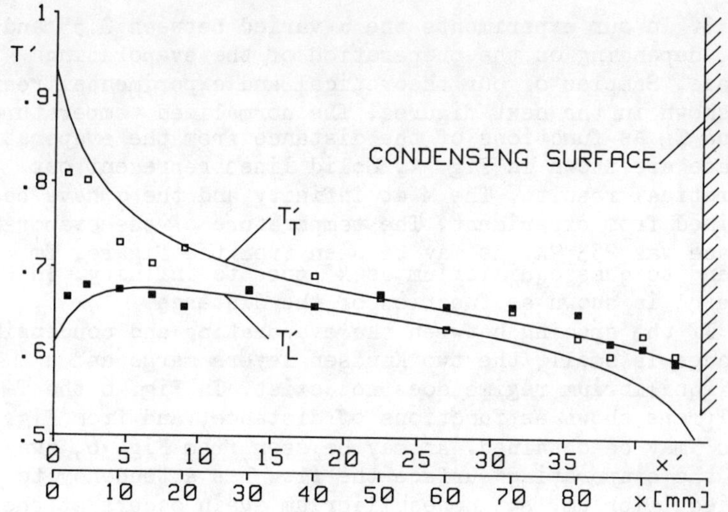

Fig. 6 Temperatures parallel to the evaporating surface T'_T and perpendicular to the surface T'_L vs distance.

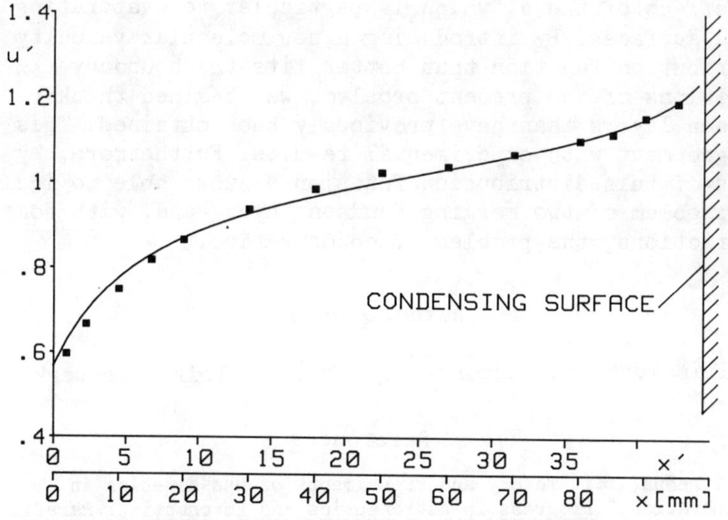

Fig. 7 Macroscopic flow velocity u' vs distance.

Fig. 3. In our experiments the α varied between 0.55 and 0.65, depending on the preparation of the evaporating surface. Samples of our theoretical and experimental results are shown in the next figures. The normalized temperatures T_T' and T_L' as functions of the distance from the evaporating surface are shown in Fig. 4. Solid lines represent our theoretical results. The M at infinity and the α have been obtained from experiment. The temperature of the evaporating surface was 253 °K. As may be seen from the figure, T_T' and T_L' tend to quasiequilibrium as x' goes to infinity. In Fig. 5 the u' is shown as function of the distance.

If the spacing between the evaporating and condensing surfaces is small, the two Knudsen layers merge and a quasiequilibrium regime does not exist. In Fig. 6 the T_T' and T_L' are shown as functions of distance, and from Fig. 7 the u' may be obtained. As may be seen from Fig. 6, away from the evaporating surface the flow has a tendency to quasiequilibrium, but nonequilibrium again occurs at the condensing surface. Furthermore, near the condensing surface the flow is accelerated again and a $M > 1$ may be obtained.

Conclusions

In comparison with our previous work we were able to improve our theoretical treatment by taking into account the effect of the α, which is particular to evaporating solid surfaces. By introducing a new molecular velocity distribution function that better fits the boundary conditions of the present problem, we obtained thicker Knudsen layers than have previously been obtained. This is in agreement with experimental results. Furthermore, by means of this distribution function we were able to include the problem of two merging Knudsen layers and, with some restrictions, the problem of condensation.

Acknowledgment

This work was supported by Stiftung Volkswagenwerk.

References

[1]Ytrehus, T., "Theory and Experiments on Gas Kinetics in Evaporation," Progress in Astronautics and Aeronautics: Rarefied Gas Dynamics, Proceedings of the 10th International Symposium, AIAA, New York, 1977, Pt. II, Vol. 51, pp. 1197-1212.

[2]Schilder, R., Adomeit, G., and Wortberg, G., "Theoretical and Experimental Investigation of the Knudsen Layer Produced by Intensive Evaporation of Iodine," Rarefied Gas Dynamics,

Proceedings of the 13th International Symposium, Plenum, New York, 1985, pp. 577-584.

[3]Schilder, R., "Theoretische und Experimentelle Untersuchung der Knudsen-Schicht bei der intensiven Verdampfung von Jod," Ph.D. Dissertation, Rheinisch-Westfälische Technische Hochschule, Aachen FRG, 1983.

[4]Koffman, L.D., Plesset, M.S., and Lees, L., "Theory of Evaporation and Condensation," Physics of Fluids, 27, April 1984, pp. 876-880.

[5]Mager, R., "Theoretische und experimentelle Untersuchung der gaskinetischen Nichtgleichgewichtsvorgänge bei der Verdampfung und Kondensation von Jod," Ph.D. Dissertation, Rheinisch-Westfälische Technische Hochschule, Aachen, FRG (to be published).

[6]Crout, P.D., "An Application of Kinetic Theory to the Problem of Evaporation and Sublimation of Monatomic Gases," Journal of Mathematical Physics, 15, January 1936, pp. 1-54.

Nonlinear Analysis for Evaporation and Condensation of a Vapor-Gas Mixture Between the Two Plane Condensed Phases. Part I: Concentration of Inert Gas $\sim O(1)$

Yoshimoto Onishi*
Tottori University, Tottori, Japan

Abstract

A weakly nonlinear behavior of a slightly rarefied vapor-gas mixture associated with the processes of evaporation and condensation between the two plane condensed phases of the vapor is investigated on the basis of the Boltzmann equation of BGK type for a gas mixture under the diffusive boundary condition. The distributions of various fluid dynamic quantities (velocity, temperature, pressure, etc.) of the mixture and its component gases over the whole flow region, including the Knudsen layers near the interfaces, are obtained for cases of moderate values of the concentration ratio of the inert gas. In this range of the values of the concentration ratio, the expression for the vapor mass flow is also obtained explicitly. The mass flow in this case is quite small since it is governed by the diffusional ability of the vapor through the inert gas, which is weak.

I. Introduction

Consider a mixture of a vapor gas (A) and an inert gas (B) between the two condensed phases (of the vapor) placed in parallel. One of the phases at $X = 0$ is kept at a uniform temperature T_0, and the other at $X = L$ kept at T_L. Let P_0^A and P_L^A be the saturated vapor pressures corresponding to T_0 and T_L, respectively, and, also, let P_0^B be the partial pressure of the inert gas at $X = 0$. Here we

---

* Professor, Department of Applied Mathematics and Physics, Faculty of Engineering.

analyze the steady behavior of the mixture and its component gases under the following assumptions: 1) the behavior of each component gas is described by the Boltzmann equation of BGK type;[1] 2) vapor and inert gas molecules leaving the interface (of the condensed phase) after the interaction with it both have Maxwellian distributions characterized by the temperature and velocity (zero here) of the interface, the number density of the vapor being given by the saturated vapor number density at that temperature, whereas that of the inert gas is determined by the condition of no net mass flow across the interface; 3) the amount of the inert gas is enough so that the mean free path l^B of the molecules is small compared with the characteristic length L of the system. In this case, the collision of the vapor molecules with those of the inert gas is so frequent that the mean free path l^A of the vapor molecules is comparable to l^B; hence, we adopt l^A for the definition of the Knudsen number Kn of the system, i.e., $Kn = l^A/L$, and here $Kn << 1$; and 4) the deviation of the system, i.e., $\epsilon = \max(|T_L - T_0|/T_0,\ |P_L^A - P_0^A|/P_0^A)$ from a reference stationary equilibrium state is small ($\epsilon << 1$), but its magnitude is of order Kn ($\epsilon \sim Kn$). In this case, the problem becomes nonlinear, and the kinetic equation and its boundary condition cannot be linearized.[2,3]

In recent years, works on the weak evaporation and condensation of a vapor-gas mixture in the half-space and/or the two-surface problems[4–7] have been carried out extensively based on either the so-called moment method applied to the Boltzmann equation[4] or the approximate analysis of the Boltzmann equation[5] as well as its BGK type.[6,7] In any method, the object was necessarily confined to finding the jump conditions and the gross quantities of constant value such as the mass flow. Reference 8 may be one of the exceptions that have treated the problem numerically, obtaining explicitly the distributions of various fluid dynamic quantities. Unfortunately, however, the results do not seem to be accurate enough because the number density distributions obtained numerically from the linearized equations are far from the correct ones theoretically predictable from the linearized theory. Incidentally, we note that there are some other numerical works[9,10] showing some of the distributions of the temperature and the number densities (Ref.9 for strong evaporation and condensation with large molecular mass ratio in the two-surface problem; Ref.10 for rather weak evaporation and condensation with small concentration of inert gas in the half-space problem). More recently, the plane two-surface problem[11] again, and subsequently the spherical droplet problem[12] and the coaxial cylindrical two-surface problem,[13] have been worked out analytically by the singular perturbation method based on the Boltzmann equation of the BGK type,[1] and the explicit solutions (distributions of various fluid dynamic quantities as well as the mass and heat flows) have been obtained that cover the whole range of the concentration ratio of the inert gas, clarifying the transition of the diffusion control to the kinetic control mechanism in the mass-transfer process as the concentration ratio decreases. However, the treatment of those problems is restricted to the linearized case in which the deviation ϵ of the system is very small, small compared to Kn^N (N: positive integer). The present study is an extension of Ref.11 to the weakly nonlinear version to account for larger deviations of the system. Unfortunately, it is impossible, at the present stage, to obtain in one single form the complete analysis that can cover the whole range of the concentration ratio of the inert gas, even if the nonlinearity of the problem is weak. The reason for this is that when the concentration of the inert gas becomes small, the full nonlinear behavior of the partial pressure and number density of the inert gas may manifest itself because its molecules are driven toward the condensing surfaces and piled up near the surfaces by collisions with those of the vapor (flowing toward the surfaces). Hence, the analysis must inevitably take full account of this behavior, which imposes on us insurmountable difficulty.[14] Therefore, two cases must be considered separately. In the first case, which is the concern of the present study, the concentration ratio of the inert gas is of order unity or larger and, hence, the variations of the fluid dynamic quantities remain within the range of weak nonlinearity; in the other, the ratio is less than unity and of order Kn (appropriately defined for the system) or less, in which case the full nonlinear variations of some of the fluid dynamic quantities should correctly be taken into account. The latter case will be treated in Part II of this study.

In the present study, we have obtained the distributions of various fluid dynamic quantities (velocity, temperature, pressure, etc.) of the mixture and its component gases over the whole flow region, including the Knudsen layers near the interfaces, as well as the expression for the mass flow, and they will be shown in Sec.V.

II. Kinetic Equations and the Boundary Conditions

Because we are assuming the deviation of the system from a stationary reference state to be small, it is convenient to introduce perturbations of various quantities of interest. Hence, the Boltzmann equation of the BGK type[1] for the present problem may be written in terms of these quantities as follows:

$$k\xi_x \frac{\partial \phi^A}{\partial x} = \frac{1}{1+a_{12}}(1+n^A)(\phi_e^A - \phi^A) + \frac{a_{12}}{1+a_{12}}(1+n^B)(\phi_e^{AB} - \phi^A) \tag{1}$$

$$k\frac{M^{1/2}}{\alpha}\xi_x \frac{\partial \phi^B}{\partial x} = \frac{a_{22}}{a_{22}+a_{21}}(1+n^B)(\phi_e^B - \phi^B) + \frac{a_{21}}{a_{22}+a_{21}}(1+n^A)(\phi_e^{BA} - \phi^B) \tag{2}$$

$$\begin{bmatrix} n^A \\ (1+n^A)u^A \\ \frac{3}{2}(1+n^A)\tau^A + (1+n^A){u^A}^2 \end{bmatrix} = \int \begin{bmatrix} 1 \\ \xi_x \\ \xi^2 - \frac{3}{2} \end{bmatrix} \phi^A \, E \, \mathrm{d}\vec{\xi} \tag{3}$$

$$\begin{bmatrix} n^B \\ (1+n^B)u^B \\ \frac{3}{2}(1+n^B)\tau^B + M(1+n^B){u^B}^2 \end{bmatrix} = \int \begin{bmatrix} 1 \\ \xi_x \\ M\xi^2 - \frac{3}{2} \end{bmatrix} \phi^B \, \tilde{E} \, \mathrm{d}\vec{\xi} \tag{4}$$

$$p^A = n^A + \tau^A + n^A\tau^A \tag{5}$$

$$p^B = n^B + \tau^B + n^B\tau^B \tag{6}$$

$$E(1+\phi_e^A) = \pi^{-3/2}\frac{1+n^A}{(1+\tau^A)^{3/2}}\exp\left\{-\frac{(\xi_x - u^A)^2 + \xi_y^2 + \xi_z^2}{1+\tau^A}\right\} \tag{7a}$$

$$\phi_e^{AB} = \phi_e^A(n^A = n^A, u^A = u^{AB}, \tau^A = \tau^{AB}) \tag{7b}$$

$$\tilde{E}(1+\phi_e^B) = \left(\frac{\pi}{M}\right)^{-3/2}\frac{1+n^B}{(1+\tau^B)^{3/2}}\exp\left\{-\frac{M[(\xi_x - u^B)^2 + \xi_y^2 + \xi_z^2]}{1+\tau^B}\right\} \tag{8a}$$

$$\phi_e^{BA} = \phi_e^{B}(n^B = n^B, u^B = u^{BA}, \tau^B = \tau^{BA}) \tag{8b}$$

$$u^{AB} = u^{BA} = \mu_A u^A + \mu_B u^B$$

$$\tau^{AB} = \tau^A + 2\mu_A\mu_B(\tau^B - \tau^A) + \frac{2}{3}\mu_B^2(u^A - u^B)^2$$

$$\tau^{BA} = \tau^B + 2\mu_A\mu_B(\tau^A - \tau^B) + \frac{2}{3}\mu_A\mu_B(u^B - u^A)^2$$

$$E = \pi^{-3/2}\exp(-\xi^2), \quad \tilde{E} = (\pi/M)^{-3/2}\exp(-M\xi^2)$$

$$M = \frac{m_B}{m_A}, \quad \mu_A = \frac{m_A}{m_A + m_B}, \quad \mu_B = \frac{m_B}{m_A + m_B}$$

$$a_{12} = \frac{N_0^B \kappa_{AB}}{N_0^A \kappa_{AA}}, \quad a_{21} = \frac{\kappa_{AB}}{\kappa_{AA}}, \quad a_{22} = \frac{N_0^B \kappa_{BB}}{N_0^A \kappa_{AA}}$$

$$\alpha = \frac{M^{1/2}(a_{22} + a_{21})}{1 + a_{12}} = \frac{l^A}{l^B}, \quad \xi^2 = \xi_x^2 + \xi_y^2 + \xi_z^2, \quad d\vec{\xi} = d\xi_x d\xi_y d\xi_z$$

and

$$k = \frac{(2R_A T_0)^{1/2}}{(N_0^A \kappa_{AA} + N_0^B \kappa_{AB})L} = \frac{\sqrt{\pi}}{2}Kn, \qquad Kn = \frac{l^A}{L} \tag{9}$$

where $N_0^A(2R_AT_0)^{-3/2}E(1+\phi^A)$ and $N_0^B(2R_AT_0)^{-3/2}\tilde{E}(1+\phi^B)$ are the molecular velocity distribution functions of gas A and gas B, respectively; $(2R_AT_0)^{1/2}(\xi_x, \xi_y, \xi_z)$, the molecular velocity; $X = Lx$, the X coordinate; $U^S = (2R_AT_0)^{1/2}u^S$, $T^S = T_0(1+\tau^S)$, $N^S = N_0^S(1+n^S)$, and $P^S = P_0^S(1+p^S)$ are, respectively the X component (the only nonzero component) of the mean flow velocity, the temperature, the number density, and the partial pressure of gas S [note that $P^S = N^S \mathrm{k} T^S$ $(S = A, B)$ holds]; m_A and m_B are the molecular masses of gas A and gas B, respectively; R_A is the gas constant per unit mass of gas A; and k is the Boltzmann constant. $N_0^A(2R_AT_0)^{-3/2}E(1+\phi_e^A)$ and $N_0^A(2R_AT_0)^{-3/2}E(1+\phi_e^{AB})$ are the local Maxwellian distributions for molecules of gas A, characterized by the local macroscopic quantities (n^A, u^A, τ^A) and (n^A, u^{AB}, τ^{AB}), respectively; and $N_0^B(2R_AT_0)^{-3/2}\tilde{E}(1+\phi_e^B)$, and $N_0^B(2R_AT_0)^{-3/2}\tilde{E}(1+\phi_e^{BA})$ are the corresponding distributions for molecules of gas B, characterized, respectively, by (n^B, u^B, τ^B) and (n^B, u^{BA}, τ^{BA}). Here L is taken as the reference length, T_0 as the reference temperature, $(2R_AT_0)^{1/2}$ as the reference velocity, N_0^S as the reference number density for S-component gas and $P_0^S(=N_0^S\mathrm{k}T_0)$ as the reference pressure for that component. $N_0^A\kappa_{AA}$ and $N_0^B\kappa_{AB}$ represent the number of collisions per unit time of a molecule of gas A with the other molecules of that component and with the molecules of gas B, respectively; thus, $N_0^A\kappa_{AA}+N_0^B\kappa_{AB}$ is the average collision frequency for a molecule of gas A, irrespective of its collision partners. The collision frequencies are related to the transport coefficients of the mixture and its component gases as[1,11,12]

$$\eta^M = \frac{P_0^A}{N_0^A\kappa_{AA}+N_0^B\kappa_{AB}} + \frac{P_0^B}{N_0^A\kappa_{AB}+N_0^B\kappa_{BB}}$$

$$D_{AB} = \frac{(m_A+m_B)\mathrm{k}T_0}{m_Am_BN_0\kappa_{AB}}, \quad \eta^A = \frac{P_0^A}{N_0^A\kappa_{AA}}, \quad \eta^B = \frac{P_0^B}{N_0^B\kappa_{BB}} \quad (10)$$

where $N_0 = N_0^A + N_0^B$, which is taken as the reference number density for the mixture; η^M and D_{AB} are the viscosity and diffusion coefficient of the mixture, η^S being the viscosity of S-component gas.

From assumption 2 in Sec.I, the kinetic boundary conditions for ϕ^A and ϕ^B at the interface are given as follows:

$$\phi^A = \phi_W^A \equiv \phi_e^A(n^A = n_W^A, u^A = 0, \tau^A = \tau_W), \quad \text{for} \pm \xi_x > 0 \quad (11)$$

$$\phi^B = \phi_W^B \equiv \phi_e^B(n^B = n_W^B, u^B = 0, \tau^B = \tau_W), \quad \text{for} \pm \xi_x > 0 \quad (12)$$

where the upper sign refers to the interface at $x=0$ and the lower to that at $x=1$. $T_W = T_0(1+\tau_W)$ is the temperature of the interface. (Here T_W stands for T_0 if the interface at $x=0$ is considered and for T_L if at $x=1$.) Note that the velocity of the interface is assumed zero. $N_W^A = N_0^A(1+n_W^A)$ and $N_W^B = N_0^B(1+n_W^B)$ are the number densities of the reflected molecules of gas A and gas B, respectively, N_W^A being the saturated vapor number density corresponding to T_W; n_W^A is a unique function of τ_W, and its explicit functional form is obtained from the Clapeyron-Clausius relation.[15] If N_0^A is so chosen as to be the saturated vapor number density corresponding to T_0, we then have

$$n_W^A = (p_W^A - \tau_W)/(1+\tau_W), \quad p_W^A = \gamma\tau_W[1+(\frac{\gamma}{2}-1)\tau_W+\ldots] \quad (13)$$

where $P_W^A = P_0^A(1 + p_W^A)$ is the saturated vapor pressure corresponding to T_W, and $\gamma = h_L/(R_A T_0)$, h_L being the latent heat of vaporization per unit mass of gas A (see, e.g., Ref.16). For different substances, γ takes different values: for example,[17] $\gamma = 13$ at 100°C, $\gamma = 19$ at 25°C for water vapor; $\gamma = 9$ at −186°C for argon; $\gamma = 2.3$ at −269°C for helium; $\gamma = 8$ at 25°C for ammonia; $\gamma = 13$ at 78°C for ethyl alcohol. On the other hand, n_W^B is so determined as to make the net mass flow of the inert gas across the interface zero, i.e.,

$$(1+n^B)u^B = \int \xi_x \phi^B \tilde{E}\, \mathrm{d}\vec{\xi} = 0, \qquad \text{at the interface} \tag{14a}$$

hence, from the consideration of Eq.(12) with Eq.(8a), it follows that

$$n_W^B = -1 + (1+\tau_W)^{-1/2}\left[1 - 2(\pi M)^{1/2}\int_{\pm\xi_x<0} \xi_x \phi^B \tilde{E}\, \mathrm{d}\vec{\xi}\,\right] \tag{14b}$$

Now, for later use, some of the important results derived from the transport equations for gas A and gas B, which are obtained from Eqs.(1) and (2), are listed:

$$\frac{\mathrm{d}}{\mathrm{d}x}\int \begin{bmatrix} \xi_x \\ \xi_x^2 \\ \xi_x(\xi^2 - \frac{3}{2}) \end{bmatrix} \phi^A E\, \mathrm{d}\vec{\xi} = \frac{1}{k}\mu_B \frac{a_{12}}{1+a_{12}}(1+n^A)(1+n^B) \times \begin{bmatrix} 0 \\ u^B - u^A \\ 3\mu_A(\tau^B - \tau^A) + 2u^{AB}(u^B - u^A) \end{bmatrix} \tag{15}$$

$$\frac{\mathrm{d}}{\mathrm{d}x}\int \begin{bmatrix} \xi_x \\ M\xi_x^2 \\ \xi_x(M\xi^2 - \frac{3}{2}) \end{bmatrix} \phi^B \tilde{E}\, \mathrm{d}\vec{\xi} = \frac{1}{k}\mu_B \frac{a_{21}}{1+a_{12}}(1+n^A)(1+n^B) \times \begin{bmatrix} 0 \\ u^A - u^B \\ 3\mu_A(\tau^A - \tau^B) + 2u^{BA}(u^A - u^B) \end{bmatrix} \tag{16}$$

The first relations in Eqs.(15) and (16) give us

$$(1+n^A)u^A = \text{const}, \qquad (1+n^B)u^B = \text{const}, \quad \text{i.e., } u^B \equiv 0 \tag{17}$$

everywhere in the flow region. The last expression for u^B is a consequence of Eq.(14a).

The fluid dynamic quantities of the mixture may often be more appropriate than those associated with the component gases and, therefore, these quantities are also introduced here:

$$(1+\sigma)u = [\rho_0^A(1+n^A)u^A + \rho_0^B(1+n^B)u^B]/\rho_0$$

$$\tau = (p-n)/(1+n), \qquad n = (N_0^A n^A + N_0^B n^B)/N_0$$

$$p = (P_0^A p^A + P_0^B p^B)/P_0, \qquad \sigma = (\rho_0^A n^A + \rho_0^B n^B)/\rho_0 \qquad (18)$$

where $U = (2R_A T_0)^{1/2}u$, $T = T_0(1+\tau)$, $N = N_0(1+n)$, $P = P_0(1+p)$, and $\rho = \rho_0(1+\sigma)$ are, respectively, the *mean mass flow velocity*, the temperature, the number density, the pressure, and the density of the mixture; $P_0 = P_0^A + P_0^B$, $\rho_0 = \rho_0^A + \rho_0^B$, $\rho_0^A = m_A N_0^A$, and $\rho_0^B = m_B N_0^B$.

The present analysis follows Refs.3 and 11. Taking into account the singular nature of Eqs.(1) and (2) owing to the existence of a small parameter k multiplying the highest derivatives of the equations, we seek the solution for the present problem in the form: the moderately varying part $[\partial/\partial x \sim O(1)]$ of the solution plus the rapidly changing part $[k\,\partial/\partial x \sim O(1)]$. The former is called the Hilbert part and the latter the Knudsen-layer correction part, which makes the correction to the Hilbert part within the Knudsen layer near the interface and vanishes quickly as the distance from it increases.

III. Hilbert Expansion and Macroscopic Equations

Here we try to seek the Hilbert part of the solution by expanding the distribution functions in terms of k as

$$\phi_H^A = k\phi_{H1}^A + k^2\phi_{H2}^A + \ldots, \qquad \phi_H^B = k\phi_{H1}^B + k^2\phi_{H2}^B + \ldots \qquad (19)$$

Substituting the preceding expansions into the original equations (1) and (2), and equating like powers of k, we obtain

$$\phi_{Hm}^A = \phi_{eHm}^A + \frac{a_{12}}{1+a_{12}}(\phi_{eHm}^{AB} - \phi_{eHm}^A) + \frac{1}{1+a_{12}}\sum_{j=1}^{m-1}[n_{Hj}^A(\phi_{eHm-j}^A - \phi_{Hm-j}^A) + a_{12}n_{Hj}^B(\phi_{eHm-j}^{AB} - \phi_{Hm-j}^A)] - \xi_x\frac{\partial\phi_{Hm-1}^A}{\partial x} \qquad (20)$$

$$\phi^B_{Hm} = \phi^B_{eHm} + \frac{a_{21}}{a_{22}+a_{21}}(\phi^{BA}_{eHm} - \phi^B_{eHm})$$
$$+ \frac{1}{a_{22}+a_{21}}\sum_{j=1}^{m-1}[a_{22}n^B_{Hj}(\phi^B_{eHm-j} - \phi^B_{Hm-j})$$
$$+ a_{21}n^A_{Hj}(\phi^{BA}_{eHm-j} - \phi^B_{Hm-j})] - \frac{\sqrt{M}}{\alpha}\xi_x\frac{\partial\phi^B_{Hm-1}}{\partial x} \tag{21}$$

for $m \geq 1$, where n^S_{Hm}, u^S_{Hm}, and τ^S_{Hm} are defined by the replacement of ϕ^S by ϕ^S_{Hm} in Eqs.(3) and (4). ϕ^A_{eHm} can be expressed in terms of these macroscopic quantities from Eq.(7a) as

$$\phi^A_{eH1} = n^A_{H1} + 2\xi_x u^A_{H1} + (\xi^2 - \frac{3}{2})\tau^A_{H1} \tag{22a}$$

$$\phi^A_{eH2} = n^A_{H2} + 2\xi_x u^A_{H2} + (\xi^2 - \frac{3}{2})\tau^A_{H2}$$
$$+ 2\xi_x u^A_{H1}[n^A_{H1} + \xi_x u^A_{H1} + (\xi^2 - \frac{5}{2})\tau^A_{H1}] - (u^A_{H1})^2$$
$$+ (\xi^2 - \frac{3}{2})n^A_{H1}\tau^A_{H1} + \frac{1}{2}(\xi^4 - 5\xi^2 - \frac{15}{4})(\tau^A_{H1})^2 \tag{22b}$$

and so on. In a similar manner, ϕ^{AB}_{eHm}, ϕ^B_{eHm}, and ϕ^{BA}_{eHm} can be obtained.

The application of these expressions for the distribution functions to the transport equations (15) and (16) with $u^B \equiv 0$ in Eq.(17) leads to the following equations governing the macroscopic or fluid dynamic quantities. It is noted that Eqs.(15) and (16) hold for the Hilbert part as well.

For the first order of approximation,

$$\frac{\mathrm{d}p_{H1}}{\mathrm{d}x} = 0 \tag{23}$$

$$u^A_{H1} \equiv 0 \tag{24}$$

$$\frac{\mathrm{d}^2 p^A_{H1}}{\mathrm{d}x^2} = 0 \tag{25}$$

$$\frac{\mathrm{d}^2 \tau^A_{H1}}{\mathrm{d}x^2} = 0 \tag{26}$$

$$n^A_{H1} = p^A_{H1} - \tau^A_{H1} \tag{27}$$

and

$$\begin{aligned} &p^B_{H1} = (P_0 p_{H1} - P^A_0 p^A_{H1})/P^B_0, \qquad n^B_{H1} = p^B_{H1} - \tau^B_{H1} \\ &\tau_{H1} = \tau^B_{H1} = \tau^A_{H1}, \qquad u_{H1} \equiv 0, \qquad n_{H1} = p_{H1} - \tau_{H1} \end{aligned} \tag{28}$$

For the second approximation,

$$\frac{\mathrm{d}p_{H2}}{\mathrm{d}x} = 0 \tag{29}$$

$$u^A_{H2} = -\frac{(1+a_{12})}{2\mu_B a_{12}} \frac{\mathrm{d}p^A_{H1}}{\mathrm{d}x} \tag{30}$$

$$u^A_{H2} \frac{\mathrm{d}p^A_{H1}}{\mathrm{d}x} = D_* \left\{ \frac{\mathrm{d}^2 p^A_{H2}}{\mathrm{d}x^2} + \frac{\mathrm{d}}{\mathrm{d}x}(\tau^A_{H1} \frac{\mathrm{d}p^A_{H1}}{\mathrm{d}x}) - (\frac{\mathrm{d}p^A_{H1}}{\mathrm{d}x})^2 \right\} \tag{31}$$

$$\begin{aligned} u^A_{H2} \frac{\mathrm{d}\tau^A_{H1}}{\mathrm{d}x} = &\frac{1}{2}\kappa_* \left\{ \frac{\mathrm{d}^2 \tau^A_{H2}}{\mathrm{d}x^2} + \frac{\mathrm{d}}{\mathrm{d}x}(\tau^A_{H1} \frac{\mathrm{d}\tau^A_{H1}}{\mathrm{d}x}) \right\} - D_* \frac{\mathrm{d}p^A_{H1}}{\mathrm{d}x} \frac{\mathrm{d}\tau^A_{H1}}{\mathrm{d}x} \\ &+ \frac{a_{12} N^A_0}{2(1+a_{12}) N_0} (1 - \frac{1}{\alpha^2}) \frac{\mathrm{d}}{\mathrm{d}x} \left[(n^A_{H1} - n^B_{H1}) \frac{\mathrm{d}\tau^A_{H1}}{\mathrm{d}x} \right] \end{aligned} \tag{32}$$

$$n^A_{H2} = p^A_{H2} - \tau^A_{H2} - n^A_{H1} \tau^A_{H1} \tag{33}$$

and for the other quantities,

$$\begin{aligned} &p^B_{H2} = (P_0 p_{H2} - P^A_0 p^A_{H2})/P^B_0, \quad n^B_{H2} = p^B_{H2} - \tau^B_{H2} - n^B_{H1} \tau^B_{H1} \\ &\tau_{H2} = \tau^B_{H2} = \tau^A_{H2}, \qquad u_{H2} = \rho^A_0 u^A_{H2} / \rho_0 \\ &n_{H2} = p_{H2} - \tau_{H2} - n_{H1} \tau_{H1} \end{aligned} \tag{34}$$

For the third approximation, we have only, for the velocities of gas A and the mixture,

$$u^A_{H3} = -\frac{(1+a_{12})}{2\mu_B a_{12}} \left[\frac{\mathrm{d}p^A_{H2}}{\mathrm{d}x} - (n^A_{H1} + n^B_{H1}) \frac{\mathrm{d}p^A_{H1}}{\mathrm{d}x} \right] \tag{35}$$

$$u_{H3} = \frac{\rho^A_0}{\rho_0} u^A_{H3} - \frac{\rho^A_0}{\rho_0} \frac{\rho^B_0}{\rho_0} \frac{(1+a_{12})}{2\mu_B a_{12}} \left(n^A_{H1} - n^B_{H1} \right) \frac{\mathrm{d}p^A_{H1}}{\mathrm{d}x} \tag{36}$$

It is noted that the relations at the first-order approximation have been taken into account in the second- and third-order equations. κ_* and D_* here are defined by

$$\kappa_* = \frac{N_0^A}{N_0} + \frac{N_0^B}{N_0}\frac{1}{\alpha\sqrt{M}}, \qquad D_* = \frac{(1+a_{12})}{2\mu_B a_{12}}\frac{N_0^B}{N_0} \tag{37}$$

and these are related to the thermal conductivity λ^M and the diffusion coefficient D_{AB} of the mixture as

$$\frac{5}{4}k\kappa_* = \frac{T_0}{L}\frac{\lambda^M}{P_0(2R_AT_0)^{1/2}}, \qquad kD_* = \frac{D_{AB}}{L(2R_AT_0)^{1/2}} \tag{38}$$

Now, Eqs.(23-36) are the desired equations governing the Hilbert part of the macroscopic or fluid dynamic quantities. With these macroscopic quantities, the Hilbert part of the distribution functions can be determined. For example, for ϕ_H^A,

$$\phi_{H1}^A = n_{H1}^A + (\xi^2 - \frac{3}{2})\tau_{H1}^A = p_{H1}^A + (\xi^2 - \frac{5}{2})\tau_{H1}^A \tag{39a}$$

$$\begin{aligned}\phi_{H2}^A = {} & n_{H2}^A + 2\xi_x u_{H2}^A + (\xi^2 - \frac{3}{2})\tau_{H2}^A + (\xi^2 - \frac{3}{2})n_{H1}^A\tau_{H1}^A \\ & + \frac{1}{2}(\xi^4 - 5\xi^2 - \frac{15}{4})(\tau_{H1}^A)^2 - \xi_x(\xi^2 - \frac{5}{2})\frac{d\tau_{H1}^A}{dx}\end{aligned} \tag{39b}$$

from Eq.(20), and similar expressions will be obtained for ϕ_H^B from Eq.(21). It will be clear, at the present stage, that ϕ_{H1}^A and ϕ_{H1}^B satisfy the given boundary conditions (11) and (12) at the first-order approximation when the Hilbert parts of the quantities are set equal to the boundary data associated with the interface. That is, at the interface,

$$n_{H1}^A = n_{W1}^A \quad (\text{or} \quad p_{H1}^A = p_{W1}^A), \quad \tau_{H1}^A = \tau_{W1}, \quad n_{H1}^B = n_{W1}^B \tag{40}$$

where n_W^A, p_W^A, τ_W, and n_W^B are expanded in terms of k just as in Eq.(19). These conditions, except the last one, will provide the boundary conditions at the interface for the macroscopic equations (25) and (26). The last relation in Eq.(40) gives the number density for the reflected molecules of gas B after the solution has been determined. The distribution functions so obtained, however, cannot, in general, be made to satisfy the given conditions owing to the singular nature of the problem. Actually, at the second-order approximation, the distribution functions cannot satisfy the conditions and, therefore, another part, which takes charge of the rapid variation of the solution, must be considered. This will be given in the next section.

IV. Knudsen-Layer Analysis

In order to obtain the solution that satisfies the given conditions (11) and (12) at the interface, another part of the solution, which varies appreciably over a small (nondimensional) length of the order of k, must be introduced. This part, called the Knudsen-layer correction part, can be analyzed in the same manner as in Refs.3, 11, and 12. We briefly mention the process of this analysis, leaving the details to the appropriate references. We again try to find the solution in the sum of the Hilbert part just discussed plus its correction in a thin layer, the Knudsen layer, near the interface, i.e.,

$$\phi^A = \phi_H^A(x) + \phi_K^A(\eta), \qquad \phi^B = \phi_H^B(x) + \phi_K^B(\eta) \tag{41}$$

where the argument of the molecular velocity in the distribution functions has been omitted. η is a stretched coordinate defined by

$$\eta = \begin{cases} x/k, & \text{near } x = 0 \\ (1-x)/k, & \text{near } x = 1 \end{cases} \tag{42}$$

ϕ_K^A and ϕ_K^B, and the correction parts of the fluid dynamic quantities (say, f) associated with these distribution functions, are also sought in the following expansion forms:

$$\phi_K^A = k\phi_{K1}^A + k^2\phi_{K2}^A + \dots, \qquad \phi_K^B = k\phi_{K1}^B + k^2\phi_{K2}^B + \dots \tag{43a}$$

$$f_K^A = kf_{K1}^A + k^2 f_{K2}^A + \dots, \qquad f_K^B = kf_{K1}^B + k^2 f_{K2}^B + \dots \tag{43b}$$

These correction parts are to vanish outside the Knudsen layer, i.e.,

$$\phi_K^A,\ \phi_K^B \text{ and } f_K^A,\ f_K^B \ \to 0, \quad \text{as } \eta \to \infty \tag{44}$$

Because the Hilbert parts of the distribution functions satisfy the boundary conditions at the first order, no Knudsen-layer correction parts will exist at this order, i.e., $\phi_{K1}^A = \phi_{K1}^B = f_{K1}^A = f_{K1}^B \equiv 0$. The analysis of the second- and higher-order approximations of the Knudsen-layer part was carried out in the same way as in Refs.3, 11, and 12, and we have obtained the Knudsen-layer correction part and, at the same time, the jump conditions to be used as the boundary conditions for the macroscopic equations (31) and (32). Only the results relevant to the present study will be listed here.

The jump conditions at the second-order approximation are

$$\begin{bmatrix} p_{H2}^A - p_{W2}^A \\ \tau_{H2}^A - \tau_{W2} \end{bmatrix} = \pm u_{H2}^A \begin{bmatrix} C_4^A \\ d_4^M \end{bmatrix} \pm \frac{\mathrm{d}\tau_{H1}^A}{\mathrm{d}x} \begin{bmatrix} C_1^A \\ d_1^M \end{bmatrix} \tag{45}$$

where C_i^A and d_i^M $(i = 4, 1)$ are the constants that depend only on m_B/m_A, N_0^B/N_0^A, κ_{AB}/κ_{AA}, and κ_{BB}/κ_{AA}. These are universal constants in the sense that they do not depend on the geometry of the problems. The values of these constants for several sets of parameters may be found in Refs.11 and 12 ($C_i^A = C_i^M$ of Ref.11). It should be understood here and hereafter that the upper sign should be adopted at $x = 0$ and the lower at $x = 1$. These conditions provide the appropriate boundary conditions for (31) and (32) at the interfaces. It is noted that u_{H3}^A does not require the condition at the interface, because it is derived from p_{H1}^A and p_{H2}^A [see Eq.(35)].

The second-order Knudsen-layer corrections are

$$u_{K2}^A \equiv 0 \tag{46a}$$

$$\begin{bmatrix} n_{K2}^A \\ \tau_{K2}^A \\ p_{K2}^A \\ n_{K2}^B \\ \tau_{K2}^B \\ p_{K2}^B \end{bmatrix} = \pm u_{H2}^A \begin{bmatrix} \Omega_4^A(\zeta) \\ \Theta_4^A(\zeta) \\ \Pi_4^A(\zeta) \\ \Omega_4^B(\zeta) \\ \Theta_4^B(\zeta) \\ \Pi_4^B(\zeta) \end{bmatrix} \pm \frac{\mathrm{d}\tau_{H1}^A}{\mathrm{d}x} \begin{bmatrix} \Omega_1^A(\zeta) \\ \Theta_1^A(\zeta) \\ \Pi_1^A(\zeta) \\ \Omega_1^B(\zeta) \\ \Theta_1^B(\zeta) \\ \Pi_1^B(\zeta) \end{bmatrix} \tag{46b}$$

$$\Pi_4^S = \Omega_4^S + \Theta_4^S, \quad \Pi_1^S = \Omega_1^S + \Theta_1^S, \quad (S = A, B) \tag{47}$$

with

$$\zeta = \begin{cases} \eta & (\alpha \geq 1) \\ \alpha\eta & (\alpha < 1) \end{cases} \quad \text{and} \quad \eta = \begin{cases} x/k & \text{near } x = 0 \\ (1-x)/k & \text{near } x = 1 \end{cases} \tag{48}$$

where Ω_i^S, Θ_i^S, and Π_i^S $(i = 4, 1)$ are universal functions of ζ, depending only on m_B/m_A, N_0^B/N_0^A, κ_{AB}/κ_{AA}, and κ_{BB}/κ_{AA}, and not on the geometry of the problems. These functions are rapidly decreasing functions of ζ. The graphs of Ω_i^S and Θ_i^S may be found in Refs.11 and 12.

The third-order Knudsen-layer correction was obtained only for the velocity of gas A, and it is

$$u^A_{K3} \equiv 0 \tag{49}$$

It is noted that the Knudsen-layer correction part of the mean mass flow velocity is also zero up to the order of the present approximation, i.e., $u_{K1} = u_{K2} = u_{K3} \equiv 0$.

V. Results and Discussion

Now we consider the present problem posed in Sec.I. Let $T_L = T_0(1+\tau_L)$ and $P^A_L = P^A_0(1+p^A_L)$, where τ_L and p^A_L are of the order of the deviation of the system. If we introduce the following parameter,

$$\Gamma = \frac{T_0}{P^A_0}\frac{P^A_L - P^A_0}{T_L - T_0} = \frac{p^A_L}{\tau_L} \tag{50}$$

we have the relation between γ and Γ from Eq.(13) as

$$\gamma = \Gamma[1-(\Gamma/2-1)\tau_L+\ldots] \quad \text{or} \quad \Gamma = \gamma[1+(\gamma/2-1)\tau_L+\ldots] \tag{51}$$

Then the parameters τ_{Wm} and p^A_{Wm} $(m = 1, 2)$ appearing in boundary conditions (40) and (45) are to be given the following values for the present case:

$$\tau_{W1} = \tau_{W2} = 0, \qquad p^A_{W1} = p^A_{W2} = 0, \qquad \text{at} \quad x = 0 \tag{52a}$$

$$\tau_{W1} = \tau_L/k, \quad \tau_{W2} = 0, \quad p^A_{W1} = \gamma\tau_{W1}, \quad p^A_{W2} = \gamma(\frac{\gamma}{2} - 1)\tau^2_{W1}, \quad \text{at} \quad x = 1 \tag{52b}$$

Furthermore, we have to specify the condition either for p^B_H or p_H. Because we have taken P^B_0 as the value of the partial pressure of gas B at the interface at $x = 0$, we have, for p^B_H at $x = 0$,

$$p^B_{H1} = 0 \tag{53a}$$

$$p^B_{H2} = -p^B_{K2}(0) = -[u^A_{H2}(0)\Pi^B_4(0) + (d\tau^A_{H1}/dx)_0\Pi^B_1(0)] \tag{53b}$$

where $\Pi^B_4(0)$ and $\Pi^B_1(0)$ are already known values [see Eq.(46b)].

Now we have the macroscopic equations (23-36) and the boundary conditions (40), (45), and (53) at the interface, with the values (52) being taken into account. It is a simple matter to solve this boundary value problem. Actually, we have obtained the distributions of various fluid dynamic quantities (velocity, temperature, pressure, etc.) of the mixture and its component gases over the whole flow region, including the Knudsen layers near the interfaces at $x = 0$ and $x = 1$. The results are:

$$\frac{U^A}{(2R_A T_0)^{1/2}} = -k\Lambda p_L^A \{1 - k\frac{\gamma}{\Gamma}[2(-\Lambda C_4^A + \frac{C_1^A}{\gamma}) - H] - \frac{\gamma}{\Gamma}(\gamma - 1)\tau_L x\}$$

$$\frac{U}{(2R_A T_0)^{1/2}} = -k\Lambda p_L^A \frac{\rho_0^A}{\rho_0}\Big\{1 - k\frac{\gamma}{\Gamma}[2(-\Lambda C_4^A + \frac{C_1^A}{\gamma}) - H] + \frac{\gamma}{\Gamma}[1 - \gamma(\frac{\rho_0^A}{\rho_0} - \frac{\rho_0^B}{\rho_0}\frac{N_0^A}{N_0^B})]\,\tau_L x\Big\}$$

$$\frac{T - T_0}{T_L - T_0} = x + k\{(1 - 2x)[-\Lambda\gamma d_4^M + d_1^M] + (x - x^2)G - \Lambda\gamma[\Theta_4^M(\zeta) - \Theta_4^M(\tilde{\zeta})] + [\Theta_1^M(\zeta) - \Theta_1^M(\tilde{\zeta})]\}$$

$$\frac{P^A - P_0^A}{P_L^A - P_0^A} = x + k\frac{\gamma}{\Gamma}\{(1 - 2x)[-\Lambda C_4^A + \frac{C_1^A}{\gamma}] + (x - x^2)H - \Lambda[\Pi_4^A(\zeta) - \Pi_4^A(\tilde{\zeta})] + \frac{1}{\gamma}[\Pi_1^A(\zeta) - \Pi_1^A(\tilde{\zeta})]\}$$

$$\frac{P^B - P_0^B}{P_L^A - P_0^A} = -x + k\frac{\gamma}{\Gamma}\{2x[-\Lambda C_4^A + \frac{C_1^A}{\gamma}] - (x - x^2)H - \Lambda\frac{N_0^B}{N_0^A}[-\Pi_4^B(0) + \Pi_4^B(\zeta) - \Pi_4^B(\tilde{\zeta})] + \frac{1}{\gamma}\frac{N_0^B}{N_0^A}[-\Pi_1^B(0) + \Pi_1^B(\zeta) - \Pi_1^B(\tilde{\zeta})]\}$$

where $\Lambda = (1+a_{12})/(2\mu_B a_{12})$, $\Theta_i^M = (N_0^A\Theta_i^A + N_0^B\Theta_i^B)/N_0$, and

$$G = \frac{\tau_L}{2k}\left[1+\frac{\gamma}{\kappa_*}\left(2\Lambda\frac{N_0^A}{N_0}+\frac{\alpha^2-1}{\alpha^2}\frac{a_{21}}{1+a_{12}}\right)\right], \quad H = \frac{\tau_L}{2k}\left(1+\gamma\frac{N_0^A}{N_0^B}\right)$$

$$\zeta = \begin{cases} x/k & (\alpha \geq 1) \\ \alpha x/k & (\alpha < 1), \end{cases} \qquad \tilde{\zeta} = \begin{cases} (1-x)/k & (\alpha \geq 1) \\ \alpha(1-x)/k & (\alpha < 1) \end{cases}$$

Note that, in the neighborhood of the interface at $x = 0$, ζ is of order unity but $\tilde{\zeta}$ is extremely large, so that $\Theta_i^M(\tilde{\zeta})$ and $\Pi_i^S(\tilde{\zeta})$ disappear completely and, near the interface at $x = 1$, the situation is reversed. The temperature distributions for gas A and gas B can be obtained by replacing Θ_i^M with Θ_i^A and Θ_i^B, respectively. We show the typical distributions of the temperature of the mixture in Fig.1 and the partial pressures of gas B and gas A in Figs.2 and 3, respectively. In all these figures, the thick solid lines indicate the total distributions with the Knudsen-layer corrections included, and the thin solid lines the Hilbert parts only. The curves of the Hilbert parts of these quantities are not straight lines because of the term associated with G in the temperature and H in the partial pressures, whereas, in the linearized theory, these are exactly stright lines (see Ref.11).

The mass flow $\dot{m}$ for the present problem now becomes

$$\dot{m} = m_A N^A U^A = \rho_0^A(2R_A T_0)^{1/2}(1+n^A)u^A \tag{54a}$$

$$= \rho_0^A(2R_A T_0)^{1/2}\frac{P_L^A - P_0^A}{P_0^A}(-k\Lambda F) \tag{54b}$$

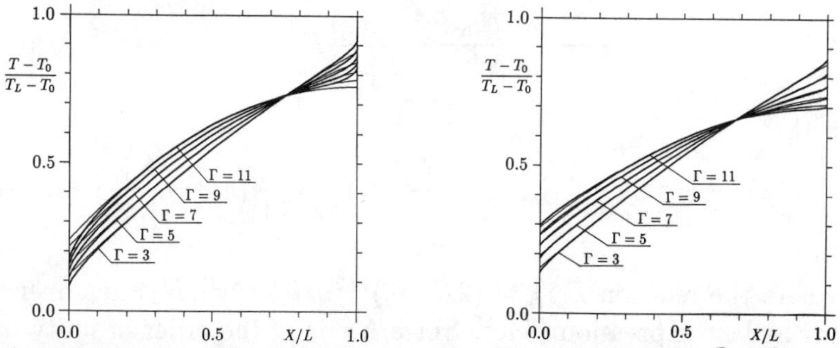

a) $m_B/m_A = 2.0$, $N_0^B/N_0^A = 1.0$, $(T_L - T_0)/T_0 = 0.05$, $k = 0.05$

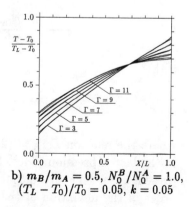

b) $m_B/m_A = 0.5$, $N_0^B/N_0^A = 1.0$, $(T_L - T_0)/T_0 = 0.05$, $k = 0.05$

Fig. 1 Temperature distributions. $\kappa_{AA} = \kappa_{AB} = \kappa_{BB}$.

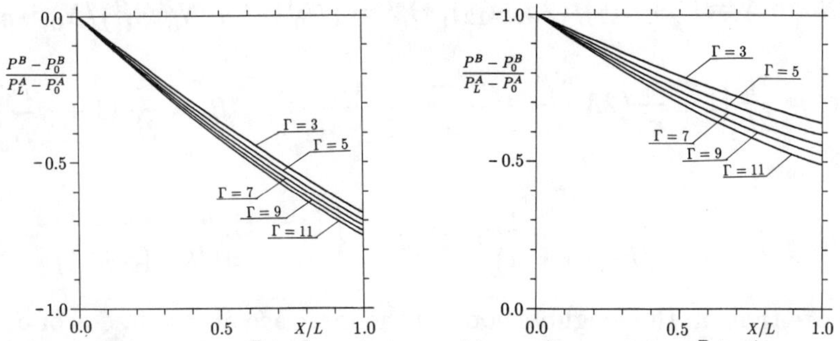

a) $m_B/m_A = 2.0$, $N^B_0/N^A_0 = 1.0$, $(T_L - T_0)/T_0 = 0.05$, $k = 0.05$

b) $m_B/m_A = 0.5$, $N^B_0/N^A_0 = 1.0$, $(T_L - T_0)/T_0 = 0.05$, $k = 0.05$

Fig. 2 Partial pressure of the inert gas. $\kappa_{AA} = \kappa_{AB} = \kappa_{BB}$.

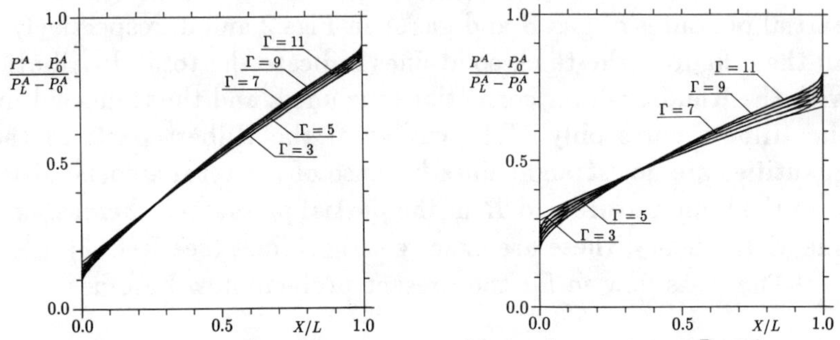

a) $m_B/m_A = 2.0$, $N^B_0/N^A_0 = 1.0$, $(T_L - T_0)/T_0 = 0.05$, $k = 0.05$

b) $m_B/m_A = 0.5$, $N^B_0/N^A_0 = 1.0$, $(T_L - T_0)/T_0 = 0.05$, $k = 0.05$

Fig. 3 Partial pressure of the vapor. $\kappa_{AA} = \kappa_{AB} = \kappa_{BB}$.

$$= -\frac{D_{AB}}{R_A T_0}\frac{N_0}{N^B_0}\frac{P^A_L - P^A_0}{L}F \tag{54c}$$

with

$$F = 1 - 2k\frac{\gamma}{\Gamma}\left(-\Lambda C^A_4 + \frac{C^A_1}{\gamma}\right) + \frac{1}{2}\tau_L\frac{\gamma}{\Gamma}\left(1 + \gamma\frac{N^A_0}{N^B_0}\right) + \ldots$$

where the relation $D_{AB} = (2R_A T_0)^{1/2}\, L\, k\Lambda\, (N^B_0/N_0)$ has been used in the last expression (54c). Since ΛF is of the order of unity in the present case, $k\Lambda F$ in Eq.(54b) is of the order of k and, hence, of the order of the deviation of the system, which is small. This case can be adequately described in terms of the diffusion coefficient D_{AB}, and the mass flow is said to be governed by the diffusional ability of the vapor through the inert gas. The mass transfer in this diffusion control case is small compared with that in the kinetic control case by a factor of k (see Refs.11-13). It is noted that, in the limit of weak evaporation and condensation ($\tau_L << k$), the mass flow coincides with that in the linearized case given in Ref.11 up to the present order of approximation (Note: The result of Ref.11 is correct to any order of k in the asymptotic sense.) In Fig.4, $\dot{m}/\dot{m}_p$ vs N_0^B/N_0^A is shown, where $\dot{m}_p$ is the mass flux in the absence of inert gas (kinetic control case) given in Ref.16, which is written in terms of the present notation as

$$\frac{\dot{m}_p}{\rho_0^A(2R_AT_0)^{1/2}} = \frac{P_L^A - P_0^A}{P_0^A}\frac{\gamma}{\Gamma}\Big\{ a_1 + \tau_L\Big[\frac{(a_2\gamma + a_3)}{e^{2U_0} - 1} + a_4\gamma + a_5\Big] + \ldots\Big\} \qquad (55)$$

where $a_1 = -0.23452$, $a_2 = 0.02574$, $a_3 = -0.12285$, $a_4 = -0.10439$, $a_5 = 0.23172$, and $U_0 = -0.23452\,\gamma\tau_L/\tilde{k}$ with $\tilde{k} = k$ (in the limit $N_0^B/N_0^A \to 0$). As can be seen from the third term in the expression for F, the mass flow increases as the nonlinearity becomes strong, i.e., $\tau_L \to$ large, and this effect becomes more and more marked as the concentration ratio N_0^B/N_0^A decreases. This is due to a nonlinear effect, although a weakly nonlinear one. Figures 4a-d show this effect. Another thing is that the mass flow becomes only slightly dependent on the ratio N_0^B/N_0^A above its certain value, which is greater than unity (roughly 10 in the present parameter ranges adopted). This can also be noticed from Eq.(54c). The figures also show that the mass flow increases with the decrease of the molecular mass ratio m_B/m_A (compare Figs.4c, 4e and 4f) because the vapor can have larger diffusional ability through the inert gas when the molecules of the vapor are heavier than those of the inert gas. This is seen from the expression

$$\frac{D_{AB}}{R_AT_0} = \Big(1 + \frac{m_A}{m_B}\Big)\frac{1}{N_0\kappa_{AB}}$$

showing the increase of $D_{AB}/(R_AT_0)$ with the decrease of m_B/m_A [see Eq.(10) for the definition of D_{AB}]. In this situation, however, if k increases, the second term including Λ in F is no longer small and, hence, the present expansion scheme becomes invalidated (see the

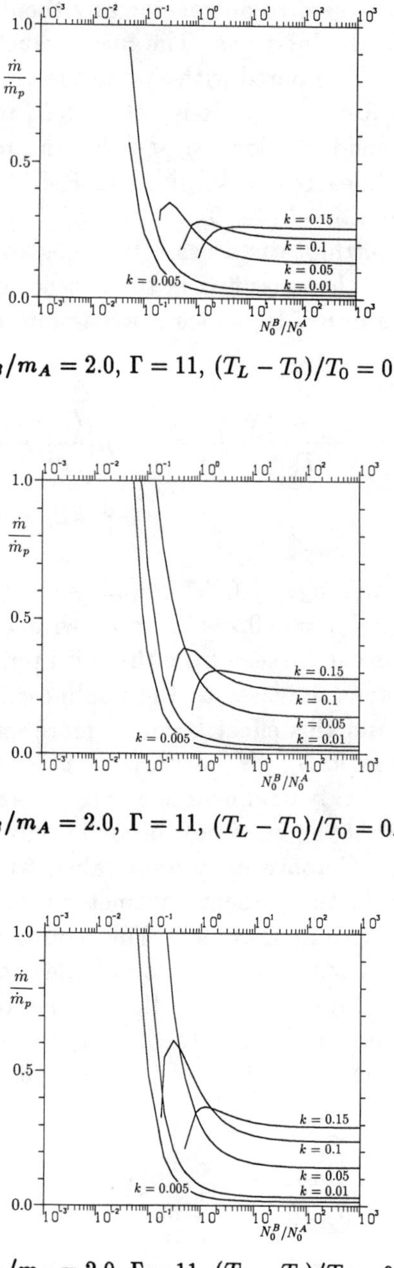

a) $m_B/m_A = 2.0,\ \Gamma = 11,\ (T_L - T_0)/T_0 = 0.01$

b) $m_B/m_A = 2.0,\ \Gamma = 11,\ (T_L - T_0)/T_0 = 0.03$

c) $m_B/m_A = 2.0,\ \Gamma = 11,\ (T_L - T_0)/T_0 = 0.05$

Fig. 4 Mass flow vs concentration ratio. $\kappa_{AA} = \kappa_{AB} = \kappa_{BB}$.

Fig. 4 continued.

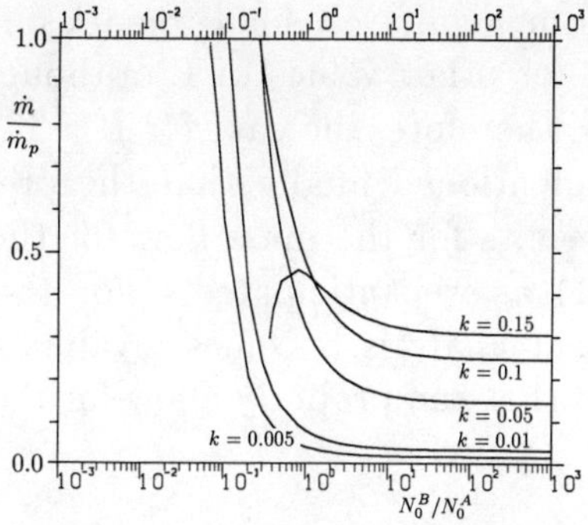

d) $m_B/m_A = 2.0$, $\Gamma = 11$, $(T_L - T_0)/T_0 = 0.08$

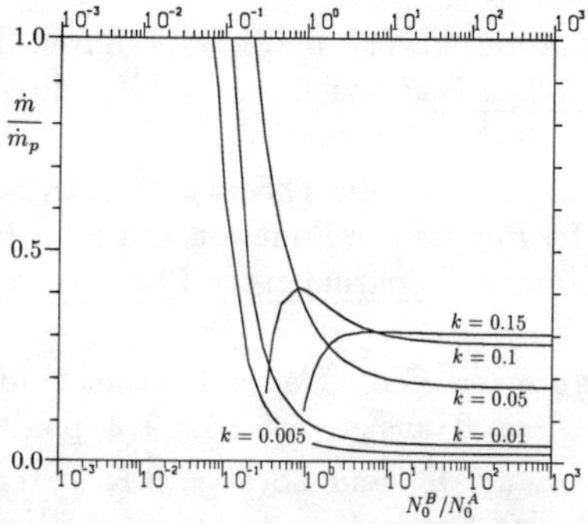

e) $m_B/m_A = 1.0$, $\Gamma = 11$, $(T_L - T_0)/T_0 = 0.05$.

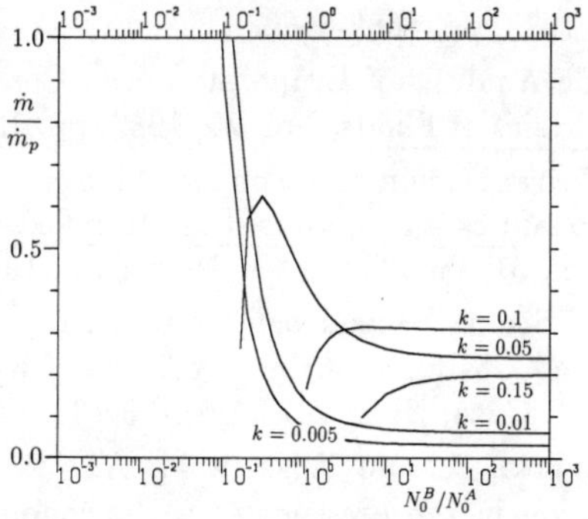

f) $m_B/m_A = 0.5$, $\Gamma = 11$, $(T_L - T_0)/T_0 = 0.05$.

Fig. 4 concluded.

curve for $k = 0.15$ in Fig.4f). A final remark on $\dot{m}/\dot{m}_p$: it is qualitatively the same for other values of Γ, although slightly different quantitatively and, therefore, the case for $\Gamma = 11$ will suffice.

It should be mentioned finally that the present analysis gives reasonably good results for the mass flow for the concentration ratio $N_0^B/N_0^A \geq O(1)$ as we anticipated. For the smaller values of N_0^B/N_0^A, the present analysis becomes invalidated, and a different analysis is required that can properly allow for the nonlinearity mentioned in Sec.I.

References

[1] Hamel, B.B., "Kinetic Model for Binary Gas Mixtures," Physics of Fluids, Vol. 8, 1965, pp. 418-425.

[2] Sone, Y., "Asymptotic Theory of Flow of Rarefied Gas over a Smooth Boundary II," Rarefied Gas Dynamics, Vol.II, edited by D. Dini, Editrice Tecnico Scientifica, Italy, 1971, pp. 737-749.

[3] Onishi, Y. and Sone, Y., "Kinetic Theory of Slightly Strong Evaporation and Condensation – Hydrodynamic Equation and Slip Boundary Condition for Finite Reynolds Number," Journal of the Physical Society of Japan, Vol. 47, 1979, pp. 1676-1685.

[4] Haas, J.C. and Springer, G.S., "Mass Transfer through Binary Gas Mixtures," Journal of Heat Transfer, Vol. 95, 1973, pp. 263-265.

[5] Cipolla, J.W.,Jr., Lang, H., and Loyalka, S.K., "Temperature and Partial Pressure Jumps during Evaporation and Condensation of a Multicomponent Gas Mixture," Rarefied Gas Dynamics, edited by M. Becker and M. Fiebig, DFVLR Press, Porz Wahn, FRG, 1974, pp. F.4-1 - F.4-10.

[6] Pao, Y.P., "Evaporation in a Vapor-Gas Mixture," Journal of Chemical Physics, Vol. 59, 1973, pp. 6688-6689.

[7] Soga, T., "Kinetic Analysis of Evaporation and Condensation in a Vapor-Gas Mixture," Physics of Fluids, Vol. 25, 1982, pp. 1978-1986.

[8] Matsushita, T., "Evaporation and Condensation in a Vapor-Gas Mixture," Progress in Astronautics and Aeronautics: Rarefied Gas Dynamics, edited by J.L. Potter, Vol. 51, Part II, AIAA, New York, 1977, pp. 1213-1225.

[9] Makashev, N.K., "Strong Recondensation in a One-component and a Two-component Rarefied Gas for an Arbitrary Value of Knudsen Number," Mekhanika Zhidkosti i Gaza, No.5, 1972, pp. 130-138.

[10] Gajewski, P., Kulicki, A. and Wisniewski, A., "A Kinetic Model of the Vapor Condensation in the presence of a Noncondensable Gas," Rarefied Gas Dynamics, edited by M. Becker and M. Fiebig, DFVLR Press, Porz Wahn, FRG, 1974, pp. F.2-1 - F.2-11.

[11] Onishi, Y., "A Two-Surface Problem of Evaporation and Condensation in a Vapor-Gas Mixture," Rarefied Gas Dynamics, Vol.II, edited by H. Oguchi, Univ. of Tokyo Press, Tokyo, 1984, pp. 875-884.

[12] Onishi, Y., "The Spherical-Droplet Problem of Evaporation and Condensation in a Vapour-Gas Mixture," Journal of Fluid Mechanics, Vol. 163, 1986, pp. 171-194.

[13] Onishi, Y., "Evaporation and Condensation of a Vapour-Gas Mixture Between the Coaxial Cylindrical Condensed Phases," Rarefied Gas Dynamics, edited by V. Boffi and C. Cercignani, Teubner, Stuttgart, 1986, pp. 251-260.

[14] Onishi, Y., "On the Behavior of a Noncondensable Gas of a Small Amount in Weakly Nonlinear Evaporation and Condensation of a Vapor," Journal of the Physical Society of Japan, Vol. 55, 1986, pp. 3080-3092.

[15] Landau, L.D. and Lifshitz, E.M., *Statistical Physics*, Pergamon Elmsford, NY, 1969, Sec.82.

[16] Onishi, Y., "Kinetic Theory of Slightly Strong Evaporation and Condensation for a Two-Surface Problem," Proceedings of the Second Asian Congress of Fluid Mechanics, edited by T.C. Lin, H. Sato, and R. Narasimha, Science Press, China, 1983, pp. 732-737.

[17] Onishi, Y., "A Nonlinear Analysis on the Behavior of a Small Amount of Inert Gas during Evaporation and Condensation of a Vapor between the Parallel Surfaces," Journal of the Physical Society of Japan, Vol. 57, 1988, pp. 2354-2364.

Nonlinear Analysis for Evaporation and Condensation of a Vapor-Gas Mixture Between the Two Plane Condensed Phases. Part II: Concentration of Inert Gas ~ *O(Kn)*

Yoshimoto Onishi*
Tottori University, Tottori, Japan

Abstract

Evaporation and condensation problem of a slightly rarefied vapor-gas mixture between the two plane condensed phases is investigated on the basis of the Boltzmann equation of BGK type under the diffusive boundary condition. The behavior of the mixture undergoing slightly strong evaporation and condensation processes is solved by the singular perturbation method when the concentration of the inert gas is small and at most of the order of the Knudsen number of the system. The large or full nonlinear variations of the partial pressure and the number density of the inert gas, which manifest themselves because of the small amount of inert gas, are shown explicitly, as well as the other quantities of interest. It is also shown that the presence of a small amount of inert gas affects considerably the evaporation and condensation rate, and its critical concentration ratio is of the order of the Knudsen number of the system.

I. Introduction

Let a mixture of a vapor gas (A) and an inert gas (B) be contained between the two plane condensed phases placed in parallel with a distance L apart. One of the phases located at $X = 0$ is kept at a uniform temperature T_0, and the other at $X = L$ kept at T_L. The saturated vapor pressures corresponding to T_0 and T_L

* Professor, Department of Applied Mathematics and Physics, Faculty of Engineering.

are denoted by P_0^A and P_L^A, respectively. Also, let P_0^B be the partial pressure of the inert gas at $X = 0$. There is a fairly large amount of vapor, so that the ratio of the mean free path of the vapor molecules l^A at T_0 and P_0^A to L, i.e., the Knudsen number of the system $Kn = l^A/L$, is small compared to unity. Also the deviation of the system ϵ, i.e., $\epsilon = \max(|T_L - T_0|/T_0,\ |P_L^A - P_0^A|/P_0^A)$ is small, but its magnitude is of the order of Kn. The problem in this case becomes nonlinear.[1,2] Furthermore, the concentration ratio P_0^B/P_0^A of the inert gas is small, being at most of the order of Kn. The present study is concerned with finding the steady behavior of the mixture and its component gases associated with the evaporation and condensation process of the problem specified. The analysis will be based on the Boltzmann equation of the BGK type[3] subject to the diffusive boundary condition, that is, both the vapor and inert gas molecules after the interaction with the condensed phase leave the interface with Maxwellian distributions characterized by its temperature and velocity (zero here), the number density of the vapor having the saturated vapor number density at that temperature, and that of the inert gas having the value determined by the condition of no net mass flow of the gas across the interface.

In Part I of this series,[2] the problem was studied in the same situation as here, except that the concentration ratio of the inert gas in Part I is of the order of unity or larger. In that case, the variations of all the fluid dynamic quantities remain within the magnitude of the order of ϵ, and the mass transfer is governed by the diffusional ability of the vapor through the inert gas. Because the diffusion coefficient of the mixture is closely related to the magnitude of Kn, which is small, the mass transfer is necessarily small in that case. However, in the present case, where the concentration ratio is small, the molecules of the inert gas are driven toward the condensing surface because of collisions with those of the vapor flowing toward it and accumulate there until the large gradients of the partial pressure and number density of the inert gas have been established. It will be expected that the smaller the concentration of the inert gas is, the larger the gradients that are formed. The accumulation of the molecules or the large gradients of the inert gas then greatly hinders the vapor flow, weakening it until a balance has been established between the ability of the diffusional flow of the inert gas and that of the vapor flow. In order to make the inert gas form these large gradients near the condensing surface, the vapor mass flow must be fairly large (weakly nonlinear at least), being naturally larger than that of Ref. 2.

Actually, it will vary greatly, down from the value given in Ref. 2 up to that for the pure-vapor case (absence of inert gas) given in Refs. 4 and 5, depending on the magnitude of the concentration ratio of the inert gas. It is noted that Ref. 6 has given the general analysis for problems in which the concentration ratio of the inert gas is negligibly small and, based on this analysis, Ref. 5 has treated the same problem as we have here in the extreme case of a negligibly small concentration. The treatment of this kind of problem must take into account this nonlinear behavior of the inert gas; this has been done in the present analysis by retaining the full nonlinear terms in the equations and solving them as in Refs. 5 and 6 by a different expansion scheme from the one used in Ref. 2.

II. Kinetic Equations and the Boundary Conditions

Because large space variations in the number density and partial pressure of the inert gas B may be expected, we introduce two types of nondimensional quantities:[6] 1) the perturbation quantities (from a reference state) of the magnitude of order ϵ or Kn, whose space variations always remain within that magnitude; 2) the quantities of order unity, such as the number density and partial pressure of gas B, and other quantities associated with them, whose space variations can become large up to $O(1)$. The Boltzmann equation of BGK type[3] for the present case may be written down in terms of these nondimensional quantities as follows:

$$\tilde{k}\xi_x \frac{\partial \phi^A}{\partial x} = (1+n^A)(\phi_e^A - \phi^A) + a_{12}\hat{N}^B(\phi_e^{AB} - \phi^A) \tag{1}$$

$$\tilde{k}\xi_x \frac{\partial \Phi^B}{\partial x} = a_{22}\hat{N}^B(\Phi_e^B - \Phi^B) + a_{21}(1+n^A)(\Phi_e^{BA} - \Phi^B) \tag{2}$$

$$\begin{bmatrix} n^A \\ (1+n^A)u^A \\ \frac{3}{2}(1+n^A)\tau^A + (1+n^A){u^A}^2 \end{bmatrix} = \int \begin{bmatrix} 1 \\ \xi_x \\ \xi^2 - \frac{3}{2} \end{bmatrix} \phi^A \, E \, \mathrm{d}\vec{\xi} \tag{3}$$

$$\begin{bmatrix} \hat{N}^B \\ \hat{N}^B u^B \\ \frac{3}{2}\hat{N}^B\tau^B + M\hat{N}^B {u^B}^2 \end{bmatrix} = \int \begin{bmatrix} 1 \\ \xi_x \\ M\xi^2 - \frac{3}{2} \end{bmatrix} \Phi^B \, \tilde{E} \, \mathrm{d}\vec{\xi} \tag{4}$$

$$p^A = n^A + \tau^A + n^A\tau^A \tag{5}$$

$$\hat{P}^B = \hat{N}^B(1+\tau^B) \tag{6}$$

$$E(1+\phi_e^A) = \pi^{-3/2}\frac{1+n^A}{(1+\tau^A)^{3/2}}\exp\Big\{-\frac{(\xi_x-u^A)^2+\xi_y^2+\xi_z^2}{1+\tau^A}\Big\} \quad (7a)$$

$$\phi_e^{AB} = \phi_e^A(n^A = n^A, u^A = u^{AB}, \tau^A = \tau^{AB}) \quad (7b)$$

$$\tilde{E}\Phi_e^B = \Big(\frac{\pi}{M}\Big)^{-3/2}\frac{\hat{N}^B}{(1+\tau^B)^{3/2}}\exp\Big\{-\frac{M[(\xi_x-u^B)^2+\xi_y^2+\xi_z^2]}{1+\tau^B}\Big\} \quad (8a)$$

$$\Phi_e^{BA} = \Phi_e^B(\hat{N}^B = \hat{N}^B, u^B = u^{BA}, \tau^B = \tau^{BA}) \quad (8b)$$

$$u^{AB} = u^{BA} = \mu_A u^A + \mu_B u^B$$

$$\tau^{AB} = \tau^A + 2\mu_A\mu_B(\tau^B - \tau^A) + \frac{2}{3}\mu_B^2(u^A - u^B)^2$$

$$\tau^{BA} = \tau^B + 2\mu_A\mu_B(\tau^A - \tau^B) + \frac{2}{3}\mu_A\mu_B(u^B - u^A)^2$$

$$E = \pi^{-3/2}\exp(-\xi^2), \quad \tilde{E} = (\pi/M)^{-3/2}\exp(-M\xi^2), \quad \xi^2 = \vec{\xi}\cdot\vec{\xi}$$

$$M = \frac{m_B}{m_A}, \quad \mu_A = \frac{m_A}{m_A+m_B}, \quad \mu_B = \frac{m_B}{m_A+m_B}$$

$$a_{12} = \frac{N_0^B\kappa_{AB}}{N_0^A\kappa_{AA}}, \quad a_{21} = \frac{\kappa_{AB}}{\kappa_{AA}}, \quad a_{22} = \frac{N_0^B\kappa_{BB}}{N_0^A\kappa_{AA}}$$

and

$$\tilde{k} = \frac{(2R_AT_0)^{1/2}}{N_0^A\kappa_{AA}L} = \frac{\sqrt{\pi}}{2}Kn, \quad Kn = \frac{l^A}{L} \quad (9)$$

where $N_0^A(2R_AT_0)^{-3/2}E(1+\phi^A)$ and $N_0^B(2R_AT_0)^{-3/2}\tilde{E}\Phi^B$ are the molecular velocity distribution functions of gas A and gas B, respectively; $(2R_AT_0)^{1/2}\vec{\xi}$, the molecular velocity vector with components (ξ_x, ξ_y, ξ_z); $X = Lx$, the X coordinate; $U^A = (2R_AT_0)^{1/2}u^A$, $T^A = T_0(1+\tau^A)$, $N^A = N_0^A(1+n^A)$, and $P^A = P_0^A(1+p^A)$ are the X component of the mean flow velocity, temperature, number density, and partial pressure of gas A, respectively; $U^B = (2R_AT_0)^{1/2}u^B$, $T^B = T_0(1+\tau^B)$, $N^B = N_0^B\hat{N}^B$, and $P^B = P_0^B\hat{P}^B$, the corresponding quantities of gas B; $P^S = N^S\mathrm{k}T^S$ $(S = A, B)$ holds; m_A and m_B are the molecular masses of gas A and gas B, respectively; R_A is the gas constant per unit mass of gas A; and k is the Boltzmann constant.

$N_0^A(2R_AT_0)^{-3/2}E(1+\phi_e^A)$ and $N_0^A(2R_AT_0)^{-3/2}E(1+\phi_e^{AB})$ are the local Maxwellian distributions for molecules of gas A, characterized by the macroscopic quantities (n^A, u^A, τ^A) and (n^A, u^{AB}, τ^{AB}), respectively; and $N_0^B(2R_AT_0)^{-3/2}\tilde{E}\Phi_e^B$ and $N_0^B(2R_AT_0)^{-3/2}\tilde{E}\Phi_e^{BA}$ are the corresponding distributions for molecules of gas B, characterized by the quantities $(\hat{N}^B, u^B, \tau^B)$ and $(\hat{N}^B, u^{BA}, \tau^{BA})$, respectively. Here L is taken as the reference length, T_0 as the reference temperature, $(2R_AT_0)^{1/2}$ as the reference velocity, N_0^S as the reference number density for S-component gas, and $P_0^S(=N_0^S\mathrm{k}T_0)$ as the reference pressure for that component. κ_{AA}, κ_{AB}, and κ_{BB} are the parameters associated with the molecular collisions,[2,3] and are related to the transport coefficients of the mixture and its component gases as

$$\eta^M = \frac{P_0^A}{N_0^A\kappa_{AA}+N_0^B\kappa_{AB}} + \frac{P_0^B}{N_0^A\kappa_{AB}+N_0^B\kappa_{BB}}$$

$$D_{AB} = \frac{(m_A+m_B)\mathrm{k}T_0}{m_Am_BN_0\kappa_{AB}}, \quad \eta^A = \frac{P_0^A}{N_0^A\kappa_{AA}}, \quad \eta^B = \frac{P_0^B}{N_0^B\kappa_{BB}} \tag{10}$$

where $N_0 = N_0^A + N_0^B$; η^M and D_{AB} are the viscosity and diffusion coefficients of the mixture, η^S being the viscosity of the S-component gas. It may be noted that the thermal conductivities of the component gases are related to their viscosities as $\lambda^A = (5/2)(\mathrm{k}/m_A)\eta^A$ and $\lambda^B = (5/2)(\mathrm{k}/m_B)\eta^B$.

The kinetic boundary conditions for ϕ^A and Φ^B at the interface as assumed take on the following form:

$$\phi^A = \phi_W^A \equiv \phi_e^A(n^A = n_W^A, u^A = 0, \tau^A = \tau_W) \tag{11}$$

$$\Phi^B = \Phi_W^B \equiv \Phi_e^{BA}(\hat{N}^B = \hat{N}_W^B, u^{BA} = 0, \tau^{BA} = \tau_W) \tag{12}$$

for the molecules leaving the interface, i.e., $\pm\xi_x > 0$, where the upper sign refers to the interface at $x=0$ and the lower to that at $x=1$. $T_W = T_0(1+\tau_W)$ is the temperature of the interface. T_W is introduced here, merely for convenience, to indicate that it takes a specified value at each interface. Other quantities with suffix W are also introduced for this purpose. $N_W^A = N_0^A(1+n_W^A)$ and $N_W^B = N_0^B\hat{N}_W^B$ represent the number densities of the reflected molecules of gas A and gas B from the interface, respectively. N_W^A is the saturated vapor number density corresponding to T_W. If N_0^A is so chosen as to be the saturated vapor number density corresponding to T_0, we then have, for the relation between n_W^A and τ_W,[2,4]

$$n_W^A = (p_W^A - \tau_W)/(1+\tau_W), \qquad p_W^A = \gamma\tau_W[1+(\frac{\gamma}{2}-1)\tau_W+\ldots] \quad (13)$$

where $P_W^A = P_0^A(1+p_W^A)$ is the saturated vapor pressure corresponding to T_W. $\gamma = h_L/(R_A T_0)$, h_L being the latent heat of vaporization per unit mass of gas A. The values of γ for several substances may be found in Ref. 2. $\hat{N}_W^B$, on the other hand, is a quantity that must be determined by the condition of no net mass flow of gas B across the interface, i.e.,

$$\hat{N}^B u^B = \int \xi_x \Phi^B \tilde{E} d\vec{\xi} = 0 \qquad (14a)$$

at the interface. Hence, from Eq.(12) with (8), $\hat{N}_W^B$ is given by

$$\hat{N}_W^B = -2(\pi M)^{1/2}(1+\tau_W)^{-1/2}\int_{\pm\xi_x<0} \xi_x \Phi^B \tilde{E}\, d\vec{\xi} \qquad (14b)$$

It may be noted that $\hat{N}_W^B$ is obtained as a part of the solution.

We now write, at the present stage, the transport equations for gas A and gas B, which are derived from Eqs.(1) and (2) and are useful in the subsequent analysis:

$$\frac{d}{dx}\int \begin{bmatrix} \xi_x \\ \xi_x^2 \\ \xi_x(\xi^2-\frac{3}{2}) \end{bmatrix} \phi^A\, E\, d\vec{\xi} = \frac{1}{\bar{k}}\mu_B a_{12}\hat{N}^B(1+n^A) \times \begin{bmatrix} 0 \\ u^B - u^A \\ 3\mu_A(\tau^B-\tau^A)+2u^{AB}(u^B-u^A) \end{bmatrix} \qquad (15)$$

$$\frac{d}{dx}\int \begin{bmatrix} \xi_x \\ M\xi_x^2 \\ \xi_x(M\xi^2-\frac{3}{2}) \end{bmatrix} \Phi^B\, \tilde{E}\, d\vec{\xi} = \frac{1}{\bar{k}}\mu_B a_{21}\hat{N}^B(1+n^A) \times \begin{bmatrix} 0 \\ u^A - u^B \\ 3\mu_A(\tau^A-\tau^B)+2u^{BA}(u^A-u^B) \end{bmatrix} \qquad (16)$$

We see from the first relations in Eqs.(15) and (16) that, everywhere in the flow region,

$$(1+n^A)u^A = \text{const}, \qquad \hat{N}^B u^B = \text{const} \tag{17}$$

hold. Hence, from Eq.(14a), we obtain a simple consequence that

$$u^B \equiv 0 \tag{18}$$

which will be taken into account in the analysis that follows.

Because the present study is restricted to small values of the concentration ratio of the inert gas, at most of the order of the Knudsen number of the system, it will be convenient to introduce a parameter b by the following relation:

$$\frac{P_0^B}{P_0^A} = \frac{N_0^B}{N_0^A} = b\,\tilde{k} \tag{19}$$

and b enters the governing equations through a_{12} and a_{22}.

III. Hilbert Expansion and Macroscopic Behavior of Gas

First of all, we try to find a solution that behaves moderately, in other words, that varies over a distance of the order of L, a characteristic length of the present problem, i.e., $[\partial/\partial x \sim O(1)]$. We call this moderately varying part of the solution the Hilbert part and attach the suffix H to it. Now we know from Ref. 2 that the following relation holds:

$$\frac{\mathrm{d}p_H^B}{\mathrm{d}x} \sim -\frac{P_0^A}{P_0^B}\frac{\mathrm{d}p_H^A}{\mathrm{d}x}$$

where p_H^A and p_H^B and their space derivatives are assumed at most of the order of ϵ, the deviation of the system. The relation between p_H^B of Ref. 2 and $\hat{P}_H^B$ here is $\hat{P}_H^B = 1 + p_H^B$. If $P_0^B/P_0^A \sim O(1)$, then $\mathrm{d}p_H^B/\mathrm{d}x$ remains within the magnitude of $O(\epsilon)$, in which case the analysis of Ref. 2 is valid. However, if P_0^B/P_0^A becomes small until it is of the order of $\tilde{k}$, $\tilde{k}$ being of the order of ϵ here (and in Ref. 2 as well), then $\mathrm{d}p_H^B/\mathrm{d}x$ must necessarily become large, up to the magnitude of the order of unity, for which case the analysis of Ref. 2 becomes invalidated. Incidentally, it may be noted that if P_0^B/P_0^A further goes down to $O(\tilde{k}^n)$ $(n \geq 2)$, $\mathrm{d}p_H^A/\mathrm{d}x$ must decrease until it is of $O(\epsilon^n)$ since $\mathrm{d}p_H^B/\mathrm{d}x$ cannot exceed the magnitude of $O(1)$, even in the full nonlinear case. This situation suggests that we should allow the large variation or full nonlinearity in the behavior of the partial pressure and the number density of the inert gas in order to treat properly the problems[5,6] with small values of P_0^B/P_0^A. It should be mentioned that this must be so, when ϵ and $\tilde{k}$ are of the same order (weakly nonlinear case),[5,6] whereas no particular attention need be paid on this point when ϵ is of a higher order than $\tilde{k}$, i.e., in the linearized case.[7,8] (In this respect, the point of view of Ref. 9 is not correct.) Therefore, the Hilbert part of the solution may be sought in the following expansion in $\tilde{k}$:

$$\phi_H^A = \tilde{k}\phi_{H1}^A + \tilde{k}^2\phi_{H2}^A + \dots, \qquad \Phi_H^B = \Phi_{H0}^B + \tilde{k}\Phi_{H1}^B + \dots \tag{20}$$

and, accordingly, the fluid dynamic quantities are also expanded

$$f_H = \tilde{k}f_{H1} + \tilde{k}^2 f_{H2} + \dots, \qquad \hat{F}_H = \hat{F}_{H0} + \tilde{k}\hat{F}_{H1} + \dots \tag{21}$$

where $f = n^A,\ u^A,\ \tau^A,\ p^A,\ \tau^B, \dots$ and $\hat{F} = \hat{N}^B, \hat{P}^B$ (and possibly the stress and heat flux vector of gas B). Substituting the expansions into the original equations (1) and (2), and equating like powers of $\tilde{k}$, we obtain the following expressions for the distribution functions at each order of approximation:

$$\phi_{Hm}^A = \phi_{eHm}^A + \sum_{j=1}^{m-1}[n_{Hj}^A(\phi_{eHm-j}^A - \phi_{Hm-j}^A) + ba_{21}\hat{N}_{Hj-1}^B(\phi_{eHm-j}^{AB} - \phi_{Hm-j}^A)] - \xi_x\frac{\partial\phi_{Hm-1}^A}{\partial x}, \quad (m \geq 1) \tag{22}$$

$$\Phi_{Hm}^B = \Phi_{eHm}^{BA} + \sum_{j=1}^{m}[n_{Hj}^A(\Phi_{eHm-j}^{BA} - \Phi_{Hm-j}^B) + b\frac{\tilde{a}_{22}}{a_{21}}\hat{N}_{Hj-1}^B(\Phi_{eHm-j}^B - \Phi_{Hm-j}^B)] - \frac{1}{a_{21}}\xi_x\frac{\partial\Phi_{Hm-1}^B}{\partial x}, \quad (m \geq 0) \tag{23}$$

where we have put $\tilde{a}_{22} = \kappa_{BB}/\kappa_{AA}$ and have assumed that κ_{AB}/κ_{AA} and κ_{BB}/κ_{AA} are quantities of the order of unity. n_{Hm}^A, u_{Hm}^A, and τ_{Hm}^A are defined in Eq.(3) by the replacement of ϕ^A by ϕ_{Hm}^A. Similarly, $\hat{N}_{Hm}^B$ and τ_{Hm}^B are defined in Eq.(4) by the replacement of Φ^B by Φ_{Hm}^B. ϕ_{eHm}^A has been calculated from Eq.(7) as

$$\phi_{eH1}^A = n_{H1}^A + 2\xi_x u_{H1}^A + (\xi^2 - \frac{3}{2})\tau_{H1}^A \tag{24a}$$

$$\phi^{A}_{eH2} = n^{A}_{H2} + 2\xi_x u^{A}_{H2} + (\xi^2 - \frac{3}{2})\tau^{A}_{H2}$$
$$+ 2\xi_x u^{A}_{H1}[n^{A}_{H1} + \xi_x u^{A}_{H1} + (\xi^2 - \frac{5}{2})\tau^{A}_{H1}] - (u^{A}_{H1})^2$$
$$+ (\xi^2 - \frac{3}{2}) n^{A}_{H1}\tau^{A}_{H1} + \frac{1}{2}(\xi^4 - 5\xi^2 - \frac{15}{4})(\tau^{A}_{H1})^2 \qquad (24b)$$

and so on. ϕ^{AB}_{eHm} is obtained by replacing $(u^{A}_{Hm},\ \tau^{A}_{Hm})$ in Eq.(24) by $(u^{AB}_{Hm},\ \tau^{AB}_{Hm})$. In a similar manner, Φ^{B}_{eHm} and Φ^{BA}_{eHm} are obtained. Only Φ^{BA}_{eHm}, which is more important here, is listed here.

$$\Phi^{BA}_{eH0} = \hat{N}^{B}_{H0} \qquad (25a)$$

$$\Phi^{BA}_{eH1} = \hat{N}^{B}_{H1} + 2M\xi_x \hat{N}^{B}_{H0} u^{BA}_{H1} + (M\xi^2 - \frac{3}{2})\hat{N}^{B}_{H0}\tau^{BA}_{H1} \qquad (25b)$$

$$\Phi^{BA}_{eH2} = \hat{N}^{B}_{H2} + 2M\xi_x(\hat{N}^{B}_{H0} u^{BA}_{H2} + \hat{N}^{B}_{H1} u^{BA}_{H1}) + (M\xi^2 - \frac{3}{2})(\hat{N}^{B}_{H0}\tau^{BA}_{H2}$$
$$+ \hat{N}^{B}_{H1}\tau^{BA}_{H1}) + 2M\xi_x \hat{N}^{B}_{H0} u^{BA}_{H1}[M\xi_x u^{BA}_{H1} + (M\xi^2 - \frac{5}{2})\tau^{BA}_{H1}]$$
$$- M\hat{N}^{B}_{H0}(u^{BA}_{H1})^2 + \frac{1}{2}\hat{N}^{B}_{H0}(\tau^{BA}_{H1})^2(M^2\xi^4 - 5M\xi^2 + \frac{15}{4}) \qquad (25c)$$

With these expressions for the distribution functions, we can find from Eqs.(15) and (16) the following equations governing the fluid dynamic quantities. For the first order of approximation,

$$\frac{\mathrm{d}u^{A}_{H1}}{\mathrm{d}x} = 0 \qquad (26)$$

$$\hat{N}^{B}_{H0} u^{A}_{H1} = \frac{1}{2\mu_B a_{21}} \frac{\mathrm{d}\hat{N}^{B}_{H0}}{\mathrm{d}x} \qquad (27)$$

$$\frac{\mathrm{d}p^{A}_{H1}}{\mathrm{d}x} = -b\frac{\mathrm{d}\hat{N}^{B}_{H0}}{\mathrm{d}x} \qquad (28)$$

$$u^{A}_{H1}\frac{\mathrm{d}\tau^{A}_{H1}}{\mathrm{d}x} = \frac{1}{2}\frac{\mathrm{d}^2\tau^{A}_{H1}}{\mathrm{d}x^2} \qquad (29)$$

$$n^A_{H1} = p^A_{H1} - \tau^A_{H1} \tag{30}$$

and

$$\hat{P}^B_{H0} = \hat{N}^B_{H0}, \qquad \tau^B_{H1} = \tau^A_{H1}, \qquad p_{H1} = p^A_{H1} + b\hat{N}^B_{H0} - b \tag{31}$$

For the second order of approximation,

$$\frac{\mathrm{d}}{\mathrm{d}x}(u^A_{H2} + n^A_{H1}u^A_{H1}) = 0 \tag{32}$$

$$\hat{P}^B_{H0}(u^A_{H2} + n^A_{H1}u^A_{H1}) = \frac{1}{2\mu_B a_{21}}\frac{\mathrm{d}\hat{P}^B_{H1}}{\mathrm{d}x} - u^A_{H1}(\hat{P}^B_{H1} - \hat{P}^B_{H0}\tau^A_{H1}) \tag{33}$$

$$\frac{\mathrm{d}p^A_{H2}}{\mathrm{d}x} = -b\frac{\mathrm{d}\hat{P}^B_{H1}}{\mathrm{d}x} \tag{34}$$

$$\begin{aligned} u^A_{H1}\frac{\mathrm{d}\tau^A_{H2}}{\mathrm{d}x} + (u^A_{H2} + n^A_{H1}u^A_{H1})\frac{\mathrm{d}\tau^A_{H1}}{\mathrm{d}x} &= \frac{1}{2}\frac{\mathrm{d}^2\tau^A_{H2}}{\mathrm{d}x^2} + \frac{1}{2}\frac{\mathrm{d}}{\mathrm{d}x}\{[(\tau^A_{H1} \\ &\quad - b(a_{21} - \frac{1}{Ma_{21}})\hat{P}^B_{H0}]\frac{\mathrm{d}\tau^A_{H1}}{\mathrm{d}x}\} \end{aligned} \tag{35}$$

$$n^A_{H2} = p^A_{H2} - \tau^A_{H2} - n^A_{H1}\tau^A_{H1} \tag{36}$$

and

$$\begin{aligned} &\hat{N}^B_{H1} = \hat{P}^B_{H1} - \hat{N}^B_{H0}\tau^B_{H1}, \qquad \tau^B_{H2} = \tau^A_{H2} + \frac{2}{3}(u^A_{H1})^2 \\ &p_{H2} = p^A_{H2} + b(\hat{P}^B_{H1} - p_{H1}) \end{aligned} \tag{37}$$

So far, we have obtained the equations governing the Hilbert part of the macroscopic or fluid dynamic quantities. It will suffice to provide the no jump conditions for the preceding equations if the boundary conditions (11) and (12) can be satisfied completely by the Hilbert part of the distribution functions so obtained. Actually, Φ^B_{H0} satisfies the condition (12) by putting

$$\hat{N}^B_{H0} = \hat{N}^B_{W0} \tag{38}$$

where $\hat{N}_W^B$ has been expanded like $\hat{F}_H$ in Eq.(21) as

$$\hat{N}_W^B = \hat{N}_{W0}^B + \tilde{k}\hat{N}_{W1}^B + \dots \tag{39}$$

Incidentally, n_W^A, p_W^A, and τ_W are also expanded as f_H in Eq.(21):

$$n_W^A = \tilde{k}n_{W1}^A + \tilde{k}^2 n_{W2}^A + \dots \tag{40a}$$

$$p_W^A = \tilde{k}p_{W1}^A + \tilde{k}^2 p_{W2}^A + \dots \tag{40b}$$

$$\tau_W = \tilde{k}\tau_{W1} + \tilde{k}^2 \tau_{W2} + \dots \tag{40c}$$

Unfortunately, it is clear that ϕ_{H1}^A, Φ_{H1}^B, ϕ_{H2}^A, and so on, cannot satisfy the boundary conditions,[6] so that their correction parts must be considered to obtain the uniformly valid solution over the whole flow region, which will be given in the next section.

IV. Knudsen-Layer Analysis

As mentioned at the end of Sec.III, another part of the solution is needed in order to obtain the complete solution that satisfies the given conditions (11) and (12) at the interface. This part, which is required to vary appreciably over a small distance of the order of the molecular mean free path, makes the correction to the Hilbert part within a thin layer, the Knudsen layer, near the interface and, therefore, it is called the Knudsen-layer correction part (suffix K is attached). The analysis for this correction part can be carried out in the same way as in Refs. 1, 2, and 6. Here we have followed Ref. 6 particularly. Leaving the details of the process of the analysis to Ref. 6, we give its brief outline. Again, we try to find the solution in the sum of two parts, the Hilbert part plus its correction:

$$\phi^A = \phi_H^A(\vec{\xi}, x) + \phi_K^A(\vec{\xi}, \eta), \qquad \Phi^B = \Phi_H^B(\vec{\xi}, x) + \Phi_K^B(\vec{\xi}, \eta) \tag{41}$$

$$f = f_H(x) + f_K(\eta), \qquad \hat{F} = \hat{F}_H(x) + \hat{F}_K(\eta) \tag{42}$$

where η is a stretched coordinate defined by

$$\eta = \begin{cases} x/\tilde{k} & \text{near } x = 0 \\ (1-x)/\tilde{k} & \text{near } x = 1 \end{cases} \tag{43}$$

ϕ_K^A, Φ_K^B, f_K, and $\hat{F}_K$ are also sought in the following expansion forms:

$$\phi_K^A = \tilde{k}\phi_{K1}^A + \tilde{k}^2\phi_{K2}^A + \dots, \qquad \Phi_K^B = \Phi_{K0}^B + \tilde{k}\Phi_{K1}^B + \dots \tag{44}$$

$$f_K = \tilde{k}f_{K1} + \tilde{k}^2 f_{K2} + \dots, \qquad \hat{F}_K = \hat{F}_{K0} + \tilde{k}\hat{F}_{K1} + \dots \tag{45}$$

and these correction parts vanish outside the Knudsen layer, i.e.,

$$\phi_K^A, \quad \Phi_K^B \quad \text{and} \quad f_K, \quad \hat{F}_K \quad \to 0, \quad \text{as} \quad \eta \to \infty \tag{46}$$

Since Φ_{H0}^B could satisfy the boundary condition, no Knudsen-layer correction part exists at this order, i.e., $\Phi_{K0}^B \equiv 0$ and, hence, $\hat{N}_{K0}^B = \hat{P}_{K0}^B \equiv 0$. The analysis for ϕ_{Km}^A and Φ_{Km}^B $(m \geq 1)$ was carried out following Ref. 6, and we have obtained the Knudsen-layer correction part and, at the same time, the jump conditions that will become the boundary conditions appropriate for the macroscopic equations (26-37). These are as follows:

1) The boundary conditions for Eqs.(26-29):

$$\begin{bmatrix} p_{H1}^A - p_{W1} \\ \tau_{H1}^A - \tau_{W1} \end{bmatrix} = \pm u_{H1}^A \begin{bmatrix} C_4^* \\ d_4^* \end{bmatrix}, \qquad \begin{aligned} C_4^* &= -2.132039 \\ d_4^* &= -0.446749 \end{aligned} \tag{47}$$

where, as before, the upper sign refers to the interface at $x = 0$ and the lower at $x = 1$. C_4^* and d_4^* are universal constants and their values are given in Ref. 10 (see also Ref. 1).

2) The boundary conditions for (32-35):

$$\begin{bmatrix} p_{H2}^A - p_{W2} \\ \tau_{H2}^A - \tau_{W2} \end{bmatrix} = \pm(u_{H2}^A + n_{H1}^A u_{H1}^A) \begin{bmatrix} C_4^* \\ d_4^* \end{bmatrix} \pm \frac{\mathrm{d}\tau_{H1}^A}{\mathrm{d}x} \begin{bmatrix} C_1 \\ d_1 \end{bmatrix}$$
$$+ (u_{H1}^A)^2 \begin{bmatrix} C_8^* \\ d_8^* \end{bmatrix} \pm \tau_{W1} u_{H1}^A \begin{bmatrix} \frac{1}{2}C_4^* \\ \frac{3}{2}d_4^* \end{bmatrix} \pm p_{W1}^A u_{H1}^A \begin{bmatrix} 0 \\ -d_4^* \end{bmatrix}$$
$$\pm\, b a_{21} \hat{N}_{H0}^B u_{H1}^A \begin{bmatrix} C_{10} \\ d_{10} \end{bmatrix} \tag{48}$$

where $C_1 = 0.558437$, $d_1 = 1.302716$, $C_8^* = C_8 - \beta_4^* C_4^* = -1.273029$, $d_8^* = d_8 - \beta_4^* d_4^* = -3.595935$, $C_8 = 2.320074$, $d_8 = -0.0028315$, and $\beta_4^* = C_4^* - d_4^* = -1.685289$. The values of C_1 and d_1 are given in Refs. 1 and 10, and those of C_8 and d_8 in Ref. 1. For the values of C_{10} and d_{10}, which depend on m_B/m_A and κ_{AB}/κ_{AA}, see Table 1.

Table 1 $\kappa_{AA} = \kappa_{AB}$

m_B/m_A	d_{10}	C_{10}	$\tilde{\Pi}_4^B(0)$
0.5	1.67983	0.28461	-0.0045825
1.0	1.33448	0.24993	-0.0054831
2.0	1.14527	0.22196	-0.0046325

Now, the Knudsen-layer corrections near the interfaces are given as follows:

$$u_{K1}^A \equiv 0 \tag{49}$$

$$\begin{bmatrix} n_{K1}^A \\ \tau_{K1}^A \\ p_{K1}^A \end{bmatrix} = \pm u_{H1}^A \begin{bmatrix} \Omega_4^*(\zeta) \\ \Theta_4^*(\zeta) \\ \Pi_4^*(\zeta) \end{bmatrix}, \qquad \Pi_4^* = \Omega_4^* + \Theta_4^* \tag{50}$$

$$\begin{bmatrix} \hat{N}_{K1}^B \\ \hat{N}_{H0}^B \tau_{K1}^B \\ \hat{P}_{K1}^B \end{bmatrix} = \pm \hat{N}_{H0}^B u_{H1}^A \begin{bmatrix} \tilde{\Omega}_4^B(\alpha\zeta) \\ \tilde{\Theta}_4^B(\alpha\zeta) \\ \tilde{\Pi}_4^B(\alpha\zeta) \end{bmatrix}, \qquad \tilde{\Pi}_4^B = \tilde{\Omega}_4^B + \tilde{\Theta}_4^B \tag{51}$$

$$u_{K2}^A = \mp (u_{H1}^A)^2 \Omega_4^*(\zeta) \tag{52}$$

$$\begin{aligned} \begin{bmatrix} n_{K2}^A \\ \tau_{K2}^A \\ p_{K2}^A \end{bmatrix} &= \pm (u_{H2}^A + n_{H1}^A u_{H1}^A) \begin{bmatrix} \Omega_4^*(\zeta) \\ \Theta_4^*(\zeta) \\ \Pi_4^*(\zeta) \end{bmatrix} \pm \frac{d\tau_{H1}^A}{dx} \begin{bmatrix} \Omega_1(\zeta) \\ \Theta_1(\zeta) \\ \Pi_1(\zeta) \end{bmatrix} \\ &+ (u_{H1}^A)^2 \begin{bmatrix} \Omega_8^*(\zeta) \\ \Theta_8^*(\zeta) \\ \Pi_8^*(\zeta) \end{bmatrix} \pm \tau_{W1} u_{H1}^A \begin{bmatrix} \Omega_9^*(\zeta) \\ \Theta_9^*(\zeta) \\ \Pi_9^*(\zeta) \end{bmatrix} \mp p_{W1}^A u_{H1}^A \begin{bmatrix} 0 \\ \Theta_4^*(\zeta) \\ 0 \end{bmatrix} \\ &\pm b a_{21} \hat{N}_{H0}^B u_{H1}^A \begin{bmatrix} \Omega_{10}(\zeta) \\ \Theta_{10}(\zeta) \\ \Pi_{10}(\zeta) \end{bmatrix} \end{aligned} \tag{53}$$

$$\Omega_8^* = \Omega_8 - \beta_4^* \Omega_4^*, \quad \Theta_8^* = \Theta_8 - \beta_4^* \Theta_4^*, \quad \Omega_9^* = \Omega_9 + \Omega_4^*$$

$$\Theta_9^* = \Theta_9 + \Theta_4^*, \quad \Pi_1 = \Omega_1 + \Theta_1, \quad \Pi_{10} = \Omega_{10} + \Theta_{10}$$

$$\Pi_8^* = \Omega_8 + \Theta_8 + (2d_4^* - C_4^* + \Theta_4^*)\Omega_4^*, \quad \Pi_9^* = \Omega_9 + \Theta_9 + 2\Omega_4^*$$

where $\alpha = \sqrt{M} a_{21}$ and ζ is defined by

$$\zeta = \int_0^\eta [1 + n^A(x')] d\eta', \quad \eta = \begin{cases} x/\tilde{k} & \text{near } x = 0 \\ (1-x)/\tilde{k} & \text{near } x = 1 \end{cases} \tag{54}$$

which is of the same order as η and, with Eq.(42), it follows that

$$\zeta = \eta + \tilde{k}(p_{W1}^A - \tau_{W1})\eta \pm \tilde{k}u_{H1}^A[\beta_4^*\eta + \int_0^\eta \Omega_4^*(\zeta_0)\mathrm{d}\zeta_0] + \dots \quad (55)$$

Ω_4^*, Θ_4^*, Ω_1, and Θ_1 are defined in Ref. 10; Ω_8, Θ_8, Ω_9, and Θ_9 in Ref. 1; and $\tilde{\Omega}_4^B$ and $\tilde{\Theta}_4^B$ in Ref. 6. Ω_{10} and Θ_{10} are the solutions of the integral equations similar to those for Ω_4^* and Θ_4^* of Ref. 10, except that the former equations have different inhomogeneous terms from those of the latter. $\tilde{\Omega}_4^B$, $\tilde{\Theta}_4^B$, Ω_{10}, and Θ_{10} depend on m_B/m_A and a_{21}. However, all these functions are universal in the sense that they are independent of the geometry of the problems.

V. Results and Discussion

Now we consider the present problem. We choose T_0, P_0^A, P_0^B, and $N_0^S = P_0^S/(\mathrm{k}T_0)$ $(S = A, B)$ as the reference quantities and put $T_L = T_0(1+\tau_L)$ and $P_L^A = P_0^A(1+p_L^A)$. The parameters τ_{Wm} and p_{Wm}^A $(m = 1, 2)$ in Eqs.(47), (48), (53), and (55) are to be given in terms of τ_L and p_L^A. That is,

$$\tau_{W1} = \tau_{W2} = 0, \qquad p_{W1}^A = p_{W2}^A = 0, \qquad \text{at} \quad x = 0 \quad (56a)$$

$$\tau_{W1} = \tau_L/\tilde{k}, \quad \tau_{W2} = 0, \quad p_{W1}^A = \gamma\tau_{W1}, \quad p_{W2}^A = \gamma(\frac{\gamma}{2} - 1)\tau_{W1}^2, \qquad \text{at} \quad x = 1 \quad (56b)$$

In addition to the conditions (47) and (48) with (56), we have to specify the condition for $\hat{P}_H^B$ in order to get a unique solution of the present problem. Since P_0^B was chosen as the value of the partial pressure of gas B at the interface at $x = 0$, the condition for $\hat{P}_H^B$ at $x = 0$ is given as follows:

$$\hat{P}_{H0}^B(0) = 1 \quad (57a)$$

$$\hat{P}_{H1}^B(0) = -\hat{P}_{K1}^B(0) = -\hat{N}_{H0}^B(0)u_{H1}^A(0)\tilde{\Pi}_4^B(0) \quad (57b)$$

where the relation (51) has been used for $\hat{P}_{K1}^B$. The values of $\tilde{\Pi}_4^B(0)$ is listed in Table 1.

By solving Eqs.(26-37), subject to the conditions (47), (48), and (57), with (56) being taken into account, and then making Knudsen-layer corrections to the Hilbert part of the solution using (49-55), we have obtained the distributions of various fluid dynamic quantities valid over the whole flow region, including the Knudsen layers near the interfaces. They are as follows:

$$\frac{U^A}{(2R_AT_0)^{1/2}} = \tilde{k}U + \tilde{k}^2\{\tilde{U}_2 - \beta_4^* U^2 - bU(1-e^{2\beta Ux}) - UD_0(1-e^{2Ux}) - U^2[\Omega_4^*(\zeta) - \Omega_4^*(\tilde{\zeta})]\}$$

$$\begin{aligned}\frac{P^A - P_0^A}{P_L^A - P_0^A} = \frac{\tilde{k}}{p_L^A}\{ & b(1-e^{2\beta Ux}) - \tilde{k}b[\beta G_0 x - U\tilde{\Pi}_4^B(0) \\ & - \beta D_0(e^{2Ux}-1)]\,e^{2\beta Ux} - \tilde{k}bU\tilde{\Pi}_4^B(0) \\ & + (U+\tilde{k}\tilde{U}_2)[C_4^* + \Pi_4^*(\zeta) - \Pi_4^*(\tilde{\zeta})] \\ & + \tilde{k}2UD_0[C_1 + \Pi_1(\zeta) - e^{2U}\Pi_1(\tilde{\zeta})] \\ & + \tilde{k}U^2[C_8^* + \Pi_8^*(\zeta) + \Pi_8^*(\tilde{\zeta})] \\ & + \tilde{k}ba_{21}U[C_{10} + \Pi_{10}(\zeta) - e^{2\beta U}\Pi_{10}(\tilde{\zeta})] - U\tau_L\Pi_9^*(\tilde{\zeta})\}\end{aligned}$$

$$\frac{P^B}{P_0^B} = \{1 + \tilde{k}[\beta G_0 x - U\tilde{\Pi}_4^B(0) - \beta D_0(e^{2Ux}-1)]\}e^{2\beta Ux} + \tilde{k}U[\tilde{\Pi}_4^B(\alpha\zeta) - e^{2\beta U}\tilde{\Pi}_4^B(\alpha\tilde{\zeta})]$$

$$\begin{aligned}\frac{T^A - T_0}{T_L - T_0} = \frac{\tilde{k}}{\tau_L}\{ & -D_0(1-e^{2Ux}) + \tilde{k}[-D_2(1-e^{2Ux}) + D_0G_0xe^{2Ux} \\ & + D_0^2(1-e^{4Ux}) - b\frac{D_0}{\beta}(a_{21} - \frac{1}{Ma_{21}})(1-e^{2U(1+\beta)x})] \\ & + (U+\tilde{k}\tilde{U}_2)[d_4^* + \Theta_4^*(\zeta) - \Theta_4^*(\tilde{\zeta})] \\ & + \tilde{k}2UD_0[d_1 + \Theta_1(\zeta) - e^{2U}\Theta_1(\tilde{\zeta})] \\ & + \tilde{k}U^2[d_8^* + \Theta_8^*(\zeta) + \Theta_8^*(\tilde{\zeta})] \\ & + \tilde{k}ba_{21}U[d_{10} + \Theta_{10}(\zeta) - e^{2\beta U}\Theta_{10}(\tilde{\zeta})] \\ & + U\tau_L[-\Theta_9^*(\tilde{\zeta}) + \gamma\Theta_4^*(\tilde{\zeta})]\}\end{aligned}$$

$$\frac{T - T_0}{T_L - T_0} = \frac{T^A - T_0}{T_L - T_0} + \frac{\tilde{k}^2}{\tau_L}bU[\tilde{\Theta}_4^B(\alpha\zeta) - e^{2\beta U}\tilde{\Theta}_4^B(\alpha\tilde{\zeta}) - \Theta_4^*(\zeta) + e^{2\beta U}\Theta_4^*(\tilde{\zeta})]$$

with $\beta = \mu_B a_{21}$ and

$$D_0 = \frac{1}{1-e^{2U}}(2d_4^* U - \frac{\tau_L}{\tilde{k}}), \qquad G_0 = 2(\tilde{U}_2 - d_4^* U^2 + U D_0)$$

$$D_2 = \frac{1}{1-e^{2U}}\Big\{2d_4^*\tilde{U}_2 + 2d_1 U D_0(1+e^{2U}) + b a_{21} d_{10} U(1+e^{2\beta U})$$
$$+ d_4^* U\big(\frac{3}{2}-\gamma\big)\frac{\tau_L}{\tilde{k}} + D_0 G_0 e^{2U} + D_0^2(1-e^{4U})$$
$$- b\frac{D_0}{\beta}(a_{21} - \frac{1}{M a_{21}})[1 - e^{2U(1+\beta)}]\Big\}$$

$$\tilde{U}_2 = \frac{1}{2(C_4^* - b\beta e^{2\beta U})}\Big\{b\beta\,[\,2U(D_0 - d_4^* U) + D_0(1-e^{2U})\,]\,e^{2\beta U}$$
$$+ bU\Pi_4^B(0)(1-e^{2\beta U}) - 2C_1 U D_0(1+e^{2U})$$
$$- b a_{21} C_{10} U(1+e^{2\beta U}) + \frac{\tau_L}{\tilde{k}}[\gamma(\frac{\gamma}{2}-1)\frac{\tau_L}{\tilde{k}} - \frac{1}{2}C_4^* U\,]\Big\}$$

$$\zeta = \eta + \tilde{k}U[\beta_4^*\eta + \int_0^{\eta}\Omega_4^*(\eta_0)\mathrm{d}\eta_0\,]$$

$$\tilde{\zeta} = \tilde{\eta} + (\gamma-1)\tau_L\tilde{\eta} - \tilde{k}U[\beta_4^*\tilde{\eta} + \int_0^{\tilde{\eta}}\Omega_4^*(\tilde{\eta}_0)\mathrm{d}\tilde{\eta}_0\,]$$

where $\eta = x/\tilde{k}$ near $x = 0$ and $\tilde{\eta} = (1-x)/\tilde{k}$ near $x = 1$. U is found by solving the following nonlinear algebraic equation

$$U = \frac{1}{2C_4^*}\left[\gamma\frac{\tau_L}{\tilde{k}} - b(1-e^{2\beta U})\right], \qquad ((2C_4^*)^{-1} = -0.23452) \tag{58}$$

Note that ζ is of order unity but $\tilde{\zeta}$ is very large near $x = 0$, so that the Knudsen-layer correction functions of $\tilde{\zeta}$ disappear completely, and so do the functions of ζ near $x = 1$. In the graphical representation of these results, we have employed, instead of the parameters $\tilde{k}$ and γ, new parameters k and Γ, which are defined, respectively, as

$$k = \tilde{k}/(1+a_{12}) = \tilde{k}/(1+b a_{21}\tilde{k}) \tag{59}$$

$$\Gamma = \frac{p_L^A}{\tau_L} = \frac{T_0}{P_0^A}\frac{(P_L^A - P_0^A)}{(T_L - T_0)} \tag{60}$$

These new parameters are equivalent to those defined in Ref. 2, and will facilitate the comparison between the present results and those in Ref. 2. It may be noted that the relation between Γ and γ is

$$\gamma = \Gamma[1-(\Gamma/2-1)\tau_L+\ldots], \quad \text{or} \quad \Gamma = \gamma[1+(\gamma/2-1)\tau_L+\ldots] \tag{61}$$

Now, with these new parameters, some of the typical distributions of the temperature of the mixture are shown in Fig.1, and the partial pressures of gas B and gas A in Figs.2 and 3, respectively. In the figures, the thick solid lines represent the total distributions, with the Knudsen-layer corrections included, and the thin solid lines the Hilbert parts only (although not so distinct in the graphs because the value of k adopted here is quite small). Figure 1 shows that as N_0^B/N_0^A becomes small, the vapor flow becomes large and, hence, the convection effect predominates over the diffusion effect in Eq.(29), flattening the temperature distributions in most of the flow region. The diffusion effect is confined very near the condensing surface. This effect becomes evident, naturally, as $(T_L - T_0)/T_0$ increases (compare Figs.1a and 1b). It is also noted that the negative temperature gradient phenomenon[10,11] can be noticed for smaller values of N_0^B/N_0^A. Figure 2 confirms the statement made in Sec.I, that is, the partial pressure of the inert gas has a steep gradient near the condensing surface and it becomes steeper as the ratio N_0^B/N_0^A becomes small and the nonlinearity (i.e., $(T_L - T_0)/T_0$ and, hence, the vapor flow velocity) increases. The partial pressure distributions of the vapor are almost flat for smaller values of N_0^B/N_0^A but have gradients of order unity for moderate values of N_0^B/N_0^A (see Fig.3). In the former case, the vapor mass flow or the evaporation and condensation rate

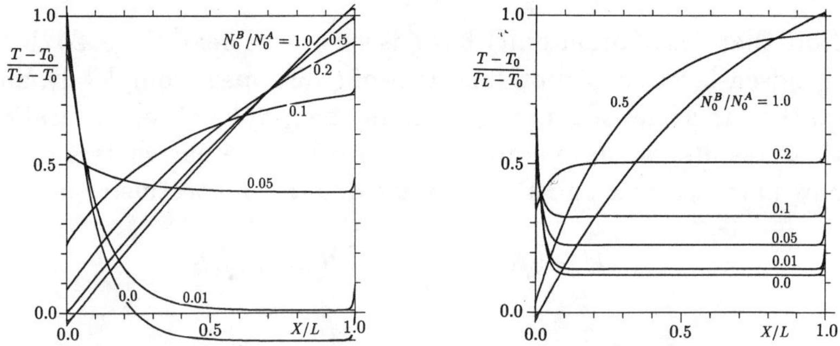

a) $m_B/m_A = 2.0$, $\Gamma = 11$, $(T_L - T_0)/T_0 = 0.01$, $k = 0.005$

b) $m_B/m_A = 2.0$, $\Gamma = 11$, $(T_L - T_0)/T_0 = 0.05$, $k = 0.005$

Fig. 1 Temperature distributions. $\kappa_{AA} = \kappa_{AB} = \kappa_{BB}$.

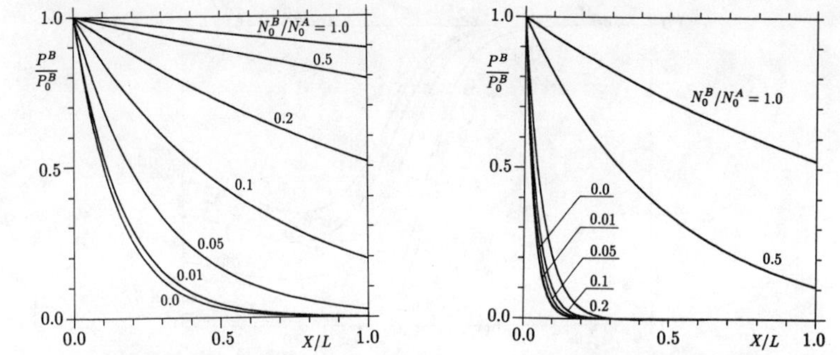

a) $m_B/m_A = 2.0$, $\Gamma = 11$, $(T_L - T_0)/T_0 = 0.01$, $k = 0.005$

b) $m_B/m_A = 2.0$, $\Gamma = 11$, $(T_L - T_0)/T_0 = 0.05$, $k = 0.005$

Fig. 2 Partial pressure of the inert gas. $\kappa_{AA} = \kappa_{AB} = \kappa_{BB}$.

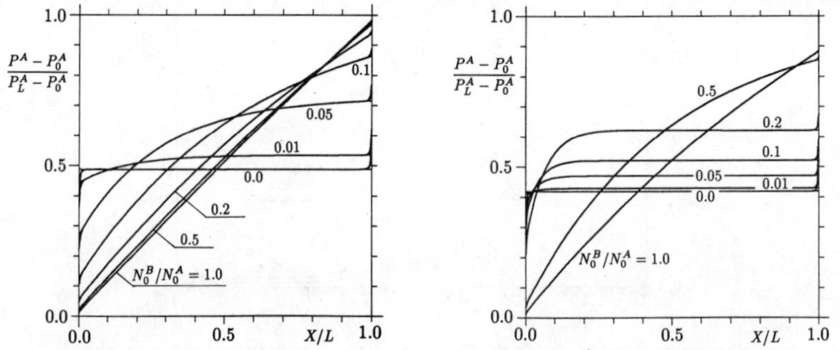

a) $m_B/m_A = 2.0$, $\Gamma = 11$, $(T_L - T_0)/T_0 = 0.01$, $k = 0.005$

b) $m_B/m_A = 2.0$, $\Gamma = 11$, $(T_L - T_0)/T_0 = 0.05$, $k = 0.005$

Fig. 3 Partial pressure of the vapor. $\kappa_{AA} = \kappa_{AB} = \kappa_{BB}$.

is determined by the kinetic control mechanism and, in the latter, by the diffusion control mechanism.[7,8] Hence, there must be a transition range where the mass-transfer mechanism shifts from one to the other. This range is somewhere around $N_0^B/N_0^A = 0.05$ in Fig.3a, whereas it is somewhere between $N_0^B/N_0^A = 0.2$ and $N_0^B/N_0^A = 0.5$ in Fig.3b (see also Fig.4a and 4d).

The mass flow $\dot{m} = m_A N^A U^A$ can be calculated as

$$\frac{\dot{m}}{m_A N_0^A (2R_A T_0)^{1/2}} = \frac{(P_L^A - P_0^A)}{P_0^A} \frac{(\tilde{k}U + \tilde{k}^2 \tilde{U}_2)}{p_L^A} \tag{62}$$

This shows that, in the limit $N_0^B/N_0^A \to 0$, $\dot{m}$ becomes $\dot{m}_p$ for the

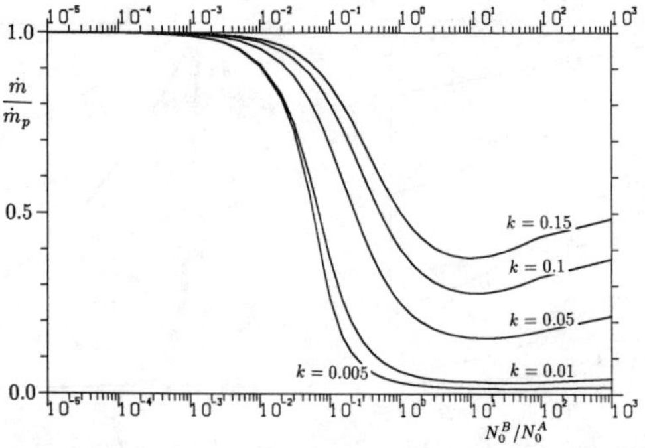

a) $m_B/m_A = 2.0$, $\Gamma = 11$, $(T_L - T_0)/T_0 = 0.01$.

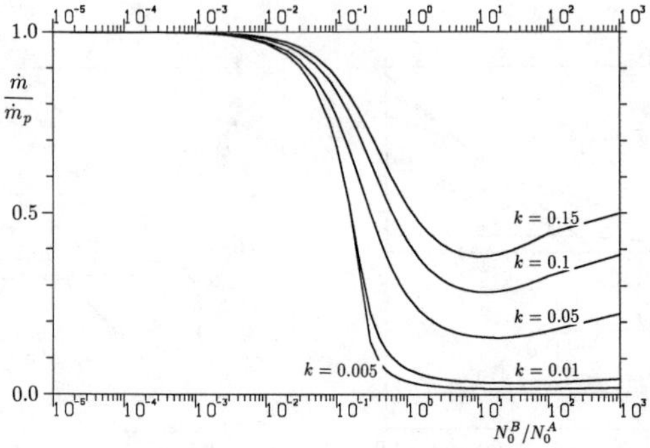

b) $m_B/m_A = 2.0$, $\Gamma = 11$, $(T_L - T_0)/T_0 = 0.03$.

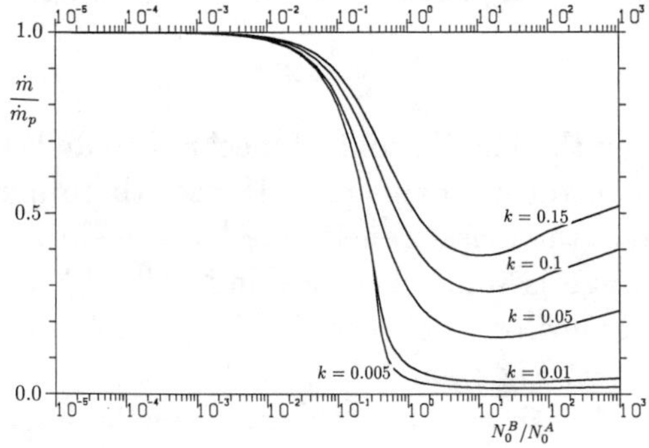

c) $m_B/m_A = 2.0$, $\Gamma = 11$, $(T_L - T_0)/T_0 = 0.05$.

Fig. 4 Mass flow vs concentration ratio. $\kappa_{AA} = \kappa_{AB} = \kappa_{BB}$.

Fig. 4 continued.

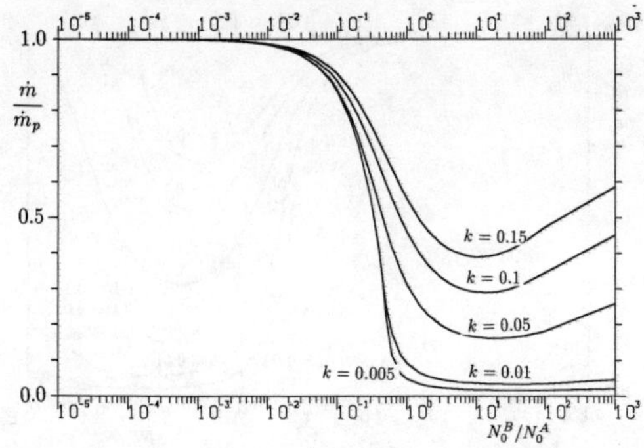

d) $m_B/m_A = 2.0$, $\Gamma = 11$, $(T_L - T_0)/T_0 = 0.08$.

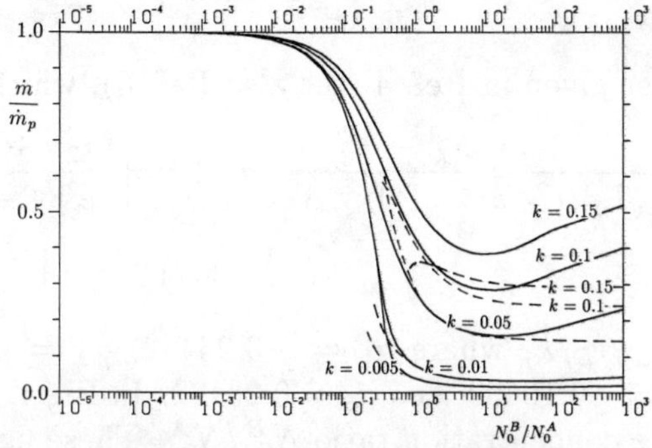

e) $m_B/m_A = 2.0$, $\Gamma = 11$, $(T_L - T_0)/T_0 = 0.05$.

Dashed lines : results of Ref. 2.

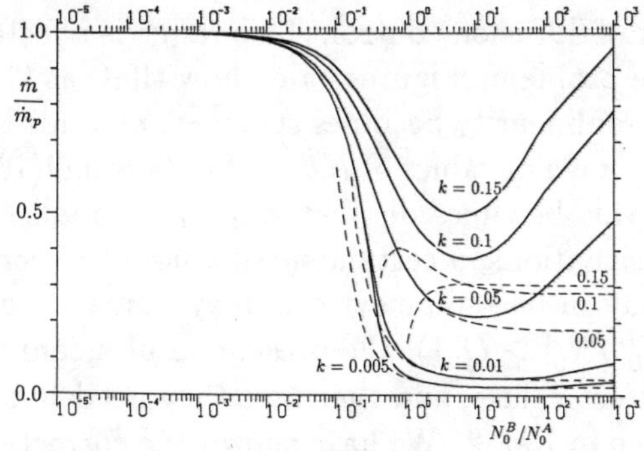

f) $m_B/m_A = 1.0$, $\Gamma = 11$, $(T_L - T_0)/T_0 = 0.05$.

Dashed lines : results of Ref. 2.

Fig. 4 continued.

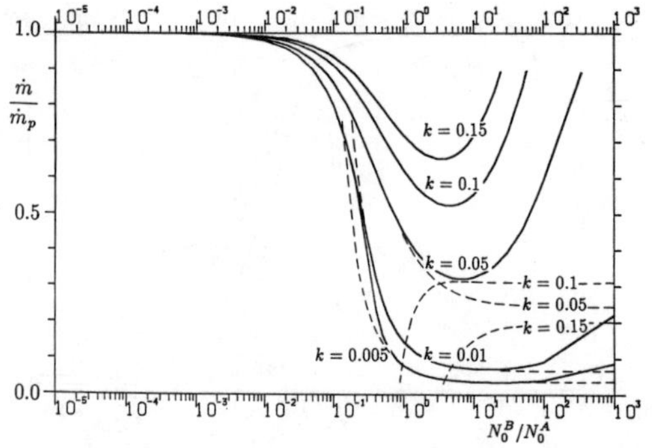

g) $m_B/m_A = 0.5,\ \Gamma = 11,\ (T_L - T_0)/T_0 = 0.05.$

Dashed lines : results of Ref. 2.

Fig. 4 concluded.

pure-vapor case given in Ref. 4 (see also Ref. 5), which is

$$\frac{\dot m_p}{m_A N_0^A (2R_A T_0)^{1/2}} = \frac{P_L^A - P_0^A}{P_0^A}\frac{\gamma}{\Gamma}\left\{ a_1 + \tau_L \left[\frac{(a_2\gamma + a_3)}{e^{2U_0} - 1} + a_4\gamma + a_5 \right] + \ldots \right\} \tag{63}$$

with $U_0 = a_1\gamma\tau_L/\tilde{k}$, where $a_1 = -0.23452$, $a_2 = 0.02574$, $a_3 = -0.12285$, $a_4 = -0.10439$, and $a_5 = 0.23172$. In Fig.4, $\dot m/\dot m_p$ is plotted against the concentration ratio N_0^B/N_0^A. These figures show the drop of the vapor mass flow $\dot m/\dot m_p$ as the concentration ratio N_0^B/N_0^A increases from its smaller values to the values of $O(1)$. The range of values of N_0^B/N_0^A in which $\dot m/\dot m_p$ drops sharply is the transition region where the mechanism of mass transfer shifts from that of kinetic control to that of diffusion control. This range is affected by the nonlineariry of the problem. Figures 4a-d show that, as $(T_L - T_0)/T_0$ increases or the nonlinearity becomes stronger, $\dot m$ tends to drop rapidly within smaller range of values of N_0^B/N_0^A between $O(10^{-1})$ and $O(1)$. In particular, this becomes distinct when k is small. This indicates that, in some situations, a certain small amount of contamination of an inert gas, say in heat pipe sytems, may damage its whole performance. For $N_0^B/N_0^A \geq O(1)$, the present results cannot be expected to give the correct values, and therefore they need to be corrected by the results given in Ref. 2. We have shown the correction in Figs.4e-g where the dashed lines indicate the results of Ref. 2 which are valid for $N_0^B/N_0^A \geq O(1)$. In general, the matching between the present results and those of Ref. 2 is good and satisfactory up to $k = 0.1$, although there is some difficulty to say so in Fig.4g when $k = 0.1$.

Note that the result of Ref. 2 in [illegible]g with $k = 0.15$ seems to be beyond the validity of the analysis of Ref. 2 (see the comment there).

Finally, both the present analysis and that of Ref. 2 are complementary, and together they give the analysis that covers the whole range of the concentration of an inert gas in a vapor undergoing slightly strong evaporation and condensation between the two interfaces.

References

[1] Onishi, Y. and Sone, Y., "Kinetic Theory of Slightly Strong Evaporation and Condensation – Hydrodynamic Equation and Slip Boundary Condition for Finite Reynolds Number," Journal of the Physical Society of Japan, Vol. 47, 1979, pp. 1676-1685.

[2] Onishi, Y., "A Nonlinear Analysis for Evaporation and Condensation of a Vapor-Gas Mixture Between the Two Plane Condensed Phases – Concentration of Inert Gas $\sim O(1)$," this volume.

[3] Hamel, B.B., "Kinetic Model for Binary Gas Mixtures," Physics of Fluids, Vol. 8, 1965, pp. 418-425.

[4] Onishi, Y., "Kinetic Theory of Slightly Strong Evaporation and Condensation for a Two-Surface Problem," Proceedings of the Second Asian Congress of Fluid Mechanics, edited by T.C. Lin, H. Sato, and R. Narasimha, Science Press, China, 1983, pp. 732-737.

[5] Onishi, Y., "A Nonlinear Analysis on the Behavior of a Small Amount of Inert Gas during Evaporation and Condensation of a Vapor between the Parallel Surfaces," Journal of the Physical Society of Japan, Vol. 57, 1988, pp. 2354-2364.

[6] Onishi, Y., "On the Behavior of a Noncondensable Gas of a Small Amount in Weakly Nonlinear Evaporation and Condensation of a Vapor," Journal of the Physical Society of Japan, Vol. 55, 1986, pp. 3080-3092.

[7] Onishi, Y., "A Two-Surface Problem of Evaporation and Condensation in a Vapor-Gas Mixture," Rarefied Gas Dynamics, edited by H. Oguchi, Univ. of Tokyo Press, Tokyo, 1984, pp. 875-884.

[8] Onishi, Y., "The Spherical-Droplet Problem of Evaporation and Condensation in a Vapour-Gas Mixture," Journal of Fluid Mechanics, Vol. 163, 1986, pp. 171-194.

[9] Soga, T., "Kinetic Analysis of Evaporation and Condensation in a Vapor-Gas Mixture," Physics of Fluids, Vol. 25, 1982, pp. 1978-1986.

[10] Sone, Y. and Onishi, Y., "Kinetic Theory of Evaporation and Condensation – Hydrodynamic Equation and Slip Boundary Condition," Journal of the Physical Society of Japan, Vol. 44, 1978, pp. 1981-1994.

[11] Pao, Y.P., "Application of Kinetic Theory to the Problem of Evaporation and Condensation," Physics of Fluids, Vol. 14, 1971, pp. 306-312.

Author Index

PROGRESS IN ASTRONAUTICS AND AERONAUTICS SERIES VOLUMES

VOLUME TITLE/EDITORS

***1. Solid Propellant Rocket Research** (1960)
Martin Summerfield
Princeton University

***2. Liquid Rockets and Propellants** (1960)
Loren E. Bollinger
The Ohio State University
Martin Goldsmith
The Rand Corporation
Alexis W. Lemmon Jr.
Battelle Memorial Institute

***3. Energy Conversion for Space Power** (1961)
Nathan W. Snyder
Institute for Defense Analyses

***4. Space Power Systems** (1961)
Nathan W. Snyder
Institute for Defense Analyses

***5. Electrostatic Propulsion** (1961)
David B. Langmuir
Space Technology Laboratories, Inc.
Ernst Stuhlinger
NASA George C. Marshall Space Flight Center
J.M. Sellen Jr.
Space Technology Laboratories, Inc.

***6. Detonation and Two-Phase Flow** (1962)
S.S. Penner
California Institute of Technology
F.A. Williams
Harvard University

***7. Hypersonic Flow Research** (1962)
Frederick R. Riddell
AVCO Corporation

***8. Guidance and Control** (1962)
Robert E. Roberson
Consultant
James S. Farrior
Lockheed Missiles and Space Company

***9. Electric Propulsion Development** (1963)
Ernst Stuhlinger
NASA George C. Marshall Space Flight Center

***10. Technology of Lunar Exploration** (1963)
Clifford I. Cummings
Harold R. Lawrence
Jet Propulsion Laboratory

***11. Power Systems for Space Flight** (1963)
Morris A. Zipkin
Russell N. Edwards
General Electric Company

12. Ionization in High-Temperature Gases (1963)
Kurt E. Shuler, Editor
National Bureau of Standards
John B. Fenn, Associate Editor
Princeton University

***13. Guidance and Control—II** (1964)
Robert C. Langford
General Precision Inc.
Charles J. Mundo
Institute of Naval Studies

***14. Celestial Mechanics and Astrodynamics** (1964)
Victor G. Szebehely
Yale University Observatory

***15. Heterogeneous Combustion** (1964)
Hans G. Wolfhard
Institute for Defense Analyses
Irvin Glassman
Princeton University
Leon Green Jr.
Air Force Systems Command

16. Space Power Systems Engineering (1966)
George C. Szego
Institute for Defense Analyses
J. Edward Taylor
TRW Inc.

17. Methods in Astrodynamics and Celestial Mechanics (1966)
Raynor L. Duncombe
U.S. Naval Observatory
Victor G. Szebehely
Yale University Observatory

18. Thermophysics and Temperature Control of Spacecraft and Entry Vehicles (1966)
Gerhard B. Heller
NASA George C. Marshall Space Flight Center

*Out of print.

***19. Communication Satellite Systems Technology** (1966)
Richard B. Marsten
Radio Corporation of America

20. Thermophysics of Spacecraft and Planetary Bodies: Radiation Properties of Solids and the Electromagnetic Radiation Environment in Space (1967)
Gerhard B. Heller
NASA George C. Marshall Space Flight Center

21. Thermal Design Principles of Spacecraft and Entry Bodies (1969)
Jerry T. Bevans
TRW Systems

22. Stratospheric Circulation (1969)
Willis L. Webb
Atmospheric Sciences Laboratory, White Sands, and University of Texas at El Paso

23. Thermophysics: Applications to Thermal Design of Spacecraft (1970)
Jerry T. Bevans
TRW Systems

24. Heat Transfer and Spacecraft Thermal Control (1971)
John W. Lucas
Jet Propulsion Laboratory

25. Communication Satellites for the 70's: Technology (1971)
Nathaniel E. Feldman
The Rand Corporation
Charles M. Kelly
The Aerospace Corporation

26. Communication Satellites for the 70's: Systems (1971)
Nathaniel E. Feldman
The Rand Corporation
Charles M. Kelly
The Aerospace Corporation

27. Thermospheric Circulation (1972)
Willis L. Webb
Atmospheric Sciences Laboratory, White Sands, and University of Texas at El Paso

28. Thermal Characteristics of the Moon (1972)
John W. Lucas
Jet Propulsion Laboratory

29. Fundamentals of Spacecraft Thermal Design (1972)
John W. Lucas
Jet Propulsion Laboratory

30. Solar Activity Observations and Predictions (1972)
Patrick S. McIntosh
Murray Dryer
Environmental Research Laboratories, National Oceanic and Atmospheric Administration

31. Thermal Control and Radiation (1973)
Chang-Lin Tien
University of California at Berkeley

32. Communications Satellite Systems (1974)
P.L. Bargellini
COMSAT Laboratories

33. Communications Satellite Technology (1974)
P.L. Bargellini
COMSAT Laboratories

34. Instrumentation for Airbreathing Propulsion (1974)
Allen E. Fuhs
Naval Postgraduate School
Marshall Kingery
Arnold Engineering Development Center

35. Thermophysics and Spacecraft Thermal Control (1974)
Robert G. Hering
University of Iowa

36. Thermal Pollution Analysis (1975)
Joseph A. Schetz
Virginia Polytechnic Institute

37. Aeroacoustics: Jet and Combustion Noise; Duct Acoustics (1975)
Henry T. Nagamatsu, Editor
General Electric Research and Development Center
Jack V. O'Keefe, Associate Editor
The Boeing Company
Ira R. Schwartz, Associate Editor
NASA Ames Research Center

38. Aeroacoustics: Fan, STOL, and Boundary Layer Noise; Sonic Boom; Aeroacoustic Instrumentation (1975)
Henry T. Nagamatsu, Editor
General Electric Research and Development Center
Jack V. O'Keefe, Associate Editor
The Boeing Company
Ira R. Schwartz, Associate Editor
NASA Ames Research Center

39. Heat Transfer with Thermal Control Applications (1975)
M. Michael Yovanovich
University of Waterloo

40. Aerodynamics of Base Combustion (1976)
S.N.B. Murthy, Editor
Purdue University
J.R. Osborn, Associate Editor
Purdue University
A.W. Barrows
J.R. Ward, Associate Editors
Ballistics Research Laboratories

41. Communications Satellite Developments: Systems (1976)
Gilbert E. LaVean
Defense Communications Agency
William G. Schmidt
CML Satellite Corporation

42. Communications Satellite Developments: Technology (1976)
William G. Schmidt
CML Satellite Corporation
Gilbert E. LaVean
Defense Communications Agency

43. Aeroacoustics: Jet Noise, Combustion and Core Engine Noise (1976)
Ira R. Schwartz, Editor
NASA Ames Research Center
Henry T. Nagamatsu, Associate Editor
General Electric Research and Development Center
Warren C. Strahle, Associate Editor
Georgia Institute of Technology

44. Aeroacoustics: Fan Noise and Control; Duct Acoustics; Rotor Noise (1976)
Ira R. Schwartz, Editor
NASA Ames Research Center
Henry T. Nagamatsu, Associate Editor
General Electric Research and Development Center
Warren C. Strahle, Associate Editor
Georgia Institute of Technology

45. Aeroacoustics: STOL Noise; Airframe and Airfoil Noise (1976)
Ira R. Schwartz, Editor
NASA Ames Research Center
Henry T. Nagamatsu, Associate Editor
General Electric Research and Development Center
Warren C. Strahle, Associate Editor
Georgia Institute of Technology

46. Aeroacoustics: Acoustic Wave Propagation; Aircraft Noise Prediction; Aeroacoustic Instrumentation (1976)
Ira R. Schwartz, Editor
NASA Ames Research Center
Henry T. Nagamatsu, Associate Editor
General Electric Research and Development Center
Warren C. Strahle, Associate Editor
Georgia Institute of Technology

47. Spacecraft Charging by Magnetospheric Plasmas (1976)
Alan Rosen
TRW Inc.

48. Scientific Investigations on the Skylab Satellite (1976)
Marion I. Kent
Ernst Stuhlinger
NASA George C. Marshall Space Flight Center
Shi-Tsan Wu
The University of Alabama

49. Radiative Transfer and Thermal Control (1976)
Allie M. Smith
ARO Inc.

50. Exploration of the Outer Solar System (1976)
Eugene W. Greenstadt
TRW Inc.
Murray Dryer
National Oceanic and Atmospheric Administration
Devrie S. Intriligator
University of Southern California

51. Rarefied Gas Dynamics, Parts I and II (two volumes) (1977)
J. Leith Potter
ARO Inc.

52. Materials Sciences in Space with Application to Space Processing (1977)
Leo Steg
General Electric Company

53. Experimental Diagnostics in Gas Phase Combustion Systems (1977)
Ben T. Zinn, Editor
Georgia Institute of Technology
Craig T. Bowman, Associate Editor
Stanford University
Daniel L. Hartley, Associate Editor
Sandia Laboratories
Edward W. Price, Associate Editor
Georgia Institute of Technology
James G. Skifstad, Associate Editor
Purdue University

54. Satellite Communications: Future Systems (1977)
David Jarett
TRW Inc.

55. Satellite Communications: Advanced Technologies (1977)
David Jarett
TRW Inc.

56. Thermophysics of Spacecraft and Outer Planet Entry Probes (1977)
Allie M. Smith
ARO Inc.

57. Space-Based Manufacturing from Nonterrestrial Materials (1977)
Gerard K. O'Neill, Editor
Princeton University
Brian O'Leary, Assistant Editor
Princeton University

58. Turbulent Combustion (1978)
Lawrence A. Kennedy
State University of New York at Buffalo

59. Aerodynamic Heating and Thermal Protection Systems (1978)
Leroy S. Fletcher
University of Virginia

60. Heat Transfer and Thermal Control Systems (1978)
Leroy S. Fletcher
University of Virginia

61. Radiation Energy Conversion in Space (1978)
Kenneth W. Billman
NASA Ames Research Center

62. Alternative Hydrocarbon Fuels: Combustion and Chemical Kinetics (1978)
Craig T. Bowman
Stanford University
Jorgen Birkeland
Department of Energy

63. Experimental Diagnostics in Combustion of Solids (1978)
Thomas L. Boggs
Naval Weapons Center
Ben T. Zinn
Georgia Institute of Technology

64. Outer Planet Entry Heating and Thermal Protection (1979)
Raymond Viskanta
Purdue University

65. Thermophysics and Thermal Control (1979)
Raymond Viskanta
Purdue University

66. Interior Ballistics of Guns (1979)
Herman Krier
University of Illinois at Urbana-Champaign
Martin Summerfield
New York University

***67. Remote Sensing of Earth from Space: Role of "Smart Sensors"** (1979)
Roger A. Breckenridge
NASA Langley Research Center

68. Injection and Mixing in Turbulent Flow (1980)
Joseph A. Schetz
Virginia Polytechnic Institute and State University

69. Entry Heating and Thermal Protection (1980)
Walter B. Olstad
NASA Headquarters

70. Heat Transfer, Thermal Control, and Heat Pipes (1980)
Walter B. Olstad
NASA Headquarters

71. Space Systems and Their Interactions with Earth's Space Environment (1980)
Henry B. Garrett
Charles P. Pike
Hanscom Air Force Base

72. Viscous Flow Drag Reduction (1980)
Gary R. Hough
Vought Advanced Technology Center

73. Combustion Experiments in a Zero-Gravity Laboratory (1981)
Thomas H. Cochran
NASA Lewis Research Center

74. Rarefied Gas Dynamics, Parts I and II (two volumes) (1981)
Sam S. Fisher
University of Virginia at Charlottesville

75. Gasdynamics of Detonations and Explosions (1981)
J.R. Bowen
University of Wisconsin at Madison
N. Manson
Université de Poitiers
A.K. Oppenheim
University of California at Berkeley
R.I. Soloukhin
Institute of Heat and Mass Transfer, BSSR Academy of Sciences

76. Combustion in Reactive Systems (1981)
J.R. Bowen
University of Wisconsin at Madison
N. Manson
Université de Poitiers
A.K. Oppenheim
University of California at Berkeley
R.I. Soloukhin
Institute of Heat and Mass Transfer, BSSR Academy of Sciences

77. Aerothermodynamics and Planetary Entry (1981)
A.L. Crosbie
University of Missouri-Rolla

78. Heat Transfer and Thermal Control (1981)
A.L. Crosbie
University of Missouri-Rolla

79. Electric Propulsion and Its Applications to Space Missions (1981)
Robert C. Finke
NASA Lewis Research Center

80. Aero-Optical Phenomena (1982)
Keith G. Gilbert
Leonard J. Otten
Air Force Weapons Laboratory

81. Transonic Aerodynamics (1982)
David Nixon
Nielsen Engineering & Research, Inc.

82. Thermophysics of Atmospheric Entry (1982)
T.E. Horton
The University of Mississippi

83. Spacecraft Radiative Transfer and Temperature Control (1982)
T.E. Horton
The University of Mississippi

84. Liquid-Metal Flows and Magnetohydrodynamics (1983)
H. Branover
Ben-Gurion University of the Negev
P.S. Lykoudis
Purdue University
A. Yakhot
Ben-Gurion University of the Negev

85. Entry Vehicle Heating and Thermal Protection Systems: Space Shuttle, Solar Starprobe, Jupiter Galileo Probe (1983)
Paul E. Bauer
McDonnell Douglas Astronautics Company
Howard E. Collicott
The Boeing Company

86. Spacecraft Thermal Control, Design, and Operation (1983)
Howard E. Collicott
The Boeing Company
Paul E. Bauer
McDonnell Douglas Astronautics Company

87. Shock Waves, Explosions, and Detonations (1983)
J.R. Bowen
University of Washington
N. Manson
Université de Poitiers
A.K. Oppenheim
University of California at Berkeley
R.I. Soloukhin
Institute of Heat and Mass Transfer, BSSR Academy of Sciences

88. Flames, Lasers, and Reactive Systems (1983)
J.R. Bowen
University of Washington
N. Manson
Université de Poitiers
A.K. Oppenheim
University of California at Berkeley
R.I. Soloukhin
Institute of Heat and Mass Transfer, BSSR Academy of Sciences

89. Orbit-Raising and Maneuvering Propulsion: Research Status and Needs (1984)
Leonard H. Caveny
Air Force Office of Scientific Research

90. Fundamentals of Solid-Propellant Combustion (1984)
Kenneth K. Kuo
The Pennsylvania State University
Martin Summerfield
Princeton Combustion Research Laboratories, Inc.

91. Spacecraft Contamination: Sources and Prevention (1984)
J.A. Roux
The University of Mississippi
T.D. McCay
NASA Marshall Space Flight Center

92. Combustion Diagnostics by Nonintrusive Methods (1984)
T.D. McCay
NASA Marshall Space Flight Center
J.A. Roux
The University of Mississippi

93. The INTELSAT Global Satellite System (1984)
Joel Alper
COMSAT Corporation
Joseph Pelton
INTELSAT

94. Dynamics of Shock Waves, Explosions, and Detonations (1984)
J.R. Bowen
University of Washington
N. Manson
Université de Poitiers
A.K. Oppenheim
University of California
R.I. Soloukhin
Institute of Heat and Mass Transfer, BSSR Academy of Sciences

95. Dynamics of Flames and Reactive Systems (1984)
J.R. Bowen
University of Washington
N. Manson
Université de Poitiers
A.K. Oppenheim
University of California
R.I. Soloukhin
Institute of Heat and Mass Transfer, BSSR Academy of Sciences

96. Thermal Design of Aeroassisted Orbital Transfer Vehicles (1985)
H.F. Nelson
University of Missouri-Rolla

97. Monitoring Earth's Ocean, Land, and Atmosphere from Space – Sensors, Systems, and Applications (1985)
Abraham Schnapf
Aerospace Systems Engineering

98. Thrust and Drag: Its Prediction and Verification (1985)
Eugene E. Covert
Massachusetts Institute of Technology
C.R. James
Vought Corporation
William F. Kimzey
Sverdrup Technology AEDC Group
George K. Richey
U.S. Air Force
Eugene C. Rooney
U.S. Navy Department of Defense

99. Space Stations and Space Platforms — Concepts, Design, Infrastructure, and Uses (1985)
Ivan Bekey
Daniel Herman
NASA Headquarters

100. Single- and Multi-Phase Flows in an Electromagnetic Field Energy, Metallurgical, and Solar Applications (1985)
Herman Branover
Ben-Gurion University of the Negev
Paul S. Lykoudis
Purdue University
Michael Mond
Ben-Gurion University of the Negev

101. MHD Energy Conversion: Physiotechnical Problems (1986)
V.A. Kirillin
A.E. Sheyndlin
Soviet Academy of Sciences

102. Numerical Methods for Engine-Airframe Integration (1986)
S.N.B. Murthy
Purdue University
Gerald C. Paynter
Boeing Airplane Company

103. Thermophysical Aspects of Re-Entry Flows (1986)
James N. Moss
NASA Langley Research Center
Carl D. Scott
NASA Johnson Space Center

104. Tactical Missile Aerodynamics (1986)
M.J. Hemsch
PRC Kentron, Inc.
J.N. Nielsen
NASA Ames Research Center

105. Dynamics of Reactive Systems Part I: Flames and Configurations; Part II: Modeling and Heterogeneous Combustion (1988)
J.R. Bowen
University of Washington
J.-C. Leyer
Université de Poitiers
R.I. Soloukhin
Institute of Heat and Mass Transfer, BSSR Academy of Sciences

106. Dynamics of Explosions (1986)
J.R. Bowen
University of Washington
J.-C. Leyer
Université de Poitiers
R.I. Soloukhin
Institute of Heat and Mass Transfer, BSSR Academy of Sciences

107. Spacecraft Dielectric Material Properties and Spacecraft Charging (1986)
A.R. Frederickson
U.S. Air Force Rome Air Development Center
D.B. Cotts
SRI International
J.A. Wall
U.S. Air Force Rome Air Development Center
F.L. Bouquet
Jet Propulsion Laboratory, California Institute of Technology

108. Opportunities for Academic Research in a Low-Gravity Environment (1986)
George A. Hazelrigg
National Science Foundation
Joseph M. Reynolds
Louisiana State University

109. Gun Propulsion Technology (1988)
Ludwig Stiefel
U.S. Army Armament Research, Development and Engineering Center

110. Commercial Opportunities in Space (1988)
F. Shahrokhi
K.E. Harwell
University of Tennessee Space Institute
C.C. Chao
National Cheng Kung University

111. Liquid-Metal Flows: Magnetohydrodynamics and Applications (1988)
Herman Branover, Michael Mond, and Yeshajahu Unger
Ben-Gurion University of the Negev

112. Current Trends in Turbulence Research (1988)
Herman Branover, Michael Mond, and Yeshajahu Unger
Ben-Gurion University of the Negev

113. Dynamics of Reactive Systems Part I: Flames; Part II: Heterogeneous Combustion and Applications (1988)
A.L. Kuhl
R & D Associates
J.R. Bowen
University of Washington
J.-C. Leyer
Université de Poitiers
A. Borisov
USSR Academy of Sciences

114. Dynamics of Explosions (1988)
A.L. Kuhl
R & D Associates
J.R. Bowen
University of Washington
J.-C. Leyer
Université de Poitiers
A. Borisov
USSR Academy of Sciences

115. Machine Intelligence and Autonomy for Aerospace (1988)
E. Heer
Heer Associates, Inc.
H. Lum
NASA Ames Research Center

116. Rarefied Gas Dynamics: Space-Related Studies (1989)
E.P. Muntz
University of Southern California
D.P. Weaver
U.S. Air Force Astronautics Laboratory (AFSC)
D.H. Campbell
The University of Dayton Research Institute

117. Rarefied Gas Dynamics: Physical Phenomena (1989)
E.P. Muntz
University of Southern California
D.P. Weaver
U.S. Air Force Astronautics Laboratory (AFSC)
D.H. Campbell
The University of Dayton Research Institute

(Other Volumes are planned.)

Other Volumes in the Rarefied Gas Dynamics Series*

Rarefied Gas Dynamics, Proceedings of the 1st International Symposium, edited by F. M. Devienne, Pergamon Press, Paris, France, 1960.

Rarefied Gas Dynamics, Proceedings of the 2nd International Symposium, edited by L. Talbot, Academic Press, New York, 1961.

Rarefied Gas Dynamics, Proceedings of the 3rd International Symposium, Vols. I and II, edited by J. A. Laurmann, Academic Press, New York, 1963.

Rarefied Gas Dynamics, Proceedings of the 4th International Symposium, Vols. I and II, edited by J. H. deLeeuw, Academic Press, New York, 1965.

Rarefied Gas Dynamics, Proceedings of the 5th International Symposium, Vols. I and II, edited by C. L. Brundin, Academic Press, London, England, 1967.

Rarefied Gas Dynamics, Proceedings of the 6th International Symposium, Vols. I and II, edited by L. Trilling and H. Y. Wachman, Academic Press, New York, 1969.

Rarefied Gas Dynamics, Proceedings of the 7th International Symposium, Vols. I and II, edited by D. Dini, C. Cercignani, and S. Nocilla, Editrice Tecnico Scientifica, Pisa, Italy, 1971.

Rarefied Gas Dynamics, Proceedings of the 8th International Symposium, Edited by K. Karamcheti, Academic Press, New York, 1974.

Rarefied Gas Dynamics, Proceedings of the 9th International Symposium, Vols. I and II, edited by M. Becker and M. Fiebig, DFVLR-Press, Porz-Wahn, Germany, 1974.

Rarefied Gas Dynamics, Proceedings of the 10th International Symposium, Parts I and II of Vol. 51 of *Progress in Astronautics and Aeronautics,* edited by L. Potter, AIAA, New York, 1977.

Rarefied Gas Dynamics, Proceedings of the 11th International Symposium, Vols. I and II, edited by R. Campargue, Commissiariat a l'Energie Atomique, Paris, France, 1979.

Rarefied Gas Dynamics, Proceedings of the 12th International Symposium, Parts I and II of Vol. 74 of *Progress in Astronautics and Aeronautics,* edited by S. Fisher, AIAA, New York, 1981.

Rarefied Gas Dynamics, Proceedings of the 13th International Symposium, Vols. I and II, edited by O. M. Belotserkovskii, M. N. Kogan, C. S. Kutateladze, and A. K. Rebrov, Plenum Press, New York, 1985.

Rarefied Gas Dynamics, Proceedings of the 14th International Symposium, Vols. I and II, edited by H. Oguchi, University of Tokyo Press, Tokyo, 1984.

Rarefied Gas Dynamics, Proceedings of the 15th International Symposium, Vols. I and II, edited by V. Boffi and C. Cercignani, B. G. Teubner, Stuttgart, 1986.

Rarefied Gas Dynamics, Proceedings of the 16th International Symposium, Vols. 116, 117, and 118 in *Progress in Astronautics and Aeronautics,* edited by E. P. Muntz, D. Weaver, and D. Campbell, AIAA, Washington, DC, 1989.

*Copies may be purchased directly from the publisher in each case. The AIAA cannot fill orders for volumes published by other publishers. This list is provided for information only.